The Pathology of Tropical Food Legumes

People of the West African Savannah where groundnuts and cowpeas are important crops

The Pathology of Tropical Food Legumes

Disease Resistance in Crop Improvement

D. J. ALLEN

Department of Applied Biology,
University of Cambridge, U.K.

A Wiley–Interscience Publication

JOHN WILEY & SONS
Chichester · New York · Brisbane · Toronto · Singapore

Library of Congress Cataloging in Publication Data:

Allen, D. J.
The pathology of tropical food legumes.
A Wiley–Interscience publication
Bibliography: p.
Includes index.
1. Legumes—Diseases and pests—Tropics. 2. Legumes—Disease and pest resistance—Tropics. 3. Legumes—Tropics—Breeding. 4. Food crops—Diseases and pests—Tropics. 5. Food crops—Disease and pest resistance—Tropics. 6. Food crops—Tropics—Breeding. I. Title.
SB608.L4A44 1983 633.3′0493′0913 82-20301

ISBN 0 471 10232 6

British Library Cataloguing in Publication Data:

Allen, D. J.
The pathology of tropical food legumes: disease resistance in crop improvement
1. Legumes—Diseases and pests—Tropics
I. Title
633′.3 SB608.L4

ISBN 0 471 10232 6

Typeset by Preface Ltd, Salisbury, Wilts.
Printed by Page Bros., (Norwich) Ltd.

Dedicated to the memory of the late Dr Alice Evans, sometime University Lecturer in Plant Genetics and Breeding in the Department of Applied Biology, Cambridge, whose dedication, infectious enthusiasm, and incisive personality made her an inspiring research supervisor

Contents

Foreword

The total output of crop protein in 1979, in the world as a whole, was 160 million tons. Of this total, three quarters came from cereals. Most of the remainder was in soya, groundnuts, and pulses, which are consequently of great importance for human diets, particularly in the poorer countries of the earth. We may expect that over the century ahead, as human population increases towards its plateau value, the demand for grain legumes will increase at least threefold.

The grain legumes are also agriculturally important, since their symbiosis with *Rhizobium* enables them to produce yield with much less assistance from purchased nitrogen fertilizers than other crops, and often with none at all. They may add to the stock of nitrogen in the soil, even though most of the nitrogen fixed by the symbiosis may be harvested in the produce. Their botanically indeterminate habit, which spreads the production of fruits over a substantial part of the life of the crops, enables them to yield something, however small, in drier conditions where a cereal, being determinate, may not be able to live long enough to produce any grain at all.

Grain legumes also complement cereals and other foods in respect of the amino-acid composition of their protein, and they supplement them as sources of dietary energy.

Perhaps most important of all, legumes add flavour and variety to diets, particularly in societies or economic classes which (because of poverty or cultural habit) eat little meat or other animal products. In this way they help poor people to consume enough starch to fuel their daily work.

Research on edible legumes in the tropics was comparatively neglected until about 1950, partly because in many tropical countries the first need was felt to be to increase the output and profitability of export crops—particularly fibres, rubber, beverages, and palm oils. Food crops were largely left to look after themselves.

When the emphasis began to move to food crops, the first priority, both nationally and internationally, went to the principal cereal crop—rice (which helps to feed half of mankind), wheat, and maize. At the international level, the International Rice Research Institute (IRRI) in the Philippines was opened in 1960, and the International Centre for the Improvement of Wheat and Maize (CIMMYT) in Mexico in 1966. But each of the next two international centres to be established, the International Centre for Tropical Agriculture (CIAT) in Colombia (1967), and the International Institute of Tropical Agriculture (IITA) in Nigeria (1968), included one or more legumes in its programmes—*Phaseolus* beans at CIAT and cowpea and tropic-adapted soya at IITA. The International Crops Research Institute for the Semi-Arid Tropics (ICRISAT), set up in India in 1972, works on pigeon pea, chickpea and groundnuts, and the programme of the International Centre for Agricultural Research in Dry Area (ICARDA) in Syria and other countries of the Middle East and North Africa (1976) includes lentils, chickpea and faba (*Vicia*) beans.

The author of this book, Dr David Allen, has been closely associated with these developments. He has devoted a number of years to research in the diseases of legumes in many parts of the tropical world, including a substantial and valuable contribution at IITA during the period when I had the honour to be chairman of the Board of Trustees. This substantial experience enables him, in the present work, to offer an authoritative professional review of the present state and future prospects of knowledge of the diseases of tropical grain legumes.

It starts, as it should, by describing the environments of the tropics, and the significance of legumes and legume diseases, in the diverse farming systems which these environments support. It then considers the general patterns of disease in legumes in the tropics. Next follows a systematic monographic treatment, crop by crop, of groundnuts, soyabean, *Phaseolus* bean, cowpea, pigeon pea and chickpea. Finally Dr Allen considers plant breeding for resistance and the general question of disease management.

It is fitting that these topics should provide the climax of the book. Reliable inherited multiple resistance to diseases (and to pests) in crops is the most valuable first gift which science can hope to offer to the hundreds of millions of small farmers in the poorer countries. Whatever other control measures they may include, successful pest and disease management systems are likely to rest as heavily as possible on inherited attributes of the crops. The nature of resistance to or tolerance of disease is thoroughly considered in relation to the different ways in which the defence systems of plants work and in which characters of defensive significance are inherited. At this point it may be acceptable for a foreword to say what a responsible author perhaps cannot—that the ultimate objective must be to incorporate into all crop plants the sort of attributes that make most plants immune to most pathogens, and (for example) totally protect potatoes from the stem rust of wheat.

This book is opportune and timely as well as scholarly and comprehensive. It will help to guide the thoughts of future workers, and it will provide them with a rich resource of access to the past information, world wide. It is a privilege to be given this opportunity to wish it well.

Reading—26th October 1982 A. H. Bunting

Preface

Increased international awareness of the consequences of world food shortage has led to increased emphasis on tropical food crop improvement, especially during the last one or two decades which have seen the establishment of the international agricultural research centres, funded through the Consultative Group on International Agricultural Research. Whereas spectacular increases in production have been made with the cereals in certain parts of Asia, the legumes remain largely unimproved. Despite their importance both in farming systems and in diets, their yields are erratic and low; farmers regard them as high-risk, low-priority crops, the more so in areas where attention has been diverted by success with high-yield cultivars of cereals. Because the food legumes are produced predominantly in subsistence systems of farming in which there are few added inputs, recent research has focused on the improvement of these crops by genetic means. Pests and pathogens are together responsible for very substantial yield losses; host-plant resistance is not only the most appropriate means of managing pests and diseases but is, in several instances, the key component of crop improvement.

Despite a large and rapidly growing literature, there is a dearth of authoritative reviews of the pathology of tropical legumes. Much that is published is insufficiently critical, too parochial of outlook, or too narrow in coverage. In this book, I have attempted to meet the need for a broader approach to the subject. The opening chapter is concerned with the tropical environment, the evolutionary diversity of the legumes, and the complex cropping systems of which they are part. This leads into a discussion of crop-pathogen evolution and the nature of cropping systems' stability, thence to germplasm exchange, new encounters between hosts and parasites, and crop vulnerability. The second and third chapters are devoted to the legume pathogens themselves, focusing on their geographical distribution, economic importance, and ecology; the third chapter ends with a discussion of pathogen interactions. The next six chapters review the major diseases by crop, each chapter opening with a brief discussion of the crop's evolution, adaptation, and production, and closing with a section on the status of disease resistance in that crop's improvement. The final chapter is devoted to breeding for disease resistance as a component of an integrated management of legume diseases. The book ends with a discussion of the potential for the translation of recent genetic advances into agricultural progress which, I suggest, may depend to a largely unrecognized degree on a better understanding of the long-established, traditional farming systems of the tropics.

The six legume species which I have chosen to discuss in detail are the groundnut (*Arachis hypogaea*), soyabean (*Glycine max*), bean (*Phaseolus vulgaris*), cowpea (*Vigna unguiculata*), pigeon pea (*Cajanus cajan*), and chickpea (*Cicer arietinum*). Examples have been drawn from other species only where this seemed pertinent. With the possible exception of the mung bean (*Vigna radiata*), none of the other tropical food legumes are yet of great economic significance though many are locally important. Nor has much breeding work been conducted on them. Should this book ever run to a second edition, I hope to extend its coverage. I have chosen to use the term 'food legume' in preference to the more common 'pulse' or 'grain legume' simply because it embraces not only the use of dry seed but also the oil-seed and vegetable uses of these crops. I have used the word 'pathology' in its widest sense to refer to diseases caused by nematodes and the parasitic angiosperms as well as to the groups more conventionally regarded as pathogens. I prefer the word 'epidemic' to the more strictly correct 'epiphytotic', if for no better reason than to avoid the clumsy 'epiphytotology'. Disease-resistance terminology, so fraught with complexity, generally follows R. A. Robinson's usage.

This book has been written principally for those actively engaged in the improvement of the legume crops, particularly plant breeders, plant pathologists, and entomologists. It is not intended as a guide to identification: several colour-illustrated field manuals are available for that purpose. I hope the book may also be useful to those concerned with resistance breeding in other crops, and I would like to think that it might even be of value to those concerned with policy-making in crop improvement. The book's extensive bibliography should make it a useful source of reference for students of applied biology, especially those with an interest in agriculture in developing countries.

Dr W. M. Steele first suggested that I write this book, and the late Dr Alice Evans gave me invaluable advice and encouragement at the outset. I have opted for single authorship in an attempt to produce a less disjointed text than is often found in volumes edited from the contributions of many authors. However, I have necessarily leant heavily on the help and advice of many others whose expertise in particular subjects far surpasses my own. As far as possible, I have sought criticism of each chapter in draft. I should like to record my particular thanks to the following for their valued suggestions: Dr S. P. S. Beniwal (Chapter 8), H. D. Franks (Chapter 1), Dr N. W. Galwey (Chapter 6), Dr R. M. Goodman (Chapter 5), C. Harkness (Chapter 4), Professor Sir Joseph Hutchinson CMG, FRS (Chapter 1), Dr R. Johnson (Chapter 10), Dr M. A. Keay (Chapter 4), Dr E. A. Kueneman (Chapter 5),

Dr C. L. A. Leakey (Chapter 6), Dr R. Michelmore (Chapter 2, *Phytophthora*), Professor G. E. Russell (Chapter 3, interactions), Dr H. F. Schwartz (Chapter 6), Professor J. B. Sinclair (Chapter 5), Dr J. B. Smithson (Chapter 9), Dr W. M. Steele (Chapters 1 and 7), Dr H. T. Tribe (Chapter 2, *Fusarium*), Dr L. J. G. van der Maesen (Chapter 8), Dr H. J. Vetten (Chapter 3, viruses), and Dr M. S. Wolfe (Chapter 10). Sir Denis Allen GCMG, CB read the entire manuscript. Photographs were kindly provided by Dr R. M. Goodman, Dr Y. L. Nene, Dr D. J. Nevill, S. G. Revell, Dr H. F. Schwartz, Professor J. B. Sinclair, Dr J. B. Smithson, Dr H. J. Vetten, and Dr R. J. Williams. I have also made extensive use of particular publications, most notably Summerfield and Buntings' *Advances in Legume Science* and Holliday's *Fungus Diseases of Tropical Crops*. Pathogen nomenclature generally follows usage at the Commonwealth Mycological Institute whose distribution maps and sets of descriptions of various groups of pathogens have been invaluable.

It is a pleasure to acknowledge my debt to Professor Sir James Beament FRS who has provided me with accommodation within his Department and access to its excellent library without which this book could not have been written. I am also most grateful to Professor A. H. Bunting CMG for agreeing to write a foreword, and to Ms Pat Reay for her expert typing from the manuscript.

D. J. ALLEN
1982

Chapter 1

Introduction

The population explosion has given rise to man's urgent search for rapid means of increasing the world's food supply. While death control has been successfully applied, attempts to reduce birth rates have been largely ineffective on a global scale, and even if they were successful, it is unlikely that they could produce significant reductions rapidly enough to avoid major food shortages. Until recently, increases in food production in the developing world (where most of the hungry live) have been due not to higher productivity but to the opening up of new land for cultivation. This will not be possible for much longer, for all the good land is already cultivated, and soon none of the remaining marginal land will be unfarmed. Current and future attempts to feed the world must therefore depend on intensification. At the centre of intensification efforts have been the recently developed, high-yielding wheat and rice cultivars which have considerably boosted yields and were the mainstay of the Green Revolution. But the Green Revolution has lost momentum, as has been widely discussed, and the problem facing us now is to bring it to the rainfed lands and the shifting cultivator, without expensive inputs (Greenland, 1975), and without the imposition of alien technology transferred from temperature agriculture. The development of 'sustained-yield tropical agro-escosystems' is required (Janzen, 1973; Greenland, 1975).

Here, we are concerned with the food legumes and their role in the development of stable cropping systems. But before discussing the diversity of legumes, we must first examine some of the characteristics of the tropical environment and its influence on crop production.

1.1 THE TROPICAL ENVIRONMENT

Strictly, the 'tropics' are those parts of the world lying between the tropics of Cancer and Capricorn at $23\frac{1}{2}°$ north and south of the equator, embracing the majority of the less developed countries. Apart from chemical and physical inadequacies of the soil, the main environmental elements limiting plant growth are solar radiation and rainfall. Only brief reference can be made here to some aspects of environmental influence on tropical crop production; the following account relies heavily on reviews given by Bunting (1961; 1975) and Cobley and Steele (1976) to which the reader is referred. For a fuller account of this vast subject, the reader is referred also to Kowal and Kassam (1978).

Solar radiation provides the source of energy used by plants in photosynthesis and, as such, is the chief determinant of dry matter production. The amount of solar energy received at the earth's surface (insolation) varies with latitude and with time of year, depending upon the intensity of the radiation and upon daylength. In turn, the intensity of radiation varies with the sun's elevation and cloud amount, while the length of day depends on latitude and season. Within the tropics, insolation is greatest in semi-arid regions during the 'winter' dry season, but is least variable in the wet equatorial tropics. Variation in temperature between locations and seasons follows the variation in insolation, and at low altitude the tropical environment is typically hot. On high plateaux and mountains, for example in East Africa, cool temperatures are associated with tropical insolation and daylength, permitting the cultivation of both temperate and tropical crops (Cobley and Steele, 1976). Though cold and frost or, on the other hand, excessive heat, may occur in typically tropical regions, they usually do so in periods of the year when plant growth is primarily restricted by drought. Temperature is important principally as a determinant of evaporation rates (Bunting, 1961). The lack of solar radiation data has hampered the mapping of potential evaporation but in East Africa relationships have been derived to assess insolation and sunshine duration from observations of cloud amount (Woodhead, 1966).

The length of day, from sunrise to sunset, varies with the seasons and with latitude; seasonal variation is greatest at high latitudes and least at the equator where the day length is constant. Many temperate crops are reproductively long-day plants, but most tropical crops are either insensitive to day length or flower only in response to short (or shortening) days. Cultivars of photosensitive crops have day-length requirements which tend to ensure that they flower and set fruit near the end of the growing season in the place where they evolved and to which, therefore, they are locally adapted (Cobley and Steele, 1976).

Of primary concern in tropical agricultural climatology is precipitation, its total amount and distribution in time and space, and its utilization by wild vegetation and crops. Taking tropical Africa as an example, the main climatic features are roughly defined, so far as plant growth is concerned, by the isohyets (rainfall contours). These have a crudely elliptical form, centred on the wet region of West Africa and Zaire, where rain falls in every month and where the wild vegetation is evergreen forest, or successional to forest. (Outside Africa, this climax vegetation is typified by the jungles of Indonesia, Malaysia, Brazil, and parts of Central America.) Away from these wet, lowland tropics is the savannah belt, of more open and usually deciduous, woody vegetation, grasses and herbs between trees and shrubs. Leaving the forest, the length of the dry season gradually increases until reaching desert.

In tropical Africa there are two main sources of precipitation: the West African source, of moist Atlantic air bringing rain on south and south-westerly winds, mainly in May to October; and the south-east Africa monsoon source, providing the rainfall of East and Central Africa between November and April. The convergencies and divergencies of these systems, together with orographic effects, account adequately, though perhaps not accurately, for the main features of the rainfall pattern (Bunting, 1961).

Rainfall data require careful interpretation. Mean total annual rainfall figures have limited significance since rainfall varies in seasonal duration as well as amount; statements of mean rainfall totals should be amplified by indications of rainfall duration. It is useful to regard as 'humid' those regions in which rainfall of at least 200 mm per month is spread over more than 7 months. Sub-humid regions receive rain in $4\frac{1}{2}$–7 months of the year whereas areas where rainfall is confined to 2–$4\frac{1}{2}$ months are considered semi-arid. In certain locations, particularly in the drier regions at the unstable frontiers of rainfall systems, rainfall is markedly variable from year to year (Bunting, 1961). A valuable method is available for the calculation of confidence limits for seasonal variation in rainfall (Manning, 1956). The concept of rainfall expectation at selected levels of productivity has been shown to provide a reliable method of assessing long-term crop risk which is not apparent when rainfall means alone are used. This helps in the formulation of the agricultural system best suited to local rainfall expectation.

Another major limitation imposed by the climate on crops is the length of the growing season. Whereas, in temperate regions, the growing season is limited at both ends by cold temperatures, in Mediterranean climates the available time is limited by either cold temperatures or drought at one end, and by drought at the other. In the seasonally arid tropics heat and drought determine both the start and close of the season while in the humid tropics weather is warm and wet more or less throughout the year (Bunting, 1975). This latter contrast is of cardinal importance in the ecology not only of crops but also of their parasites, as we shall see in Chapter 2.

1.2 LEGUMES IN TROPICAL CROPPING SYSTEMS

1.2.1 The evolutionary diversity of the food legumes

Within the very small portion of the plant kingdom that includes the world's crop plants, the legumes are a large group second only to the cereals as a source of human food. They belong to the family Leguminosae which has the characteristic of being able to form a symbiotic relationship with a group of bacteria (*Rhizobium*) which can utilize atmospheric nitrogen. This symbiosis, which occurs in nodules on the roots, may be the key to the success of the legumes, for they have a worldwide distribution; the family includes about 18,000 species, characterized by their fruits, which are pods, and by their (usually) alternate, compound, pinnate or trifoliate leaves (Cobley and

Steele, 1976). Diversity of form within Leguminosae ranges from hardwood trees of the tropical rain forests to the herbs of temperate pastures.

The Leguminosae have three sub-families, the Caesalpinioideae, the Mimosoideae, and the Papilionoideae; the first two consist mainly of tropical trees and shrubs with few economically important species and it is the Papilionoideae that are of agricultural importance. Members of the Papilionoideae are of value, firstly, for their nitrogen-rich plant material for consumption by man and his animals, and secondly for the nitrogen-rich residues they leave in the soil, thus enhancing the productivity of other crops grown in association with them. The human food they supply is of three kinds: edible tubers; leaf, green pods, and unripe seed (vegetable legumes); and ripe dry seeds (grain legumes, pulses). Together these will be referred to as food legumes, as distinct from forage.

The Papilionoideae are classified in ten botanical tribes, three of which together contain the major vegetable and grain legumes. They are the Fabeae which includes the genera *Vicia, Pisum, Lens,* and *Cicer;* the Phaseoleae which includes *Phaseolus, Vigna, Glycine,* and *Cajanus;* and the Hedysareae which contains *Arachis.* The principal food legume species in tropical agriculture are shown in Table 1 (the most important shown with an asterisk), sub-divided according to their region of origin (Evans, 1979). This book is about these species.

The grain legumes were domesticated very early in history, in the major centres of origin of agriculture (Hutchinson, 1970). Peas (*Pisum*), faba beans (*Vicia*), chickpeas, and lentils are among the early domestications in the Old World, dating from 7500 BC in the Mediterranean and temperate Eurasian region (Evans, 1979). The Afro-Asian group is indigenous in areas to which agriculture was introduced from the Middle East and it is probable that they were taken into cultivation by the early colonists as farming peoples spread outwards from the centre of origin of Old World agriculture (Hutchinson, 1970). Cowpeas appear to have evolved with, or perhaps were domesticated into, the West African cereal farming system of sorghum and pearl millet (Steele and Mehra, 1980), while the pigeon pea has been domesticated in India which is the present centre of diversity of *Cajanus* and its relative, *Atylosia* (De, 1974). The soyabean is apparently more recently domesticated than other species of the Afro-Asian group, domesticates emerging in north-eastern China in the eleventh century BC (Hymowitz, 1970). In the New World, the use of the bean (in this book, used strictly to refer to *Phaseolus vulgaris*) and its relatives has been reported throughout pre-history from the beginnings of agriculture in the Americas. The earliest domesticated types of *Phaseolus vulgaris* were discovered in Peru and are dated between 8000 and 5500 BC (Kaplan *et al.,* 1973). Remains of *P. vulgaris* nearly as old have been found also in Latin America, as have domesticates of other *Phaseolus* species (Evans, 1980), while the groundnut has been found in archaeological remains dated 2000–3000 BC in Peru (Krapovickas, 1968 cited by Evans, 1979). With the exception of the groundnut, these pulses are known to have been domesti-

TABLE 1 The Region of Origin and of Major Production of Tropical Food Legumes (after Evans, 1979)

Region of origin	Species	Common name	Region of major production
Mediterranean and temperate Eurasia	*Lens culinaris* Medik	Lentil	India, Near East
	[a]*Cicer arietinum* L.	Chickpea, gram	India
Afro-Asia	[a]*Vigna unguiculata* (L.) Walp.	Cowpea	W Africa, India, NE Brazil
	Vigna subterranea (Thouars) Verdc. (= *Voandzeia subterranea* (L.) Thouars.)	Bambarra groundnut	Africa
	Vigna radiata (L.) Wilcz. (= *Phaseolus aureus* Roxb.)	Mung bean, green gram	Asia
	Vigna mungo (L.) Hepper (= *Phaseolus mungo* L.)	Black gram, urd	Asia
	[a]*Cajanus cajan* (L.) Millsp.	Pigeon pea, red gram	India
	[a]*Glycine max* (L.) Merr.	Soyabean, soybean	N America, China, Brazil
	Psophocarpus tetragonolobus (L.) DC.	Winged bean, Goa bean	Papua New Guinea
	Sphenostylis stenocarpa (Hochst. ex Rich.) Harms.	African yam bean	W Africa
	Cyamoposis tetragonoloba (L.) Taub.	Cluster bean, guar	India
	Lablab purpureus (L.) Sweet (= *Dolichos lablab* L.)	Hyacinth bean, bonavist bean	India
	Macrotyloma uniflorum (Lam.) Verdc. (= *Dolichos biflorus* auct., non L.)	Horse gram	S India
	Canavalia gladiata (Jacq.) DC.	Sword bean	India, SE Asia
America	[a]*Phaseolus vulgaris* L.	Bean	Latin America, N America, E Africa
	Phaseolus lunatus L.	Lima bean	Latin America
	Phaseolus acutifolius A. Gray	Tepary bean	N Mexico, SW USA
	[a]*Arachis hypogaea* L.	Groundnut, peanut	India, Africa, USA
	Canavalia ensiformis (L.) DC.	Jack bean	Pan tropical
	Pachyrrhizus erosus (L.) Urban	Mexican yam bean	Asia

[a]Principal species covered in this book.

cated from the wild by agricultural man who both cultivated these species and made selections within them to suit his evolving environment.

Since these early origins, the grain legumes have spread widely so that, in some cases, the region of major production is far removed from the region of origin, as exemplified by the soyabean and groundnut (Table 1). Such migration to new areas, often associated with partial or complete failure of nodulation, tends to accentuate the changes which occur as a result of domestication. The most obvious effects of pulse domestication, as reviewed by Hutchinson (1970), Smartt (1976b), and Evans (1979), involve the modification of growth habit. Stems tend to be thicker, leaves larger, branches fewer, the node number less, and internode length shorter. This process culminates in the evolution of self-supporting plants well adapted to monocrop farming systems (Smartt, 1976b). Other consequences of domestication include a change from perennial to annual life form, earlier flowering, increasing self-pollination, loss of seed dormancy and pod dehiscence mechanisms, and increased seed size at the expense of seed numbers per pod. Such changes have led to an increase in yield potential. Following domestication, seed coats became more readily permeable to water (almost certainly as a consequence of human selection), thus decreasing their dormancy and leading to uniform germination as well as ease and speed of cooking. Local consumer preference may lead to selection of a particular seed-coat colour and consequently remarkable diversity, especially in *Phaseolus vulgaris, Phaseolus lunatus,* and *Vigna unguiculata.*

The Leguminosae as a whole are notorious for their wide range of biochemical substances that are not of nutritive value. While some (e.g. dyes from *Indigofera,* fish poisons and insecticides from *Derris* and *Tephrosia*) are of economic importance, others are considered deleterious and tend to be eliminated by plant breeders (Hutchinson, 1970). This may have dangerous consequences. Certain heat-labile anti-nutritional factors appear to have roles in the ecology of their hosts. For instance, phytohaemagglutinins (or lectins) are involved not only in the recognition of *Rhizobium* by the host-plant root (Hamblin and Kent, 1973) but may play some role in host-resistance and specificity of legume pests (Janzen *et al.,* 1976; 1977) and pathogens (Sequeira and Graham, 1976; Callow, 1977). Similarly, trypsin inhibitors govern resistance to certain pests (Gatehouse *et al.,* 1979). As pointed out by Hamblin (1980), there is a danger that if plant breeders select against these toxic substances in legumes before their roles are fully elucidated, they may inadvertently select against crop vigour.

1.2.2 The place of legumes in agriculture

We have seen that the utilization and domestication of legumes is of great antiquity. The importance of legumes lies in their food value as major sources of protein of high biological value, energy, minerals, vitamins, and roughage in addition to their role in the maintenance of soil fertility. In many tropical developing countries, legumes supply most of the plant proteins which are not only the main, but also the cheapest source of dietary proteins in areas where meat is scarce and too expensive for a large proportion of the population. The seed protein contents, ranging from 20 to 38 per cent, usually exceed those of other food grains; thus, legume seeds are the chief source of essential amino acids in tropical diets which consist predominantly of starchy cereals, roots, and tubers (Okigbo, 1978), though they are notably deficient in the sulphur amino acids.

The quantities of legumes produced and consumed vary considerably within the tropical world. The average daily legume seed consumption per head is in the range of 30–50g in most areas though much higher estimates have been given for some areas, notably Rwanda and Uganda (Aykroyd and Doughty, 1964). Seasonal fluctuations in consumption are also pronounced.

In the traditional farming systems of the tropics such as the rotational bush fallows of tropical Africa, the wet rice-farming systems of south-east Asia, and the compound gardens of both areas, legumes are found as subordinate components in mixture with the starchy major cereal or root and tuber staples. In the humid tropics, grain legumes such as lima beans, cowpeas, and yam beans (*Sphenostylis stenocarpa*) are high risk crops which are frequently grown scattered among other crops or in strategic positions on fences and hedgerows (Okigbo, 1978). In the semi-arid tropics, legumes are also relay-cropped through cereals, maturing on residual moisture after cereal harvest. As a result of their ability to mature rapidly, and because of their nitrogen-fixing ability, grain legumes feature both in the intensive multiple-cropping systems of south-east Asia and in crop rotations patterned on the modern cropping systems typical of temperate agriculture (Okigbo, 1978). Sole cropping of legumes in the tropics is chiefly confined to commercial production, perhaps particularly of soyabeans and groundnuts which are principally commercial oil seeds but beans for canning in eastern Africa, and cowpeas in NE Brazil are also grown as sole crops on a relatively large scale. Thus, changes from subsistence to commercial farming of legumes are usually associated with changes in both husbandry and farming systems.

Legumes in traditional farming systems are notoriously low-yielding. While their low productivity is apparently responsible in some cases for farmers' diverted interest to other crops, often with a consequent disruption of the balance of the local food supply (Stanton, 1966), it is evident that low productivity is not inherent in pulses and that there is considerable potential for their improvement (Rachie and Roberts, 1974); hitherto they have been comparatively neglected. Furthermore, it has been suggested (Bunting, 1975) that low yield in Nigerian pulses is a farming compromise resulting from the inefficient use either of time or of space; farmers generally lack both the resources and the methods to manage the environment properly. But here there seems to be a dilemma: evidence suggests that as the harvestable yield is increased through breeding, the value of these plants as suppliers of biologically fixed nitrogen to agricultural systems decreases.

In view of the external limitations imposed by the climate on the length of the growing season, the timing of field operations is always critical in tropical agriculture. So far as labour constraints allow, there is considerable scope for the development of new and more productive systems of cultivation, including mixed, sequential, and multiple cropping (Bunting, 1975), probably including an important legume component if the system is to be self-sustaining. Recent attempts at modifying traditional practices are producing encouraging results (Baker, 1978; 1980), and underline the importance of appreciating that subsistence agriculture is a balanced, self-supporting tropical agro-ecosystem, as cogently stressed by Igbozurike (1971) and Janzen

Figure 1 Cowpeas intercropped with millet in the Sudan savannah zone of West Arica

(1973). The complexity of the evolution of traditional agricultural systems is best brought out by discussing an example of a farming system involving indigenous crop species. The system we shall consider in detail is the cereal-legume farming system in West Africa (Figure 1) which is comparatively well known. Others will be mentioned more briefly.

1.2.3 Legumes in traditional farming systems

Our example comes from the traditional, largely subsistence farming of northern Nigeria, and nicely shows some of the ways in which traditional cultivars of crops, and traditional farming systems, are adapted with surprising precision both to their social uses and to the time-tables imposed by the environment. The following is based on the account given by Bunting (1975). In semi-arid regions of Nigeria, the effective start of the rains is uncertain but their end is predictable. Here, the farmer has two overriding purposes when the rains begin: to secure the earliest possible crop from part of his land, to break the 'hungry gap'; and to sow and weed as large an area as possible of the staple cereal crops, on which he and his family depend for survival, before newly mineralized nitrogen is leached. The first crop to be sown, on the first light rains, is day-neutral types of bulrush millet; these mature quickly and, being photoinsensitive, independently of the calendar. The main crop, sown when the rains are established, is sorghum. Whenever they are sown, the sorghum cultivars flower at dates precisely related to the average date of the end of the rains in their home locality. If they were to flower too soon, insects and fungi would damage the grains; if they were to flower too late, there would be too little water left in the profile for the grains to fill. The mechanism is short-day photoperiodic.

Into these crops of sorghum, after weeding, farmers usually sow a long-season, spreading, short-day photoperiodic cowpea. These spreading inter-sown legumes help to control late weeds; they may protect the soil from erosion during beating rain; and they supply edible leaves and protein-rich seeds for the diet of the farm family. It is also possible that the cereals may derive some nitrogen from the associated cowpea late in the season to help fill the ears at a time when nitrogen is scarce in the soil; but this 'interspecific symbiosis' has not been proved, although there is evidence (Pulver *et al.*, 1978) that growth habit influences the amount of residual nitrogen turned back into the soil: with traditional, spreading cowpeas more nitrogen (up to 50 kg/ha) may be turned back than with erect cultivars. Photoperiodic sensitivity initiates flowering at a time when sorghum is heading, and its leaf area index begins to decline. So the phenological time-table of the sorghum and of the associated cowpea are closely, even elegantly, coupled (Curtis, 1968; Steele, 1972; Bunting, 1975; Steele and Mehra, 1980).

In Latin America, long-established and evidently stable associations of indigenous cereal-legume crop mixtures are also known. Wild, indeterminate types of *Phaseolus vulgaris* may be found intertwined with a wild relative of maize, teosinte (*Euchlaena mexicana*) (Miranda Colin, 1967), and maize-beans mixtures were an established feature of Amerindian agriculture well before the Spanish conquest (Weatherwax, 1954; Davis *et al.,* 1981). Now some 60 per cent of the maize and 80 per cent of the beans are still produced in crop mixtures (Figure 2; Francis, 1978). In south-east Asia, too, grain legumes are important components of traditional farming systems, of which mung beans (*Vigna radiata*) in rice-based cropping in the Philippines are an example. Although mung beans were domesticated in India by about 1500 BC, their subordinate place in agriculture has meant that traditional cultivars are still grown in edaphic conditions not much different from their native habitats. In lowland rice areas of the Philippines, mung beans are typically broadcast after rice harvest so that they mature into the dry season on residual soil moisture with few inputs. Traditional systems are giving way to more intensive agriculture based upon new, earlier maturing cultivars (Herrera *et al.,* 1978; Litsinger *et al.,* 1978; Jain and Mehra, 1980).

There are few detailed studies of the physiology of cultivar adaptation in traditional systems of mixed cropping. But we assume that traditional cultivars, themselves often products of thousands of years of natural selection during which stress tolerance accumulates, have become part of a balanced agro-ecosystem. The disruption of such long-established systems will have important consequences if the nature of the system is not first understood.

Figure 2 Beans intercropped with maize, sown on the same hill, Colombia (photograph courtesy H. F. Schwartz)

1.3 DISEASE IN LEGUME FARMING SYSTEMS

1.3.1 Systems' balance in tropical subsistence agriculture

Subsistence agriculture tends to select continuously for 'fitness' (in the genetic sense), since dependability is at a higher premium than high yield; crop failure can mean famine to the farmer (Harlan, 1976). This suggests that co-evolved crops in traditional farming systems have achieved an equilibrium not only with one another and with their environment, as we saw in the previous section, but also with their parasites. Any agent that selectively destroys a component of an ecosystem must play a role in the evolution of that system. It is not 'evolutionarily stable strategy' for a parasite to kill its host; in an endemic situation, host and parasite accommodate to each other in a dynamic balance. That this is so is borne out by genetic analysis of obligate parasites and their hosts, suggesting that such systems are the relics of an ancient balanced polymorphism which was a necessary condition for their co-evolution (Mode, 1958). Furthermore, supporting evidence is also obtained from the concentration of resistance genes present in centres of origin of hosts and parasites where they have co-evolved (Leppik, 1970). Examples include certain viruses (Holmes, 1951) and the rust fungi which, as obligate parasites, have evolved in interdependence with their hosts. In centres of origin and genetic diversity, the host population contains a wide spectrum of protective mechanism which is matched by broad spectra of pathogenicity in the parasite population (Browning, 1974; Anikster and Wahl, 1979). Plant pathogens in their most diverse and primitive forms are likely to be found where the ancestors of cultivated crops still exist as wild species, thereby making it possible to trace the evolution of plant pathogens (Watson, 1970b). Such information can help us to understand the origin of epidemics, a point to which we shall return.

Studies on disease in plant communities have shown that the interactions which occur between co-evolved plants and pathogens in undisturbed ecosystems usually involve low disease levels; the continuous genetic interaction that occurs between host and pathogen populations results in the pathogen being held in check by a combination of natural resistance and spatial isolation of hosts (Burdon and Shattock, 1980). It is important to recognize that certain traditional farming systems in the tropics, especially those involving complex crop mixtures, are essentially similar to semi-natural or natural ecosystems in the low levels of disease normally found in them (Buddenhagen, 1978). Robinson's (1976) advocacy of the systems' concept has done much to enhance understanding in this direction, as noted by Putter (1978).

1.3.2 The nature of systems' stability

Genetic diversity within a species tends to increase the stability of its reproductive success (Allard, 1961) as well as contributing to the control of pests and diseases (Suneson, 1960). The use of multiline cultivars has been proposed for disease control in inbreeding cereals (Browning and Frey, 1969). Such cultivars consist of a mixture of lines with similar agronomic phenotypes but with differing vertical (Section 10.2.1) resistance genes. Where some of the component lines are susceptible, disease develops more slowly on the susceptible plants than in pure stands of the susceptible lines (Leonard, 1969). Reviewing the mechanisms by which disease incidence is reduced in multiline cultivars, Browning and Frey (1969) suggested that a proportion of the initial inoculum falls on to resistant lines, thus reducing the effective dose. Of the spores produced by the infected susceptible lines, again only a proportion will land on the susceptible plants, the remainder being deposited on resistant ones; this reduces the rate of increase of the pathogen. Furthermore, the resistant individuals may act as a mechanical barrier to dispersal (the 'fly-paper' effect; Trenbath, 1977). Resistance induced by infection with an avirulent race of a pathogen and effective against subsequent infection with a virulent race may further contribute to resistance in multilines (Johnson and Allen, 1975; Johnson and Taylor, 1976). Another approach is to use mixtures of unrelated cultivars of a crop, each possessing different genes for resistance. Although the approach to disease control is essentially similar to that in multilines, cultivar blends offer greater diversity and flexibility in that their components may be readily varied with both site and season. A problem central to both these approaches, however, is the effect that multilines and cultivar mixtures might have on the evolution of the pathogen. Will 'super-races' arise? Concern over the lack of theoretical understanding of the long-term fate of multilines has generated the development of models and debate (Groth, 1976, 1978; Barrett and Wolfe, 1978; Marshall and Prior, 1979). Current conclusions (Barrett, 1980) are that disease levels will be less in cultivar mixtures or multilines; that disease control increases with time, and with increasing exchange of spores; that 'super-races' will not necessarily evolve; and that pathogen genotypes able to attack more than one component may undergo a rapid increase in frequency at the beginning of the season and then decline later in the season. The behaviour of pathogen populations on mixed hosts is a complex function of the reproduction rate, time, the amount of propagule transfer between host components, the proportions of the components and the fitness of the pathogen genotypes on each of the host components.

To what extent, then, are these conclusions relevant to the behaviour of pathogens in mixtures of different species of crop? While the influence of mixed cropping on pest populations has received some attention (Perrin, 1977; Altieri *et al.,* 1978), very few studies have been made on the effects of cropping system on disease, and little attempt has yet been made to draw parallels between natural plant communities and agricultural ecosystems in this respect (Browning, 1974; Burdon, 1978). The few studies that have been made on the incidence of disease and the rates of spread of pathogens in pure stands and mixed crops do suggest that the outcome depends upon complex interactions between several factors (Aiyer, 1949; Mukiibi, 1976b). For example, in cowpea in association with maize, the severity and rate of spread of *Ascochyta phaseolorum* was greater in sole-cropped cowpea than in the intercrop, apparently because the maize foliage impeded spore dispersal (Moreno, 1975). Powdery mildew (*Erysiphe polygoni*) may initially develop more rapidly in sole-crop cowpea than when intercropped with maize (Moreno, 1979) but when maize populations are sufficiently dense to shade the cowpea, powdery mildew incidence may be much higher in the intercrop (Allen, 1977a). That bean rust (*Uromyces appendiculatus*) severity may be lessened by intercropping *Phaseolus vulgaris* and maize (Moreno, 1979), is accounted for by impeded spore dispersal, and possibly also by induced resistance (Allen, 1975; 1976). The severity of angular leaf spot (*Phaeoisariopsis griseola*) may increase on beans in maize, probably through a more favourable microclimate in the mixture (Moreno, 1977) but when beans are intercropped with cassava, angular leaf spot may be slightly reduced. Chickpea blight (*Ascochyta rabiei*) is also said to be less severe when the crop is intersown with a cereal (Luthra *et al*., 1935).

Mixed cropping may affect the development not only of fungal diseases, as discussed above, but also of other groups of pathogens. Thus, the intercropping of cassava with maize may significantly decrease the incidence of cassava bacterial blight (*Xanthomonas campestris* pv. *manihotis*), apparently by providing shade (Arene, 1976). Similarly, the severity of halo blight (*Pseudomonas syringae* pv. *phaseolicola*) and common blight (*Xanthomonas campestris* pv. *phaseoli*) has been shown to be less when beans are grown in association with maize (Van Rheenen *et al*., 1981; Msuku and Edje, 1982). Beetle-transmitted virus diseases of cowpeas (Shoyinka, 1976; Allen, 1977a; Moreno, 1979) may be reduced in maize mixtures, either through lower vector populations in the mixture (Altieri *et al.,* 1978) or through their reduced mobility (Kayumbo *et al.,* 1976). Similarly, the incidence of groundnut rosette may be reduced in groundnut-maize mixtures, apparently because the aphid vector reproduces less rapidly in the intercrop (Guillemin, 1952). In mixtures of groundnuts and beans, the spread of rosette is decreased because the aphids are trapped by the hooked trichomes of the beans (Farrell, 1976c). And, finally, the incidence of the parasitic witchweed (*Striga hermonthica*) may be reduced in sorghum when intercropped with cowpea, groundnut (C. Parker, personal communication 1977), or mung bean (L. Ohlander, personal communication 1976), presumably because these associated crops stimulate the germination of *Striga* seeds (Johnson *et al.,* 1976). Some of these observed effects are summarized in Table 2. As more work is conducted, the relative importance of the underlying mechanisms involved may become apparent. Much is likely to depend on the temporal and spatial arrangement of the cropping system. For instance, whether or not maize pollen may induce susceptibility to disease in an intercropped legume is liable to depend in part on the time of maize anthesis relative to the period of peak susceptibility of the companion legume (Allen and Skipp, 1982). With pathogens favoured by

TABLE 2 The Influence of Maize as a Companion Crop on Disease Development in Legumes

Crop combination	Pathogen	Parameter	Sole	Mixed	% Difference	Reference
Cowpea/maize	*Ascochyta phaseolorum*	Severity (% area affected)	0.0499	0.0183	−63.3	Moreno, 1975; 1979
Cowpea/maize	*Erysiphe polygoni*	Days to onset of epidemic	46	57	−23.9	Moreno, 1979
Cowpea/maize	*Erysiphe polygoni*	Incidence (mean score, 1–7)	2.8	4.4	+57.1	Allen, 1977a
Cowpea/maize	Cowpea (yellow) mosaic virus	Incidence (mean no. infected plants)	64	25	−60.9	Allen, 1977a
Cowpea/maize	Cowpea severe mosaic virus + cowpea chlorotic mottle virus	Incidence at 38 days after planting	68.6	28.4	−58.6	Moreno, 1979
Bean/maize	*Uromyces appendiculatus*	Severity at flowering time	15.1	10.1	−33.1	Moreno, 1979
Bean/maize	*Phaeoisariopsis griseola*	Severity at flowering time	14.4	17.8	+23.6	Moreno, 1979
Bean/maize	*Colletotrichum lindemuthianum*	Severity (disease index, %)	42.4	28.8	−32.1	Msuku and Edje, 1982
Bean/maize	*Pseudomonas syringae* pv. *phaseolicola*	Severity (disease index, %)	37.7	19.2	−49.1	Msuku and Edje, 1982
Bean/maize	*Xanthomonas campestris* pv. *phaseoli*	Severity (mean score, 0–5)	1.2	0.9	−25.0	van Rheenen *et al.*, 1981

increasing shade, plant density influences the outcome. It has been shown that powdery mildew of cowpea increases with decreasing light intensity, varied both by maize spacing and by artificial shade (Allen, 1977a). In cases where humidity is the principal determinant of disease severity, plant arrangement may be important. It has been shown (Allen, 1977a) that web blight (*Thanatephorus cucumeris*) may be less severe where single rows of cowpeas alternate with rows of maize than in sole cowpea, yet disease severity can be greater in climbing cowpeas supported by maize than when sole-cropped. The variation in protection afforded by different companion crops (Moreno, 1979) may relate to their architecture, but no data are available.

We saw at the outset of this section that systems of traditional agriculture were often stable but that they tended to be low-yielding. Thus, peasant agriculture, in the face of world food shortages, is under pressure to change and to intensify. If intensification is attempted by the continued imposition of alien technology from temperate countries, the results are unlikely to be any improvement over earlier efforts (Janzen, 1973). It may be more prudent to attempt first to analyse existing balanced systems, through experimentation and the development of mathematical models (Trenbath, 1977) which may then be used to predict the long-term effects of a wide range of strategies; the most evolutionarily stable yet productive strategy for crop production might be so identified.

1.3.3 Crop-pathogen co-evolution and the origin of epidemics

During the last decade, an unprecedented amount of attention has been devoted to tropical food crop improvement, and the success of such programmes has often relied initially upon international exchange of plant genetic resources. While germplasm collection, conservation, and evaluation are vital prerequisites of successful plant breeding, such efforts disrupt the systems balance which we have discussed. It seems that most cases of extreme susceptibility and extreme virulence are consequences of man's activities, occurring when host and pathogen are introduced for the first time (Harlan, 1976). Buddenhagen (1977) has examined the nature of epidemic imbalance in tropical crops, classifying epidemics according to their origin. He shows that epidemics, resulting from an intensified union of crop and pathogen following separation, may be grouped according to the type of separation and reunion (Table 3). Type I epidemics are those with a long-established, co-evolved host-pathogen relationship which has either remained relatively constant (IA), or has been recently interrupted and rejoined spatially (IB). Cowpea (yellow) mosaic in Africa is an example of a type IA disease which may become serious through intensification of the cropping system. An example of a type IB disease is groundnut rust in Africa; *Puccinia arachidis* has spread across Africa only during the last decade while groundnuts have been cultivated in West Africa since the sixteenth century (Grigg, 1974). Both host and pathogen are thought to have co-evolved in Latin America (Bromfield, 1974;

TABLE 3 The Origin of Epidemics Relative to Crop-pathogen Evolution (after Buddenhagen, 1977)

I. Long-term crop-pathogen co-evolution
  (A) Smooth or cyclical intensification of relationship
  (B) Reunion of co-evolved systems following recent spatial separation

II. New encounters after ancient (geologic) separation
  (A) Host moved to new region
  (B) Pathogen moved to new region
  (C) New-encounter-developed pathogen moved to separate host location

III. Man-created genetic separation followed by genetic reunion

Rothwell, 1975). A similar situation has occurred with soyabean and *Rhizobium japonicum* which co-evolved in Asia yet, through separate introduction into America, their host-parasite relationship has changed (Devine and Breithaupt, 1980). Of relevance to type IB diseases is the risk of accidental introduction of new races of pathogens on seed from gene centres (Leppik, 1970), especially where intensive cultivation of genetically uniform crops is involved, such as beans, soyabeans (Anon, 1972), and groundnuts (Hammons, 1976) in the USA.

Type II diseases are genetically new-encounter diseases, and follow reunion of two long-separated components of isolated evolutionary systems. These may be a result of intercontinental or regional movement of a crop into a new environment (IIA) or the movement of a pathogen into a new region where it infects a related host (IIB) (Buddenhagen, 1977). Examples of type IIA diseases in legumes include various virus diseases of cowpea in the New-World, to which *Vigna unguiculata* was introduced with the slave trade. Cowpea severe mosaic virus (CSMV) and cowpea chlorotic mottle virus (CCMV), both apparently confined to America, would appear to have become adapted to the introduced cowpea from other legume species. Wild reservoirs of CSMV include the American native *Macroptilium lathyroides* (de Jager, 1979), while the American natives *Desmodium laevigatum* and/or *Phaseolus vulgaris* are likely to be the original hosts of CCMV (Walters and Dodd, 1969; Fulton *et al.,* 1975). This speculation is supported in part by the recognized variation within CCMV. The five naturally occurring isolates so far obtained apparently represent three distinct serological groups; two are confined to beans while the third has been obtained from *Desmodium,* as well as from the exotics, cowpea, and soyabean (Kuhn and Wyatt, 1979). An example of a new-encounter fungal disease is the Australian stem rot of cowpea caused by *Phytophthora vignae* (Purss, 1957) which is apparently absent from Africa where cowpeas originate. Southern bean mosaic in *Phaseolus vulgaris* may perhaps afford an example of a type IIB disease, resulting from the movement of a pathogen into a new region, and thence to a new host. Although southern bean mosaic virus (SBMV) was first described from beans in the USA (Shepherd, 1971), Lamptey and Hamilton (1974) speculate that the wild-type SBMV is of West African origin; comparison of their Ghana isolate with other isolates of SBMV suggests such an evolutionary trend. SBMV has since been obtained from both beans and cowpeas in Togo (Steiner, 1975) and from cowpeas in Nigeria (Shoyinka and Okusanya, 1975) and the Ivory Coast (Givord, 1981). If SBMV is, indeed, of West African origin, then it seems likely that the virus was carried to the New World, perhaps in cowpea seed, thence spreading to *Phaseolus*. Once a new-encounter disease occurs, it may be transferred to a region from which the disease was absent. A hypothetical example might be the introduction of SBMV on bean seed into East Africa, where the virus is at present unknown.

The third type of epidemic disease (III) is a result of overcoming a man-created genetic separation, through plant breeding. It is this group that has caused the greatest concern and includes the classic examples of 'boom and bust' which characterize vertical resistance and its breakdown by emergent matching pathotypes. Cases amongst legumes of type III epidemics have occurred with bean rust, bean anthracnose, and the *Phytophthora* diseases of soyabean and lima bean in the USA.

Not enough is known of pathogen distribution and spread in tropical food crops for accurate reconstruction of the course of host-pathogen co-evolution. Pathogens of the legume crops of concern to us here are best documented in areas to which the crops are exotic (e.g. in the USA), and more complete territorial checklists of plant diseases are urgently required, especially before extensive international exchange of germplasm takes place. Afterwards, it may be more difficult to distinguish potentially dangerous alien introductions from endemic, and often overlooked, pathogens, thus complicating the analysis of the origin of epidemics. The paucity of information is especially acute in the case of virus diseases. The relatively recent origins of virology as a science, and the difficulties in detecting and identifying viruses (Thresh, 1980), make assessment of virus distribution especially difficult. Yet the considerable dependence of agriculture on introduced crops led Bennett (1952) to conclude that newly recognized virus diseases frequently had their origins in new encounters, viruses spreading to the new crop plants from native hosts, or from cultivated species with which the viruses had co-evolved. A recent review (Jones, 1981) shows that this is true of the viruses which infect potatoes in the Andean region of South America. The cosmopolitan distrubution of some legume viruses is probably due to their transmission in seed, and Bos (1977) points out that long association of a virus with its host through seed passage may lead to host adaptation; bean yellow mosaic virus and its allies may be examples.

Chapter 2

Legume Pathogens I

2.1 THE GEOGRAPHICAL DISTRIBUTION AND ECONOMIC IMPORTANCE OF PATHOGENS

The importance of a plant pathogen depends on the economic loss it causes. In turn, the extent of crop loss depends on both the intensity of the disease and its distribution. In the last chapter we discussed the co-evolution of crop species and their parasites since such an approach can assist in the analysis of the origin of epidemics, of relevance to plant-breeding strategy. Similarly, information on the geographical distribution of plant disease can be useful both in setting priorities for disease control and in the possible prediction of disease spread into new areas. Although transport through man's activities, and natural spread, have often led to the introduction of pathogens into new habitats, many remain restricted to specific areas often defined by ecological boundaries. The study of the geography of plant disease, or 'geophytopathology' (Weltzien, 1972), is in its infancy and territorial checklists of plant pathogens (CMI, 1970; 1971; 1979) are scanty (biased toward the distribution of pathologists), incomplete (biased toward economic importance of the host), and occasionally contain erroneous records, especially in the tropics. Nevertheless, examination of distribution maps of plant diseases such as those published by the Commonwealth Mycological Institute, can be instructive. Taking examples from among the diseases of legumes, it is apparent that while pathogens such as bean common mosaic virus, *Colletotrichum lindemuthianum* and *Mycosphaerella berkeleyi,* are widespread and occur virtually wherever their hosts are cultivated, other pathogens have restricted distributions. Examples may be categorized as follows:

(i) Confined to centre of origin of host, e.g. *Sphaceloma arachidis;*
(ii) Localized but absent from centre of origin of host e.g. *Phytophthora vignae;*
(iii) Present in centre of crop origin and locally distributed elsewhere, e.g. *Septoria glycines, Corynebacterium flaccumfaciens;*
(iv) Rapidly increasing distribution, e.g. *Puccinia arachidis*.

While these groups are rather arbitrarily selected, such examples illustrate some of the striking features of pathogen distribution. *S. arachidis* was first reported from Brazil in 1940 since when it has not spread elsewhere. This contrasts with those pathogens (iii above) which are present in the centre of crop origin from where some spread has occurred; these may present a threat of further spread, especially if known to be seedborne, as are both *Septoria glycines* (Punithalingam and Holliday, 1972) and *Corynebacterium flaccumfaciens* (Hayward and Waterston, 1965a). While *Phytophthora vignae* is also very restricted in its known distribution, it represents a new-encounter disease, as we saw in Chapter 1, being absent from the centre of its host's origin.

The general paucity of disease recording in tropical regions often leads to the sudden appearance of new records. This warrants cautious interpretation, if real spread of a pathogen is to be distinguished from new reports of endemic but overlooked pathogens, as will be the inevitable outcome of intensified search. Yet there can be no doubt of the recent dramatic intercontinental spread of *Puccinia arachidis*, as shown by the distribution maps issued between 1948 and 1980 (Figure 3). In such cases, distribution maps rapidly become obsolete.

Special difficulties arise with virus diseases, which we shall consider in Section 3.1. Because viruses are difficult to identify, maps of their distribution in the tropics are prone to inaccuracy, especially when identification has been based on symptomatology alone. An example is the misrecording of soyabean mosaic virus in eastern Africa, where the virus has not yet been critically identified. Since mosaics of soyabean (caused by other viruses) are known in East Africa, it is probable that the early records are referable to these and not to soyabean mosaic virus. With the more widespread use of serological assays (Clark and Adams, 1977; Talley *et al.*, 1980) for detection of plant viruses, we can expect a rapid improvement in the mapping of virus distribution.

Another problem in the interpretation of distribution maps is the frequent difficulty in distinguishing between the distribution of a pathogen and the distribution of a disease. They are not synonymous. For instance, the rust fungus *Phakopsora pachyrhizi* is widespread throughout tropical Africa yet the soyabean rust which it causes is not reliably reported (see Section 2.2.5) from that continent, where soyabeans are increasing in importance. What this means is that mapping of pathogens with relatively wide host ranges (the principal host of *P. pachyrhizi* in Africa is cowpea, although lima bean and black gram are also natural hosts; Allen, 1979a) may give little information on the geographical distribution of a particular disease. Where *formae speciales* are clearly distinguished, these are best mapped separately.

Obviously the incidence and severity of diseases are heavily dependent on ecological factors, and observed distribution patterns need ecological interpretation. While the absence of a disease from a region may merely indicate the absence of the pathogen (thus affording opportunities for its further exclusion through quarantine procedures), the local absence of a disease may suggest the pathogen's epidemiological incompetence. While in the

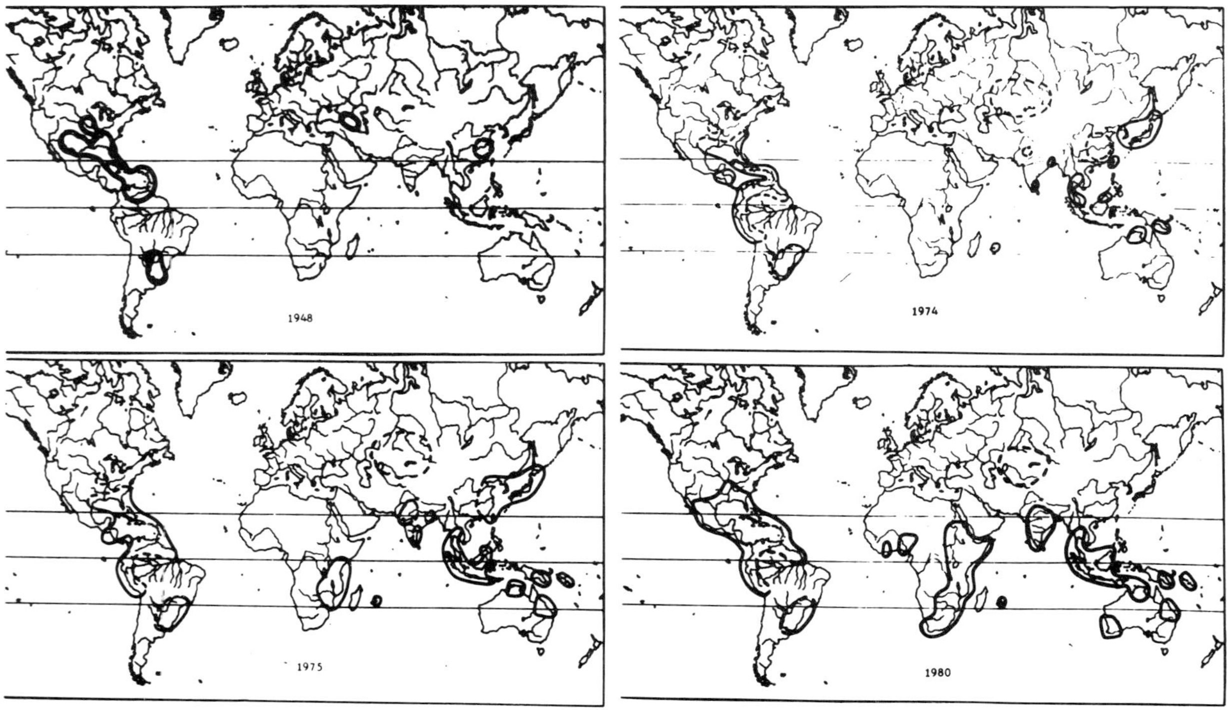

Figure 3 The geographical distribution of *Puccinia arachidis* in 1948, 1974, 1975 and 1980 (reproduced with permission from the Commonwealth Mycological Institute's Distribution Maps)

humid tropics there may be no well-defined dry season, in the savannahs heat and drought determine both the beginning and end of the growing season, as we saw in Chapter 1. The presence or lack of a well-defined close season for plant growth has a profound effect on the epidemiology of many plant diseases. This is reflected in the variation in pathogen priority with ecological zone, as noted with cowpeas in West Africa (IITA, 1980) and beans in Latin America (Schwartz and Gálvez, 1980). In those regions where a pronounced dry season occurs, pathogens must carry over on seed, on alternative (often woody) hosts, or as drought-resistant structures in infected plant debris or soil, if they are to survive. Bacteria, which do not produce resistant spores, often tolerate desiccation and survive for long periods under dry conditions (Schuster and Coyne, 1974). Species producing drought-resistant structures often survive less well under moist conditions, presumably through increased rates of decomposition of debris, and perhaps increased antagonism. When irrigation programmes become widely established in semi-arid regions, breaking the long dry fallow between crops, a change in pathogen populations can be expected.

Waller (1976) points out that some pathogens with airborne spores may not need to survive the dry season in some tropical areas because of the cropping pattern, with the seasons moving to and fro across the equator following the movement of the intertropical convergence zone. As the prevailing winds blow towards this, they can carry airborne inoculum from areas where crops are mature and therefore perhaps heavily diseased, to areas of young crops where the spores can initiate new epidemics. A similar situation may help to explain the annual recurrence of groundnut rosette in northern Nigeria, where no alternative hosts are known and the virus is not seedborne. It has been suggested that viruliferous aphids may be windborne from semi-permanent virus reservoirs in the more humid south (Rossel, 1977).

There is also a close relationship between altitude and disease incidence, principally through the influence of temperature (Waller, 1976). Bacterial blight of soyabean (*Pseudomonas syringae* pv. *glycinea*) is apparently seldom found in the lowland humid regions of Africa, and this presumably relates to the bacterium's intolerance of high temperatures (Sinclair and Shurtleff, 1975), unlike the bacterial pustule organism (*Xanthomonas campestris* pv. *glycines*) which is more widespread. Soyabean downy mildew (*Peronospora manshurica*) and *Ascochyta* blight of cowpea (*A. phaseolorum*) are also apparently restricted to cooler environments at medium elevations in Africa (Allen, 1979a). Soil factors may also have a pronounced influence on plant disease distribution, and Whitehead (1969) has shown that where hosts occur the abundance of root knot nematodes (*Meloidogyne* spp.) depends on soil factors, especially soil moisture and temperature. *Meloidogyne javanica* was found to be more abundant than *M. incognita* in wetter areas of intermediate altitude in East Africa, whereas the two species were equally abundant in drier areas of similar altitude. *M. incognita* was more successful in hotter soils and *M. javanica* in cooler, wetter soils. When more data on pathogen distribution become available, there would be considerable merit in plotting records of occurrence on a grid system, such as used for other biological records (e.g. Perring and Walters, 1962), in preference to the delineation of pathogen ranges as used at present in the CMI Distribution Maps. The former would permit easier ecological analysis of pathogen distribution.

In discussing the geographical distribution of plant disease, we have considered only the presence or absence of that disease, or of its causal agent, within a given area. Weltzien (1972) has emphasized that the value of disease distribution maps can be further increased by attempting to distinguish degrees of disease intensity within the area of its occurrence, distinguishing areas of main damage, marginal damage, and sporadic attack. We shall now consider some aspects of the measurement of disease intensity, the second factor determining the economic importance of a pathogen.

Although it is firmly established that plant diseases constitute major constraints to crop production, there are very few reliable data quantifying these crops losses (James and Teng, 1979). Fewer still have attempted to analyse the relative importance of production constraints in tropical food crops. An exception is the study made by Pinstrup-Andersen *et al.* (1976) who examined data from field surveys of bean production in the Cauca Valley of Colombia. Their crop loss profile revealed that among the six yield constraints identified, three diseases (rust, bacterial blight, and angular leaf spot) together accounted for 50 per cent of the yield loss. More commonly, field surveys have been employed to assess the incidence of a particular disease, or diseases, within a region. For example, Nene (1972) has surveyed pulse crops in Uttar Pradesh, measuring the incidence of certain virus diseases within districts. Aerial surveys, using remote sensing techniques to quantify disease severity from aerial photographs, have been used for the assessment of bacterial blight of beans (Jackson and Wallen, 1975; Wallen and Jackson, 1975), but such methods are likely to remain largely inappropriate to tropical food crops.

Estimates of yield losses may be based on experimental data from the study of a single disease, from which the relationship between disease progress and crop loss is established. The study of cercospora leaf spot in cowpea affords a good example (Schneider *et al.,* 1976). But such studies have seldom been made of the legume diseases and yield loss estimates are probably rather inaccurate. Furthermore, it is difficult to extrapolate from such estimates, made under experimental conditions, to those likely to be incurred in agricultural practice where crop yields are limited by a complex of stress factors of which diseases are but part. Until more studies have been carried out, systems of field rating of legume diseases must continue to rely on largely arbitrary scales whose relationship to crop loss is uncertain.

We shall now review the major groups of legume pathogens, starting with the fungi. Emphasis will be placed on their biology and epidemiology and no attention will be given to their morphology or physiology of parasitism as these are beyond the scope of this book.

2.2 FUNGAL PATHOGENS OF LEGUMES

The fungi are the most important group of pathogens and a very wide range of species are recognized as parasites of the legume crops with which we are concerned. Relative to the bacteria and viruses, fungi are comparatively easy to identify and mycological records are relatively well documented, to which the territorial checklists (CMI, 1970, 1979) will testify. The availability of identification services, such as those provided by the Commonwealth Mycological Institute at Kew and the Centraalbureau voor Schimmelcultures at Baarn, has vastly contributed to the accurate recording of plant pathogenic fungi in the tropics.

The legume pathogens have representatives among most of the major groups of the fungi. Their diversity is illustrated in Table 4 which lists selected species of the majority of genera parasitic on the legumes. Some genera contain numerous species of importance as legume pathogens and, for present purposes, it seems preferable to focus attention on certain selected groups which between them include the majority of the more important pathogens while providing useful contrasts in biology and modes of spread. Thus, the next section will discuss the ecology of legume pathogenic species of *Phytophthora, Fusarium, Cercospora* and allied genera, the anthracnose and scab fungi, and the rust fungi. But, before doing so, let us briefly consider the nature of diseases caused by fungi, drawing attention to cases where further aetiological investigation appears warranted.

Certain fungi including species of *Pythium*, *Fusarium*, and *Rhizoctonia* are commonly responsible for seed and seedling diseases in a wide range of legumes, including cowpea (Williams, 1975a), chickpea (Singh and Mehrotra, 1980), winged bean (Price and Monroe, 1978a), bean (Gay, 1969a), and soyabean (Sinclair, 1977). Species of *Phomopsis* are involved in soyabean seed decay (Ndimande *et al.*, 1981; Shortt *et al.*, 1981) and *Aspergillus* spp. cause seedling and pod disease in groundnuts (Gibson, 1953; McDonald, 1969). Collar and root rots, characterized by a progressive rotting of the root system or basal portion of the stem leading to chlorosis and wilting, are also caused by species of *Pythium* (Lehman and Wolf, 1926a; Onuorah, 1973; Hoch *et al.*, 1975), *Fusarium* (French and Kennedy, 1963; Steadman *et al.*, 1975; Singh and Allen, 1980), and *Rhizoctonia*, as well as by *Calonectria, Corticium, Mycoleptodiscus, Operculella, Phymatotrichum,* and *Thielaviopsis* (Sinclair and Shurtleff, 1975; Nene *et al.*, 1979; Schwartz and Gálvez, 1980). Diseases of the stem include brown stem rot of soyabeans (*Cephalosporium*), charcoal rot of various legumes caused by *Macrophomina phaseolina*, *Diaporthe* stem blight and *Neocosmospora* stem rot of soyabean, and various species of *Colletotrichum* and *Sphaceloma* which will be considered in Section 2.2.4. True wilts are those in which wilting occurs before there is extensive damage to the root system. Vascular tissues of affected plants are typically discoloured. Two principal genera are involved: *Fusarium* and *Verticillium*, of which the first is the more important in the legumes.

TABLE 4 Some Representative Species of Legume Pathogens Among the Orders of Fungi

Order	Species	Selected synonyms and/or imperfect states	Natural hosts	References
Chytridiales				
	Synchytrium psophocarpi (Rac.) Gäumann	—	Winged bean	Drinkall, 1978
Peronosporales				
	Pythium aphanidermatum (Edson) Fitz.	—	Cowpea, and many others	Onuorah, 1973
	Phytophthora drechsleri Tucker f.sp. *cajani* Kannaiyan *et al.*	—	Pigeon pea	Kannaiyan *et al.* 1980
	Peronospora manshurica (Naoum.) Syd. ex Gaum.	—	Soyabean	Sinclair and Shurtleff, 1975
Mucorales				
	Choanephora cucurbitarum (Berk. and Rav.) Thaxt.	—	Cowpea, pigeon pea, and many others	Lefebvre and Weimer, 1939; Wilson and Jose, 1968; Misra and Mehra, 1969
Erysiphales				
	Erysiphe polygoni DC ex Merat	*Oidium* sp.	Bean, cowpea, and many others	Schwartz and Gálvez, 1980
	Leveillula taurica (Lév.) Arn.	*Oidiopsis taurica* (Lév.) Salm.	Pigeon pea, cluster bean, and many others	Mukerji, 1968
Myringiales				
	Elsinoë phaseoli Jenk.	*Sphaceloma* sp.	Lima bean, bean, and others	Sivanesan and Holliday, 1971
Helotiales				
	Sclerotinia sclerotiorum (Lib.) de Bary	—	Bean, soyabean, chickpea, and many others	Kaiser, 1972a; Sinclair and Shurtleff, 1975; Mordue and Holliday, 1976; Schwartz and Gálvez, 1980
Hypocreales				
	Calonectria crotalariae (Loos) Bell and Sobers	*Cylindrocladium crotalariae* (Loos) Bell and Sobers	Groundnut, soyabean, and others	Bell and Sobers, 1966; Sinclair and Shurtleff, 1975
	Nectria haematococca Berk. and Br.	*Fusarium solani* (Mart.) Sacc.	Bean, cowpea, chickpea, pigeon pea, and many others	Booth, 1971
Sphaeriales				
	Diaporthe phaseolorum (Cke. and Ell.) Sacc.	*Phomopsis sojae* Leh.	Soyabean, bean, cowpea, lima bean, and others	Athow and Caldwell, 1954; Punithalingham and Holliday, 1972a

TABLE 4 (*continued*)

Order	Species	Selected Synonyms and/or imperfect states	Natural hosts	References
	Glomerella glycines (Hori) Lehman and Wolf	*Colletotrichum glycines* Hori ex Hemmi	Soyabean	Lehman and Wolf, 1926b
	Guignardia arachidis Punith.	*Phyllostictina* sp.	Groundnut	Punithalingham, 1974
	Leptosphaerulina trifolii (Rost.) Petr.	—	Cowpea, soyabean, and other hosts	Booth and Pirozynsky, 1967; Anahosur *et al.*, 1974
	Mycosphaerella cruenta Latham	*Pseudocercospora cruenta* (Sacc.) Deighton	Cowpea, and others	Mulder and Holliday, 1975; Deighton, 1976
	Neocosmospora vasinfecta Smith	—	Soyabean, chickpea, pigeon pea	Sarojini, 1954; Phillips, 1972; Haware and Nene, 1976; Oyekan, 1978
Uredinales				
	Phakospora pachyrhizi Syd.	see Table 9		
	Puccinia arachidis Speg.	see Table 9		
	Uromyces appendiculatus (Pers.) Ung.	see Table 9		
Ustilaginales				
	Entyloma vignae Batista *et al.*	—	Bean, cowpea	Vakili, 1978; Prabhu and Albuquerque, 1980
Aphyllophorales				
	Thanatephorus cucumeris (Frank) Donk	*Rhizoctonia solani* Kuhn	Bean, cowpea, soyabean, and very many other hosts	Mordue, 1974a
	Corticium rolfsii Curzi	*Sclerotium rolfsi* Sacc.	Groundnut, bean, cowpea, soyabean, and others	Mordue, 1974b
Sphaeropsidales				
	Ascochyta rabiei (Pass.) Labr.	—	Chickpea	Punithalingham and Holliday, 1972b; Kaiser, 1973
	Septoria glycines Hemmi	—	Soyabean	Punithalingam and Holliday, 1972c
	Chaetoseptoria wellmanni Stev.	—	Bean, cowpea, and other legumes	Stevenson, 1946; Muller, 1953
	Phaeoseptoria sp.	—	Soyabean, winged bean	Allen, 1979a
	Aristastoma guttulosum Sutton	—	Cowpea, lima bean	Allen, 1979a

TABLE 4 (*continued*)

Order	Species	Selected Synonyms and/or imperfect states	Natural hosts	References
	Pyrenochaeta glycines R.D. Stewart	—	Soyabean	Stewart, 1957
	Macrophomina phaseolina (Tassi) Goid.	*Rhizoctonia bataticola* (Taub.) Butl.	Bean, cowpea, groundnut, soyabean, and many others	Holliday and Punithalingham, 1970
Melanconiales				
	Colletotrichum lindemuthianum (Sacc. and Magn.) Br. and Cav.	—	Bean, cowpea, and other legumes	Mordue, 1971; Onesirosan and Barker, 1971
	Sphaceloma arachidis Bitanc. and Jenk.	—	Groundnut	Bitancourt and Jenkins, 1940
	Pestalotiopsis versicolor (Speg.) Steyaert	—	Lima bean, groundnut, and other hosts	Allen, 1979a
Moniliales				
	Cercospora canescens Ell. & Mart	—	Soyabean, bean, cowpea, and other legume hosts	Mulder and Holliday, 1975a
	Fusarium udum Butl.	—	Pigeon pea	Booth, 1978
	Verticillium dahliae Kleb.	—	Groundnut	Purss, 1961
	Cephalosporium gregatum All. and Chamb.	*Phialophora gregata* All. and Chamb.	Soyabean	Allington and Chamberlain, 1948
	Botrytis cinerea Pers. ex Fries	*Sclerotinia fuckeliana* (de Bary) Fuckel	Bean, and many other hosts	Ellis and Waller, 1974
	Aspergillus flavus Link ex Fries	—	Groundnut, and other hosts	Onions, 1966; McDonald, 1969
	Mycovellosiella phaseoli (Drummond) Deighton	*Ramularia phaseoli* (Drummond) Deighton	Bean	Deighton, 1974
	Thielaviopsis basicola (Berk. and Br.) Ferr.	—	Soyabean, bean	Lockwood *et al.*, 1970
	Cladosporium vignae Gardn.	—	Cowpea	Gardner, 1925
	Corynespora cassiicola (Burk. and Curt.) Wei.	—	Bean, cowpea, soyabean, and numerous other hosts	Ellis and Holliday, 1971
	Alternaria spp.	—	Bean, soyabean, groundnut	Aulakh, 1969; Sinclair and Shurtleff, 1975; Schwartz and Gálvez, 1980
	Stemphylium botryosum Wallr.	*Pleospora herbarum* (Fr.) Raebenh.	Soyabean	Anahosur and Fazalnoor, 1972
	Phaeoisariopsis griseola (Sacc.) Ferraris	—	Bean	Schwartz and Gálvez, 1980
Mycelia sterilia				
	Dactuliophora tarrii Leakey	—	Cowpea, mung bean, lima bean, bean	Leakey, 1964; Mukiibi, 1969a

Diseases characterized by sudden and serious damage to all aerial parts of the plant are often referred to as blight. Examples of fungal blights are the phytophthora blights of cowpea (Purss, 1957) and pigeon pea (Kannaiyan *et al.*, 1980), the ascochyta blights of chickpea (Punithalingham and Holliday, 1972b), bean and cowpea (Sutton and Waterston, 1966; Singh and Allen, 1979), rhizoctonia web blight and white mould (*Sclerotinia sclerotiorum*) of bean (Schwartz and Gálvez, 1980). In many leaf diseases the area of necrosis is limited and the main symptom is simply a spot, of varying size and shape. Numerous leaf spot fungi occur on legumes, the most important group being *Cercospora* and allied genera; these fungi will be the subject of a subsequent section. Other locally important leaf spots are caused by species of *Septoria* on cowpea and soyabean, *Chaetoseptoria wellmanii* on bean and cowpea, and various species of *Alternaria* and *Phyllosticta* on many legume hosts. *Pyrenochaeta glycines* is locally important on soyabeans in eastern Africa, and *Dactuliophora* species cause zonate leaf lesions on legumes in tropical Africa. *Phaeoisariopsis griseola* is amongst the most widespread and important leaf pathogens of bean on which it causes angular leaf spot in both Latin America and eastern Africa. *Corynespora cassiicola* is the cause of target spot on a very wide range of hosts, including cowpea, soyabean, and cluster bean.

In some diseases, the presence of a pathogen is more obvious than any change it induces in the plant. The mildews and rusts are examples in which their visible growth are signs of the disease, as distinct from true symptoms which are usually a consequence of host reactions. In the downy mildews, a white or grey 'bloom' develops on lesions under humid conditions owing to the production of sporangiophores while, in the powdery mildews, the presence of a superficial white mycelium and fruiting bodies on the host surface is characteristic. Downy mildews are represented by *Peronospora manshurica* which infects soyabean, and by *P. lentis* of lentil (Beniwal and Srivastava, 1968). A disease caused by *Phytophthora phaseoli* in beans and lima beans is often referred to as 'downy mildew' but this usage is misleading; the term downy mildew is normally reserved for diseases caused by the Peronosporaceae (Wheeler, 1969). Powdery mildews, which generally are not well represented in the tropics (Weltzien, 1978), are reported for most of the legume crops, the majority being attributed to *Erysiphe polygoni*. *Microsphaera diffusa* is the cause of powdery mildew on soyabean (Paxton and Rogers, 1974; McLaughlin *et al.*, 1977) and *Erysiphe cichoracearum* is apparently responsible for powdery mildew of winged bean (Price, 1977). *Leveillula taurica*, in which the mycelium is largely intercellular rather than superficial, causes powdery mildew of pigeon pea and cluster bean (Mukerji, 1968; Nene, 1978).

The rust fungi which, like the powdery mildews, are airborne, biotrophic pathogens, derive their name from the orange-brown spore masses which many of these fungi produce. Each species of rust fungus may have from one to five spore forms and associated fruiting structures which may, or may not, occur on widely different host species. The rusts will be dealt with in detail in a subsequent section, using terms as defined by Laundon (1967). The smut fungi, to which the rusts are closely related, are so-called because of the soot-like masses of spores which many of them produce. This group is well reviewed by Ainsworth and Sampson (1950). Although the smuts are of considerable importance to many crops, especially the cereals, very few are known to affect the legumes. Smut galls have been found in soyabean seed lots but, despite an earlier claim (Whitehead and Thirumalachar, 1960), it appears that the smut is merely a contaminant not a soyabean pathogen (Schiller and Sinclair, 1979). A notable exception is *Entyloma vignae* which causes an important black leaf smut of beans and cowpeas (Figure 4) in Central and South America (da Ponte, 1966; Schieber and Zentmyer, 1971; Vakili, 1972; Vakili, 1978a). Since the taxonomy of this fungus is controversial, it is appropriate to discuss it further.

Figure 4 Leaf smut (*Entyloma vignae*) of cowpea, Grosso, Pernambuco, NE Brazil

E. vignae was first described from Brazil by Batista *et al.* (1966). More recent studies have demonstrated that the product of chlamydospore germination is, indeed, an aseptate promycelium bearing primary sporidia; the sporidia after fusion in pairs give rise to a mycelium which forms secondary, H-shaped structures (Prabhu and Albuquerque, 1980). In 1975, an apparently similar disease was observed for the first time on cowpea in western Nigeria; collections were subsequently identified at the Commonwealth Mycological Institute as belonging to *Protomycopsis phaseoli* Ramakrishnan and Subraman, known previously from mung bean, black gram, and cowpea in India (Prasad *et al.*, 1962). The cytology of chlamydospore development and germination in *Protomycopsis* species has also been studied (Haware and Pavgi, 1976a, b). With the advent, in Brazil, of a cowpea improvement programme involving the exchange of breeding material with Nigeria, it became important to determine whether the diseases in Africa and tropical America were the same; to this end, collections from Brazil, Nigeria and elsewhere in Africa were compared at the Commonwealth Mycological Institute. All were identified as *Protomycopsis phaseoli*, suggesting that *P. phaseoli* and *E. vignae* were synonyms (Allen, 1979a). However, Dennis (1968) has warned that the two genera are easily confused since both induce similar symptoms in their hosts, the tissues of which are filled with spherical chlamydospores. While the vital distinction between the two depends on the different products of chlamydospore germination, Dennis suggests that a useful supplementary distinction can be found in the size of the chlamydospores. It would be unexpected to find an *Entyloma* with chlamydospores exceeding 25 μm in diameter. Early studies on *Protomycopsis* in India led to its being confused with *Synchytrium* (Patel *et al.*, 1949; Pavgi and Thirumalachar, 1953). While *Synchytrium* is a chytrid in the class Phycomycetes, *Protomycopsis* is a hemiascomycete genus, and *Entyloma* species are true smuts in the Basidiomycetes. Thus, there is need for comparative studies in this important group of legume pathogens. But let us now return to consider the ecology of certain selected groups of fungal pathogens.

2.2.1 *Phytophthora* species as legume pathogens

The fungus *Phytophthora infestans* causes late blight of potatoes, which played a central role in the Irish potato famine of the mid-nineteenth century, and it is for this that the genus *Phytophthora* is notorious. Several species are responsible for important diseases in legume crops: *P. megasperma* f.sp. *glycinea* causes a root and stem rot of soyabean in North America (Figure 5); *P. drechsleri* f.sp. *cajani* is the cause of blight in pigeon pea in India (Figure 6); *P. vignae* causes a stem rot of cowpea in Australia; and *P. phaseoli* causes a rot, often erroneously referred to as 'downy mildew', of bean (Figure 7) and lima bean in the Americas. The distribution and hosts of these and other species pathogenic to legumes are shown in Table 5; these species represent five of the six taxonomic groups of the genus, as summarized by Newhook *et*

Figure 5 Phytophthora rot of soyabean caused by *P. megasperma* f.sp. *glycinea* (photograph courtesy J. B. Sinclair)

al. (1978), to which reference should be made for details of morphology and taxonomy. At the sub-specific level, host-specialized variants have generally been assigned to varieties in *Phytophthora,* but Kuan and Erwin (1980) suggest that varietal status may be invalid and their use of *formae speciales* is followed in taxa where this has been proposed. Ecotypes within species have been distinguished (Shepherd and Pratt, 1973), and pathogenic races have also been recognized (Wester, 1968; Purss, 1972; Schmitthenner, 1972). Some features of the ecology of *Phytophthora* pathogens of legumes will be considered in the following section.

Species of *Phytophthora* may persist in soil, or in infected debris, for considerable periods. *P. megasperma* f.sp. *glycinea* may carry-over in field soil in the USA (Slusher and Sinclair, 1973), *P. phaseoli* overwinters in infected lima bean pods in the USA (Cox and Hyre, 1951); *P. vignae* survives for at least

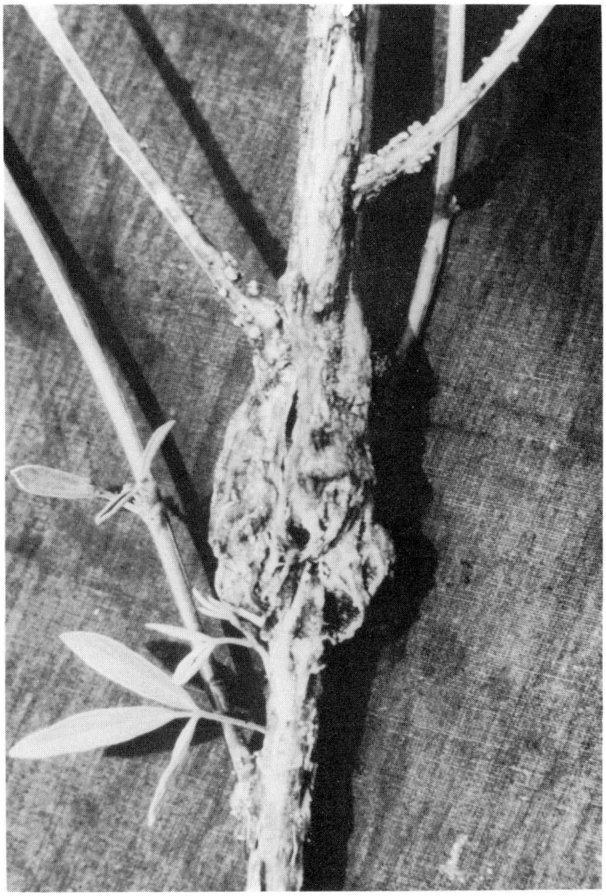

Figure 6 Phytophthora blight of pigeon pea caused by *P. drechsleri* f.sp. *cajani* (photograph courtesy Y. L. Nene)

two years under Queensland conditions (Purss, 1953), and *P. drechsleri* f.sp. *cajani* apparently survives for at least four years in the absence of its host in India (Nene, 1980). While successive cropping with susceptible hosts often builds up the pathogen population in soil (Purss, 1957), rotation with non-host crops for four years may be insufficient to reduce the initial inoculum to sub-economic levels (Apple, 1963). Species of *Phytophthora* may differ widely in their ability to survive in soil, and care needs to be taken in extrapolating from one species to another within the genus (Park, 1965). While the production of chlamydospores may be the primary means of survival of some species (Mircetich and Zentmyer, 1966; Hwang and Ko, 1978), none of the legume pathogens has yet been shown to produce chlamydospores (Newhook *et al.,* 1978) so that oospores probably afford the chief

Figure 7 Phytophthora pod rot of bean caused by *P. phaseoli* (photograph courtesy of H. F. Schwartz)

means of carry-over. Large numbers of oospores are formed in the roots and stems of soyabean in which *P. megasperma* f.sp. *glycinea* survives long periods (Sinclair and Shurtleff, 1975). However, owing to the difficulties in germinating oospores experimentally, precise data on the role oospores play in perpetuation of *Phytophthora* spp. are generally sparse (Erwin and McCormick, 1971). There is evidence (Blackwell, 1943 cited by Garrett, 1970) that individual oospores may vary considerably in their period of dormancy, and such physiological heterogeneity within a population of oospores (of *P. cactorum*) promotes maximum longevity of the population, because no single set of environmental conditions will induce more than a small proportion of the population to germinate at any one time. Thus, the oospore population provides against wide environmental variation.

Since oospores are products of sexual reproduction in *Phytophthora,* it is clearly of cardinal significance as to whether a given species is homothallic or

TABLE 5 *Phytophthora* Species as Legume Pathogens

Species	Taxonomic group[a]	Distribution	Natural hosts[b]	References
P. cactorum (Leb. and Cohn) Schroet.	I	USA, widespread on other hosts	Cowpea, many others	Weimer, 1949
P. citrophthora (Sm. and Sm.) Leon.	II	Argentina, widespread on other hosts	Chickpea, [citrus], and others	Frezzi, 1950
P. drechsleri Tucker	VI	Australia, Zimbabwe, and elsewhere	Velvet bean, various others	Anon, 1956; Sturgess and Egan, 1960
P. drechsleri Tuck. f.sp. *cajani* Kannaiyan *et al.* (= *P. drechsleri* var. *cajani* Pal *et al.* = *P. cajani* Amin *et al.*)	VI	India	Pigeon pea, *Atylosia* spp.	Pal *et al.*, 1970; Williams *et al.*, 1975; Amin *et al.*, 1978; Kannaiyan *et al.*, 1980
P. megasperma Drechs.	V	Argentina, India, widespread on other hosts	Chickpea, and various other hosts	Frezzi, 1950; Suryanarayana and Pathak, 1968
P. megasperma Drechs. f.sp. *glycinea* Kuan and Erwin (= *P. megasperma* Drechs. var. *sojae* Hildebrand)	V	USA, Canada	Soyabean, lupin, and others	Hildebrand, 1959; Waterhouse and Waterston, 1966; Sinclair and Shurtleff, 1975; Kuan and Erwin, 1980
P. nicotianae Breda de Haan var. *parasitica* (Dast.) Waterh. (= *P. parasitica* Dastur)	II	India, Puerto Rica and elsewhere, widespread on other hosts	Bean, pigeon pea, and many others	Venkatakrishnaiya, 1950; Kaiser and Melendez, 1978; Schwartz and Gálvez, 1980
P. phaseoli Thaxt.	IV	Temperate and tropical America, Zaire and elsewhere	Bean, lima bean and others	Thaxter, 1889; Hendrickx, 1939; Schwartz and Gálvez, 1980
P. vignae Purss	VI	Australia, India, Japan	Cowpea, *Vigna angularis*	Purss, 1957; Nirwan and Upadhyaya, 1972; Kitazawa *et al.*, 1978

[a] Newhook *et al.*, 1978.
[b] Food legume hosts only.

heterothallic. In homothallic species, the formation of sex organs (oogonia and antheridia) occurs in cultures of a single isolate, while heterothallic species normally depend on the presence of both compatibility types (called A1 and A2 mating types) for the formation of oospores. Amongst the major legume pathogenic species, *P. megasperma* f.sp. *glycinea, P. phaseoli* and *P. vignae* are homothallic (Purss, 1957; Savage *et al.,* 1968) and *P. drechsleri f.sp. cajani* is heterothallic (Kannaiyan *et al*., 1980). It often appears that one or other of the mating types of heterothallic species is relatively uncommon in nature. This appears to be the case in *P. drechsleri* f.sp. *cajani* since all the isolates examined by Kannaiyan *et al.* (1980) were found to belong to the A1 mating type. Thus, if successful carry-over depends on the production of oospores, rather than on chlamydospores (Cother and Griffin, 1973), it seems probable that oospore production relies more on selfing of single isolates than on chance pairing. Under certain conditions, some isolates of *P. drechsleri* f.sp. *cajani* may be homothallic (Kannaiyan *et al.,* 1980), possibly due to the presence of an extra chromosome (Mortimer *et al.,* 1977), and in other circumstances oospores may be induced to form by other species of *Phytophthora* (Shepherd, 1978; Kannaiyan *et al.,* 1980; Boccas, 1981; Skidmore *et al.,* 1981) in crosses between sexually compatible isolates. Sexual reproduction, and thus oospore production, in single A2 mating type isolates may also be induced by *Trichoderma viride* (Brasier, 1971) which is a widespread and abundant soil inhabitant. It has been suggested that the '*Trichoderma* effect' may have evolved as a specific defence mechanism, since the formation of oospores by *Phytophthora* at the approach of such an antagonist would be of survival value (Brasier, 1975), despite the number of other micro-organisms able to parasitize oospores (Sneh *et al.,* 1977). Although it appears that the heterothallic *Phytophthora* species common on roots of woody hosts (e.g. *P. cinnamomi*) may be more ecologically specialized for response to *Trichoderma* than other species, Brasier (1975) has suggested that such specialization could lead to speciation in the heterothallic species, possibly accounting for the occurrence of ecotypes as recognized in *P. drechsleri* (Shepherd and Pratt, 1973).

Other means of carry-over may include alternate host reservoirs, perhaps particularly on woody perennials, and in species of *Phytophthora* with wide natural host ranges such as *P. cactorum* and *P. nicotianae* var. *parasitica.* But the principal legume pathogens apparently have narrow host ranges and alternate hosts are not recognized as playing a significant role in the epidemiology of the diseases they cause. *Phytophthora* species are seldom reported to be seed-transmitted in legumes (Purss, 1957; Nene, 1980) and carry-over on, or in, seed is probably an unimportant means of their survival. The few reports of seed transmission include the overwintering of *P. phaseoli* on lima bean seed (Wester *et al.,* 1966), the transmission of *P. drechsleri* with velvet bean seed (Bates, 1961), and the isolation of *P. megasperma* f.sp. *glycinea* from immature soyabean seed, as well as from soil inadvertently mixed with seed samples following harvest (Klein, 1959).

Field infection and spread is favoured by moisture. Phytophthora rot of soyabean is usually more severe on the heavier clays than on lighter soils (Hildebrand, 1959), and poorly drained sites and local waterlogging increase the incidence also of the phytophthora blights of cowpea (Purss, 1957) and pigeon pea (Nene, 1980). Soil chemical factors influence the development of phytophthora diseases in non-legumes (Kincaid and Gammon, 1954) through affecting sporangium production (Gooding and Lucas, 1959; Allen and Nandra, 1975). Typically, sporangia are borne on sporangiophores which may be produced on infected debris or on diseased plant parts. In *P. vignae,* sporangiophores often emerge from stomata on disease lesions, leading to aerial spread and infection (Purss, 1957). Sporangia germinate, releasing abundant motile zoospores which are dispersed in soil water, along drainage channels and in flooded soil. The production of sporangia and subsequent release and motility of zoospores is regulated by moisture, optimum conditions varying with the species. Thus, *P. megasperma* produces sporangia in flooded soil, with production decreasing with decreasing water potential. Sporangia germinate both directly (via a germ tube) and indirectly (via zoospores) at −0.05 bars water potential, but direct germination may also occur at −0.1 bars (Pfender *et al.*, 1977). In marked contrast, *P. drechsleri* may produce abundant sporangia in soil at lower potentials (between −0.3 and −0.4 bars), none being formed in flooded soil. Zoospores may be released in soil at −0.3 bars water potential (Duniway, 1975). This suggests that different *Phytophthora* species may have inherently different water requirements for the production of sporangia, and this, in turn, could have important ecological implications. Temperature also influences asexual reproduction in *Phytophthora*. Zoospores of *P. megasperma* f.sp. *glycinea* remain motile longest at 15 °C, a temperature at which repeated emergence of zoospores is favoured, whereas encysted zoospore germination and germ-tube growth may be greatest at a high temperature (25 °C) (Ho and Hickman, 1967a). Such effects of temperature may possibly relate to the relatively narrow temperature range over which aggressive parasitism by *P. vignae* occurs, relative to the much wider range of temperature for fungal growth (Purss, 1957).

The motility of zoospores clearly provides an advantage over other fungal propagules since it enables them to collect in areas favouring infection, thus increasing their inoculum potential and subsequent disease incidence. Although zoospores, and encysted zoospores, do not retain infectivity for long periods, there is evidence that species vary in the survival of zoospores. Zoospores and cysts of *P. drechsleri* have been found to survive and remain infective for as long as 15 days while the infectivity of zoospores of *P. megasperma* f.sp. *glycinea* may apparently be lost within one day. This suggests that *P. drechsleri* possesses superior survival ability: not only may its propagules persist longer as transportable units in soil, but also its per centage cyst germination may be higher and hyphal persistence longer (Mehrotra, 1972), in addition to tolerating drier soils, as we have seen. Since there is evidence that zoospores of *P. megasperma* f.sp. *glycinea* do not persist in soil as germ tubes, and that encysted zoospores generally fail to germinate in nature (Ho, 1969), it would seem that such a species might gain advantage from its rapid finding of a suitable host. While the swimming capacity of zoospores would provide such an advantage, further benefit would accrue if swimming were directional. It has been shown that zoospores are, indeed, attracted to, and trapped in the immediate vicinity of, the root tip where they accumulate, encyst, and germinate at the side closest to the root surface towards which the germ tube subsequently grows (Ho and Hickman, 1967b). However, with one possible exception, the chemotaxis of zoospores of *Phytophthora* spp., including *P. megasperma* f.sp. *glycinea,* appears to be non-specific with respect to host and non-host, as well as to resistant and susceptible cultivar (Zeutmyer, 1966; Ho and Hickman, 1967b).

Discussion of spread so far has been concerned only with local spread of *Phytophthora* pathogens. Let us now turn to consider possible mechanisms of long-distance dissemination. Evidence for the introduction of *P. cinnamomi* into the Americas (Zentmyer, 1977) and Australia (Weste, 1974) suggests that the fungus is of Asian origin (Ko *et al.*, 1978). The Latin American origin of *P. infestans* is better established. In each case, the pathogen appears to have spread with a vegetatively propagated host. Although *Phytophthora* species are occasionally reported to be seedborne in legumes (Klein, 1959; Bates, 1961; Wester *et al.,* 1966), there is no evidence that seed transmission is responsible for new outbreaks, nor is there evidence that long-distance dissemination has occurred. On present evidence, the most probable explanation for the appearance of *P. vignae* on cowpea in Asia and Australia, and *P.*

Figure 8 A suspected phytophthora rot of cowpea; Ilonga, Tanzania (photograph courtesy J. B. Smithson)

megasperma f.sp. *glycinea* on soyabean in North America, in areas remote from the centre of origin of their respective hosts, is that an alternative native plant species is the original host. This is supported by the findings that native lupins are hosts of *P. megasperma* f.sp. *glycinea* in the USA (Sinclair and Shurtleff, 1975) and that the Asiatic *Vigna angularis* is a host of *P. vignae* (Kitazawa *et al.,* 1978). Whether or not this hypothesis has validity will depend partly on the identity of a *Phytophthora* sp. thought to be the cause of a blight of cowpea in Tanzania (Figure 8) (P. N. Patel and C. Kuwite, personal communication, 1979), the first record of *Phytophthora* on cowpea in Africa. If *P. vignae* is involved, seed transmission may indeed have played a role in long-distance dissemination. But it is plausible that, again, the fungus is endemic. The disease first appeared in 1976, soon after the establishment of a legume-improvement project on land seasonally prone to waterlogging, at a site where *P. nicotianae* is known to occur (Wienk, 1968). This species has a wide host range including legumes (Ruppel and Barnes, 1965; Allen, 1970). Only time will tell which, if either, of these hypotheses is correct.

2.2.2 *Fusarium* species as legume pathogens

Fusarium is a large genus within the Moniliales of the Fungi Imperfecti containing various plant pathogens of very great economic importance to agriculture. The taxonomy of the genus is complex and controversial; there are perhaps two basic schools of thought on classification in *Fusarium,* nicely summarized in two adjacent papers (Booth, 1975; Toussoun and Nelson, 1975). For present purposes, the system used by Booth (1971) is followed. *F. acuminatum, F. avenaceum, F. culmorum, F. equiseti, F. semitectum* (all members of the '*F. roseum* concept', see Booth, 1975) and *F. moniliforme* are each sometimes associated with the legume crops with which we are concerned (Norse, 1974; Russell and Warburg, 1976; Lin and Cook, 1977; Price and Munroe, 1978a; Kannaiyan and Nene, 1979b), but they are saprophytes or weak parasites and are not amongst the major pathogens of legumes. The principal species are *F. oxysporum* and *F. solani,* the latter having a perfect state (*Nectria haematococca*). These two species are ubiquitous in soil, occurring both as saprophytic populations and as more specialized parasites distinguished as *formae speciales. Formae speciales* of fusaria are physiologic strains that are indistinguishable from saprophytic strains of the same species; they show different physiological properties in their ability to parasitize specific hosts (Snyder and Hansen, 1940; 1941). Initially it was believed that *formae speciales* were specific to one host, but this concept of highly selective pathogenicity has been shown to be untrue in certain cases. Extensive cross-inoculations conducted in the USA by the Armstrongs have shown the existence of complex patterns of pathogenicity, and of host resistance, especially with *F. oxysporum.* For instance, *F. oxysporum* f.sp. *tracheiphilum* from soyabean and cowpea will also infect chrysanthemum, while *F. oxysporum* f.sp. *vasinfectum* from cotton can naturally infect soyabeans and tobacco

(Armstrong and Armstrong, 1958; 1965b). The pigeon pea wilt pathogen, treated here as *F. udum* (Booth, 1978), is sometimes regarded as a *forma specialis* of *F. oxysporum* (Snyder and Hansen, 1940), and the latter treatment is supported by the application of serology and electrophoresis used in investigations of *formae speciales* of *F. oxysporum* by Kaiser and Sen Gupta (1976). Although *formae speciales* are also recognized within *F. solani*, the degree of host specificity amongst isolates of *F. solani* from cowpea (da Ponte and da Silva, 1977; Singh and Allen, 1979), chickpea (Nene *et al*., 1979) and pigeon pea (Kannaiyan and Nene, 1979b) requires further investigation; it seems possible that further *formae speciales* may await description. Races have also been distinguished within *formae speciales*, on the basis of specific pathogenicity to host cultivars (Armstrong and Armstrong, 1968; 1975). Taxa in which races have been described include *F. oxysporum* f.sp. *tracheiphilum* (Armstrong and Armstrong, 1950) and f.sp. *ciceris* (Chauhan, 1962a; Nene, 1980a), and there is some evidence that pathogenic variation occurs also within *F. udum* (Baldev and Amin, 1974).

The principal fusaria (Table 6) cause two main types of disease in the legumes: foot, or collar, and root rots; and vascular wilts. The collar and root rots are characterized by a progressive rotting of the root system or basal portion of the stem leading to chlorosis and wilting; whereas, in the true wilts, wilting occurs before extensive damage occurs to the root system, and vascular tissues are usually discoloured (Figure 9; Wheeler, 1969). Various fusaria are also often associated with legume seed and seedlings, and with the subterranean pod of groundnuts (McDonald, 1969), but these are comparatively unimportant, unlike the seedling diseases of cereals. While the foot and root rot pathogens are usually forms of *F. solani*, the vascular wilt fungi comprise various *formae speciales* of *F. oxysporum* and *F. udum*. But these groups are not entirely distinct: the two species may occur in combination, as in 'St. John's Disease' of pea (Buxton, 1955) and, occasionally, isolates of *F. oxysporum* are responsible for root rots (French and Kennedy, 1963).

F. solani f.sp. *phaseoli* causes a dry root rot of beans which is the best known disease of this group. Until more information becomes available on similar diseases, such as the root rot of cowpea (*F. solani*), the extent to which the principles derived from studies of bean root rot may be applied to similar diseases of other legumes remains uncertain. In beans, the rot starts at the tips of the main and lateral roots, and there is a gradual drying-out of the tissues to the stem bases. Affected plants often produce adventitious roots above the diseased area (Wheeler, 1969), in a manner similar to cowpea plants affected by *Pythium* wet stem rot, and as a result they often survive. The fungus only invades the vascular tissue in the later stages of disease development, and it seldom extends in the stem much above soil level. Chlamydospores are produced within the root cortex and when the roots decay these spores are released into the soil, where they survive to provide the primary inoculum for infection of the subsequent crop. The chlamydospore inoculum in the soil may also be replenished by the conversion of conidia washed into the soil

TABLE 6 *Fusarium* Species as Legume Pathogens

Species	Distribution	Natural hosts	References
F. oxysporum Schl. ex Fr. f.sp. *ciceris* (Padw.) Matuo and Sato (= *F. lateritium* (Nees) Snyder and Hansen f.sp. *ciceri* (Padwick) Erwin)	India, Pakistan, Burma, Ethiopia, Malawi, USA, Mexico, Peru	Chickpea	McKerral, 1923; Padwick, 1940a; Erwin, 1958b; Stewart and Yirgou, 1967; Echandi, 1970; Nene, 1978; Kannaiyan, 1981
F. oxysporum f.sp. *glycines* Armst. and Armst.	USA	Soyabean	Armstrong and Armstrong, 1965a
F. oxysporum f.sp. *lentis* (Vasudeva and Srinivasan) Gordon	India	Lentil	Vasudeva and Srinivasan, 1952
F. oxysporum f.sp. *phaseoli* Kendrick and Snyder	Temperate and tropical America	Bean	Kendrick and Snyder, 1942
F. oxysporum f.sp. *tracheiphilum* (E. F. Smith) Snyder and Hansen	USA, Colombia, Nigeria, ? Uganda, India, Taiwan, Australia, USSR	Cowpea, soyabean (chrysanthemum)	Armstrong and Armstrong, 1950; Singh and Sinha, 1955; Armstrong and Armstrong, 1965b; Holliday, 1970; Oyekan, 1975; Armstrong and Armstrong, 1980; J. Mukiibi, pers. comm.
F. oxysporum f.sp. *vasinfectum* (Atk.) Snyder and Hansen	USA	Soyabean (cotton)	Armstrong and Armstrong, 1958
F. oxysporum f.sp. *voandzeiae* Armst., Armst. and Bill.	Tanzania	Bambarra groundnut	Ebbels and Billington, 1972; Armstrong *et al.*, 1975
F. udum Butl. (= *F. oxysporum* Schl. ex Fr. f.sp. *udum* (Butl.) Snyd. and Hans.)	India, Indonesia, Thailand, Vietnam, ? Kenya, Tanzania, Uganda, Europe, Trinidad	Pigeon pea	Butler, 1910; Hansford, 1937; Riley, 1960; Nattrass, 1961; Booth, 1971; 1978; Nene, 1978
F. solani f.sp. *phaseoli* (Burk.) Snyder and Hansen	Temperate and tropical America, Caribbean, Europe, Kenya, ? Malawi, ? Fiji, and presumably elsewhere	Bean	Firman, 1972; Mukunya, 1974; Norse, 1974; Allen, 1975b; Steadman *et al.*, 1975; Schwartz and Gálvez, 1980
F. solani f.sp. *pisi* (Jones) Snyder and Hansen	USA, ? India	Chickpea (pea)	Kraft, 1969; Grewal *et al.*, 1974a; Westerlund *et al.*, 1974

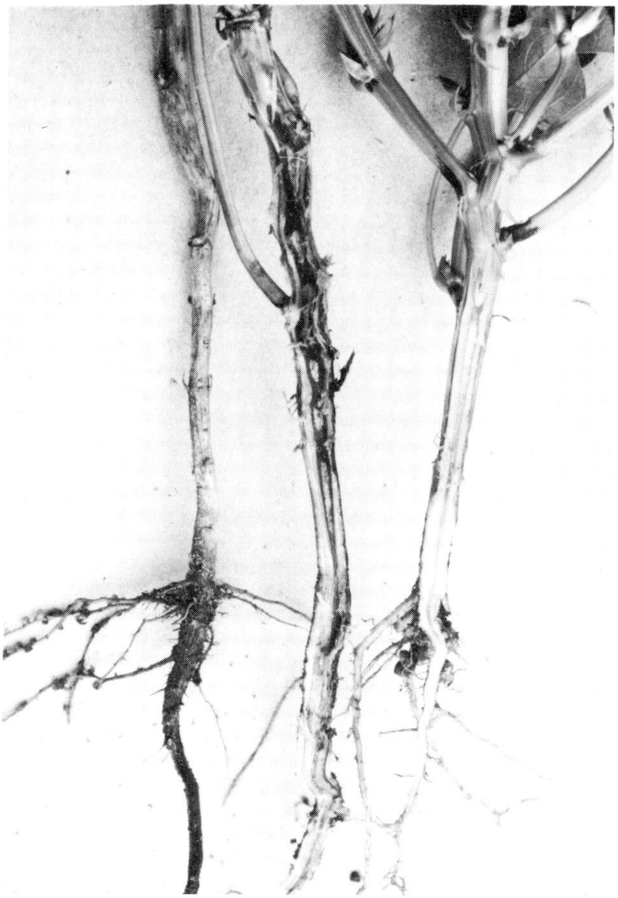

Figure 9 Vascular discoloration in cowpea caused by *Fusarium oxysporum* f.sp. *tracheiphilum*, Ibadan, Nigeria

(Nash *et al*., 1961); their formation can also be induced by soil bacteria (Ford, *et al*., 1970). The soil survival of *F. solani* f.sp. *phaseoli* may be enhanced by temporary supplies of nutrients provided by diffusates from non-host plants and crop residues, enabling the fungus to vegetate briefly and form new chlamydospores (Schroth and Hendrix, 1962; Papavizas *et al.,* 1968). Chlamydospore germination is stimulated by exudates from both seed and hypocotyls of beans (Schroth and Snyder, 1961) but, in the absence of living plant residues, substances produced during the natural decomposition of crop residues can also influence the germination of chlamydospores of *F. solani* f.sp. *phaseoli* (Toussoun *et al.,* 1963). However, a high degree of spore germination around roots does not necessarily lead to a high incidence of disease. Nutritional conditions favouring chlamydospore germination may not be correlated with conditions favouring growth and survival of *F. solani* f.sp.

phaseoli, and some germination stimulants cause lysis of germ tubes. Those that develop in the vicinity of non-hosts, if not lysed, may return to the chlamydospore form without any infection occurring. Thus, the overall development of *F. solani* f.sp. *phaseoli* in the rhizosphere appears to be governed by the balance of nutrients available in exudates as well as in the soil solution (Schroth *et al.,* 1963; Cook and Snyder, 1965). Furthermore, it is evident that nutrients, especially nitrogen and certain sugars, govern the pathogenicity of the bean root rot fungus, perhaps particularly where the soilborne inoculum load is relatively low (Toussoun *et al.,* 1960).

Fusarium root rot of bean is more important in its influence on the development of lateral roots than on the hypocotyl or tap root (Burke and Barker, 1966), and this may have relevance to the effect of plant spacing (Burke, 1965a), and the mechanical impeding of root growth (Burke, 1968), as factors in root rot development. Temporary waterlogging of soil, in which oxygen diffusion rates may be low, is liable also to aggravate root rot (Miller and Burke, 1977). The pathogen is spread in soil as chlamydospores or conidia in surface water, and by drainage and irrigation. Rates of spread are very low, and no appreciable spread occurs by capillary movement of water through the soil (Burke, 1965b). The fungus is not internally seedborne, but it may accompany seed in infested soil particles (Nash and Snyder, 1964).

Turning now to consider the vascular wilt fusaria, this group of specialized, soilborne pathogens probably do not survive indefinitely in soil as free-living saprophytes although they do survive for long periods as chlamydospores (Garrett, 1970). Soil type and nutrient status influence the survival of *F. udum* (Sarojini, 1951; Nene, 1980a), while variation in the persistence of chlamydospore populations in soil may also possibly relate to morphological differences in the chlamydospore wall (Nash and Alexander, 1965). It is possible that alternate hosts may play some role in the carry-over of these pathogens, since it is known that certain non-host plants may harbour vascular wilt fusaria as latent infections (Armstrong and Armstrong, 1948; Katan, 1971), and several legume wilt pathogens may share a common host (Armstrong and Armstrong, 1964). There is evidence that *F. oxysporum* f.sp. *tracheiphum* (Kendrick, 1931) and f.sp. *phaseoli* (Kendrick, 1934), and *F. udum* (Nene, 1980a), may carry over as seed contaminants, although none appears to be internally seedborne. However, *F. oxysporum* f.sp. *ciceris* may be transmitted internally through seed of chickpea, the rate varying with the cultivar (Haware *et al*., 1978).

The development of vascular wilts caused by *F. oxysporum* is, in general, favoured by temperatures of 25 °C or above, and by soil acidity, high available nitrogen content of the soil, and sometimes by a deficiency in soil potassium supply (Garrett, 1970). The available information on the legume wilts confirms that their incidence depends in part on temperature and soil pH as well as on moisture and plant maturity (Mundkur, 1935; Chauhan, 1962b; 1963a, b). Field studies reveal that the prevalence of fusarium wilt varies with the depth, and timing, of sowing (Raheja and Das, 1957; Kannaiyan and

Nene, 1979c), probably relating to variation in soil temperature and moisture (Padwick and Bhagawagar, 1943). The effects of nitrogen and potassium status of soils appear not to have been investigated, although it is well established that disease development in legumes is favoured by relatively sandy soils (Vasudeva and Srinivasan, 1952; Echandi, 1970; Shukla, 1975). Recent studies in India have shown that *F. udum* not only survives longer but wilt of pigeon pea develops more rapidly on the red soils (alfisols) than on the deep black soils (vertisols) which together cover much of the semi-arid tropics. Alfisols comprise some 60 per cent sand while vertisols are only about 40 per cent sand (Nene, 1980a). Observations in south-western Nigeria suggest that the incidence of cowpea wilt is greater during the dry season than in either of the two rainy seasons (D. J. Allen, unpublished), but whether this relates to the plants' predisposition by low soil moisture (Foster and Walker, 1947), or to some other factor, is not known. It is well established that soils infested with nematodes may predispose crops to vascular wilt, as occurs in soyabean (Cromwell, 1917), and it is also apparent that certain soils may suppress wilt development, suggesting an antagonism (Burke, 1965c). It has been shown that the low incidence of pigeon pea wilt in soils containing *Bacillus subtilis* is due to the production of an antibiotic, bulbiformin, which suppresses *F. udum*, the inhibitory effect being enhanced by the presence of root residues (Vasudeva and Roy, 1950; Vasudeva *et al*., 1962). Interactions involving the fusaria are further discussed in section 3.6.

2.2.3 *Cercospora* and allies as legume pathogens

It has been customary for plant pathologists to describe as new any species of *Cercospora* found on a host plant for the first time, and this has led inexorably to extensive synonymy (Johnson and Valleau, 1949) and so to chaos in the classification of the genus. Some 2000 names have been published for *Cercospora* (Chupp, 1954) which has become one of the largest genera within the Moniliales. In view of the considerable heterogeneity of the genus, recent studies have attempted to classify the species described, together with similar species which have been described in other genera, and extensive taxonomic revisions have been made (Deighton, 1967b; 1973; 1974; 1976; 1979; Ellis, 1971; 1976). The family Dematiaceae, in which the conidia and conidiophores are typically dark, includes the genera *Cercoseptoria*, *Cercosporidium*, *Corynespora*, *Mycovellosiella*, *Phaeoisariopsis*, *Pseudocercospora*, and *Stenella*, all of which contain legume pathogenic species at one time described as species of *Cercospora*. For present purposes, these are treated together with 'true' *Cercospora* as a group, a treatment warranted not only on taxonomic grounds (Deighton, 1967b *et seq*.) but also on a pathological basis; for all are essentially splash-dispersed, leaf spot pathogens. Some of the species have perfect (*Mycosphaerella*) states but most are known only as imperfect fungi. Within species, variants have been distinguished on the bases of morphology (Deighton, 1974), colony characteristics (Miller, 1949), and pathogenicity (Hocking, 1967; Spencer and Walters, 1969; Phillips and Boerma, 1981).

Of the numerous *Cercospora*-like fungi reported from the legumes, only the more reliably named species will be considered; these are given in Table 7. *Cercospora arachidicola* and *Cercosporidium personatum* are the most intensively studied, each causing a serious disease of groundnut. Other species of major economic importance are: *Phaeoisariopsis griseola*, the cause of angular leaf spot of bean; *Pseudocercospora cruenta*, which causes a leaf spot of cowpea and other legumes; and *Cercospora kikuchii*, the cause of purple seed stain in soyabean. These fungi are all widespread, occurring effectively wherever their respective hosts are cultivated. There is considerable variation between species in the degree of their specificity. Many are known to make rather slow growth in culture (Cardona-Alvarez and Walker, 1956; Smith, 1971), and this carries with it the risk of inadvertent selection of mycelial variants (Ryker, 1942). Such species often appear to be relatively specialized parasites, and may be confined to a single host genus, as for instance are both the groundnut leaf spot pathogens. Other species are comparatively weak parasites with wide host ranges, such as *Corynespora cassiicola* and, to a lesser extent, *Cercospora canescens*. These two species often occur in mixed infections (Wei, 1950).

As legume pathogens, the *Cercospora*-like fungi induce a range of leaf spots which may or may not be characteristic. *P. griseola* and *Cercospora vanderysti* each cause discrete, angular lesions on bean leaves (Skiles and Cardona-Alvarez, 1959), while those induced by *C. canescens* in various hosts are usually circular. Zonate lesions are characteristic of foliar infection by *C. cassiicola*, often termed 'target spot' (Figure 10). In other cases, lesions tend to be somewhat diffuse, as in floury leaf spot of bean caused by *Mycovellosiella phaseoli* (Cardona-Alvarez and Skiles, 1958) and often also in cowpea leaves infected with *Pseudocercospora cruenta* (Williams, 1975b). Severe leaf infections are often accompanied by stem and petiole spotting, and the development of pod infections. The latter can lead to the discoloration of seed, as commonly occurs in soyabean infected with *C. kikuchii* (Figure 11) (Murakishi, 1951), *Cercospora sojina* (Sherwin and Kreitlow, 1952), and other *Cercospora* species (Kilpatrick and Johnson, 1956). Foliar symptoms typically develop late in the growth of the crop, often after flowering, as reported for the leaf spots of cowpea (Verma and Patel, 1969; Schneider and Sinclair, 1975; Fery *et al*., 1977; Vakili, 1977), mung bean (Mew *et al*., 1975), and groundnut (Hemingway, 1955; Fowler, 1970b). Because of their late development, it has often been assumed that these *Cercospora*-like leaf spots are relatively unimportant, but recent studies on their epidemiology and potential crop loss have shown that as much as 40–60 per cent loss may occur (Barros *et al*., 1958; Fowler, 1970b; 1971; Schneider *et al*., 1976), infections adversely affecting various components of yield (Mercer, 1973; Fery *et al*., 1977). In most cases the principal effect is through premature defoliation (Hemingway, 1954; Cardona-Alvarez and Walker, 1956; Hartwig, 1959;

TABLE 7 *Cercospora* and Allies as Legume Pathogens

Species	Distribution	Natural hosts	References
Cercoseptoria cajanicola Pavgi and Singh	India, Kenya, Tanzania, Sierra Leone	Pigeon pea	Pavgi and Singh, 1965; Deighton, 1974; 1976; Ebbels and Allen, 1979
Cercospora arachidicola Hori, conidial state of *Mycosphaerella arachidis* Deighton	Widespread	Groundnut	Jenkins, 1938; Deighton, 1967c; Mulder and Holliday, 1974a
Cercospora canescens Ell. and Mart. (= *C. vignicaulis* Tehon)	Widespread	Soyabean, bean, lima bean, cowpea, groundnut, bambarra groundnut, mung bean, hyacinth bean, and others	Fowler, 1970b; Mew *et al.*, 1975; Mulder and Holliday, 1975a
Cercospora cercosporelloides Deighton (= *Cercosporella canavaliae* Sawada)	Taiwan	Jack bean	Deighton, 1973
Cercospora kikuchii (Matsumato and Tomoyasu) M. W. Gardner	Widespread	Soyabean, bean, cowpea, cluster bean	Murakishi, 1951; Johnson and Jones, 1962; Mulder and Holliday, 1975c
Cercospora sojina Hara (= *C. daizu* Miura)	Temperate and tropical America, Europe, Manchuria, Japan	Soyabean (and other *Glycine* spp.)	Lehman, 1928
Cercospora vanderysti P. Henn.	Latin America	Bean	Skiles and Cardona-Alvarez, 1959; Schwartz and Gálvez, 1980
Cercosporidium personatum (B and C) Deighton, conidial state of *Mycosphaerella berkeleyi* W. A. Jenkins (= *Cercospora personata* (B and C) Ell. and Ev.)	Widespread	Groundnut	Jenkins, 1938; Deighton, 1967b; Mulder and Holliday, 1974b
Corynespora cassiicola (Berk. and Curt.) Wei (= *Cercospora vignicola* Kawamura)	Worldwide	Soyabean, bean, lima bean, cowpea, mung bean, winged bean, cluster bean, and very many others	Wei, 1950; Ellis, 1957; Ellis and Holliday, 1971
Mycovellosiella cajani (P. Henn.) Rangel ex Trotter (= *Cercospora cajani* P. Henn.) var. *cajani*	Tropical America, Africa	Pigeon pea	Deighton, 1974
Mycovellosiella cajani var. *indica* (U. B. Singh) Deighton	India, Pakistan, Bangladesh, Burma	Pigeon pea	Deighton, 1974

Species	Distribution	Natural hosts	References
Mycovellosiella cajani var. *trichophila* (Curzi) Deighton	Kenya, Somalia, Taiwan	Pigeon pea	Deighton, 1974
Mycovellosiella phaseoli (Drummond) Deighton (= *Ramularia phaseoli* (Drummond) Deighton)	Tropical America, eastern Africa, NW India, Malaysia, Papua New Guinea	Bean	Deighton, 1967a; 1974; Schwartz and Gálvez, 1980
Phaeoisariopsis griseola (Sacc.) Ferraris (= *Isariopsis griseola* Sacc. = *Cercospora stuhlmanni* P. Henn. = *Cercospora solimani* Speg. = *Cercospora columnare* Ell. and Ev.)	Widespread	Bean, and occasionally other legumes	Deighton, 1977; Schwartz and Gálvez, 1980
Pseudocercospora cruenta (Sacc.) Deighton, conidial state of *Mycospharella cruenta* Latham (= *Cercospora cruenta* Sacc.)	Widespread	Bean, cowpea, mung bean, hyacinth bean, and others	Latham, 1934; Mulder and Holliday, 1975b; Deighton, 1976
Pseudocercospora mucunae-ferrugineae (Yamam.) Deighton	Taiwan	Velvet bean	Yamamoto, 1934; Deighton, 1976
Pseudocercospora mungo Deighton	Tanzania, Kenya, Sudan, Nigeria	Mung bean, black gram	Deighton, 1979
Pseudocercospora psophocarpi (Yen) Deighton (= *Cercospora psophocarpi* Yen = ? *Cercospora aranetae* Borlaza and Roldan)	SE Asia, Papua New Guinea, Nigeria	Winged bean	Deighton, 1976; Price and Munro, 1978; Allen, 1979a
Pseudocercospora stizolobii (H. and P. Syd.) Deighton (= *Cercospora stizolobii* H. and P. Syd. = *Cercospora lussoniensis* Sacc.)	Philippines; Zambia, Malawi, Nigeria, Togo; Trinidad, and presumably elsewhere	Velvet bean (and other *Mucuna* spp.)	Sydow and Sydow, 1913; Baker and Dale, 1951; Angus, 1962–66; Bailey, 1966; Peregrine and Siddiqi, 1972; Steiner, 1975
Pseudocercospora vignae-reticulatae Deighton	Sierra Leone, Ghana, Zaire, Kenya	Mung bean (and *Vigna reticulata*)	Deighton, 1976; Allen, 1979a
Pseudocercosporella albida (Matta and Bell.) Yoshii and Aamodt	Colombia, Guatemala	Bean	Schwartz *et al.*, 1981
Stenella canavaliae (H. and P. Syd.) Deighton (= *Cercospora canavaliae* H. and P. Syd. = *Cercospora canavaliae-rosae* Yen and Gilles)	Tropical Asia, Africa and America	Jack bean (and other *Canavalia* spp.)	Deighton, 1971; 1979

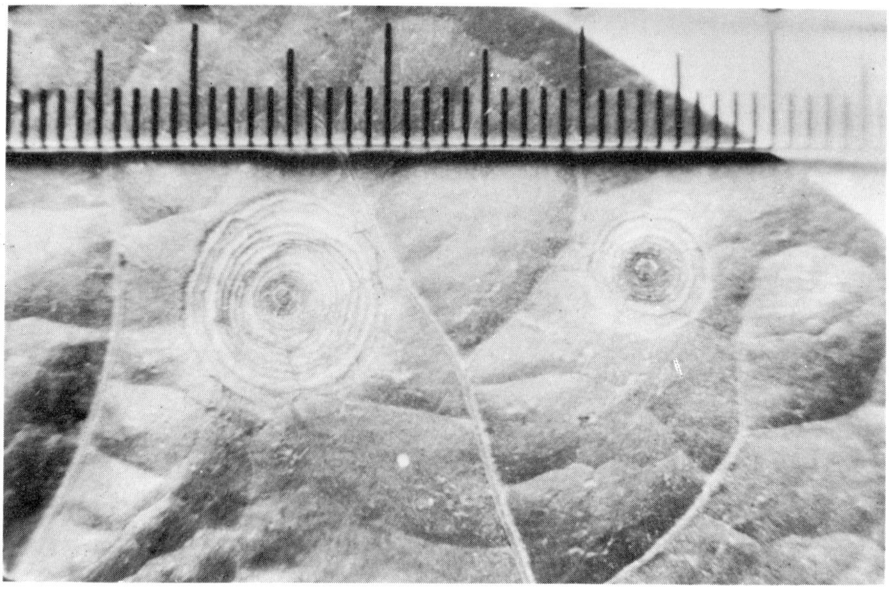

Figure 10 Zonate lesions on cowpea leaves caused by *Corynespora cassiicola*, Ibadan, Nigeria (photograph courtesy R. J. Williams)

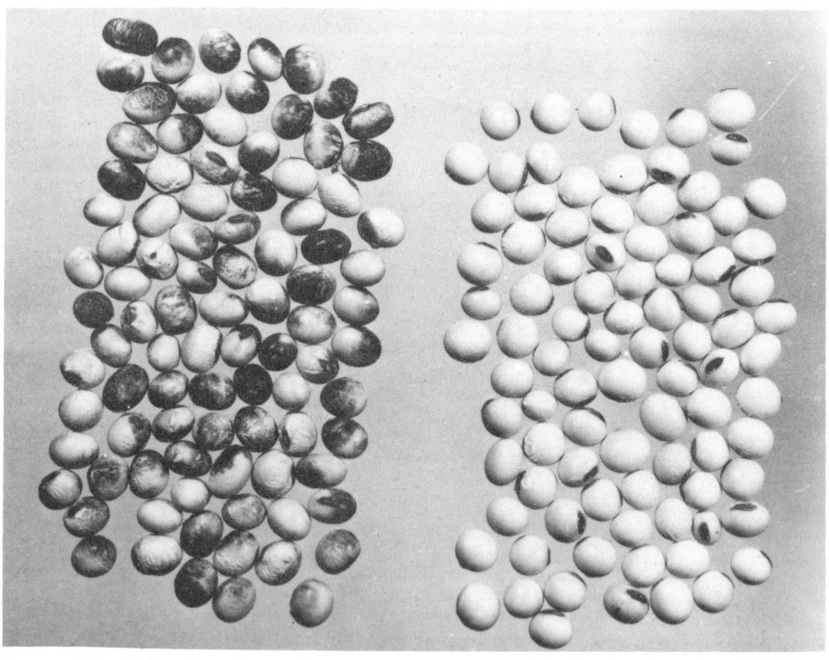

Figure 11 Discoloration of soyabean seed caused by *Cercospora kikuchii* (photograph courtesy J. B. Sinclair)

Cole, 1966; Deighton, 1967a; Elston *et al*., 1976; Schneider *et al*., 1976), and it has been found that groundnut leaflets fall when between 6 and 34 per cent of their leaf area is diseased (Fowler, 1970b). It is pertinent to note that Enyi (1975) has shown that grain yield losses in various legumes are greatest when defoliation occurs early in the pod-filling period. The relatively late development of symptoms may, in some instances, relate to biochemical changes in the host with age. It has been shown that the extent to which conidial germination is enhanced by leaf extracts is age-dependent (Ekpo and Esuruoso, 1977), young tissue being inhibitory (Schneider and Sinclair, 1975).

The relative prevalence of the *Cercospora*-like pathogens varies with ecological zone, as discussed more generally in Section 2.1. Some species like *Pseudocercospora psophocarpi* (Price and Munroe, 1978b) appear well adapted to hot, lowland tropical conditions, while others are essentially confined to the cooler conditions which prevail at high altitude within the tropics. Examples of the latter include *Mycovellosiella phaseoli* (Cardona-Alvarez and Skiles, 1958) and *Cercospora vanderysti* (Skiles and Cardona-Alvarez, 1959). Similarly, the relative importance of the two groundnut leaf spot pathogens also varies with location (Gibbons, 1966). While *Cercospora arachidicola*, the cause of 'early' leaf spot, is the more important in Malawi (Mercer, 1973), *Cercosporidium personatum*, which causes 'late' leaf spot, is more damaging in Tanzania (Hemingway, 1955) and Nigeria (Fowler, 1970b). However, the prevalence of these two pathogens is determined not only by climatic factors but also by the maturity type of groundnut grown (Gibbons, 1966).

Turning now to consider certain aspects of their epidemiology, it is evident that seed transmission has played an important role in the distribution of the *Cercospora*-like pathogens of legumes. The introduction of soyabeans into the USA from China and Japan led to the establishment of various oriental species, including *Cercospora sojina*, in North America where their seedborne nature was confirmed (Wolf and Lehman, 1926). *Cercospora kikuchii* was first found in the USA rather later (Murakishi, 1951). Similarly, bean pathogens such as *Phaeoisariopsis griseola* and *Mycovellosiella phaseoli* have newly appeared in India (Srinivasan, 1953; Sohi *et al*., 1965). *P. griseola* is known to be seedborne (Orozco-Sarria and Cardona-Alvarez, 1959), and various pathogens of this group are seed-transmitted also in cowpea, including *Cercospora canescens, Pseudocercospora cruenta* (Williams, 1975b), and *Corynespora cassiicola* (Esuruoso, 1975). There is evidence that this group of fungi are internally seedborne (Sherwin and Kreitlow, 1952; Orozco-Sarria and Cardona-Alvarez, 1959; Ilyas *et al*., 1975).

The length of the flowering period may influence the extent to which *C. kikuchii* is seedborne (Crane and Crittenden, 1966), and such seedborne infections may provide sources of primary inoculum under certain conditions (Sherwin and Kreitlow, 1952; Orozco-Sarria and Cardona-Alvarez, 1959). However, infected host debris constitutes the most important means of carry-over of these leaf spot pathogens (Lehman, 1928; Cardona-Alvarez and

Walker, 1956; Jones, 1968; Fowler, 1970b). The role of perithecia in survival is uncertain. Although the cowpea and groundnut leaf spot fungi produce perithecia (Latham, 1934; Jenkins, 1938), their perfect stages have apparently never been found under African field conditions (Hemingway, 1955; Smartt, 1961; Gibbons, 1966; Fowler, 1970b). It seems likely that alternate hosts may provide sources of primary infection of those species having wide host ranges, but no evidence is available to support this. Under certain conditions, volunteer seedlings may play a role in pathogen carry-over (Hemingway, 1954; Orozco-Sarria and Cardona-Alvarez, 1959), but where the dry season is too long, volunteers may not survive (Fowler, 1970b).

Secondary spread of the *Cercospora*-like pathogens occurs typically by means of rain-splash. *P. griseola* is also readily disseminated with water- or windborne soil particles as well as in dry air currents (Cardona-Alvarez and Walker, 1956), epidemic development being enhanced by intensive cultivation (Barros *et al*., 1958). The build-up of groundnut leaf spot varies considerably with the pathogen species, the epidemic potential of *Cercosporidium personatum* being much greater than that of *Cercospora arachidicola* (Hemingway, 1955). Epidemic development evidently depends on the duration of high humidity and the minimum temperature prevailing during such periods (Jensen and Boyle, 1966). High humidity appears to be more critical than temperature in governing conidial germination in *C. arachidicola*. Under certain conditions, the 'early' leaf spot pathogen may produce secondary conidia, and it has been suggested that their production may have survival value in a nutrient-poor environment, permitting further opportunity for dissemination (Oso, 1972). Studies on the relationship between disease development and expected crop loss have seldom been made, although a model is available for cercospora leaf spot of cowpea (Schneider *et al*., 1976).

2.2.4 Anthracnose and scab pathogens of legumes

The term anthracnose literally means 'like coal', but it is now used in an aetiological sense for diseases caused by fungi in the Melanconiales with *Sphaceloma*, *Colletotrichum*, and *Gloeosporium* conidial states (Wheeler, 1969). Some of these diseases are also called 'scab', and it has been suggested that these terms be used in a symptomatic sense (Jenkins, 1933a). For present purposes, a compromise usage is warranted: legume diseases caused by species of *Sphaceloma*, some of which have *Elsinoë* perfect states, are often associated with the development of raised corky lesions on some plant parts so may be called scab; diseases in legumes caused by *Colletotrichum* species, some having *Glomerella* perfect states, do not involve hyperplasia and will be termed anthracnose. Occasionally, other fungi induce scab-like symptoms in legumes, as does *Cladosporium vignae* (Gardner, 1925; Weber, 1941), but these are not discussed here. Anthracnose and scab diseases, as defined here, constitute a small group of morphologically similar fungi (Jenkins and Bitancourt, 1941; Tiffany and Gilman, 1954) some of which are of considerable economic importance. The principal pathogens affecting tropical legumes are shown in Table 8. While the *Colletotrichum* species are relatively well known, and *C. lindemuthianum* in particular has been extensively studied, the scab fungi deserve much greater attention. Species appear to have been described more on the basis of pathogenicity and colony characters than on morphology

TABLE 8 Pathogens Causing Anthracnose and Scab Diseases in Legumes

Species	Distribution	Natural hosts	References
Colletotrichum lindemuthianum (Sacc. and Magn.) Bri. and Cav.	Worldwide	Bean, cowpea, hyacinth bean, and others	Mordue, 1971a; Onesirosan and Barker, 1971; Nowsher *et al.*, 1975
Colletotrichum phaseolorum Tak.	Japan	Adzuki bean	Takimoto, 1934; Sasaki, 1936
Colletotrichum capsici (Syd.) Butler and Bisby	Pan-tropical	Groundnut, cowpea, cluster bean, and many others	Desai and Prasad, 1955; Riley, 1960; Mordue, 1971b; Emechebe, in press
Colletotrichum truncatum (Schw.) Andrus and Moore (= *C. dematium* (Pers. ex Fr.) Grovt var. *truncata* (Schw.) Arx)	Pan-tropical	Soyabean, lima bean, cowpea, pigeon pea, and others	Andrus and Moore, 1935; Hemmi, 1955; Khan and Singh, 1974; Sinclair and Shurtleff, 1975; Emechebe, in press
Glomerella cingulata (Stonem.) Spauld. and Schrenk (incl. *Colletotrichum cajani* Rangel)	Brazil, Puerto Rico, India (as *C. cajani*)	Pigeon pea	Tucker, 1927; McRae, 1932; Mordue, 1971c
Glomerella glycines (Hori) Lehman and Wolf	USA, China	Soyabean	Lehman and Wolf, 1926b; Ling, 1940; Sinclair and Shurtleff, 1975
Sphaceloma arachidis Bitanc. and Jenk.	Brazil	Groundnut	Bitancourt and Jenkins, 1940
Sphaceloma glycines Kurata and Kuribayashi	Japan	Soyabean	Jenkins, 1951; Kurata and Kuribayashi, 1954
Elsinoë canavaliae Racib.	Eastern Asia and Africa	Jack bean, sword bean	Jenkins, 1931a; Hansford, 1937; Sivanesan and Holliday, 1971a
Elsinoë dolichi Jenk., Bitanc. and Cheo	China, Africa	Hyacinth bean	Cheo and Jenkins, 1945
Elsinoë iwatae Kajiwara and Mukelar	Indonesia	Mung bean	Mukelar, 1978
Elsinoë phaseoli Jenk.	Tropical America, Africa, ? India	Lima bean, bean, ? cowpea and mung bean	Jenkins, 1931b; Bruner and Jenkins, 1933; Sivanesan and Holliday, 1971b; Emechebe, 1980

(Jenkins, 1931b; Bruner and Jenkins, 1933) and it is probable some of the species described are synonymous. There is considerable host specificity within the scab pathogens (Bruner and Jenkins, 1933; Cheo and Jenkins, 1945; Bates, 1957; Mukelar, 1978; Emechebe, 1980) and it is likely that *formae speciales* should be established, as proposed by Staples (1958). For example, bean scab (Bates, 1957; Mutitu and Mukunya, 1979) and cowpea scab (Van Hoof, 1963; Emechebe, 1980) isolates may have host ranges quite distinct from the type collection of *Elsinoë phaseoli* (Jenkins, 1931b; Bruner and Jenkins, 1933) to which they are usually ascribed. Conversely, collections of mung bean scab in Africa (Nattrass, 1961; and IMI 237694, *leg*. D. J. Allen) have been tentatively assigned also to *E. phaseoli* while mung bean scab in Indonesia is referred to *E. iwatae* (Mukelar, 1978). Further confusion perhaps may result from the premature usage of *Elsinoë* before the perfect stage has been demonstrated; apparently no perfect state has been described for isolates of cowpea scab (Emechebe, 1980) although the pathogen is often referred to as an *Elsinoë* (Bates, 1955; Angus, 1962–66; Van Hoof, 1963; Singh and Allen, 1979). In general, progress with the study of the legume scab diseases has been hampered by their frequent confusion with anthracnose (Jenkins, 1931b; Singh and Allen, 1979; Emechebe, 1980), compounded by the relative difficulty encountered in the isolation of their causal agents (Emechebe, 1980), and much remains to be clarified in this group, whose economic importance is often underrated.

Species of *Colletotrichum* from legumes fall into two clearly defined groups based on spore shape (Tiffany and Gilman, 1954). Those with straight, cigar-shaped spores include *C. lindemuthianum, C. cajani*, and *C. glycines*, while species with curved (falcate) spores include *C. truncatum* and *C. capsici*. The perfect (perithecial) stage is known only for those in the straight-spored group: *C. cajani* (*Glomerella cingulata*) and *C. glycines* (*Glomerella glycines*). *C. lindemuthianum* is a major pathogen of bean and as such has been intensively studied, and thoroughly reviewed (Wheeler, 1969; Mordue, 1971a; Schwartz and Gálvez, 1980); it is also the cause of a locally important stem anthracnose of cowpea (Onesirosan and Barker, 1971; Williams, 1975b) and hyacinth bean (Nowsher *et al*., 1975). Both *C. truncatum* and *C. capsici*, unlike *C. lindemuthianum*, are relatively unspecialized parasites, having wide host ranges and inducing a variety of symptoms. *C. truncatum* causes an important stem and pod anthracnose of lima bean in southern USA (Toler and Wester, 1966) and in western Nigeria (D. J. Allen, unpublished), as well as soyabean anthracnose in the USA (Sinclair and Shurtleff, 1975). Both *C. truncatum* and *C. capsici* and responsible for a disease of cowpea in Nigeria, called 'brown blotch' (Singh and Allen, 1979; Emechebe, in press) to distinguish it from anthracnose caused by *C. lindemuthianum*. In both lima bean and cowpea, symptoms induced by *C. truncatum* involve the development of purplish, or reddish, brown blotches of petioles, leaf veins, stems, peduncles and, especially, pods (Figure 12), without the formation of definite lesions (Andrus and Moore, 1935; Cox, 1949; Tiffany and Gilman, 1954; Singh and

Figure 12 'Brown blotch' symptoms in cowpea caused by *Colletotrichum capsici* and *C. truncatum*, Nigeria

Allen, 1979). Although vascular necrosis of leaves is a characteristic also of bean anthracnose (Figure 13), *C. lindemuthianum* does induce definite lesions in its hosts and, in this respect, more closely resembles the scab fungi discussed above. There is considerable physiologic specialization in *C. lindemuthianum* and numerous races of bean anthracnose have been defined (Kruger *et al*., 1977). There is evidence that a cowpea isolate is pathogenically distinct from the bean anthracnose organism (Williams, 1975b) so there may be a case for the erection of *formae speciales* in *C. lindemuthianum*, as has been proposed for a cluster bean isolate of *C. capsici* (Desai and Prasad, 1955).

Most of the anthracnose and scab pathogens of legumes are known to be seedborne (Noble and Richardson, 1968), both as contaminants and as internal infections. Although some species, such as *Sphaceloma arachidis* (Section 2.1 and *S. glycines* (Figure 14), have restricted distributions, most are widespread. *Colletotrichum capsici* and *C. truncatum* occur widely throughout the

Figure 13 Vascular necrosis of bean leaves caused by *Colletotrichum lindemuthianum*, Southern Highlands, Tanzania

tropics and sub-tropics; *C. lindemuthianum* extends further into the cool temperate zones than the other species. Seed transmission probably accounts for the long-distance dispersal of these pathogens; Sitterly and Epps (1958) consider that lima bean scab was introduced into the USA on seed. Seed-borne infections and infected crop residues together probably account for the carry-over of the scab and anthracnose pathogens (Chambers and Andes, 1962; Onesirosan and Sagay, 1975; Emechebe, 1980), although, in certain instances, volunteers and alternate hosts may be significant. In Colombia, sphaceloma scab of a tropical forage legume (*Zornia latifolia*) has been found on native *Zornia* species (Lenné, 1981), and scab-like lesions are common on a woody perennial *Cassia* sp. in northern Nigeria where scab of cowpea is a major disease (D. J. Allen and A. M. Emechebe, unpublished), but in neither case has their epidemiological significance been established. Weeds such as *Abutilon theophrasti* may act as alternate hosts of *C. truncatum*, a soyabean

Figure 14 Pod lesions in soyabean caused by *Sphaceloma glycines* (photograph courtesy J. B. Sinclair)

seed pathogen (Dhingra and da Silva, 1978; Hepperly *et al*., 1980). Anthracnose and scab diseases are typically splash-dispersed, and their prevalence generally increases during periods of heavy rainfall (Toler and Wester, 1966; Wheeler, 1969; Emechebe, 1980). In Kenya, bean scab appears to be disseminated principally by wind and plant contact, spreading at the rate of 7.5 m from an infection focus over a period of six weeks (Mutitu and Mukunya, 1979) while, in Canada, it has been estimated that bean anthracnose spreads from an infection focus at the rate of 3–4.6 m per rainstorm (Tu, 1981).

The relative importance of anthracnose and scab diseases varies with ecological zone, and this is nicely illustrated by cowpea in Nigeria where it is susceptible to scab and two anthracnose diseases. In the south-west of Nigeria, under comparatively humid, forest conditions, *C. lindemuthianum* causes a locally devastating disease, especially under experimental conditions; stem anthracnose is seldom encountered further north, under drier conditions. *C. truncatum* and *C. capsici* together cause a 'brown blotch' of cowpea which is occasionally damaging in southern Nigeria; however, the disease is of greater importance in the drier savannahs further north extending into the Sahel (14° N). Sphaceloma scab is absent from southern Nigeria but is the principal disease of cowpea in the intermediate savannahs (approximately 10°–12° N), though it is uncommon in the northern Sudan savannah and Sahel belt (12°–14° N). Although it is possible that cowpea scab, with an ecologically restricted distribution, may in time spread into more humid zones, as has been suggested (Emechebe, 1980), it is also conceivable that *Sphaceloma* cannot survive in the forest belt. Field surveys in eastern Africa and in north-east Brazil have confirmed the importance of cowpea scab under savannah conditions, extending up to 1500 m in Kenya (D. J. Allen, unpublished).

2.2.5 The rust fungi as legume pathogens

Many rusts have been reported on the tropical food legumes; the more important and better known species are shown in Table 9. All these species are autoecious, or apparently so, although there are examples of heteroecism amongst temperate species of *Uromyces* on the Leguminosae, alternating predominantly between a legume and *Euphorbia* (Olive, 1911; Butler, 1958; Wilson and Henderson, 1966). An example of a legume-grass alternation is shown by *Puccinia tristachyae* Doidge the accia of which occur on *Vigna angustifolia* and the uredia and telia on *Tristachya* in South Africa (Doidge, 1926). Some tropical rusts, such as *U. appendiculatus*, are macrocyclic, although aecia are rare in nature (Jones, 1960; Groth and Mogen, 1978). Furthermore, little is known about the many species of *Aecidium* described from legumes in the tropics, and it seems possible that some may represent hitherto unknown aecial stages of otherwise well-known species. *Aecidium caulicola* P. Henn. and A. *vignae* Cke from cowpea in Africa (Snowdon,

TABLE 9 Rust Fungi as Legume pathogens

Species	Distribution	Natural hosts	References
Puccinia arachidis Speg.	Pan-tropical	Groundnut and other *Arachis* spp.	Laundon and Waterston, 1965a
Uromyces appendiculatus (Pers.) Ung. (= *U. phaseoli* (Pers.) Wint. = *U. vignae* Barcl.)	Worldwide	Bean, lima bean, cowpea, bambarra groundnut, mung bean, hyacinth bean, and others	Laundon and Waterston, 1965b
U. ciceris-arietini Jacz. apud Boy. and Jacz.	India, Europe, N and E Africa, Mexico	Chickpea (and *Trigonella*)	Punithalingham, 1968
U. mucunae Rabenh.	Africa, Asia	Velvet bean	Laundon and Rainbow, 1971
U. viciae-fabae (Pers.) Schroet.	Worldwide	Lentil, and other legumes	Laundon and Waterston, 1965c
Uredo cajani Syd.	Africa, Asia, Caribbean	Pigeon pea	Anahosur and Waller, 1978b
Phakopsora pachyrhizi Syd. (= *P. vignae* Arth. = *Uromyces sojae* (P. Henn.) Syd. = *Uredo vignae* Bres. = *Uredo concors* Arth. and other synonyms)	Widespread in tropics and sub-tropics	Soyabean, bean, lima bean, cowpea, mung bean, black gram, hyacinth bean, yam bean, and others	Bromfield, 1976; Anahosur and Waller, 1978a

1921; Allen, 1979a) are recognized synonyms of *U. appendiculatus* (Laundon and Waterston, 1965b). For present purposes, let us consider in greater detail the three most important species, which provide useful contrasts: *P. arachidis*, the cause of groundnut rust, *U. appendiculatus* which causes bean and cowpea rust, and *Phakopsora pachyrhizi*, the soyabean rust pathogen, which we shall discuss first.

With the growing concern over the threat of soyabean rust in the Western Hemisphere, considerable attention is being devoted to *P. pachyrhizi*. At present, the disease is principally of economic importance in Asia and Australia, but the pathogen has a much wider distribution. The numerous synonyms of *P. pachyrhizi* complicate the analysis of its geographical distribution and host range, and thus of its possible origin. Some of the names applied to *P. pachyrhizi* include *Uredo vignae* Bres., *Uredo sojae* Hennings, *Uredo concors* Arthur, *Physopella concors* Arthur, *Phakopsora vignae* Arthur, *Phakopsora sojae* Fujikuro, *Phakopsora sojae* Sawada, and *Uromyces sojae* (P. Henn.) Syd. (Hiratsuka, 1936; Sathe, 1972; Meléndez, 1978). The species was originally described, in 1914, in Taiwan from the yam bean (*Pachyrrhizus erosus*) (Sydow and Sydow, 1914), which is an American native. At the same time (1913), the fungus was collected in Puerto Rico from hyacinth bean (*Lablab purpureus*) (Meléndez, 1978), which is an Asian native species. The natural host range of *P. pachyrhizi* is wide, including more than a dozen legume genera (Anahosur and Waller, 1978a; Deslandes, 1979; Vakili, 1979). In tropical America, the hyacinth bean appears to be the most common host (Norse, 1974; Vakili and Bromfield, 1976; Meléndez, 1978; Deslandes, 1979; Vakili, 1981), although *P. pachyrhizi* has been found on soyabean in the Americas (Vakili and Bromfield, 1976; Deslandes, 1979). In Africa, cowpeas are much the most common host (Figure 15) although other recorded hosts include lima bean in Nigeria (Allen, 1979a) and both black gram (Ebbels and Allen, 1979) and mung bean (IMI 254520, *fide* P. N. Patel) in Tanzania. Records of *P. pachyrhizi* from soyabean from Zaire (Hendrickx, 1939), Zambia (Javaid and Ashraf, 1978) and Ethiopia (J. B. Sinclair cited by Bromfield, 1978) all warrant confirmation; apparently none is supported by herbarium collections. If identifications were based solely on field symptoms, confusion with other diseases such as bacterial pustule is always possible, as emphasized by Vakili (1978b).

Pathogenicity tests are beginning to suggest that host-specific *formae speciales* should be established within *P. pachyrhizi* (Lin, 1966; Bromfield, 1977), as long recognized in the cereal rusts. The extent to which pathogenicity to soyabean is within, or beyond, the capacity for micro-evolutionary change (Robinson, 1976) of a particular population of *P. pachyrhizi* is clearly of cardinal significance, and is at the nub of the current debate on the extent to which the US soyabean crop is threatened by rust. Field observations over four years in Western Nigeria have shown that when heavily rusted (*P. pachyrhizi*) cowpeas were grown in close proximity to soyabeans, soyabeans did not develop rust (D. J. Allen, unpublished). There is now evidence that a

Figure 15 The rust fungus. *Phakopsora pachyrhizi*, on cowpea, Ibadan, Nigeria

Puerto Rican isolate of *P. pachyrhizi* lacks virulence on soyabean (Bromfield *et al*., 1980), and histological studies show there may be a degree of incompatibility between an Australian isolate and soyabean cultivars (Keogh *et al*., 1980). Indeed, *P. pachyrhizi* usually causes relatively extensive host necrosis around the point of penetration, and it is only later that productive uredia appear within the necrotic zone (Melching *et al*., 1979). This apparent lack of compatibility between rust- 'susceptible' soyabean cultivars and *P. pachyrhizi*, despite the economic importance of the disease, seems to suggest that soyabeans were not the original host of *P. pachyrhizi*.

Pycnia and aecia of *P. pachyrhizi* have not been found, and nothing is known of the role that sexual reproduction might play in pathogenic variation. But, since *Neonotonia wightii* is a host of *P. pachyrhizi* in Australia (Deverall *et al*., 1977), it is pertinent to note that an aecial rust (*Aecidium glycines* P. Henn.) has been known to occur on this host, an African native, since 1895 in Tanzania (Ebbels and Allen, 1979); it is also known from Uganda (Snowdon, 1921) and Kenya (Ondieki, 1973). But whether or not *A. glycines* is the aecial state of *P. pachyrhizi* appears to be unknown. Nor is it known with certainty that *P. pachyrhizi* is not heteroecious. Preliminary studies in Taiwan (Yang, 1978) on the possible relationship of *Aecidium paedariae* Diet. (Sydow and Sydow, 1914) to soyabean rust warrant further investigation.

Turning now to consider the other two species of rust fungi with which we are to deal, the origin and current distribution of *P. arachidis* provides an impressive contrast to the situation with *P. pachyrhizi*. Unlike the latter, the host range of *P. arachidis* is genus-specific, being confined to *Arachis* to which the groundnut belongs. Not surprisingly, the rust was first reported within the centre of origin of its host, in Argentina in 1881 (Spegazzini, 1881). While the crop spread widely, its rust effectively remained confined to tropical and sub-tropical America until the early 1960s, as shown by the pathogen's distribution maps (see Figure 3). Groundnut rust then began to increase in importance in the USA (Wells, 1962; Harrison, 1972) and in the period 1969–72 the world situation changed dramatically, as traced by Bromfield (1974) and McDonald and Raheja (1980). *P. arachidis* reached India in 1969 (Chalal and Chohan, 1971) where it spread widely (Goswami, 1975; Misra and Misra, 1975; Mayee *et al*., 1977; Subrahmanyam *et al*., 1979); and in 1970 the rust reached Thailand, Japan and elsewhere in the Far East. In 1971, it was recorded in Brunei (Peregrine, 1971b) and, in 1973, in Australia (Pitkethley, 1973; O'Brien, 1974). The first record of groundnut rust from the African mainland came in March 1974 from Rhodesia (now Zimbabwe) (Rothwell, 1975); thence it spread to Kenya and Tanzania (Keswani and Ondieki, 1976), Zambia (Raemaekers and Preston, 1977) and elsewhere in southern and eastern Africa. It reached West Africa in 1976 (McDonald and Emechebe, 1978), and is now widespread in virtually all regions where groundnuts are cultivated.

U. appendiculatus is similarly widespread throughout the tropics and sub-tropics, but extends further also into the temperate regions than either of the other two rust fungi as yet. Its host range is tribe-specific, being effectively confined to the genera *Phaseolus*, *Vigna* and *Lablab* of the Phaseoleae. Between them, these genera contain species native to the Americas (bean and lima bean), Africa (cowpea and bambarra groundnut) and Asia (mung bean and hyacinth bean); and no single ancestral host, nor region of origin, is readily discernible. The host-specific forms within *U. appendiculatus*, given specific or varietal status by some authors (Fromme, 1924; Arthur, 1934), may possibly reflect a New World origin of the bean rust fungus versus an Old World origin of the cowpea rust fungus, each of course being composed of pathogenic variants (races).

The international dissemination of legume rusts, as with the cereal rusts, is presumed to be predominantly by means of windborne urediospores. Intercontinental spread may possibly be assisted by man's air transport, as may have occurred with maize rust (Robinson, 1976), particularly where short-lived propagules are concerned. While *P. arachidis* has spectacularly demonstrated its ability to be spread, Vakili (1978b) has pointed to the surprising lack of spread of *P. pachyrhizi* from Puerto Rico, despite many years of tourism, and of vegetable commerce, with the USA. While its absence from the North American soyabean crop may be due to effective avirulence of the Puerto Rican rust population (Bromfield *et al*., 1980), it is also probable that the hyaline or lightly pigmented nature of the urediospores of *P. pachyrhizi* may limit its survival (Dillon-Weston, 1931), relative to the darker pigmented urediospores of both *P. arachidis* and *U. appendiculatus*. Seed transmission probably plays a negligible role in the long-distance dissemination of rusts, although occasionally indirect evidence suggests that *U. appendiculatus* (Deighton, 1945) and *P. arachidis* (Peregrine, 1971b) may be seedborne; *P. pachyrhizi* may be dispersed on infected debris mixed with seed.

Teliospores may play a role in the survival of rusts during the close season, but there is a variation within and between the rust fungi in their ability to produce telia, and in the relative importance of teliospores in carry-over. The production of teliospores of *U. appendiculatus* is in part under environmental control and partly genetically regulated. Bean rust collections from ecologically distinct localities may differ in their propensity to produce teliospores (Harter *et al*., 1935; Allen, 1975c). While teliospores may be the major source of overwintering inoculum in some areas (Milbrath, 1944), in areas where *U. appendiculatus* does not produce teliospores, bean rust epidemics may depend on the transport of urediospores from elsewhere (Townsend, 1939), or on the direct overwintering of urediospores (Fromme and Wingard, 1921; Marcus, 1952). Teliospore production in *P. arachidis* appears to be uncommon, there being but a single report (Chalal and Chohan, 1971) since the species was first described a century ago, so that urediospores from volunteer plants or crop debris (Rothwell, 1975; Hammons, 1977) are the primary source of inoculum for groundnuts. However, in the absence of such sources, epidemic development may depend on the annual introduction of windborne urediospores (Van Arsdel and Harrison, 1972). *P. pachyrhizi* also produces teliospores, the development of the telial stage apparently being triggered by temperatures below 20 °C (Bromfield, 1978; Yeh *et al*. 1981), but their role in the epidemiology of soyabean rust is unknown. Alternate hosts may also play a role in the seasonal carry-over of legume rusts. In Australia, *P. pachyrhizi* survives throughout the year on pasture legumes, including *Neonotonia wightii*, which may act as reservoirs of inoculum for soyabeans (Bromfield, 1976) and, in India, *U. ciceris-arietini* survives the summer as uredia on *Trigonella polycerata* in the hills from where it then spreads to the plains (Payak, 1962).

Secondary dispersal occurs as urediospores, in some tropical rusts assisted by insects (Turner, 1974). The aerial dissemination of urediospores of *P. arachidis* has been studied in India where it has been shown that urediospore concentrations vary markedly with the season and time of day. The incidence of urediospores in air was greatest when temperatures were 29–31 °C and relative humidity 75–85 per cent (Mallaiah and Rao, 1982). Urediospore deposition, and their retention by leaves, is governed by various factors. The selective retention of bean rust urediospores on bean leaves may depend upon the release of film-forming substances from the spores, the amount produced being related to host cuticle wettability (Yarwood, 1968; Woodbury and Stahmann, 1970). Changes in the wettability of groundnut leaves with age similarly influences the retention of urediospores of *P. arachidis* (Cook, 1980). Spores in clumps tend not to germinate, and germination inhibitors have been identified in both bean rust (Macko *et al*., 1970) and groundnut rust urediospores (Foudin and Macko, 1974). Laboratory studies suggest that a brief period at high humidity is required to stimulate germination and subsequent infection by *P. arachidis*; 16–24 hours in a saturated atmosphere is sufficient and surface moisture is unnecessary (McVey, 1965; Bromfield and Cevario, 1970; Cook, 1972). Urediospores of *P. pachyrhizi* germinate between 10 ° and 28.5 °C, and maximum infection of soyabean occurs at 20–25 °C with 10–12 hours of dew; a longer dew period is required at lower temperatures. The minimum dew period for infection was found to be 6 hours at 20–25 °C. Infection may not occur above 27.5 °C (Marchetti *et al*., 1976). Kochman (1979) has confirmed that high temperature may reduce both the survival of *P. pachyrhizi* urediospores and the rate of development of soyabean rust, possibly in part explaining the late development of the disease in Queensland. The optimum temperature for urediospore germination of *U. appendiculatus* is at about 14.5 °C and germ tubes are longer at temperatures where the most germination takes place; maximum germ-tube growth occurs at 20 °C. At temperatures between 10 ° and 20 °C, germination is completed within four hours. The optimum temperature for infection of bean leaves is 17° C (Harter *et al*., 1935; Naito, 1951). The growth of germ tubes of *U. appendiculatus* and *P. arachidis* is directional, and appressoria form over stomata in response to a surface-contact stimulus leading to stomatal penetration by infection pegs (Figure 16; Wynn, 1976; Cook, 1980). The stomata need not be open to allow successful infection, and the process can occur in the dark (Maheshwari *et al*., 1967; Cook, 1980). The infection process in *P. pachyrhizi* is quite different. Appressoria develop promptly following urediospore germination, many being sessile to the parent spore. Typically, they form over junctions between epidermal cells, penetration occurring directly through the cuticle (Bonde *et al*., 1976); cuticular penetration by rusts is comparatively uncommon (Hunt, 1968).

To what extent, then, can these findings be used to explain field observations? Given that far more data are required in order to make valid comparisons between the legume rusts, such speculation may later prove incorrect. Nevertheless, some ecological trends are apparent. For instance, *Phakopsora pachyrhizi* appears to depend on high humidity for its development. Soyabean rust is most severe during the wettest months in Thailand (Bromfield, 1976), and the rust is also prevalent during the rainy season in Puerto Rico, particularly in the cool humid highlands (500–850 m), where shading may permit longer dew periods (Vakili, 1978). In Nigeria, *P. pachyrhizi* is most prevalent during the heavy rains of September–November and is apparently confined to the humid forest zone (D. J. Allen, unpublished), contrasting with *U. appendiculatus* which, in Nigeria, is common in the savannahs and under dry season irrigation but seldom develops significantly during periods of heavy rainfall (Williams, 1975b). Under Puerto Rican conditions, Vakili (1978b) has noted

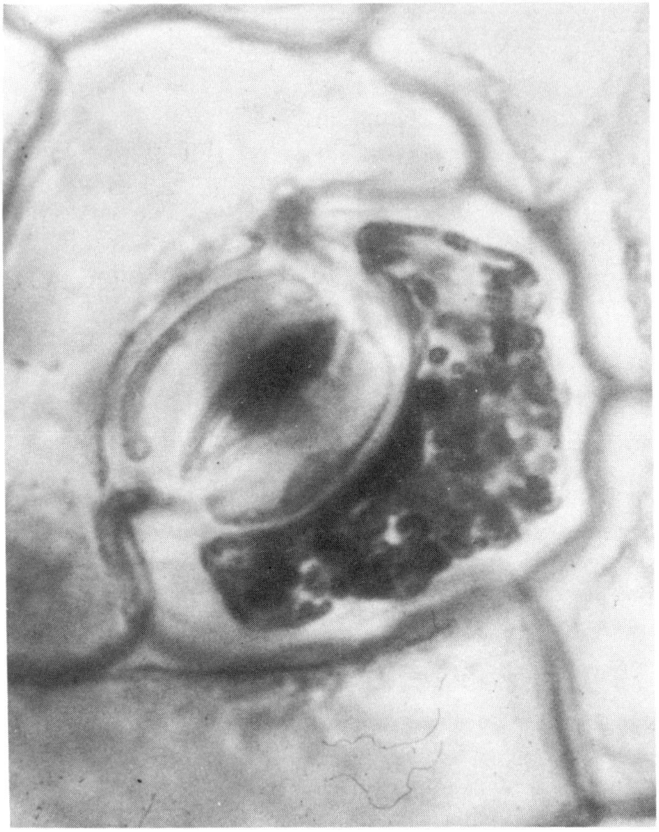

Figure 16 Stomatal penetration and development of *Puccinia arachidis* in groundnut (photograph courtesy D. J. Nevill)

the very much faster rate of spread of *U. appendiculatus* relative to *P. pachyrhizi*, and it is tempting to suggest, as Vakili has done, that the urediospores' tendency to clump around the ostiole of the uredium may favour the splash dispersal of *P. pachyrhizi* rather than the windborne spread of more pulverulent rusts like *U. appendiculatus*. However, another factor is involved. Melching *et al* (1979) have drawn attention to the comparatively low sporulating capacity of *P. pachyrhizi* relative to other rust fungi, including *U. appendiculatus*. They found that whereas a Taiwan isolate of *P. pachyrhizi* produced about 0.7×10^4 urediospores per lesion, the bean rust fungus produced ten times that number (Yarwood, 1961), probably accounted for by the smaller proportion of lesion area which produces spores, or the smaller lesion size in soyabean rust. Presumably this would also influence the rate of secondary spread of these two rusts.

The evolutionary significance of the contrasting infection processes of *P. pachyrhizi*, on the one hand, and *P. arachidis* and *U. appendiculatus* on the other, is yet harder to speculate upon. The hyaline, and thus presumably vulnerable, nature of the urediospores of *P. pachyrhizi* would seem to put a premium on relatively rapid infection. The very wide environmental range over which the legume rusts occur might be expected to exert a considerable influence on their biology. Too little is known of their environmental adaptation, and 'environmental races' have not been described, as they have with cereal rusts (Browning and Bustamante, 1973; Green, 1973; Sharp, 1973). Geographically diverse isolates of *P. pachyrhizi* appear not to have significantly different temperature ranges (Marchetti *et al*., 1976), and current evidence suggests that temperature optima for both germination and infection of *P. pachyrhizi* are rather higher than those of *U. appendiculatus*.

Chapter 3

Legume Pathogens II

3.1 VIRUSES AS LEGUME PATHOGENS

3.1.1 Virus identification and nomenclature

The viruses which naturally infect legumes are in no way unusual, having representatives from almost all of the major groups of plant viruses. The majority have been isolated from northern temperate, herbaceous crops, and few have been obtained from tropical, wild or woody species, presumably reflecting the distribution of virologists and the importance of host crops rather than any fundamental difference in the preferences of viruses (Gibbs and Watson, 1980). Although experimental host ranges of viruses are usually broader than their natural host ranges, the adoption of a taxonomically more balanced range of test plants might indicate whether there is any taxonomic component in virus host ranges. Appropriate plant species have been suggested (Gibbs and Watson, 1980) and the need for host plant standardization stressed (Bos *et al.*, 1960). Although certain host reactions may be diagnostic (e.g. Milbrath and Soong, 1976) and others useful in schemes for the separation of virus mixtures which commonly occur under field conditions (Bos *et al.*, 1960; Kuhn, 1964), legume virus identification based on host plant reactions alone is seldom reliable. More sophisticated methods must usually be used for the accurate identification of legume viruses (Bos, 1976; Hamilton *et al.*, 1981) although a dichotomous key based on host reaction has been drawn up to aid diagnosis of some legume viruses (Hampton *et al.*, 1978).

Inadequate identification has tended to be compounded with a lack of any internationally accepted classification of plant viruses. Appreciating this, Bos (1964) suggested the use of standardized vernacular names of legume viruses. Subsequent schemes of nomenclature (Gibbs *et al.*, 1966) and grouping (Harrison *et al.*, 1971; Kurstak, 1981) of viruses, and the advent of the CMI/AAB Descriptions, have done much to clarify the situation. And now there is scope for the determination of virus relationships within groups by serological means (Fulton and Scott, 1979). Nevertheless, nomenclatural confusion and incomplete characterization remain rife among the viruses infecting the tropical food legumes with which we are concerned here. Of the many published reports of virus diseases in these crops, only about 29 viruses have been formally described to date; of these, five are potyviruses, four are comoviruses and two each belong to the cucumovirus and geminivirus groups (Table 10). The aetiology of several important legume virus diseases, such as groundnut rosette and pigeon pea sterility mosaic, remains imperfectly understood, while in other cases (e.g. potyviruses) comparative studies are required to determine relationships between viruses. Geminiviruses (Harrison *et al.*, 1977; Goodman, 1981) are being implicated as causal agents of some of the leafhopper- and whitefly-transmitted diseases, such as beet curly top, bean golden mosaic, cowpea golden mosaic, and mung bean yellow mosaic, the characterization of which is only just beginning (Goodman and Bird, 1978); and it is certain that many other viruses naturally infecting the food legumes await discovery.

3.1.2 Economic importance of virus diseases

Although common viruses such as cucumber mosaic virus may have little effect on legume hosts, some virus diseases can devastate legume crops. An epidemic of rosette in 1975 led to almost total loss of the groundnut crop in northern Nigeria, with an estimated loss of about $300m (Rossel, 1977; Yayock, 1977). And groundnut yields may be reduced by as much as 90 per cent by infection with tomato spotted wilt virus in Australia (Helms *et al.*, 1961). Brazilian bean production has been reduced considerably in recent years by bean golden mosaic virus (Costa, 1975), while cowpea severe mosaic virus is a major production constraint in north-eastern Brazil where cowpeas are an important food legume (Lima and Nelson, 1977). Seedborne viruses, often having cosmopolitan distributions, may cause extensive losses in seed yield. For instance, seed losses of up to 68 per cent have been estimated from beans naturally infected with bean common mosaic virus (Hampton, 1975), and cowpea aphidborne mosaic virus can decrease cowpea seed yields by as much as 87 per cent from natural field infection (Kaiser and Mossahebi, 1975). Natural aphid transmission of soyabean mosaic virus can result in up to 35 per cent loss of seed yield in susceptible soyabean cultivars (Ross, 1977), while mechanical inoculation of seedlings can cause reductions of up to 86 per cent (Goodman and Oard, 1980). Such seed yield losses, which are usually greatest when virus infection occurs early, may result from adverse effects of infection on the components of yield, such as pod number per plant (Hampton, 1975), as well as affecting nodulation (Tu *et al.*, 1970; Adnew *et al.*, 1977). Mixed infections of two or more viruses can cause synergistic responses and lead to greater yield losses than may result from either virus alone (Ross, 1968a; Pio-Ribeiro *et al.*, 1978).

3.1.3 Seed transmission of viruses in legumes

In 1919, Reddick and Stewart showed that bean common mosaic virus was transmitted through seed of *Phaseolus vulgaris* and, in 1924, the seed trans-

TABLE 10 Viruses Naturally Infecting Tropical Legumes[a]

Virus	Distribution	Hosts[b]	Seed transmission (%)	References
I. *Aphid-transmitted viruses*				
(i) *Potyviruses*				
Bean common mosaic (BCMV)	Worldwide	Bean, mung bean, tepary bean	<83	Bos, 1971; Kaiser and Mossahebi, 1975; Provvidenti and Cobb, 1975
Bean yellow mosaic (BYMV)	Worldwide	Bean, soyabean, chickpea, lentil, mung bean, and others	0?	Kaiser *et al.*, 1968; Bos, 1970; Sinclair and Shurtleff, 1975; Iwaki and Auzay, 1978
Soyabean mosaic (SMV)	Widespread	Soyabean	30	Bos, 1972
Peanut mottle (PnMV)	Europe, USA, Venezuela, E Africa, Malaysia, India, Philippines, Japan, Australia	Groundnut, soyabean, lima bean, bambarra groundnut, other legumes	<2 (g'nut) 0 (soya)	Bock and Kuhn, 1975; Bock *et al.*, 1978; Reddy *et al.*, 1978; Ghanekar, 1980; Demski *et al.*, 1981
Cowpea aphidborne mosaic (CAMV)	Widespread	Cowpea, adzuki bean	0–40	Tsuchizaki *et al.*, 1970; Bock and Conti, 1974; Kaiser and Mossahebi, 1975; Ladipo, 1977
(ii) *Cucumoviruses*				
Cucumber mosaic (CMV)	Worldwide	Cowpea, mung bean, bean, chickpea, soyabean, winged bean, and many others	4–28 (cowpea) 0.3 (mung) <30 (bean)	Gibbs and Harrison, 1970; Kaiser and Danesh, 1971a; Bos, 1977; Purivirojkul and Poehlman, 1977; Tamada and Kojima, 1977; Schwartz and Gálvez, 1980; Brunt and Phillips, 1981b

Peanut stunt (PSV)	Japan, USA, Morocco	Groundnut, soyabean, bean	<0.1	Mink, 1972; Milbrath and Tolin, 1977; Fischer and Lockhart, 1978
(iii) *Other groups*				
Alfalfa mosaic (AMV)	Worldwide	Bean, soyabean, cowpea, mung bean, chickpea, lentil, hyacinth bean, and many others	?	Allington *et al.*, 1960; Nour and Nour, 1962; Kaiser *et al.*, 1968; Kaiser, 1979; Jaspars and Bos, 1980
Pea enation mosaic (PEMV)	Widespread in northern temperate regions	Chickpea, lentil, and other legumes	0	Shepherd, 1970; Vovlas and Rana, 1972
Lettuce necrotic yellows (LNYV)	Australia	Chickpea and others	?0	Francki and Randles, 1970; Corbin, 1975
Soyabean dwarf (SDV)	Japan	Soyabean	0	Tamada and Kojima, 1977
II. *Beetle-transmitted viruses*				
(i) *Comoviruses*				
Cowpea (yellow) mosaic (CYMV)	Africa (America)	Cowpea, winged bean, pigeon pea	1–5	K. R. Bock, pers. comm. 1976; R. F. Bozarth, pers. comm. 1977; McLaughlin *et al.*, 1977; van Kammen and de Jager, 1978
Cowpea severe mosaic (CSMV)	Warm temperate and tropical America	Cowpea, soyabean, pigeon pea, mung bean, black gram, bean, and other legumes	0–10	Dale, 1949; Alconero and Santiago, 1973; de Jager, 1979
Bean pod mottle (BPMV)	USA	Bean, soyabean	0	Semancik, 1972
Quail pea mosaic (QpMV)	USA, Central America	Bean, soyabean (and *Strophostyles helvola*)	0	Moore and Scott, 1981
(ii) *Other groups*				
Cowpea chlorotic mottle *(CCMV)*	Warm temperate and tropical America	Cowpea, bean, soyabean, and other legumes	0	Bancroft, 1971; Fulton *et al.*, 1975

TABLE 10 (*Continued*)

Virus	Distribution	Hosts[b]	Seed transmission (%)	References
Southern bean mosaic (SBMV)	Warm temperate and tropical America; West Africa	Bean, cowpea	3–4	Shepherd, 1971
Cowpea mottle (CMeV)	Nigeria	Cowpea, bambarra groundnut	<10	Bozarth and Shoyinka, 1979
Blackgram mottle (BgMeV)	India, Thailand	Black gram	8	Scott and Hoy, 1981
Bean mild mosaic (BMMV)	El Salvador, Colombia	Bean	?	Waterworth, 1981
III. *Whitefly-transmitted viruses*				
(i) *Geminiviruses*				
Bean golden mosaic (BGMV)	Tropical America	Bean, lima bean, and other legumes	0	Goodman and Bird, 1978
(ii) *Carlaviruses*				
Cowpea mild mottle (CMMV)	Africa, India, Thailand	Cowpea, groundnut, soyabean, and other legume hosts	2–90	Brunt and Kenten, 1974; Brunt, 1975; Bock *et al.*, 1976; Brunt and Phillips, 1978; 1981a; ICRISAT, 1978b; Iwaki *et al.*, 1981; Thouvenel *et al.*, 1982
IV. *Leafhopper-transmitted viruses*				
(i) *Geminiviruses*				
Beet curly top (BCTV)	North America	Bean, and many other hosts	0	Thomas and Mink, 1979

V. *Thrips-transmitted viruses*				
Tomato spotted wilt (TSWV)	Worldwide	Groundnut, and many other hosts	0	Ie, 1970; Ghanekar *et al.*, 1979
VI. *Nematode-transmitted viruses*				
Tobacco ringspot (TRSV)	Widespread	Soyabean, mung bean, black gram, cluster bean, and many other hosts	? –	Orellana, 1966; Stace-Smith, 1970; Vignarajah, 1978
VII. *Fungus-transmitted viruses*				
Tobacco necrosis (TNV)	Worldwide	Bean, and many other hosts	0	Kassanis, 1970
Peanut clump (PCV)	West Africa, India	Groundnut (and *Sorghum arundinaceum*)	5–6	Ghanekar, 1980; Thouvenel and Fauquet, 1981b
VIII. *Viruses with no known vector*				
(i) *Tobamoviruses*				
Sunnhemp mosaic (ShMV)	India, Nigeria and USA	Cowpea, hyacinth bean, and other legumes	0–20	Kassanis and Varma, 1975
(ii) *Ilarviruses*				
Tobacco streak (TSV)	Europe, temperate and tropical America and New Zealand	Soyabean, bean, groundnut, and many other hosts	?	Mink *et al.*, 1966; Fagbenle and Ford, 1970; Fulton 1971

[a]Formally described viruses only (CMI/AAB Descriptions of Plant Viruses).
[b]Natural infections.

mission of soyabean mosaic virus was reported (Kendrick and Gardner, 1924). Since these early studies, which were amongst the first published records of seed transmission of plant viruses, there has been a rapid increase in the number of known seed-transmitted viruses (Phatak, 1974; Bos, 1977). Seed transmission plays a vital role in the ecology of crop virus diseases, in providing foci of primary infection, carrying the virus from one season to the next, and in the long-distance transport of virus, of concern to the international exchange of plant genetic resources.

The incidence of seed transmission depends on the virus and its strain (or isolate), on the host plant and its cultivar, and on the environment. Although natural infection of a susceptible host usually leads to the systemic spread of virus throughout the plant, seeds of infected plants do not invariably become virus infected and viruses in some plants are not seedborne. Bean pod mottle, bean golden mosaic, pigeon pea sterility mosaic, groundnut rosette, and cowpea chlorotic mottle viruses are examples of non-seed-transmitted viruses (Bancroft, 1971; Nene, 1972; Semancik, 1972; Goodman and Bird, 1978). Peanut mottle virus is seed transmitted in groundnut but not through cowpea or soyabean (Bock and Kuhn, 1975). While seed transmission of bean yellow mosaic virus is uncommon, the closely related bean common mosaic virus is transmitted through bean seed, often at high rates (Medina and Grogan, 1961). Other viruses which are transmitted through seed at high incidences include cowpea mild mottle (Brunt and Kenten, 1973) and tobacco ringspot (up to 100 per cent in soyabean; Athow and Bancroft, 1959).

Seed transmission may vary with virus isolate, as shown with tobacco streak virus in soyabean (Ghanekar and Schwenk, 1974), and with host cultivar. Ladipo (1977) has shown that cowpea cultivars differ in their ability to transmit cowpea aphidborne mosaic virus through seed, the incidence of seed transmission ranging from zero to more than 20 per cent. Similarly, Goodman *et al.* (1979) identified soyabean accessions which were apparently 'non-transmitters' of soyabean mosaic virus through seed, although subsequent work (Goodman and Oard, 1980) showed that the accessions with resistance to seed transmission could transmit the virus, but at very low rates (<0.2 per cent).

It is generally considered that the rate of seed transmission depends on the time of infection. Plants infected after flowering do not produce infected seed, and transmission is successful only before cytoplasmic separation of the developing embryo from maternal tissue (Bos, 1977). That this is not invariably so has been suggested by Bowers and Goodman (1979). They showed that soyabean plants inoculated with soyabean mosaic virus while bearing immature green seeds produced infected seed, suggesting that post-flower infection can result in seed transmission. However, they pointed out that their observations were open to another interpretation: that outcrossing may have occurred with virus-infected pollen from early infected plants growing nearby, since Iizuka (1973) has shown that soyabean mosaic virus may be transmitted in pollen produced on infected plants, in a manner similar to bean common mosaic virus transmission in bean pollen (Medina and Grogan, 1961). The incidence of seed transmission often appears to vary according to whether reported rates refer to natural or to artificial infection, but whether or not such effects are attributable to experimental conditions or merely to the time of infection is uncertain. Discrepancies in the reported incidence of seed transmission of cowpea mottle virus (CMeV) provide an example. Although Robertson (1966) failed to detect seed transmission of CMeV through many thousands of seeds harvested from infected cowpea plants, Shoyinka *et al.* (1978) recorded up to 10 per cent seed transmission from naturally infected plants, and up to 7 per cent from manually inoculated plants. While confirming that CMeV may be seedborne, subsequent studies with five cowpea cultivars failed to detect an incidence higher than 0.4 per cent seed transmission, despite the use of serological assays, and then only in a very susceptible cultivar following manual inoculation. Only in one case ($^1/_{4305}$ = 0.02 per cent) was there evidence of seed transmission of CMeV from naturally infected plants (Allen *et al.*, 1982).

The rate of seed transmission of a virus may also be influenced by the presence of other viruses in mixed infections. Kuhn and Dawson (1973) found that southern bean mosaic virus was seedborne at 12 per cent in cowpea but the rate increased to 20 per cent in the presence of cowpea chlorotic mottle virus.

Having considered some of the factors that influence the rate of virus transmission through seed, let us now briefly examine the mechanisms of seed transmission, reviewed by Bos (1977). Viruses can be seedborne in several ways. Embryo infection leads to direct infection of progeny, and this is apparently the commonest mode of virus transmission. Virus can also be carried outside the embryo, in or on the seed coat, but this is ecologically precarious, exposing the virus to dessication and to the problem of finding a means of infecting the seedling, so that few viruses are seedborne in this manner. Bean common mosaic virus (BCMV) and cowpea aphidborne mosaic virus are examples of internally seedborne viruses. Each has been detected internally in cotyledons and embryos but not from seed coats of their respective hosts (Ekpo and Saettler, 1974; Ladipo, 1977); there is evidence (Pierce and Hungerford, 1929) that BCMV may survive in infected seed for at least 30 years. Southern bean mosaic virus (SBMV) was initially thought to be confined to the seed coat of cowpea (Lamptey and Hamilton, 1974) and bean (McDonald and Hamilton, 1972), suggesting that the mechanism of seed transmission was through infection of the embryo at germination with virus from the seed coat. However, subsequent studies (Uyemoto and Grogan, 1977) have shown that SBMV is, in fact, transmitted through the embryo of beans, not by fortuitous infection by surfaceborne virus. Whether the same applies to SBMV transmission through seed of cowpea remains uncertain. Cowpea chlorotic mottle virus (CCMV), on the other hand, is absent from the host embryo. CCMV can be recovered from the coats of immature seed as well as from mature, but not dry, seed. Since the virus is not seed-transmitted in cowpea, CCMV probably becomes inactivated during seed maturation (Gay, 1969b). The same probably applies to tomato spotted wilt virus on groundnut seed (Helms *et al*., 1961).

In practice, what is of consequence is whether or not a virus survives on, or in, seed and becomes established in an infected seedling, initiating an epidemic. We have seen there is considerable variation in the incidence of seed transmission as well as in the type of mechanism involved. Viruses confined to the testa of seed may or may not be successfully seedborne. The rate of seed transmission required to initiate an epidemic must obviously depend on many factors, but what is evident is that too little is known of that critical threshold. Zink *et al*. (1956) found that if the rate of seedborne infection exceeded 0.1 per cent the degree of control (of lettuce mosaic) was unsatisfactory. This demonstrates that even very low rates of seed transmission can be sufficient to establish an epidemic.

In the humid tropics, crop growth may be almost continuous so that seasons overlap; such dry season as occurs may be too short to destroy groundkeepers and volunteer seedlings so as to threaten the survival of inoculum. In semi-arid and arid environments, there are clearly defined seasons of growth separated by a long dry season, as discussed in Chapter 1. Since the principal crops in such regions are rainfed annuals such as cereals and legumes, there is a long period without crops. If a virus has no alternative host on which it can survive the close season, seed transmission is likely to be the chief means of survival (woody perennials, such as pigeon pea, may possibly provide exceptions).

Bean common mosaic and soyabean mosaic viruses survive in this manner (Fajardo, 1930; Halbert *et al*., 1981) as probably do many other legume viruses. Cowpea aphidborne mosaic virus (CAMV), unlike other common cowpea viruses, is present in naturally infected cowpea crops in the northern savannah belt of Nigeria (Rossel, 1977) where no alternative hosts are known. This suggests that seedborne infection may be involved in CAMV carry-over, although it is evident that increasing use of irrigation, in addition to the perenially damp areas (the 'fadamas' of Nigeria and the 'dambos' of eastern Africa), may provide virus reservoirs (Raheja and Leleji, 1974; Rossel, 1977). In other cases, certain seedborne viruses apparently have relatively restricted distributions and this suggests that seed transmissibility under experimental conditions may not always correspond closely with the incidence of seed transmission in nature. For instance, if southern bean mosaic virus (SBMV) is of African origin as suggested by Lamptey and Hamilton (1974), it is surprising that the virus has been isolated relatively infrequently from a few widely scattered locations in the humid and sub-humid parts of West Africa and from nowhere else in Africa (Ladipo and Allen, 1979). This suggests that the seed transmission of SBMV may not provide a means of survival under dry savannah conditions, but the reason for this is uncertain. A further example is afforded by cowpea mottle virus. Despite its reported seed transmission to up to 10 per cent of seedlings

(Shoyinka *et al.*, 1978; Bozarth and Shoyinka, 1979), the virus is apparently confined to the more humid regions of Nigeria, being absent from cowpeas in farmers' fields in the northern savannahs (Rossel, 1977). This, too, suggests that its seed transmissibility contributes little to close season survival and is ecologically inefficient, a supposition reinforced by recent studies (Allen *et al.*, 1982).

3.1.4 Weeds and the ecology of virus diseases

Weeds may play an important role in the ecology of plant viruses, acting as virus reservoirs, vector reservoirs, or both (Duffus, 1971). Despite the dearth of knowledge of viruses in wild plants, a few examples are available from the legumes. Tomato spotted wilt virus occurs in naturally infected weeds (*Bidens pilosa* and *Tagetes minuta*) in the vicinity of groundnut crops in Australia (Helms *et al.*, 1961), and the woody leguminous herb *Tephrosia villosa* is a common host of cowpea mild mottle virus in coastal Kenya (Bock *et al.*, 1981). *Macroptilium lathyroides* (= *Phaseolus lathyroides*) is a major reservoir of cowpea severe mosaic virus in Puerto Rico and Brazil (Alconero and Santiago, 1973; Lima and Nelson, 1977). Forage legumes (*Trifolium* spp. and *Lupinus* spp.) and a leguminous weed (*Desmodium canum*) are alternate hosts of peanut mottle virus in the USA (Demski *et al.*, 1981). *M. lathyroides* and *Calopogonium mucunoides*, as well as various Malvaceae including *Sida* spp., are sources of bean golden mosaic virus and other whitefly-transmitted disease agents in tropical America (Costa, 1976). Some weeds may harbour several different viruses (Alconero and Santiago, 1973; Costa, 1976) of which they may be not merely an alternative but the original host (Section 1.3.3). An example of a weed acting as an alternative host of a virus vector is *Euphorbia hirta* on which the vector of groundnut rosette, *Aphis craccivora*, survives the dry season in West Africa (Booker, 1963). Furthermore, weeds may increase vector populations (Altieri *et al.*, 1977), or alter vector behaviour (Williams, 1976), within a crop. *Melilotus alba* is a major reservoir of both bean yellow mosaic virus (BYMV) and its aphid vectors in Iran, where BYMV is an important chickpea pathogen (Kaiser and Danesh, 1971b). Similarly, it appears that *Sorghum arundinaceum* is a natural host of peanut clump virus as well as of the fungus *Polymyxa graminis* which, being resistant to air-drying, is thought to be the virus vector (Thouvenel and Fauquet, 1981a).

3.1.5 The role of vectors in the epidemiology of legume viruses

While alternative host reservoirs and seedborne infections are of great importance to the carry-over of virus between seasons, vectors are of paramount importance as agents of secondary spread. Viruses that are spread widely by natural means are chiefly those with efficient vectors; other viruses have a limited distribution or fail to survive unless they are disseminated by vegetative propagation or other human activities. Thresh (1974) has emphasized that the effectiveness of transmission necessary to ensure survival of the virus depends on the nature and distribution of host plants as well as on the environment. While occasional transmission by a rare, immobile, or inefficient vector may ensure perpetuation of a virus in common perennial hosts, it is unlikely to be adequate for a virus with a restricted host range, or widely scattered hosts. Consequently, various relationships have evolved between viruses and vectors, varying in degree of specificity and mechanism of transmission. The rate of spread depends on both the density and activity of the vector population, while the optimum feeding conditions for transmission depend upon the nature of the mechanism, each having features that facilitate spread and others that make for unreliability: rapid transmission in short probes and long persistence in the vector are mutually exclusive attributes (Thresh, 1974).

Aphids are much the most important group of virus vectors, including more vector species and transmitting many more viruses than any other group, and as such have been extensively studied. The diverse viruses with aphid vectors are referred to as styletborne or circulative according to the mode of transmission, approximating the earlier grouping of non-persistent and persistent based on retention in the vector. Of the many aphidborne viruses of food legumes, the majority, including the cucumovirus and potyvirus groups, are transmitted in the non-persistent manner. Examples of legume viruses transmitted by aphids in the persistent manner include soyabean dwarf (Tamada and Kojima, 1977), groundnut rosette (Watson and Okunsanya, 1967) and pea leafroll (Kaiser and Danesh, 1971a). The non-persistent viruses are rapidly acquired and transmitted, by transient as well as resident aphids, leading to the early development and rapid spread of virus diseases, if local sources of infection are available. This contrasts with the persistent viruses, vectors of which take longer to become infective but, once having done so, retain virus sometimes for life. Prolonged retention compensates for the usually greater specificity of the relationship, as well as for the longer acquisition period required. Nevertheless, persistent viruses are less dependent on the presence of nearby sources of infection, and are more readily controlled by insecticides, than non-persistent viruses (Thresh, 1974).

Several important legume viruses, especially those of the comovirus group, are transmitted by beetles, principally of the family Chrysomelidae. Although viruses transmitted by beetles cannot be transmitted by insects with sucking mouth parts, it has been generally assumed that when a biting insect behaves as a vector, it only does so by carrying infective sap on its mouth parts (Walters, 1969). However, since viruses such as cowpea severe mosaic (Dale, 1953) and southern bean mosaic (Walters and Henry, 1970; Allen *et al.*, 1981a) may be retained by beetles for periods of two to three weeks, other mechanisms must also be involved. Regurgitation from the foregut may perhaps account for the transmission of cowpea severe mosaic (Dale, 1953) and southern bean mosaic viruses (Kopek and Scott, 1970); and Slack and

Scott (1971) showed that beetle haemolymph could act as a virus reservoir. The efficiency of transmission, and the length of time a beetle vector remains infective may depend more on the beetle species than on the virus (Fulton and Scott, 1974), correlated with feeding behaviour (Jansen and Staples, 1971). The virus source plant can also influence vector efficiency (Jansen and Staples, 1970a; Hobbs and Fulton, 1979). However, in other virus-vector combinations, the nature of the virus apparently determines transmission efficiency more than the beetle species does (Hobbs and Fulton, 1979; Allen *et al.*, 1981a). Nevertheless, beetleborne viruses appear to have certain characteristics in common: they are usually relatively stable, develop a high concentration in infected plants, are readily transmitted by manual inoculation, and have spherical or polyhedral particles (Walters, 1969). They may occur in mixed infections (Kuhn and Dawson, 1973; Shoyinka *et al.*, 1978) from which they may be simultaneously transmitted (Allen *et al.*, 1981a). The degree of virus-vector specificity is generally not pronounced. Some species of beetle may transmit several distinct viruses and certain viruses may have numerous beetle vectors (Jansen and Staples, 1971; de Jager, 1979). Grasshoppers are also known to transmit certain beetleborne viruses but their transmission by other groups of insect requires further investigation (Whitney and Gilmer, 1974).

Whiteflies are important vectors of several virus diseases of tropical legumes, the chief species being *Bemisia tabaci* which is the vector of more than 25 diseases (Costa, 1976) including several in the food legumes. Despite their importance throughout the tropics, very little is known of the aetiology of the whitefly-transmitted diseases of legumes. Although there is increasing evidence that geminiviruses are involved (Thongmeearkom *et al.*, 1981; H. J. Vetten, unpublished), so far only bean golden mosaic virus (BGMV) has been satisfactorily described (Goodman and Bird, 1978). Recently, it has been shown that cowpea mild mottle virus (CMMV), a carlavirus, is also whitefly-transmitted (D. V. R. Reddy, personal communication, 1981; Iwaki *et al.*, 1981). Investigations have been made of the virus-vector relationships of mung bean yellow mosaic in India (Nene, 1972), cowpea golden mosaic in Nigeria (Anno-Nyako, 1980) as well as of BGMV (Gámez, 1971) and many other similar diseases in non-legume hosts. Since the disease agent in most cases has been reported to be persistent, or semi-persistent, in *B. tabaci*, a certain degree of specificity could be expected, yet available evidence does not support this (Costa, 1976). The flight activity of the adults contrasts with the immobility of larvae. Consequently only the adults are important as vectors, although larvae can acquire virus which persists through pupation and which can be transmitted immediately they become adults. The transmission efficiency of adults is usually erratic, infectivity often being lost. Adult whiteflies generally require a minimum of 10–15 minutes to acquire virus since the insects apparently need this time to reach the phloem tissues in which these viruses are concentrated, their stylets following an intercellular course (Pollard, 1955). Whiteflies are not always immediately infectious and there is a definite latent period of 4–20 hours in many cases. Virus usually persists for several days and in some instances for life. Individual adult whiteflies transmit BGMV and cowpea golden mosaic intermittently for up to 16 and 21 days after acquisition, respectively (Goodman and Bird, 1978; Anno-Nyako, 1980). There is no evidence of passage to the egg or of multiplication in the vector.

There may be marked seasonal fluctuations in population density of *B. tabaci*, corresponding closely with periodicity in disease incidence, peak catches of adult whiteflies following the onset of rains after the dry season at a time of year when temperature is high. Virus transmission may apparently depend on either a transient or a resident vector population, depending on environment (Vetten and Allen, in press). Mixed infections of different whitefly-transmitted diseases are not uncommon (Costa, 1976) and the simultaneous transmission of two viruses by *B. tabaci* is known (Varma, 1955; Anno-Nyako, 1980).

As is the case with the whitefly-transmitted viruses, the status of many of the pathogens transmitted by leafhoppers remains uncertain. Recent evidence suggests that while some are truly of virus aetiology, the 'yellows' and other diseases are apparently caused by mycoplasmas and will not be discussed further here. Examples from among the legumes are beet curly top and bean summer death, both of which are diseases of bean caused by geminiviruses (Thomas and Mink, 1979; Thomas and Bowyer, 1980), transmitted by *Circulifer tenellus* and *Orosius argentatus* respectively. Beet curly top virus circulates but does not multiply in its vector nor is the virus transmitted through the egg. Leafhoppers transmit occasionally after an acquisition time of one minute but efficiency is greatest after acquisition access periods of 2–3 days. The minimum latent period in the vector is about 4 hours. Highly infective vectors transmit the virus for life (Thomas and Mink, 1979). Although *O. argentatus* transmits mycoplasmas as well as bean summer death virus (Bowyer, 1974), there appears to be a relatively high degree of specificity among leafhopper vectors of plant viruses, and this has a considerable influence on the epidemiology of the diseases induced by leafhopper-transmitted viruses.

The only other group of insect vectors of legume viruses are thrips. *Frankliniella schultzei* and *Scirtothrips dorsalis* are vectors of bud necrosis of groundnut in India, caused by tomato spotted wilt virus (Ghanekar *et al.*, 1979). The virus is acquired by the larvae but not by the adults, whereas only adults transmit. Thus, transmission is only by adults that feed on infected plants in the larval stage. The minimum acquisition access period is 15 minutes; the latent period is 4–10 days, depending on the vector species. Vectors are maximally infective 22–30 days after acquisition but sometimes retain the virus for life. They do not transmit virus to their progeny (Ie, 1970); and there is no evidence of multiplication in the vector. The adults' inability to acquire virus may account for the lack of secondary spread within crops. Newly infected plants tend to be randomly distributed, and the rate of spread depends upon environmental factors influencing the number of incoming thrips rather than on the pattern and incidence of infected plants (Thresh, 1974). While thrips are the only known vectors of tomato spotted wilt virus, certain viruses with other vectors may also be thrips-transmitted. Tobacco ringspot virus, transmitted by nematodes, may also to some extent be transmitted in soyabean by nymphs but not adults of *Thrips tabaci* in which the virus may be retained for 14 days (Messieha, 1969). And the beetle-transmitted cowpea (yellow) mosaic virus is also reported to be transmitted by both *Sericothrips occipitalis* and *Megalurothrips sjostedti* (Whitney and Gilmer, 1974), but recent results with *M. sjostedti* refute earlier conclusions, and suggest that thrips are unlikely to be significant vectors of cowpea (yellow) mosaic in the field (Allen and Van Damme, 1981).

Several important plant viruses are transmitted by eriophyid mites which have piercing and sucking mouthparts (Oldfield, 1970); the only example relevant to us is the transmission of pigeon pea sterility mosaic by *Aceria cajani* in India (Seth, 1962; see Section 8.2.3). A single mite is sufficient to transmit the disease agent but at a low efficiency. More than ten mites per seedling are required to achieve 100 per cent infection, while a minimum inoculation access period of 5 minutes is necessary for transmission (ICRISAT, 1977). It is probable that a high degree of specificity between the agent and its vector is involved, suggested by the difficulty of sap transmission and absence of other vectors (Nene, 1972).

Soilborne virus diseases appear to be uncommon in annual crops under tropical conditions. Although nematodes are important vectors of many plant viruses, only tobacco ringspot virus (TRSV), the type member of the nepovirus group, is transmitted by nematodes in the legumes with which we are concerned. The nematode *Xiphenema amercianum* transmits TRSV to and from soyabean in which the virus causes bud blight in the USA, but it appears that the importance of this nematode in the spread of the disease may be negligible (Bergeson *et al*., 1964). Other vectors of TRSV in soyabean include grasshoppers (Dunleavy, 1957), thrips (Messieha, 1969) and possibly other groups, including mites (Thomas, 1969; Stace-Smith, 1970). Recent work on groundnut in West Africa has shown that peanut clump virus (PCV) is soilborne. Failure to find nematodes belonging to virus vector genera in infective soil, coupled with the finding that soil remained infective after prolonged air-drying, suggested that the vector was a fungus. It is now thought that the fungus *Polymyxa graminis* is the vector of PCV. The frequency of soil transmission of PCV was seasonal, being greatest during the dry, windy period ('the harmattan') (Thouvenel and Fauquet, 1981a). Since there is some evidence that peanut clump virus also occurs in India where its natural vector has not yet been identified (Ghanekar, 1980), it is of considerable interest that the fungus *Olpidium brassicae*, a known virus vector and a close relative of *Polymyxa*, has recently been isolated from groundnut roots in India (Subrahmanyam and McDonald, 1980).

Some legume viruses, such as sunnhemp mosaic, have no known vector (Kassanis and Varma, 1975). While in some cases this may merely reflect incomplete search, it is also possible that some mechanism obstructs insect transmission. Scott and Fulton (1978) examined the relations between sunn-hemp mosaic virus and the beetle, *Ceratoma trifurcata*, and found that although the virus was present in the beetle's haemolymph and regurgitant, transmission did not occur. Since the beetle fed extensively, and the virus was present near damaged tissue, they concluded that the virus was not deposited at the proper site for infection to occur. In other cases, insect transmission may depend on the presence of an assistor virus (Hull and Adams, 1968; Cockbain and Jones, 1980). Mixed infections may play some role in such dependent transmission of viruses (Rochow, 1972).

As we have seen from the above examples, the degree of virus-vector specificity varies considerably. At one extreme, different vector biotypes may transmit different strains of a virus, as in the aphid transmission of groundnut rosette (Watson and Okusanya, 1967); at the opposite extreme, a single virus such as tobacco ringspot may be transmitted by very diverse vectors including species of nematode, thrips, spider mite, grasshopper, and flea beetle (Stace-Smith, 1970). In between these extremes, there are examples of a single virus being transmitted by a single vector species, as is the case with pigeon pea sterility mosaic (Nene, 1972); a single virus being transmitted by several species of a single group of insects, as with soyabean mosaic virus (Bos, 1972); and a single insect species transmitting a wide range of viruses, as with *Bemisia tabaci* and geminiviruses (Costa, 1976).

Vector biology and behaviour have a profound influence on the patterns of virus spread, as well discussed by Thresh (1974a; 1976). The economic importance of an insect species as a virus vector depends on many factors including its abundance, activity, and virus transmission efficiency. With aphids, the time of emergence and abundance of winged forms in the population may be significant, especially since virus vectors are often transient rather than resident in the crop (Kennedy, 1950; Halbert *et al*., 1981). Factors governing probing behaviour include alarm pheromones (Yang and Zettler, 1975) and host plant resistance (Kennedy, 1976) but neither need necessarily affect virus transmission efficiency. The yellow-sensitivity of aphids (Kennedy *et al*., 1961) and of whiteflies (Mound, 1962) plays a role in host-finding and preference and thus in part governs the numbers of potential vectors and consequent disease incidence, apparently through affecting the alighting response of the insect (A'Brook, 1968).

3.2 MYCOPLASMAS AND LEGUMES

The 'yellows' diseases of plants, once thought to be caused by viruses, are now known to be associated with mycoplasma-like organisms (MLOs), spiroplasmas or probably also rickettsia-like bacteria. Symptoms, which appear to be due to hormonal changes, include yellowing of foliage, growth abnormalities and changes in flowers and fruit (Markham, 1980). Although about 60 plant diseases comprise the yellows-type group, only three have been satisfactorily characterized, essentially because of the difficulty in culturing the MLOs.

Most of the 'yellows' diseases, which often have wide host ranges, are transmitted by leafhoppers.

Various MLOs are known to infect legumes and incite symptoms described as legume little-leaf, witches' broom, phyllody, and virescence. However, too little is known of their aetiology to assess their probable relationships; much more work is required to clarify the position. In 1956, Hutton and Grylls described legume 'little leaf' disease of pasture legumes in Australia where pigeon pea also became naturally infected. Mycoplasma-like bodies were later found in sieve tube elements of infected plants (Bowyer and Atherton, 1970) and shown to be similar but distinct from MLOs associated with lucerne witches' broom and tomato big bud (Bowyer, 1974). Witches' broom diseases of various legumes have been described from various parts of Asia; examples include witches' brooms of soyabean in Taiwan (Lo and Han, 1969), of cowpea in India (Varma *et al*., 1978), and of soyabeans, groundnuts, beans, and other legumes in Japan (Murayama, 1966). A pigeon pea rosette is also associated with MLOs in India (Maramorosch *et al*., 1976). Other examples are given by Kitajima (1980). In Africa, phyllody of groundnuts (C. Harkness, pers. comm.), soyabeans and, occasionally, of cowpeas have been observed (D. J. Allen, unpublished) but no information is available on their aetiology. Witches' broom of pigeon pea is known to be associated with MLOs in Puerto Rico (Maramorosch *et al*., 1974; Vakili and Maramorosch, 1974) and a leafhopper-transmitted shoot proliferation of soyabean ('Machismo' disease) is suspected also to be induced by MLO in Colombia (Granada, 1979a,b).

The general characteristics of MLOs, well reviewed by Maramorosch (1974), are that they are pleiomorphic micro-organisms very sensitive to tetracycline antibiotics. Most are transmitted by leafhoppers in which they usually have long incubation periods. MLOs are not passed through the egg to the progeny of infected vectors, nor are they seedborne in their host plants in which they are confined to the phloem. Relatively little is known of the epidemiology of the yellows-type diseases, but current evidence suggests that MLOs are of relatively minor importance as pathogens of tropical food legumes.

3.3 BACTERIA AS LEGUME PATHOGENS

The principal bacterial pathogens of the legume crops with which we are concerned belong to the genera *Pseudomonas, Xanthomonas*, and *Corynebacterium*. The first two of these genera are gram-negative and the third gram-positive. Virtually all plant pathogenic bacteria can be artifically cultured, and it is only by their cultural and biochemical behaviour that they can be satisfactorily characterized. There has been considerable controversy over the taxonomy of plant pathogenic bacteria. Physiological criteria have often been found inadequate, and insufficient cross-inoculations have been conducted to minimize synonymy. This has meant that species have been separated largely on the basis of their pathogenicity and host ranges. Early serological studies did not help, and there was apparently a complete lack of correspondence between pathogenicity and serological groupings (Burkholder and Starr, 1948; Wernham, 1948). Recent taxonomic revisions have tended to lump the so-called 'nomenspecies' into a few taxa clearly distinguishable by usual bacteriological procedures, while recognizing that, for plant pathological purposes, at the sub-specific level strains with distinctive pathogenicity required differentiation. These pathogenic strains which roughly correspond with the better characterized nomenspecies are now referred to as 'pathovars' for which a set of international standards has been drawn up (Dye *et al*., 1980). Following this classification, the principal bacterial pathogens of food legumes are shown in Table 11. Various other bacteria are reported to be associated with the legumes, such as *Bacillus subtilis* on soyabean seed (Ellis *et al*., 1977), but these are of minor importance and therefore not discussed here. Coloured pigment variants of certain pathogens have been described (Schuster and Christiansen, 1957; Hayward and Waterston, 1965c; Schuster *et al*., 1968), and pathogenic variants within the newly erected pathovars are also known (Schuster and Coyne, 1975); though variants may be serologically indistinguishable (Cross *et al*., 1966; Guthrie, 1968), agglutination tests can differentiate races (Slusarenko and Wood, 1981).

Unlike the fungi and viruses, there are comparatively few bacterial diseases of plants, and bacterial infection less often leads to outbreaks of epidemic proportions. Nevertheless, bacterial pathogens induce a range of blights, cankers, wilts, leaf spots, and pustales in susceptible legume hosts, often leading to significant economic crop loss. The more important diseases include halo blight of bean, bacterial blight of soyabean, and bacterial wilt of groundnut, each caused by a pseudomonad; xanthomonads are responsible for the common and fuscous blights of bean, bacterial pustule, blight and canker of cowpea, and bacterial pustule of soyabean. Crop loss estimates range from less than 15 per cent in bacterial pustule of soyabean (Hartwig and Johnson, 1953; Laviolette *et al*., 1970), and 26 per cent in bacterial blight of cowpea (Ekpo, 1979), up to 45 per cent from common blight of bean (Yoshii *et al*., 1976), the extent of crop loss depending in each case on the host cultivar, the time (and method) of infection, and locality.

Bacteria invade plants only through wounds and natural openings and once inside susceptible hosts they divide and multiply rapidly in intercellular spaces. Some move systemically while others cause only local infections. After multiplication in the host, bacterial pathogens are spread in various ways. Bacterial exudates from leaves may be spread in rain, as are the pseudomonads of bean and soyabean, disease spread being favoured by mechanical injury caused by wind during rainstorms (Walker and Patel, 1964a; Daft and Leben, 1972). Naturally produced aerosols may also play a significant role in the epidemiology of legume pseudomonads (Venette and Kennedy, 1975). Wind-blown soil may assist the dissemination and infection of bacterial pathogens of beans under dry conditions (Burke, 1957; Claflin *et al.*, 1973), while insects may be important vectors of legume xanthonomads elsewhere. Kaiser and Vakili (1978) noticed that bacterial blight lesions on beans and cowpeas in Puerto Rico were frequently associated with pest damage, and xanthomonads were recovered from washings of five leaf-feeding

TABLE 11 Bacterial Pathogens of Food Legumes

Bacterium	Distribution	Natural hosts	References
Corynebacterium			
C. flaccumfaciens	North America, Europe, Australia, and southern Africa	Bean, soyabean	Dunleavy, 1963; Hayward and Waterston, 1965a; Schwartz and Gálvez, 1980; Leakey, personal communication, 1982
Pseudomonas			
Ps. syringae			
pv. *syringae*	Temperate and tropical America, Kenya, Australia	Bean, soyabean, cowpea, lima bean, and many other hosts	Gardner and Kendrick, 1923; Tisdale and Williamson, 1923; Wilson, 1936; Ondieki, 1973; Kaiser and Ramos, 1980; Schwartz and Gálvez, 1980
pv. *glycinea*	Widespread	Soyabean, other *Glycine* spp.	Allen, 1975b; Sinclair and Shurtleff, 1975
pv. *phaseolicola*	Worldwide	Bean, lima bean, hyacinth bean, mung bean, ? pigeon pea, yam bean	Patel and Jindal, 1972b; Allen, 1979a; Ebbels and Allen, 1979; Schwartz and Gálvez, 1980; Birch *et al.*, 1981
pv. *tabaci*	Widespread	Soyabean, bean, cowpea, and other hosts	Tisdale, 1924; Allington, 1945; Ribeiro and Hagedorn, 1976
Ps. solanacearum	Worldwide	Groundnut, bean, soyabean, winged bean, and many other hosts	Smith and McCulloch, 1919; Sinclair and Shurtleff, 1975; Abdullah, 1980; McDonald and Raheja, 1980
Xanthomonas			
X. campestris			
pv. *cajani*	India, Sudan	Pigeon pea	Kulkarni *et al.*, 1950; Sabet *et al.*, 1969
pv. *cassiae*	India	Chickpea, and other legumes	Rangaswami and Prasad, 1959; Nene, 1978
pv. *cyamopsidis*	India	Cluster bean	Patel *et al.*, 1953
pv. *glycines*	Worldwide	Soyabean, and other hosts	Sinclair and Shurtleff, 1975
pv. *phaseoli* (including var. *fuscans*)	Widespread	Bean, hyacinth bean, moth bean, mung bean	Hayward and Waterston, 1965b,c; Sabet and Ishag, 1969; Patel and Jindal, 1972a,b
pv. *vignae-radiatae*	Sudan	Mung bean	Sabet *et al.*, 1969
pv. *vignicola*	Widespread	Cowpea, bean, mung bean	Burkholder, 1944; Allen, unpubl., 1976

species, including three beetles, a leafhopper, and a sucking bug. Naturally infested beetles transmitted the bacteria to beans, and several xanthomonads survived for up to 19 days on the bodies of beetles. From their data, Kaiser and Vakili (1978) concluded that the beetle *Ceratoma ruficornis* was the predominant vector of pathogenic xanthomonads of bean and cowpea under Puerto Rican conditions. Heavy infestations of the whitefly, *Bemisia tabaci*, may also lead to transmission of xanthomonads in legumes (Sabet and Ishag, 1969). Other vectors of bacterial pathogens in legumes may possibly include sheep (Starr and Kercher, 1969), as well as man through his cultivation and irrigation practices.

Although escape to the plant surface may be necessary for the spread of plant pathogenic bacteria, it is not necessary for their perpetuation. Many survive through intimate association with the resting host (Buddenhagen, 1965), in buds and in seed. Leben *et al*. (1968) found *Ps. syringae* pv. *glycinea* in buds of mature soyabean and suggested that this pathogen may have a resident phase of ecological advantage to the pathogen, the buds providing a continual source of inoculum. Soyabeans may also support epiphytic colonies of this bacterium (Mew and Kennedy, 1971; Kennedy and Ercolani, 1978). Similarly, *Ps. syringae* pv. *syringae* survives throughout the year as epiphytic populations on a weed host which apparently is the source for infection of bean (Hagedorn *et al*., 1972), while weed hosts for bacterial blight of bean, caused by *X. campestris* pv. *phaseoli*, and bacterial pustule of soyabean (*X. campestris* pv. *glycines*) have also been recognized (Gardner, 1924; Jones, 1961).

Like the fungi and viruses, many of the bacterial pathogens of legumes are seedborne. Bacteria may retain their viability for as long as or longer than that of the host seed. Both *C. flaccumfaciens* and *X. campestris* pv. *phaseoli* have been recovered from 15 year-old bean seed (Schuster and Coyne, 1974). It has been suggested that the manner of seed germination may affect transmission of seedborne bacteria. Seeds with epigeal germination could favour transmission of bacteria that infect the aerial parts of plants, but may limit pathogens infecting subterranean plant organs; and vice versa (Schuster and Coyne, 1974). Seed transmission may result from both infected (internally-borne) and contaminated seed (Grogan and Kimble, 1967; Guthrie, 1970), often at very low incidence. An initial inoculum load of 1 per cent infected seed led to an outbreak of bacterial blight of cowpea in India (Shekhawat and Patel, 1977), while in Ontario a serious epidemic of fuscous blight of bean developed from a seed infection incidence of 0.5 per cent (Wallen and Sutton, 1965). As few as 12 infected seed per acre (*c*. 25/ha) may lead to an epidemic of halo blight of bean under Wisconsin conditions (Walker and Patel, 1964a). Zero tolerance of seed transmission appears to be a necessity where rates of disease spread are likely to be high, although reduction of primary inocula from infected seed may give effective disease control under certain conditions (Taylor *et al*., 1979a). Under British conditions, a tolerance level of 10 infected seed per cwt ($\equiv$ 1 seed/5 kg) has been estimated for halo blight

(Wharton, 1967). In tests of disease transmission from seed to seedling, Taylor *et al*. (1979b) found a relationship between seed symptoms and bacterial numbers in *Ps. syringae* pv. *phaseolicola*; heavily infected seeds usually failed to produce seedlings. Most infections developed from seeds with slight, or no symptoms. The nature of symptoms in the seedling may also depend on whether or not infection came from the seed. Seed transmission of *X. campestris* pv. *vignicola* often causes seedling mortality and stem canker in cowpea, while secondary infection leads to leaf blight only (Shekhawat and Patel, 1977).

It is probable that infected, or contaminated, seed is the primary inoculum source of the majority of bacterial pathogens of legumes; but bacteria may also persist in infected crop debris. *C. flaccumfaciens*, and both pseudomonads and xanthomonads of bean and soyabean, may overwinter in residues, particularly when the residues remain on the soil surface (Schuster and Coyne, 1974; Schuster, 1977). When buried, survival is often longer under dry conditions than when soil is saturated. Since it has been found that *X. campestris* pv. *phaseoli* may survive in moist soil longer when the soil has been autoclaved than in untreated soil (Sabet and Ishag, 1969), it appears that competitive saprophytes may play a role in bacterial survival. This has relevance to moist, lowland tropical conditions, where seasonal changes are minor and biological activity continuously high; here, one might expect a greater attrition of plant pathogenic bacteria in the soil (Buddenhagen, 1965). The extent to which plant pathogenic bacteria may survive in the free living state in soil is uncertain, partly owing to the difficulty of distinguishing the presence of host debris in the soil. In attempting to assess the role of the soil in the biology of bacterial plant pathogens, Buddenhagen (1965) has divided the pathogens into three groups, based upon the place of origin of their populations and upon their survival ability in soil. Such a treatment focuses attention on the non-host phase of the pathogen, and thus on the evolution and ecology of bacterial diseases. The three groups may be summarized as follows:

(i) Pathogens whose populations are developed almost exclusively in the plant host.
(ii) Pathogens that build up populations within hosts, but whose populations only gradually decline in the soil.
(iii) Bacteria whose populations are largely produced in the soil, and whose relation to plant disease is transitory.

The first category, of transient visitors to the soil, probably embraces the majority of bacterial plant pathogens, including most of the xanthomonads and some pseudomonads, while the second group includes pseudomonads, such as *Ps. syringae* pv. *tabaci*, and other species. The third category contains the true soil saprophytes.

In considering the types of host-parasite relationships that have evolved,

Buddenhagen (1965) has suggested that bacterial pathogens evolving further away from the requirement of a soil saprophytic existence have tended to develop sustained plant-to-plant infection cycles, often through insect transmission. Coupled with seed transmission, they have thus lost the more complex requirements of survival in the soil. Bacterial diseases, such as halo blight of bean and bacterial blight of soyabean, are caused by pathogens which have been essentially liberated from the requirement of a soil phase, through seed or insect transmission, or through association with a perennial host. In the humid tropics, where, as we have ssen, sustained biological activity is liable to deplete rapidly bacterial populations in the soil, the near continuous presence of host plants, and the rapid succession of crops, would appear to provide pathogenic bacteria with ample opportunity for development exclusively in the host phase.

Bacteria do not produce resistant spores, and bacterial cells are susceptible to dessication, light, and other external factors, thus emphasizing the need for their rapid and direct dispersal. There is some evidence that legume pseudomonads are more susceptible to high temperatures than are some of the xanthomonads (Goss, 1940; Patel and Walker, 1963; Sinclair and Shurtleff, 1975). This is borne out by the apparent ecological distribution of certain legume diseases in Africa: halo blight, and bacterial blight of soyabeans, which are each caused by pathovars of *Ps. syringae*, are common only under cooler conditions at higher elevations in tropical Africa, whereas soyabean bacterial pustule and the xanthomonad blights of bean and cowpea are also abundant in the hot lowlands (D. J. Allen, unpublished data). It appears that bacterial ooze may contribute to the high temperature tolerance of *X. campestris* pv. *phaseoli*, and may also afford some protection against light injury to bacterial cells (Leach *et al*., 1957).

3.4 NEMATODES AS LEGUME PATHOGENS

Amongst the wide range of plant-parasitic nematodes, or eelworms, associated with legume crops, the root knot nematodes (*Meloidogyne* spp.) are of paramount importance in tropical regions. Of the four principal species, three are cosmopolitan and the fourth, *M. hapla,* has a more northerly distribution (Table 12). The cyst nematodes (*Heterodera* spp.) which can cause severe damage to soyabeans, and locally to pigeon peas (Koshy *et al.,* 1970), are generally more important in temperate climates. The root lesion nematodes (*Pratylenchus* spp.) are among the most commonly occurring nematodes, causing extensive damage to a wide range of hosts throughout the world. *P. brachyurus* is apparently the most widely distributed species. *Pratylenchus* is probably the second most important genus, after *Meloidogyne*. The reniform nematode, *Rotylenchulus reniformis*, is also widely distributed in warm temperate and tropical regions where its very broad host range includes all the major food legume crops (Ayala and Ramirez, 1964). Other nematodes of economic significance to legume crops have more restricted geographical

TABLE 12 Nematode Parasites of Tropical Food Legumes

Species	Distribution	Hosts	References
Sedentary endoparasites			
Meloidogyne (root knot nematodes)			
M. arenaria	Cosmopolitan, tropical to warm temperate regions	Groundnut, soyabean, cowpea, mung bean	Sinclair and Shurtleff, 1975; Castillo and Litsinger, 1978; McDonald and Raheja, 1980; Schwartz and Gálvez, 1980; Singh and Allen, 1980
M. hapla	N Europe, Japan, USA, Canada (a northern range) but also in warmer regions of Africa and Middle East	Groundnut, soyabean, bean	Sinclair and Shurtleff, 1975; McDonald and Raheja, 1980; Schwartz and Gálvez, 1980
M. incognita	Cosmopolitan, tropical to warm temperate regions	Soyabean, bean, lima bean, cowpea, mung bean, winged bean	Allard, 1954; Caveness, 1975; Sinclair and Shurtleff, 1975; Price, 1976; Castillo and Litsinger, 1978; Schwartz and Gálvez, 1980
M. javanica	Cosmopolitan, tropical to warm temperate regions	Soyabean, bean, cowpea, mung bean, pigeon pea, winged bean	Ayala, 1962; Sinclair and Shurtleff, 1975; Castillo and Litsinger, 1978; Duncan *et al.*, 1979; Schwartz and Gálvez, 1980; Singh and Allen, 1980
M. spp.	—	Chickpea, hyacinth bean, bambarra groundnut, and various other legumes	Doku and Karikari, 1971; Anon., 1979; Nene *et al.*, 1979; Bridge, 1981
Heterodera (cyst nematodes)			
H. cajani	India	Pigeon pea, and other hosts	Koshy *et al.*, 1970; Bridge, 1981
H. glycines	Japan, China (Manchuria), Korea, USA	Soyabean, and other hosts	Winstead *et al.*, 1955; Sinclair and Shurtleff, 1975; Bridge, 1981
Migratory endoparasites			
Pratylenchus (root lesion nematodes)			
P. brachyurus	Cosmopolitan	Groundnut, soyabean, bean, cowpea	Bridge, 1973; Sinclair and Shurtleff, 1975; McDonald and Raheja, 1980; Schwartz and Gálvez, 1980

TABLE 12 (*Continued*)

Species	Distribution	Hosts	References
P. spp.	—	Soyabean, bean, pigeon pea	Sinclair and Shurtleff, 1975; Singh, 1975; Schwartz and Gálvez, 1980
Radopholus (burrowing nematodes)			
R. similis	Cosmopolitan in tropical to warm temperate regions	Groundnut (banana)	Taylor, 1976
Ditylenchus (stem and bulb nematodes)			
D. dipsaci	Cosmopolitan, but more important in cool, moist climates, and on heavy soils	Soyabean, and numerous other hosts	Sinclair and Shurtleff, 1975
Aphelenchoides (bud and leaf nematodes)			
A. arachidis	Nigeria	Groundnut	Bridge *et al.*, 1977
A. spp.	—	Bean	Schwartz and Gálvez, 1980
Sedentary semi-endoparasites			
Rotylenchulus (reniform nematode)			
R. reniformis	W Africa, USA, Hawaii, Indonesia, Philippines	Groundnut, soyabean, bean, lima bean, cowpea, pigeon pea, chickpea, mung bean, hyacinth bean, winged bean, and numerous other hosts	Ayala and Ramirez, 1964; Singh and Farrell, 1972; Bridge, 1973; Sinclair and Shurtleff, 1975; Castillo and Litsinger, 1978; IITA, 1979; McDonald and Raheja, 1980; Schwartz and Gálvez, 1980
Migratory semi-endoparasites			
Helicotylenchus (spiral nematodes)			
H. spp.	—	Soyabean, bean, cowpea	Bridge, 1973; Hague, 1980; Schwartz and Gálvez, 1980
Hoplolaimus (lance nematodes)			
H. seinhorsti	Nigeria	Cowpea	Bridge, 1973
H. spp.	—	Soyabean	Sinclair and Shurtleff, 1975

TABLE 12 (*Continued*)

Species	Distribution	Hosts	References
Scutellonema (false spiral nematode)			
S. cavenessi	W Africa	Groundnut (millet, sorghum)	Taylor, 1976
Sedentary ectoparasites			
Criconemoides (ring nematodes)			
C. spp.	Widespread	Groundnut, soyabean, bean	Feakin, 1973; Sinclair and Shurtleff, 1975; Schwartz and Gálvez, 1980
Migratory ectoparasites			
Belonolaimus (sting nematodes)			
B. gracilis	USA	Groundnut, soyabean, bean	Feakin, 1973; Sinclair and Shurtleff, 1975; Schwartz and Gálvez, 1980
B. longicaudatus	USA	Groundnut, soyabean, bean	Sasser *et al.*, 1968; Feakin, 1973; Sinclair and Shurtleff, 1975; Schwartz and Gálvez, 1980
Trichodorus (stubby root nematodes)			
T. christiei	Widespread	Groundnut, and many other hosts	Sasser *et al.*, 1968; Feakin, 1973
T. spp.	—	Soyabean, bean	Sinclair and Shurtleff, 1975; Schwartz and Gálvez, 1980
Xiphenema (dagger nematodes)			
X. americanum	Widespread	Groundnut, soyabean	Feakin, 1973; Sinclair and Shurtleff, 1980
X. diversicaudatum	Widespread	Groundnut, soyabean	Feakin, 1973; Sinclair and Shurtleff, 1980
X. spp.	—	Bean, cowpea	Caveness *et al.*, 1975; Schwartz and Gálvez, 1980

distributions. *Aphasmatylenchus straturatus* has been found associated with chlorosis of groundnut, as well as of soyabean, pigeon pea, and other legume hosts in Upper Volta where considerable crop loss was recorded (Germani and Dheri, 1973). *Hoplolaimus seinhorsti* may damage cowpea (Bridge, 1973), and *Aphelenchoides arachidis* is potentially important on groundnuts (Bridge *et al.,* 1977); both are known only from these hosts in Nigeria. The sting nematodes (*Belonolaimus* spp.), which may cause damage to soyabeans and groundnuts, are confined to the southeastern coast area of the USA. The distribution and legume host range of various other species are given by Bridge (1981).

The relative economic importance of plant-parasitic nematodes is most often assessed by the use of soil fumigants, although crop loss estimates determined in this way tend to underrate the complexity of the soil environment, as stressed by Mountain (1965). Although crop yield may be inversely related to nematode counts (Sasser *et al.,* 1968), as might be expected, this is not invariably the case (Robbins *et al.,* 1972). Furthermore, significant negative correlations with yield may occur without accounting for much of the yield variation (Sasser *et al.,* 1968), and direct assessment of symptoms may be a more accurate index of damage, as shown by the use of root knot indices in soyabeans (Minton and Parker, 1974) and counts of kernel damage in groundnuts (Boswell, 1968). Chlorosis, such as is associated with *Aphasmatylenchus* on groundnut (Taylor, 1976) and *Heterodera glycines* on soyabean (Sinclair and Shurtleff, 1975), may also be correlated with nematode population level, but such symptoms are seldom sufficiently diagnostic to provide useful damage indices. Such difficulties have tended to lead to the equating of nematodes' pathogenicity with association, partly justified by the fact that all known plant pathogenic nematodes are highly adapted morphologically for plant parasitism. The difficulties in applying Koch's postulates to plant nematology have been discussed (Mountain, 1965; Oostenbrink, 1969; Seinhorst, 1980).

The importance of nematodes is not limited to their individual effects on crop yield; infestations may also lead to reduced commercial value, as in the case of *Aphelenchoides* on confectionery groundnuts (Bridge *et al.,* 1977). Nematodes are also well known to be involved in disease complexes, such as *Meloidogyne–Fusarium* synergism in cowpea (Thomason *et al.,* 1959). These and other nematode-pathogen interactions are discussed later (See section 3.6). The role of nematodes as virus vectors in food legumes is a minor one, an example being the transmission of tobacco ringspot virus by *Xiphenema americanum* in soyabean (Stace-Smith, 1970). The claimed transmission of cowpea (yellow) mosaic virus by *X. basiri* (Caveness *et al.,* 1975) requires confirmation.

Having discussed the geographical distribution and economic importance of nematodes as parasites of legumes, we shall now turn to consider selected aspects of eelworm ecology in relation to the food legumes. For more detailed treatment of this and other features of plant parasitic nematodes, the reader is referred to the reviews of Williams (1968), Peachey (1969) and Taylor (1976), which deal with the nematodes of tropical crops in general, and to Hague (1980) and Bridge (1981), who deal specifically with the nematodes of legume crops.

For taxonomic purposes, nematodes are classified on the basis of their morphology and anatomy, which are beyond the scope of this book (see the literature cited in the reviews mentioned above). But nematodes may also be classified on an ecological basis, according to the degree of penetration and mobility of the parasite. Such a scheme is largely arbitrary, as pointed out by Williams (1968), because there is often considerable variation between species within genera in habit. Examples of such generic variation in habit are *Ditylenchus* and *Aphelenchoides,* each of which exhibits a full range from ectoparasitism to endoparasitism. The grouping used in Table 12 roughly follows those given by Mountain (1965) and Williams (1968). In discussing the evolution of plant parasitism, Mountain (1965) considers that there are two main trends in the development of plant parasitism among nematodes. First, there is a trend from ectoparasitism to migratory endoparasitism to sedentary endoparasitism, the last being restricted to the female. Secondly, there is a trend from more-or-less indiscriminate feeding to discrete feeding sites with specific plant responses. Thus, the sedentary endoparasitic genera *Meloidogyne* and *Heterodera* represent the most highly evolved plant parasites in which one might expect considerable specificity to have developed. Despite the wide host ranges of the root knot and cyst nematodes (Table 12), there is considerable pathogenic variation within species of each of the two genera (Dropkin, 1959; Minton, 1963; Golden *et al.,* 1970), but whether or not host-adapted races occur is less clear.

Virtually all plant parasitic nematodes have a soil phase, and spread by soil is the major means by which nematodes are locally dispersed. Over short distances, nematodes may be spread on soil adhering to farm implements, in flood and irrigation water, and in wind-blown soil. Established introductions of nematodes over long distances are often also via soil (Taylor, 1977); the introduction of *Heterodera glycines* into North America may be an example. A few species invade aerial parts of plants, occasionally entering the seeds. Examples of seedborne nematodes in legume crops include *Aphelenchoides arachidis* (Bridge *et al.,* 1977) and *Ditylenchus dipsaci* (Diercks and Klewitz, 1962). Some cyst-forming species have also been dispersed over long distances by seed, but this is uncommon and the few available examples apparently do not involve legumes. This, and other aspects of the international dispersal of nematodes, are discussed by Bingefors (1967). Despite such opportunities for introduction into new areas, the successful establishment of a nematode depends on its ability to overcome certain obstacles, amongst which are the search for a suitable host and the ability to survive in the soil (Bingefors, 1967; Taylor, 1977). Nematodes stored under dry conditions survive for long periods, which may exceed 20 years in plant tissue, and for 10 years in dried mud (Van Gundy, 1965), but the conditions under which such studies were conducted may not occur in nature. Nevertheless, it is evident that species vary considerably in their ability to withstand adverse conditions. The nematode *Scutellonema cavenessi,* which is commonly associated with groundnut, sorghum, and millet in the Sahel zone of West Africa, can survive the 8-month-long dry season in the soil; juveniles and adults are equally drought-resistant (Demeure, 1975). Conversely, species of *Meloidogyne* and *Rotylenchulus* are responsible for extensive damage to mung bean in upland areas of the Philippines; but in lowland areas where mung bean follows paddy rice, very low populations of nematodes are found, apparently relating to their intolerance of flooding (Castillo and Litsinger, 1978). In other cases, survival under dry savannah conditions may depend on host trees (Taylor *et al.*, 1978) or on volunteer plants. While *Aphelenchoides arachidis* can survive dessication in stored groundnut pods under certain conditions, viability is low after storage. Since volunteer plants in infested fields may contain many adult nematodes, it seems probable that they continue to develop to maturity under natural conditions in pods left in the ground during the dry season in northern Nigera (Bridge *et al.*, 1977). Species of *Heterodera* may carry over in soil as cysts from which the emergence of larvae is stimulated by root leachates from susceptible host plants, as in the case of *H. cajani* (Koshy and Swarup, 1971c), in a manner analogous to the plant parasitic *Striga* discussed in Section 3.5. Such a mechanism is presumably an adaptation to host finding in relatively narrow host range soilborne pathogens, such as *H. cajani* (Koshy and Swarup, 1972).

Despite the evolution of such soil survival mechanisms, it is often evident that nematode populations in soil decline rapidly in the absence of suitable hosts, and rotations with resistant crops have provided good control (Ross, 1962; Rhoades, 1976). Continuous cropping with susceptible species, or perhaps the presence of nematode-susceptible weeds, has the opposite effect (Riggs and Hamblen, 1966; Birat, 1969). However, certain weeds such as species of *Tagetes,* many of which are common in the tropics, appear to have a cleansing action in soil, perhaps related to the presence of a natural nematicide which the weed contains (Daulton and Curtis, 1964). Soil-tillage practices may also affect nematode populations, depending on the species. In south-western Nigeria, where zero-tillage systems show considerable promise, the numbers of root lesion nematodes are less under zero-tillage than conventional tillage regimes, but the converse is true of root knot nematodes (Caveness, 1974).

3.5 ANGIOSPERM PARASITES OF LEGUMES

Numerous species of higher plants are parasitic on other plants to varying degrees. Those parasitizing the food legumes belong to three families: the Scrophulariaceae and the closely related Orobanchaceae (the broomrapes), and the Convolvulaceae, which contains the dodders. The most important group under tropical conditions are the witchweeds (*Striga*) in the Scrophulariaceae, on which there is a large literature (Wilson-Jones, 1953b; Parker, 1965; Musselman, 1980). The legume parasites, summarized in Table 13, fall into two broad categories: root parasites (*Striga, Alectra, Orobanche*), and parasites of aerial structures (*Cuscuta*). The genus *Striga* and its close relative, *Alectra,* are entirely Old World and tropical, while *Orobanche* and *Cuscuta*, which are more widespread, are predominantly temperate genera. *Cassytha*, which is reported to parasitize beans under artificial conditions

TABLE 13 Higher Plant Parasites of Food Legumes

Species	Distribution	Hosts	References
Scrophulariaceae			
Striga gesnerioides	Africa, India, (USA)	Cowpea, groundnut (*Tephrosia*, tobacco, sweet potato and others)	Parker, 1965; Musselman, 1980
Alectra vogelii	Africa	Cowpea, groundnut bean, soyabean, mung bean	Rattray, 1932; Wild, 1954
Orobanchaceae			
Orobanche aegyptiaca	SW Asia, Middle East	Lentil (and other hosts)	Hawtin *et al.*, 1980
O. crenata	S Europe, Mediterranean	Lentil (and other hosts)	Hawtin *et al.*, 1980
O. minor	Europe, USA, E Africa	Bean (and various other hosts)	A. H. Ramos, pers. comm., 1979
Convolvulaceae			
Cuscuta ? campestris	Widespread (on cowpea in Nigeria and Brazil)	Cowpea, bean (and various other hosts)	D. J. Allen, unpubl.; Schwartz and Gálvez, 1980
Cuscuta spp.	China	Soyabean	R. M. Goodman, pers. comm., 1981
C. hyalina	India	Chickpea, pigeon pea	Vyas and Joshi, 1975; Kotasthane *et al.*, 1980

(Schwartz and Gálvez, 1980), is a member of Lauraceae although it superficially closely resembles dodder; it has a palaeotropical distribution and is common in coastal areas of West Africa. We shall first discuss the biology of the root parasites.

The host ranges of the root-parasitic weeds are typically broad but physiologic specialization often occurs within species, and there is evidence that host exudates may play a role in such specialization (Musselman, 1980). All species of *Orobanche* (Figure 17) lack chlorophyll and are obligate parasites. They are herbaceous, and are either annuals or perennials with underground tubers attached to the roots of the host plant. The witchweeds, which are chiefly annuals, vary in their degree of parasitism; the two chief species parasitic on cereals (*S. asiatica* and *S. hermonthica*) have relatively well-developed foliage, as does *Alectra vogelii,* while *S. gesnerioides* (Figure 18) bears only scale leaves. Although the shoots of *S. gesnerioides* are green and so presumably photosynthesize (Okonkwo and Nwoke, 1975), this species is probably rather more completely parasitic than its relatives we have mentioned. Both the broomrapes and witchweeds are spread by seeds which are produced in enormous numbers (in the case of *Striga,* perhaps up to 50,000 per plant; Wilson-Jones, 1953b), remaining viable in soil for many years. *Orobanche* seed viability may be retained for 2–12 years (Kasasian, 1971),

Figure 17 *Orobanche* species lack chlorophyll and are obligate parasites

Striga for 10 years (Wilson-Jones, 1953b), and *Alectra* for more than 6 years (Rattray, 1932). Viability may be lost more rapidly under high soil moisture (Parker, 1965; Kasasian, 1971). Seed germination occurs, after a period of dormancy, only in the presence of a stimulant ('strigol') secreted by the host root. More is known about seed germination in *Striga* than in *Orobanche*, although strigol, as well as ethylene and various other chemicals, affects germination of each genus. Germination inhibitors are produced by some certain non-hosts of broomrapes (Whitney, 1978). It has been shown (Williams, 1961b) that the host rhizosphere strongly influences the growth and development of *Striga* seedlings. Host roots exert a positive chemotropic response of the parasite's radicle, and subsequent branching of the root, and shoot production, depend on the secretion of growth substances from the host (Williams, 1961a). The latter is apparently reflected in the known changes which occur in the concentration of growth substances in the parasitized host,

Figure 18 The witchweed, *Striga gesnerioides*, on cowpea, Nigeria (photograph courtesy J. B. Smithson)

and perhaps in their higher concentrations in *Striga*-resistant cultivars (Musselman, 1980). A vital feature of parasitic weeds is the development of a haustorium, the fine structure of which is discussed by Kuijt (1977). In *S. gesnerioides,* the tip of the radicle swells at the site of host penetration, a primary haustorial bulb develops and a shoot with scale leaves appears. Parasite seedlings later produce secondary roots which, in turn, establish secondary haustoria (Okonkwo and Nwoke, 1975). Once vascular contact with the host is established, the parasite draws water and nutrients from the host, and the shoot grows, emerging after several weeks. Competition for nutrients in heavily infested hosts may limit the proportion of *Striga* shoots which reach the soil surface. Those that do emerge, flower and set seed which is again spread widely by wind, water, animals, and man (Parker, 1965). There is evidence that the flow of water from the host to the parasite depends on a strong transpirational pull, and it is possible that the preponderance of *Striga* and *Orobanche* species in open sunny habitats that favour increased transpiration may relate to this, borne out perhaps by the inhibition of growth of witchweeds under conditions of high humidity (Musselman, 1980). At one time, it was considered that witchweeds were only important under conditions of low soil fertility, but it is thought that *Striga* infestation is more often the cause, not the effect, of low productivity (Wilson-Jones, 1953a). Cowpea plants infested with *S. gesnerioides* may be stunted and chlorotic, and seed yield may be considerably reduced.

The dodders (*Cuscuta* spp.) are slender, twining annuals which are attached to the aerial parts of their host plants by suckers (Figure 19). Like the broomrapes, dodders are obligate parasites, and are devoid of chlorophyll although there may be small, rudimentary leaves. Dodder seeds germinate in the soil and produce a hair-like shoot which makes movements until it contacts a suitable host plant, after which it twines around the host stem which it penetrates by means of sucker-like haustoria. The stem of the parasite then dies away at its lower end, so losing connection with the soil (Tarr, 1972). Occasionally, dodders will turn back on their own stems and invade them via haustoria (Lackey, 1946), in a manner similar to the autoparasitism of *Striga* roots (Okonkwo and Nwoke, 1975). Dodders often have wide host ranges which may include food legumes. They are known to be vectors of plant

Figure 19 Dodder (*Cuscuta* sp.) on soyabean, Shandong Province, China (photograph courtesy R. M. Goodman)

viruses under experimental, and occasionally also natural, conditions (Lackey, 1966) but their role in the ecology of plant viruses is likely to be minor. Similarly, the economic importance of the dodders as agents of crop loss is probably only of local, or seasonal significance. *Cuscuta* spp. occasionally invade crops of cowpeas in Nigeria and Brazil, haustoria being produced on green pods as well as on leaves and stems (D. J. Allen, unpublished), and other dodder species are known to parasitize beans in the New World (Schwartz and Gálvez, 1980), chickpeas and pigeon peas in India (Vyas and Joshi, 1975; Kotasthane *et al*., 1980), and soyabeans in China (R. M. Goodman, personal communication, 1981).

3.6 INTERACTIONS BETWEEN PATHOGENS

Plant pathogens seldom occur in isolation under natural conditions. On leaf surfaces, and perhaps particularly in the soil rhizosphere, pathogens occur in mixed populations with other parasites, and with saprophytes, with which they may interact in various ways. Synergistic interactions are said to occur whenever the association of two or more organisms effects a change in pathogenicity that neither alone is able to make. Inhibitory interactions may be the consequence of the production of specific toxins, or antibiotics, by an antagonistic micro-organism. Both synergism and inhibition may result not only from direct interactions external to the host but may also occur within the host. Established infection by one pathogen may assist, or impede, the development of a second pathogen, principally through altered host physiology. These phenomena are variously referred to respectively as predisposition, induced susceptibility, or internal synergism; and as cross-protection, acquired immunity, or induced resistance. The nature and consequences of interactions between micro-organisms are complex, and largely beyond the scope of this book. For detailed accounts of synergistic interactions, the reader is referred to Dickinson (1979) and Powell (1979); soil antagonism is covered by Garrett (1970), and host resistance induced by previous invaders is discussed in Horsfall and Cowling (1980). In the following section, we shall briefly examine some of the principles which underlie interactions, as far as possible exemplified by the legume pathogens.

When cucumber mosaic virus occurs in mixed infections of cowpea with blackeye cowpea mosaic virus, they may interact in such a way as to cause a serious, distinct disease known as cowpea stunt. The extent of crop loss caused by stunt far exceeds that resulting from infection with either virus alone (Pio-Ribeiro *et al.,* 1978). This case, which nicely underlines the potential economic importance of pathogen interactions, is an example of synergism. The reliance of viruses on vectors for their effective dispersal, as discussed in Section 3.1.5, becomes especially complex in situations where insect transmission occurs only in the presence of a helper virus, as in the case of groundnut rosette and its assistor, or 'helper', virus (Hull and Adams, 1968). In other cases, virus mixtures may enhance the efficiency of transmission by a vector (Pio-Ribeiro *et al.*, 1978) or through seed (Kuhn and Dawson, 1973). While the vector's role is primarily to ensure that infection is successful, a vector may also provide a refuge for a pathogen from adverse conditions, as may be the case with the fungal vector of peanut clump virus which survives the long dry season of the Sahel (Thouvenel and Fauquet, 1981). These, too, are examples of synergism.

Fifty years ago, Fawcett (1931) predicted that a number of plant diseases would be shown to be a consequence of an association of organisms. Several diseases of complex aetiology are now well-known, and perhaps best illustrated by the interaction between nematodes and fungi (Powell, 1971). Plants previously invaded may predispose a host to infection by a secondary pathogen. Increased penetration by the latter may be a consequence of wounding caused by the primary parasite, and examples include the colonization of cowpea pods by the fungus *Choanephora cucurbitarum* following weevil damage (Fery and Cuthbert, 1972); the increased incidence of southern blight, caused by *Sclerotium rolfsii,* in soyabeans infested with root knot nematodes (Minton *et al.,* 1975); and the incidence of bean bacterial wilt, caused by *Corynebacterium flaccumfaciens,* depending on root knot nematode damage (Schuster, 1959). In other cases, metabolic leakage from an invaded host may enhance the development of a secondary pathogen, as in the case of infection by *Rhizoctonia solani* in the presence of exudates from galls caused by *Meloidogyne incognita* (Van Gundy *et al.,* 1977). Similarly, prior infection with virus can lead to increases in the concentration of amino acids in root exudates thus increasing the inoculum potential of *Fusarium solani*, and the root rot which it causes (Beute and Lockwood, 1968). And prior infection with soyabean mosaic virus may reduce soyabean seed germination and quality by predisposing plants to infection by *Phomopsis sojae* (Hepperly *et al.,* 1979). Premature senescence of host tissue, which is a common consequence of foliar infection, can increase the development of secondary pathogens by providing a more suitable substrate for its development; a possible example is the synergistic interaction between *Pseudomonas syringae* pv. *glycinea* and *Septoria glycines* on soyabean (Benedict, 1964). A somewhat similar situation may apply to the epidemic development of wildfire, caused by *Pseudomonas syringae* pv. *tabaci*, on soyabean, which is invariably associated with lesions of bacterial pustule (*Xanthomonas campestris* pv. *glycines*) but is independent of wounding (Graham, 1953; Chamberlain, 1956). Water soaking, such as occurs in storms, may be involved (Clayton, 1936) and it seems possible that bacterial pustule may predispose the soyabean to wildfire in a similar manner. Dual infections can also lead to increased multiplication of viruses. Rusted bean leaves may be more susceptible to virus infection (Yarwood, 1951) and may enhance virus replication, relative to healthy bean seedlings (Gill, 1965). Similarly, infestation of soyabean with the reniform nematode (*Rotylenchulus reniformis*) can increase the replication of cowpea chlorotic mottle virus (Bustillo, 1972).

Non-pathogenic micro-organisms associated with the rhizosphere, or as epiphytes on foliage, are well known to play a role in the host's defence through their acting as antagonists (e.g. Vasudeva and Roy, 1950); but their role in moderating the microenvironment within which pathogens germinate is less well known (Dickinson, 1979). Such assistance provided by epiphytes in the establishment of pathogens may be rendered directly, by acting as a source of nutrients, or perhaps by their production of chelating agents, since it appears that iron deficiency may enhance the aggressiveness of *Botrytis cinerea* on *Vicia faba* (Swinburne, 1980), in a manner analogous to the effects of pollen (Chu Chou and Preece, 1968). A further means of direct assistance might be detoxification of chemicals produced by the host (Dickinson, 1979). A species of pseudomonad can apparently reduce the phytoalexin response of bean leaves challenged with *Botrytis cinerea,* which is more aggressive as a result (McCarthy and Swinburne, 1980); thus this is one of the few documented cases of induced susceptibility between species.

Turning now to interactions which hinder pathogen development, with obvious relevance to biological control, there are many examples amongst legume pathogens of antagonism, or inhibition, the mechanisms of which are complex, varied, and incompletely understood. Legume pathogens which are themselves parasitic on other pathogens of the same host plant species include the nematode *Aphelenchoides arachidis,* which feeds ectoparasitically on the fungus *Macrophomina phaseolina* (Bridge *et al*., 1977), and probably *Rhizoctonia solani* on species of *Pythium* (Butler, 1957). Although in the latter case the nature of parasitism apparently may not involve antibiotic production (Butler, 1957), there are several more recent reports of antagonism between *R. solani* and *Pythium* spp. (Flowers and Littrell, 1973; Garcia and Mitchell, 1975; Pieczarka and Abawi, 1978), perhaps having relevance to seedling diseases and root rots of legumes in which these two fungal genera are part of a complex aetiology (Williams, 1975a). The non-pathogen *Gliocladium* sp. is an antagonist of *Colletotrichum lindemuthianum* (Porter, 1924) while *G. virens* is a hyperparasite of *R. solani* on beans (Tu and Vaartaja, 1981). The soil antagonist *Trichoderma viride* has the ability to induce sexual reproduction and thus oospore production, in heterothallic species of *Phytophthora* (Brasier, 1971), and certain soil bacteria can induce chlamydospore formation in *Fusarium* (Ford *et al.,* 1970). Other examples of antagonism between legume pathogens include the inhibition of *Macrophomina phaseolina* by *Aspergillus flavus* in groundnut pods (Jackson, 1965) and the antagonism between *Cercospora kikuchii* and *Diaporthe phaseolorum* in pods and seeds of soyabean (Roy and Abney, 1977). Non-pathogenic bacteria may play a role in retarding the development of *Pseudomonas syringae* pv. *phaseolicola* on bean (Adam and Pugsley, 1935), and may inhibit lesion development of *Ps. syringae* pv. *glycinea* in soyabean (Scherff, 1973). The antagonist apparently competes successfully with the bacterial blight organism on soyabean buds on which it is one of the most common epiphytes; under suitable conditions, the non-pathogen increases at greater rates than other bacteria, possibly influencing the development of bacterial blight (Scherff, 1973). Such an example has obvious bearing on the ecological relationships of bacterial pathogens, discussed in Section 3.3. In other cases, non-pathogenic bacteria may inhibit pathogenic fungi through production of an antibiotic, as shown by Vasudeva *et al*. (1962) in the interaction between *Bacillus subtilis* and *Fusarium* wilt of pigeon pea.

The above examples concern interactions which occur more or less directly between micro-organisms, without significant involvement of the host plant. Prior infection of a host often leads to internal changes which in turn inhibit the subsequent development of a secondary pathogen or 'challenger'. We shall refer to this phenomenon as induced resistance. Of the many examples in which some measure of protection against a challenger pathogen is afforded by an established infection, the mechanism underlying this effect remains unknown. Examples include the inhibition of fusarium wilt symptoms in cowpea by prior inoculation of a non-pathogen, *Cephalosporium* sp. (Long, 1963); the protection afforded against fusarium wilt in pigeon pea by prior inoculation with sterility mosaic virus (Chadha and Raychaudhuri, 1966), or with non-pathogenic *formae speciales* of *F. oxysporum* (Kaiser and Sen Gupta, 1977); the mutual inhibition of fusaria and root knot nematode development in soyabean (Goswami and Agarwal, 1978); the inhibition of cyst nematode populations by established infections of *Phytophthora megasperma* in soyabean (Adeniji *et al.,* 1975); the inhibition of tobacco necrosis virus infection of bean infested with the nematode *Ditylenchus dipsaci* (Caspary and Nienhaus, 1976); and the retarded synthesis of challenging virus in virus-infected cowpea (Kuhn and Dawson, 1973). Amongst the cases in which the mechanism of induced resistance has been investigated, studies on the rust fungi, including bean rust (*Uromyces appendiculatus*), have suggested that either a gaseous metabolite or an aqueous diffusate from the urediospores may be responsible for the protection which is restricted to the germination and penetration phases of infection (Yarwood, 1954; Wilson, 1958). Littlefield (1969), however, concluded that, in addition to a stomatal blockage mechanism, diffusates from the inducing inoculum triggered the production of phytoalexins in the host, and studies with bean anthracnose have also provided evidence that phytoalexins are implicated in induced resistance (Rahe *et al.,* 1969; Elliston *et al.,* 1971; Skipp and Deverall, 1973). If phytoalexins are, indeed, involved, then induced resistance should occur between widely differing organisms; the examples of cross-protection cited above would support this. The effects of induced resistance to bean rust have been quantified (Yarwood, 1956b; Johnson and Allen, 1975), and it has been suggested that induced resistance may play a role in the epidemiology of rusts in multilines (Johnson and Allen, 1975; Johnson and Taylor, 1976) and perhaps also in crop mixtures (Allen, 1975a; 1976).

Finally, mention should be made of the interactions between legume pathogens and nitrogen-fixing bacteria. In certain instances, *Rhizobium* may afford protection against soilborne pathogens, such as *Phytophthora megasperma* which causes a root rot of soyabean, apparently by parasitizing the hyphae of the fungus (Tu, 1978). In other cases, it appears that *Rhizobium* may have an indirect influence by increasing nitrogen concentration, which may reduce the activity of the soyabean cyst nematode (Barker *et al.,* 1971). There are also cases of the converse situation, wherein virus infection reduces nodulation (Tu *et al.,* 1970a; Adnew *et al.,* 1977). Although the underlying mechanism is unclear, in view of the dual role that certain host toxins may play in *Rhizobium*-host recognition and in disease resistance (Hamblin, 1980), it seems possible that such interactions, like the cases of induced resistance we have discussed, may involve disturbance of host defence systems. This is perhaps partly reinforced by the greater susceptibility of healthy soyabeans to *Rhizobium* infection than plants infected with soyabean mosaic virus (Tu *et al.,* 1970b).

Chapter 4

Groundnut Diseases

4.1 EVOLUTION, ADAPTATION, AND PRODUCTION OF GROUNDNUTS

The genus *Arachis* to which the groundnut, or peanut, belongs is of South American origin. The genus as a whole has a centre of diversity in the Matto Grosso of Brazil. The origin of the individual species ranges, east of the Andes, from north-eastern Brazil, near the mouth of the Amazon (0°), southwards into north-west Argentina and southern Uruguay (34° S). The genus contains some 40–50 wild species (Gregory and Gregory, 1976) but only 22 species are formally described (Resslar, 1980). All are geocarpic herbs: species in areas of high rainfall are perennial, those in semi-arid areas, annual. Most species are diploid (2n = 2x = 20), but *A. hypogaea* L., the groundnut, and *A. monticola* Krap. and Rig. are tetraploids (2n = 4x = 40). *A. monticola,* from north-western Argentina, is an annual that crosses freely with cultivated *A. hypogaea,* and is presumed to be the current wild descendant of the amphidiploid species ancestral to the cultigen. Wild species can sometimes be crossed with each other but the hybrids are usually sterile. As a rule, crosses made between species within a section are more successful than in hybridization between sections of the genus (Gregory and Gregory, 1976; Gregory *et al.,* 1980).

Two sub-species are recognized within *A. hypogaea,* based on branching pattern, which is either alternate or sequential. In ssp. *hypogaea,* the Virginia group, the main axis bears only vegetative branches, while on lateral branches on the main axis there is a regular alternation, usually in pairs, of vegetative and reproductive branches. In ssp. *fastigiata* Waldron, there may be a short proximal sequence of vegetative branches followed by a distal sequence of reproductive branches. While the sequentially branched varieties are susceptible to the *Mycosphaerella* leaf spots, the alternately branched types are less susceptible (Feakin, 1973). Within sub-species *fastigiata,* Valencia-type cultivars are referred to var. *fastigiata,* and the Spanish type to var. *vulgaris* Harz.; three varieties (*hypogaea, hirsuta* Kohler, and *nambyquarae* (Hoehne) Burk.) are recognized within sub-species *hypogaea* (Gibbons *et al.,* 1972;

Resslar, 1980). Details of the structure, variation, and classification, as well as of the evolution of *Arachis,* are reviewed by Gregory *et al.* (1980). Other useful reviews are given by Cobley and Steele (1976) and Smartt (1976a).

The earliest archaeological records of the groundnut are from Peru, 2000–3000 BC. The crop was, however, domesticated well to the south-east and is presumed to have had a much longer history. Six geographical groups of cultivated groundnuts in South America have been distinguished, these seeming to represent secondary centres of diversity that have developed from a primary centre of domestication in southern Bolivia and northern Argentina (Gregory and Gregory, 1976). The groundnut was probably brought to Africa from Brazil by the Portuguese in the sixteenth century, when they established regular communication with the Indian sub-continent, and with Sri Lanka and the Far East, by way of the southern and eastern coasts of Africa. It was taken somewhat later from the west coast of South America to Asia. Stocks of groundnuts from widely separated regions of the New World appear to have been assembled in Africa where they have given rise to a new and distinct pattern of variation, so that the African continent may be regarded as a secondary centre of variation of the crop. Groundnuts may have reached the USA by way of slave ships from West Africa (Higgins, 1951; Gibbons *et al.,* 1972; Gibbons, 1980a).

The groundnut is now grown on nearly 20 million hectares in a very wide range of environmental conditions between 40° N and S of the equator in all continents. Although essentially a tropical crop, the climates in which groundnuts are produced range from humid equatorial through seasonally arid to Mediterranean and continental temperature. Groundnuts behave as day-neutral plants, their phenology being determined primarily by temperature (Bunting and Elston, 1980). Annual world production of fruits (seeds, or kernels, 'in shell') is about 18.5 million tonnes, of which some 70 per cent consists of seeds. More than 70 per cent of the total production comes from crops in the developing countries. Although experimental farm yields may reach 4–5 tonnes/hectare, typical farmers' yields in the tropics vary between 0.5 and 1.5 tonnes/hectare. India produces 5.7 million tonnes; other major centres of production are Nigeria, Senegal, Sudan, the USA, Brazil, and Argentina (Gibbons, 1980a). The ideal soil for groundnut production is a well-drained, light, sandy loam, well supplied with calcium and a moderate amount of organic matter. But in practice groundnuts are produced on a wide range of soils. When grown on black clay soils (vertisols), harvesting can be a problem. Under rainfed conditions, the choice of cultivar depends principally on the duration of the growing season, the Spanish and Valencia types being cultivated under short-season conditions, and where the rainy season is longer, Virginia types are used. The adaptation of cultivars can vary with differences in environment within a country, as in Nigeria where production extends from the Sudan savannah to the derived savannah zone, where commercial production ceases. Now that cultivars resistant to rosette are available, there is potential for the extension of production into the more humid areas where crop failures due to this disease were once common (Gillier, 1978; Gibbons, 1980a).

In developed countries, and many of the developing ones, groundnuts are grown as a sole crop under rainfed conditions. In developed countries like the USA, production is intensely mechanized and specialized, but much of the crop in the developing world is grown on relatively small plots by farmers who use only hand tools, or simple animal-drawn implements. Crops may be sown at too wide a spacing, or too late, because of time and labour constraints, exacerbated by the low priority groundnuts often receive (Gibbons, 1980a). Both late sowing and gappy stands are known to predispose the crop to rosette (A'Brook, 1964; Hull, 1964). Much of the land on which groundnuts are grown is of inherently poor fertility. In parts of India and Africa, both phosphorus and sulphur deficiency may limit productivity (Kanwar, 1963; Goldsworthy and Heathcote, 1964). However, in traditional agricultural systems fertilizers are seldom applied. Similarly, disease control measures are seldom practised; but indirect benefits from the use of fungicides may occasionally occur. It has been shown that the application of sulphur dust to control [cercospora] leaf spots is economical on the sulphur-deficient sandy soils of Malawi (Mercer, 1973; Laurence *et al.,* 1976). It remains to be seen whether the same would apply to other areas in which both leaf spot and sulphur deficiency occur. In developing countries, groundnuts are usually harvested and cured by hand, or with simple machinery. Rapid curing and drying, in addition to prompt harvesting, is important if the production of aflatoxin by *Aspergillus flavus* is to be minimized (McDonald, 1970b; Gibbons *et al.,* 1975).

There is potential for the use of irrigation in the production of groundnuts, but so far only a small proportion of the total crop is irrigated in developing countries in the tropics. In parts of south-east Asia, groundnuts are often components of sequential cropping systems based on rice (Gibbons, 1980a). However, the majority of groundnuts produced in the tropics are grown in complex mixtures of two or more crops, both in humid areas as in Cameroun (Mutsaers *et al.,* 1981) and in the savannahs. In Nigeria, 95 per cent of the groundnuts are intercropped, especially with cereals, in a system described in detail previously (Section 1.2.3). In Zambia, such traditional practices appear to contribute to the normally low incidence of disease in the local cultivation of groundnuts (Smartt, 1961).

The uses of the crop have been reviewed by Woodroof (1966) and, recently, by Gibbons (1980a). About two-thirds of the total groundnut harvest is crushed for oil and the crop provides about 20 per cent of the total world vegetable oil production, used primarily for cooking. The residual cake is an animal feed. Thus, to the traditional farmer, the groundnut is both a food crop for home consumption as a pulse, and a cash crop. Often it may be considered too valuable to consume, the total crop being sold. Shelled seeds (kernels) are exported, especially from China, India, parts of Africa, and the USA, for confectionery use in Japan, Canada, and Europe. Groundnuts are exported in shell from Brazil, principally to the Mediterranean region. Groundnut haulms are used as fodder for cattle in developing countries, their quality depending on the extent of prior defoliation caused by foliar pathogens (Gibbons, 1980a).

The extent to which diseases limit the production of the groundnut crop will be discussed in the next section. But a conservative estimate of the annual loss of kernels caused by the cercospora leaf spot fungi alone is about 3 million tonnes (Bunting *et al.,* 1974 cited by Gibbons, 1980b), which emphasizes the need in groundnut improvement to give priority to disease management.

4.2 THE PATHOLOGY OF GROUNDNUTS

Groundnut diseases have been reviewed most notably by Loden and Hildebrand (1950), Garren and Wilson (1951), Feakin (1973), Garren and Jackson (1973), and McDonald and Raheja (1980). Regional reviews include those available for Australia (Morwood, 1945; Purss, 1962), Africa (Smartt, 1961; Rothwell, 1962; Mercer, 1977; 1978; McDonald, 1978), and the USA (Dickson, 1956; Higgins, 1963). In the early years of its cultivation in the southern states of the USA, the groundnut was relatively disease-free. As production increased, diseases attracted more attention. Much of the substance of the literature preceding 1970 referred to the American work; since then, research has been carried out on the pathology of groundnuts, especially in Asia and Africa (McDonald and Raheja, 1980). Here, we shall consider some of the many diseases reported from the crop, paying particular attention to those of greatest economic importance to production in developing countries. Amongst the fungal diseases, three are of particular importance, effectively occurring wherever groundnuts are cultivated; these are the two cercospora leaf spots and rust. The aflatoxin problem associated with the fungus *Aspergillus flavus* is also of great importance. These will be discussed more fully below. Other fungal pathogens of the crop may be of local or sporadic importance and these are summarized in Table 14.

Groundnuts are susceptible also to a wide range of virus diseases, the most important of which are rosette, bud necrosis, and peanut mottle; these will be discussed in detail later. Other virus diseases are stunt, caused by peanut stunt virus, and clump, caused by a soilborne virus known from West Africa and apparently also from India. Cowpea mild mottle virus causes an important disease in groundnuts in coastal areas of East Africa where it has been known as 'Ngomeni mottle' at least since 1957 (Storey and Ryland, 1957; Bock *et al.,* 1977). Other viruses which have been described more recently include groundnut eyespot virus in the Ivory Coast (Dubern and Dollet, 1980) and peanut green mosaic virus in India (Sreenivasulu *et al.*, 1981). Tobacco streak and mung bean yellow mosaic viruses may also naturally infect groundnuts. The geographical distribution, mode of transmission, and economic importance of each of these virus diseases is summerized in Table 15. Various other, less satisfactorily characterized, virus diseases have been

TABLE 14 Fungal Diseases of Groundnut

Disease	Causal fungi	Distribution	Importance/crop loss	References
1. Seed and seedling diseases				
Pre-emergence rot	Associated with spp. of *Fusarium, Aspergillus, Penicillium, Rhizopus, Pythium myriotylum, Rhizoctonia solani, Macrophomina phaseolina, Botryodiplodia theobromae, Sclerotium rolfsii*	Widespread	Up to 75% seedbed losses recorded	Gibson and Clinton, 1953; Tarr, 1958; Clinton, 1960; Purss, 1960; Garcia and Mitchell, 1975; McDonald, 1978
Crown and collar rot	*Aspergillus niger* Van Tiegh.	Widespread	Major	Gibson, 1953; Nema *et al.*, 1955; Purss, 1960; Jackson, 1962; Ashworth *et al.*, 1964; Chohan, 1965; Hsi, 1966
Aflaroot	*Aspergillus flavus* Link. ex Fr.	India	Minor	Chohan and Gupta, 1968
2. Stem, collar and root rots, and wilts				
Diplodia collar rot	*Botryodiplodia theobromae* Pat. (perfect state = *Physalospora rhodina* Berk. and Curt. apud Cooke) = *Diplodia gossypina* Cooke = *Diplodia natalenis* Pole Evans	Widespread	Of sporadic and local importance	Jacoway and Owen, 1951; Boyle, 1953; Chorin and Frank, 1961; Purss, 1962; McGuire and Cooper, 1965; Porter and Hammons, 1975; Punithalingham, 1976
Cylindrocladium black rot	*Cylindrocladium crotalariae* (Loos) Bell and Sobers (perfect state = *Calonectria crotalariae* (Loos) Bell and Sobers)	USA, Japan	Notable yield loss occasionally occurs	Bell and Sobers, 1966; Bell, 1967; Garren *et al.*, 1971; Beute, 1980
Texas root rot	*Phymatotrichum omnivorum* (Shear) Dugg. (perfect state = *Trechispora brinkmanii* (Bres.)) = *Ozonium omnivorum* Shear	Southern USA	Locally important; losses up to 83% recorded	Kenknight, 1942; Baniecki and Bloss, 1969; Feakin, 1973

Rhizoctonia root rot	*Rhizoctonia solani* Kühn (perfect state = *Thanatephorus cucumeris* (Frank) Donk)	Worldwide	Yield loss potential = 25% in USA	Ashworth *et al.*, 1961; Feakin, 1973
Macrophomina root and stem rot	*Macrophomina phaseolina* (Tassi) Goid. (sclerotial state = *Rhizoctonia bataticola* (Taub.) Butl.)	Widespread in the tropics and sub-tropics	Minor	Feakin, 1973; McDonald, 1978
Sclerotium blight (southern blight)	*Sclerotium rolfsii* Sacc. (perfect state = *Corticium rolfsii* Curzi)	Widespread	Locally important	Garren and Wilson, 1951; Ashworth *et al.*, 1961; Purss, 1962; Garren, 1964a; Feakin, 1973; McDonald, 1978
Sclerotinia blight	*Sclerotinia sclerotiorum* (Lib.) de Bary and *S. minor* Jagger	China, Taiwan, Japan, Australia, Argentina, USA	Locally of major importance	Porter and Beute, 1974; Porter, 1980
Pythium wilt	*Pythium myriotylum* Dresch.	Nigeria, USA	Of sporadic and local importance	Perry, 1967; Porter, 1970; McDonald, 1978
Fusarium root rot and wilt	*Fusarium oxysporum* Schl. ex Fr. and *F. solani* (Mart.) Sacc.	Widespread	Minor	McDonald, 1978
Verticillium wilt	*Verticillium dahliae* Kleb. = microsclerotial form of *V. albo-atrum* Reinke and Berth	Australia, USA	Seasonally of local importance; individual plant yield loss 14–60%	Smith, 1960; Purss, 1961
3. Foliar diseases				
Early leaf spot	*Mycosphaerella arachidis* Deighton (conidial state = *Cercospora arachidicola* Hori)	Widespread	Major } together may cause losses in pod yield of up to 60%	Gibbons, 1966; Fowler, 1970b; Mulder and Holliday, 1974a
Late leaf spot	*Mycosphaerella berkeleyi* W. A. Jenkins (conidial state = *Cercosporidium personatum* B and C) Deighton	Widespread	Major }	Gibbons, 1966; Fowler, 1970b; Mulder and Holliday 1974b
Rust	*Puccinia arachidis* Speg.	Pantropical	Major; pod yield losses exceeding 50%	McDonald, 1978; Subrahmanyam *et al.*, 1980

TABLE 14 (*Continued*)

Disease	Causal fungi	Distribution	Importance/Crop loss	References
Net blotch (web blotch)	*Didymosphaeria arachidicola* (Chochrjakov) Alcorn, Punith. and McCarthy = *Mycosphaerella arachidicola* Chochrjakov = *M. argentinensis* Frezzi = *Phoma arachidicola* Marasas, Pauer and Boerema = *Ascochyta adzamethica* Schoschiaschvili = ? *A. arachidis* Woronichin	Widespread	Generally minor, but severe infections can lead to 40–60% defoliation	Frezzi, 1969; Pettit *et al.*, 1973; Marasas *et al.*, 1974; Alcorn *et al.*, 1976; Blamey *et al.*, 1977; Cole and Grimmer, 1977; McDonald, 1978
Pepper spot and leaf scorch	*Leptosphaerulina trifolii* (Rost.) Petr. = *Pseudoplea trifolii* (Rost.) Petr. = *Leptosphaerulina arachidicola* Yen, Chen and Huang = ? *L. crassiasca* (Sechet) Jackson and Bell = ? *Pleospora crassiasca* Sechet	Widespread. Reported from Taiwan, Sabah, India, Mauritius, Madagascar, Malawi, Nigeria, Argentina	Minor	Sechet, 1955; Yen *et al.*, 1956; Luttrell and Boyle, 1960; Nayudu, 1963; Booth and Pirozynski, 1967; Jackson and Bell, 1968; Karan and Rao, 1968; Pettit *et al.*, 1968; Mallaiah and Rao, 1976; McDonald, 1978; Holliday, 1980
Phyllosticta leaf spot	*Phyllosticta arachidis-hypogaeae* Vasant Rao	India, Zimbabwe, Nigeria, Togo, Argentina, USA	Minor	Rao, 1962; Rothwell, 1962; Steiner, 1975; McDonald, 1978; Smith, 1980
Pestalotiopsis leaf spot	*Pestalotiopsis adusta* (Ell. and Ev.) Stey., *P. arachidis* Satya, and *P. versicolor* (Speg.) Stey	India, Nigeria	Minor	Satya, 1964; McDonald, 1978
Alternaria leaf blight	*Alternaria alternata* (Fr.) Keissler (= *A. tenuis* C. G. Nees)	Widespread	Minor	Aulakh, 1969; McDonald, 1978; Balasubramanian, 1979; Smith, 1980

Myrothecium leaf blight	*Myrothecium roridum* Tode ex Fr.	India	Minor	Subrahmanyam, 1979
Phomopsis leaf spot	*Phomopsis* sp.	India	Minor	Sharma, 1974
Curvularia leaf spot	*Curvularia* spp.	Nigeria, Togo, Malaysia	Minor	Steiner, 1975; Williams and Liu, 1976; McDonald, 1978
Anthracnose	*Colletotrichum capsici* (Syd.) Butl. and Bisb., *C. truncatum* (Schw.) Andrus and Moore (= *C. dematium* (Pers. ex Fr.) Grove var. *truncata* (Schw.) Arx)	Widespread	Minor	Saksena *et al.*, 1967; McDonald, 1978; Smith, 1980
Scab	*Sphaceloma arachidis* Bitanc. and Jenk.	Brazil	Locally important	Bitancourt and Jenkins, 1940
Powdery mildew	*Erysiphe polygoni* DC and *Oidium arachidis* Chorin	Indonesia, Israel, Tanzania	Occasional	Thung, 1937; Riley, 1960; Chorin, 1961
Grey mould	*Botrytis cinerea* Pers. ex Pers. (perfect state = *Sclerotinia fuckeliana* (de Bary) Fuckel	Worldwide, but most prevalent in humid temperate and sub-tropical areas	Minor	Ellis and Waller, 1974
4. Pod diseases				
Pre-harvest pod rot	*Rhizoctonia solani, Sclerotium rolfsii, Fusarium* spp., *Pythium myriotylum, Macrophomina phaseolina, Botrytis* spp. and *Botryodiplodia theobromae*	Widespread	Important under some conditions	McDonald, 1969; 1970a; Garren, 1970; Frank, 1972; Mercer, 1978
Concealed damage	*Botryodiplodia theobromae* Pat., *Macrophomina phaseolina* (Tassi) Goid	Widespread	Important under some conditions	Garren and Higgins, 1947; Wilson, 1947; McDonald, 1969
Aflatoxin problem	*Aspergillus flavus* Link. ex Fr.	Widespread	Major	Sargeant *et al.*, 1961; McDonald, 1969; Martin and Gilman, 1976; Coker, 1980

TABLE 15 Virus Diseases of Groundnut

Disease	Causal virus	Distribution	Transmission		Importance/Crop loss	References
			Vector	Seed (%)		
Rosette	(A complex of incompletely characterized viruses)	Africa	Aphid	0	Major. Total crop loss occurs	Storey and Ryland, 1955; Okusanya and Watson, 1966; Hull and Adams, 1968; Bock *et al.*, 1970; Yayock, 1977
Mottle	Peanut mottle virus (PnMV)	Widespread	Aphid	0–8.5	Locally important. 30% crop loss reported	Kuhn, 1965; Bock, 1973b; Bock and Kuhn, 1975; Adams and Kuhn, 1977; Reddy *et al.*, 1978
Mottle (Ngomeni mottle)	Cowpea mild mottle virus (CMMV)	Africa, India	Whitefly	?	Important in coastal E Africa where 64–80% crop loss may occur	Storey and Ryland, 1957; Bock *et al.*, 1977; ICRISAT, 1978b; D. V. R. Reddy, pers. comm., 1981
Bud necrosis	Tomato spotted wilt virus (TSWV)	Widespread	Thrips	0	Major. Up to 90% yield losses reported	Helms *et al.*, 1961; Ie, 1970; Ghanekar *et al.*, 1979

Stunt	Peanut stunt virus (PSV)	Japan, USA, Morocco	Aphid	<0.1	Local. 70–80% crop loss may occur in USA	Culp and Troutman, 1967; Mink, 1972; Fischer and Lockhart, 1978
Clump	Peanut clump virus (PCV)	Upper Volta, Senegal, ? India	Fungus	14 under natural conditions	Locally important. 43% and >60% crop loss reported for W Africa and India respectively	Germani and Dhery, 1973; Thouvenel *et al.*, 1976; Thouvenel *et al.*, 1978; Ghanekar, 1980; McDonald and Raheja, 1980; Thouvenel and Fauquet, 1981
Eyespot	Groundnut eyespot virus (GEV)	Ivory Coast	Aphid	?	Of no economic importance	Dubern and Dollet, 1980
Mosaic	Peanut green mosaic (PGMV)	India	Aphid	0	?	Sreenivasulu *et al.*, 1981
Streak	Tobacco streak virus (TSV)	Europe, temperate and tropical America, New Zealand	?	?	Minor ?	Fulton, 1971
Yellow mosaic	Mungbean yellow mosaic virus (MYMV)	India	Whitefly	0	Minor ?	Rao *et al.*, 1980

reported but these are omitted from the table. These include a mosaic (Nariani and Dhingra, 1963), a 'bunchy top' chlorosis and ring mottle (Sharma, 1966), and chlorotic spot diseases (Haragopal and Nayudu, 1971), all from India. Ringspot diseases have also been reported from South Africa (Klesser, 1966) and the USA (Kuhn *et al.,* 1964), while a tymovirus, named peanut yellow mottle virus, has been obtained from groundnuts in Nigeria (Lana, 1980).

The only bacterial disease of groundnut of any importance is a wilt caused by certain pathotypes of *Pseudomonas solanacearum* (Kelman and Person, 1961). Infected plants are typically chlorotic, and wilt under conditions of water stress, the vascular tissue becoming discoloured. Sources of immunity, and of moderate resistance, have been identified in the USA and Uganda (Jenkins *et al.,* 1966; Simbwa-Bunnya, 1972).

Groundnuts are also prone to attack by various nematodes. Amongst the root knot nematodes (Section 3.4), *Meloidogyne arenaria* and *M. hapla* are the most important. The development of *M. hapla* in groundnut has been investigated by Castillo *et al.* (1973b). Damage results from larval feeding which causes the formation of giant cells that disrupt the vascular system, leading to stunting and wilt. Hypertrophy of cortical tissue causes the galls which typify root knot. In groundnut, galling is not confined to the root system but may also occur on the pegs, pods and, in some cultivars, the runners (Feakin, 1973). Estimates of the crop loss caused by *M. arenaria* and *M. hapla* are about 30 per cent and 70 per cent respectively (Cooper, 1952; McDonald and Raheja, 1980), although *M. arenaria* is generally regarded as the potentially more important species. The interaction of root knot nematodes with fungi, especially species of *Fusarium* and *Aspergillus,* may lead to losses exceeding those caused by either pathogen alone. These aspects are discussed more fully elsewhere (Section 3.6). Crop rotation and the use of nematicides may provide means of control. Attempts to find sources of resistance to the root knot nematodes have met with limited success (Castillo *et al.,* 1973a; Minton and Hammons, 1975), perhaps complicated by the existence of races within species (Minton, 1963). Other nematodes which may cause damage include the root lesion nematodes (*Pratylenchus brachyurus*), the sting nematodes (*Belonolaimus* spp.), the ring nematodes (*Criconemoides* spp.), the burrowing nematode (*Radopholus similis),* the reniform nematode (*Rotylenchulus reniformis*), the stubby root nematode (*Trichodorus christiei*), dagger nematodes (*Xiphenema* spp.), and a bud and leaf nematode (*Aphelenchoides arachidis*) (Endo, 1959; Boswell, 1968; Sasser *et al.,* 1968; Bridge *et al.,* 1977). Further details of the nematode diseases of groundnut are given in Section 3.4; other useful reviews are given by Feakin (1973) and McDonald and Raheja (1980). Two higher plants, *Alectra vogelii* and *Striga gesnerioides,* are also parasites of groundnuts and these may cause damage in some localities in Africa.

Let us now consider the major diseases in greater detail, focusing first on the leaf spot diseases.

4.2.1 Cercospora leaf spots

The cercospora, or 'tikka', leaf spots are generally regarded as the most important fungal diseases of groundnut, occurring wherever the crop is cultivated. Two principal species are involved: *Cercospora arachidicola,* the cause of 'early' leaf spot, and *Cercosporidium personatum,* which causes 'late' leaf spot. A third species, *Cercospora canescens*, also occasionally infects groundnut leaves but is of no economic importance (McDonald, 1978). *C. personatum* was first described in the USA in 1875 while *C. arachidicola* was reported first from Japan in 1919 (Gibbons, 1966); Woodroof (1933) was the first to recognize the existence of two distinct species. Their perfect stages (*Mycosphaerella arachidicola* and *M. berkeleyi* for *C. arachidicola*, *C. personatum,* respectively) were described by Jenkins (1938). Strictly, these fungi should be referred to by the names of their perfect stages, as Gibbons (1966) has done in his useful review. However, since the diseases are more commonly referred to as cercospora leaf spots, for it is their conidial stages which are normally found in the field, this usage is followed here. Symptoms of the two leaf spots are effectively indistinguishable in the early stages of their development. Later, lesions caused by *C. arachidicola* tend to be lighter brown, less regular in outline, and more often surrounded by a yellow halo than those caused by *C. personatum.* Symptom development has been described in detail by Gibbons (1966), Feakin (1973), and McDonald (1978). *C. arachidicola* produces conidia from scattered groups of conidiophores predominantly on the upper leaf surface, while *C. personatum* sporulates chiefly on the lower leaf surface, the conidia being produced from groups of concentrically arranged conidiophores (McDonald, 1978). Lesions induced by *C. arachidicola* normally first appear 3–4 weeks after sowing, *C. personatum* appearing some 2–4 weeks later. However, the relative importance of the two species varies considerably, depending both on the type of groundnuts grown and the prevailing weather conditions. Despite its later development, *C. personatum* is potentially the more destructive species because of its much higher rate of spread leading to more rapid defoliation (Hemingway, 1955). If early-maturing groundnut cultivars are grown, the crop may be nearly mature before the development of a late leaf spot epidemic (Gibbons, 1966).

The principal effect of both leaf spots is to cause premature defoliation (Hemingway, 1954; Elston *et al.,* 1976), and it has been estimated (Fowler, 1970b) that defoliation of leaflets may begin when 6 per cent of their leaf area is diseased. Severe infections may cause complete defoliation. However, lesions are not confined to the leaves but may occur also on the stems and pegs, leading to direct deterioration of the developing pods (Reyes and Romasanta, 1940). Losses to the groundnut crop may be as high as 60 per cent under experimental conditions. It has been estimated that, under farmers' field conditions in northern Nigeria, losses of about 23 per cent for pods and about 56 per cent for haulms may occur regularly, suggesting that some 200,000 tonnes of groundnuts may be lost annually in Nigeria (Fowler, 1970a;

1971). Control may lead not only to increases in kernel yield but also to improved kernel quality, and to increased haulm yield, *in vitro* digestibility and protein content (Fowler, 1970a; Cummins and Smith, 1973; Mercer, 1973). The epidemiology of the cercospora leaf spots of groundnuts, discussed in general terms in Section 2.2.3, has been investigated by Hemingway (1954; 1955) and Fowler (1970b); a system of forecasting has been devised in the USA (Jensen and Boyle, 1966).

Cercospora leaf spots are effectively controlled by various fungicides, though tolerance to benomyl has developed (Clark *et al.,* 1974; Littrell, 1974). The economic benefit of their use depends on climatic conditions, the cultivar and its rate of maturity, the nature of the farming system as well as the local importance of the diseases. In northern Nigeria, fungicide spraying is recommended for the control of cercospora leaf spots in high-value breeding material and on seed-multiplication schemes but not for the main crop (McDonald and Fowler, 1976). Small-scale farmers in Africa generally can ill afford the high cost of spraying equipment, and the large amounts of water which are required may be difficult to obtain and to handle. Recent work in Nigeria with ultra-low volume (ULV) application of fungicides is promising (McDonald, 1978); the use of sulphur and copper fungicide dusts is effective, though often too expensive (McDonald, 1973). However, Mercer (1973) has found that sulphur dusting is sufficiently cheap to warrant recommendation to the smallholder farmer in Malawi. Since tropical soils are commonly sulphur-deficient, the fungicide may also act locally as a fertilizer. Cultural control measures such as crop rotation, clean weeding, and destruction of volunteer groundnuts, on which these pathogens often survive the dry season, may all assist in delaying the onset of infection and reducing its severity (Hemingway, 1954; Fowler, 1970b). There may also be scope for adjusting the time of sowing and spacing to reduce disease incidence (Farrell *et al.,* 1967).

It has long been appreciated that groundnut cultivars differ in susceptibility to the cercospora leaf spots (Higgins, 1935; Reyes and Romasanta, 1940; Hemingway, 1957). Early-maturing cultivars, which are usually of the erect bunch type, sequentially branched with light green foliage, are generally considered highly susceptible. Late-maturing types are either runner or spreading bunch in habit, alternately branched with dark green foliage; these late cultivars are thought to have varying degrees of resistance (Hemingway, 1957; Gibbons, 1966; Feakin, 1973). Although alternately-branched and sequentially-branched cultivars, derivatives of ssp. *hypogaea* and ssp. *fastigiata* respectively, may respond differently to fungicidal control of cercospora leaf spot (Elston *et al.,* 1976), there is now evidence that the field assessment of resistance has tended to be confounded by variations in host development, perhaps leading to spurious conclusions in the identification of resistance to the cercospora leaf spots (Nevill and Evans, 1980). Nevertheless, despite the failure to find immunity within *A. hypogaea,* sufficient resistance is present both within the species and in related species of *Arachis* to indicate considerable scope for breeding (Abdou *et al.*, 1974; Sowell *et al.*, 1976; Nevill, 1980). Further aspects of the expression of resistance and its use in groundnut improvement are discussed in Section 4.3.

4.2.2 Rust

The geographical distribution and, with it, the potential economic importance of groundnut rust (*Puccinia arachidis*) have dramatically increased over the last decade. Details of the recent intercontinental spread of rust, and some features of its epidemiology, have been described previously (Section 2.2.5). Here, we are concerned first with the symptomatology of rust and its effects on the groundnut crop, then on the principal means of rust control.

The first symptons of rust normally appear some 6–10 days after infection, but the incubation period varies with season (Mallaiah, 1976). Whitish flecks are seen on the lower surface of leaves, then yellow-green spots appear on the upper surface. Minute orange-red pustules develop within the flecks, enlarging rapidly before rupturing to expose the reddish-brown urediospores. The tissue surrounding the pustules becomes necrotic and, in severe infections, the entire leaflet may dry up and fall prematurely. Outbreaks commonly occur in patches, often termed 'hot spots', but severe epidemics can damage entire crops (Feakin, 1973; McDonald, 1978). Yield losses from rust can be substantial, particularly where the cercospora leaf spots are also involved. In India, the extent of crop loss has been estimated to be about 27 per cent from rust infection alone, and up to 70 per cent when leaf spot is also present (Nevill, 1980). In Zambia, losses of up to 50 per cent have been recorded from combined infections of rust and leaf spot (Raemaekers and Preston, 1977) and in the USA groundnut yields have been reduced by as much as 70 per cent by rust alone (Harrison, 1973). Chemical control of rust is feasible and fungicide applications usually attempt simultaneous control also of the leaf spots (Arneson, 1970; Harrison, 1973; Raemaekers and Preston, 1977; McDonald, 1978). However, the same limitations to the use of fungicides apply to rust control as to leaf spot, discussed above, and the development of integrated methods of disease management is more appropriate for developing countries. In areas where rust is not yet established, strict observance of quarantine regulations may help to prevent the introduction of *P. arachidis* as a seed contaminant (Peregrine, 1971b), but eradication is seldom feasible (Pitkethley, 1973; McDonald and Emechebe, 1978). Other cultural practices, such as crop rotation and the destruction of volunteer groundnut plants and infected debris, may limit sources of primary inoculum, and thus delay the onset of the epidemic, but such measures alone are unlikely to be fully satisfactory (Hammons, 1977). Certain hyperparasitic fungi including *Eudarluca* spp. and *Tuberculina costaricana* are often found on uredia of *P. arachidis* (Misra and Misra, 1975; Rothwell, 1975; Hammons, 1977; McDonald and Emechebe, 1978) but their role in the biological control of groundnut rust is unknown.

Field screening of groundnut cultivars under Venezuelan conditions first led to the identification of a Tarapoto selection as possessing rust resistance (Mazzani and Hinojosa, 1961), which appears to be effective also against a Jamaican isolate (Cook, 1972) and under field conditions in northern Nigeria (McDonald, 1978) and India (Subrahmanyam *et al.,* 1980). Other sources of resistance have since been identified in the USA by Bromfield and Cevario (1970), Bromfield and Bailey (1972) and Cook (1972), as well as in India (Subrahmanyam *et al*., 1980) where they are being used in breeding programmes.

4.2.3 *Aspergillus flavus* and aflatoxin

Species of *Aspergillus* are universal components of the soil mycoflora. *A niger* causes seed and seedling diseases of groundnut which may be locally important (Gibson, 1953; Nema *et al.,* 1955; Purss, 1960; Jackson, 1962), and *A. flavus* occasionally causes a seedling root rot (Chohan and Gupta, 1968). However, *A. flavus* is especially important in view of its ability to produce aflatoxin which is thought to cause liver damage in livestock and man (Kraybill and Shimkin, 1964). In 1960, widespread fatalities occurred in poultry fed on groundnut meal which had been imported into the UK, the source of the toxin being traced to *A. flavus* (Sargeant *et al.,* 1961). Although groundnuts for human consumption tend to be hand-picked and selected to eliminate discoloured kernels, it seems probable in developing countries that consumption of groundnuts (e.g. as cake, after oil extraction) is less discriminating and that local exposure to aflatoxin contamination is relatively common. The consequences are hard to appraise, but there seems like doubt that much aflatoxin must have been consumed in the past in contaminated groundnuts, thus exerting long-term, particularly carcinogenic, effects many years later (Barnes, 1970). This subject has been reviewed by Martin and Gilman (1976) and Coker (1980).

A considerable body of information is available on the conditions favouring the growth and development of *A. flavus,* and the subject is reviewed elsewhere (McDonald and Harkness, 1965; McDonald, 1969; Feakin, 1973; McDonald and Raheja, 1980). Undamaged pods at maturity are commonly infected with *A. flavus* but kernels are normally free of the fungus. Damage to pods, caused either by mechanical means or by parasitic attack, tends to increase seed infection (Ashworth and Langley, 1964). Since *A. flavus* will grow only when the moisture content of groundnuts exceeds 9 per cent, infection usually occurs after lifting but before drying reduces the moisture content below this critical level. Thus, post-harvest drying conditions tend to influence the degree of seed infection and aflatoxin contamination. Delayed harvesting also leads to increased fungal invasion (McDonald, 1970b).

Such information has been used to develop harvest and subsequent husbandry methods which minimize the risk of infection and aflatoxin production by *A. flavus.* Prompt harvesting at crop maturity followed by rapid drying, and subsequent maintenance in the dry state, are the essential measures for the cultural control of the aflatoxin problem (McDonald and Raheja, 1980). In Nigeria, it was found that the traditional windrow-drying methods permitted sufficiently rapid post-harvest drying to avoid substantial fungal invasion in the drier savannahs, while under more humid conditions initial windrow-drying followed by sun-drying on mats provided a suitably rapid method of drying. Although appropriate recommendations have been made to farmers, there is little evidence that such advice is followed, at least in Nigeria, and cultural control consequently appears to have restricted value in developing countries (McDonald, 1978).

There is evidence that groundnut cultivars differ in their susceptibility to invasion by *A flavus,* and both fungal isolates and, perhaps, host genotypes vary in their abilities to produce aflatoxin. Resistance to penetration of pods (Kushalappa *et al*., 1979) and of the dry seed by *A. flavus* has been identified (La Prade and Bartz, 1972; Mixon and Rogers, 1973; Zambettakis *et al.,* 1977; Subrahmanyam *et al.,* 1980). The intact testa of resistant genotypes apparently provides a mechanical barrier to infection, determined by the ultrastructure of the seed coat (La Prade and Bartz, 1972; Zambettakis and Bockelee-Morvan, 1976). The pigmentation of the testa may also influence penetration by *A. flavus* (Carter, 1973). Resistant genotypes imbibe water at slower rates and dry out more rapidly than accessions susceptible to infection (Ketring *et al.,* 1976). The expression of resistance to penetration is dependent on the environment and resistant lines do produce aflatoxin under certain conditions. Although accessions with this dry seed resistance may have some advantage in the field, they are nevertheless prone to fungal deterioration when stored at high humidity (Mixon and Rogers, 1975; Wilson *et al.,* 1977). Furthermore, there is no evidence that genotypes with dry seed resistance possess resistance to pod infection before lifting, and this requires further investigation (Subrahmanyam *et al.,* 1980). There are indications that cultivars may differ in their production of aflatoxin, but the existence of sources of resistance to aflatoxin production (Rao and Tulpule, 1967) has not been confirmed in subsequent studies (Doupnik, 1969).

4.2.4 Rosette

First reported by Zimmerman (1907), rosette disease is now known to be widespread in Africa, south of the Sahara. Apparently similar diseases also occur in Argentina, India, and Fiji (Feakin, 1973). Early studies (Storey and Bottomley, 1925; 1928) identified the aphid vector of rosette, later shown to be transmitted in a persistent manner (Storey and Ryland, 1955; Watson and Okusanya, 1967), and showed that infected plants bore virtually no seed. The only yield obtained from diseased plants was the seed formed before infection; where infection was early, the crop was a total loss. Recent experience in Nigeria (Yayock, 1977) has emphasized the potentially devastating effect of early infection on the groundnut crop. An increase in irrigated crops may also be expected to exacerbate rosette.

Three distinct types of rosette (chlorotic, green, and mosaic rosette) were recognized by Storey and Ryland (1957). Two of these types are now widely known, 'chlorotic mottle' predominating in eastern and southern Africa, while 'green rosette' is the more common syndrome in West Africa (Feakin, 1973). However, despite the intensity of investigation, the aetiology of groundnut rosette remains obscure. There has long been evidence that more than one virus component was involved (Hull and Adams, 1968). Attempts at isolation of virus from each of the types of symptom has met with only limited success, certain components proving to be extremely refractory (Bock *et al.*, 1976). However, it seems certain that rosetting is induced by co-infection of at least two viruses, and the concept of a groundnut rosette virus is thus open to question (Bock and Ngugi, 1972).

There have been many studies on the ecology of rosette and its principal vector, *Aphis craccivora*. Primary infection appears to depend on the presence of infected volunteer groundnuts from which winged aphids transmit rosette to the new crop. Early sowing and close spacing may reduce rosette incidence, probably by inhibiting the landing response of the vector. In trapping experiments, the effect of plant population on the number of aphids caught was thought to be due, in part, to the colour sensitivity of aphids and perhaps also to the contrast of plants and soil in widely spaced crops. At crop height, an alighting response was shown only by species of aphid which feed on groundnuts, *A. craccivora* and *A. gossypii* (A'Brook, 1964; 1968; Hull, 1964), the latter also being a vector of rosette (Adams, 1967). It has also been found that weeding tends to increase the incidence of rosette (Hayes, 1932; Mukiibi, 1978), perhaps also through influencing vector behaviour. Aphids may carry over on groundnut volunteers (Evans, 1954), although in areas were volunteers do not survive the dry season alternate hosts such as *Euphorbia* spp. may be important (Booker, 1963; Davies, 1972). At other times of year, legumes such as *Gliricidia sepium* and certain pasture species may support large populations of *A. craccivora* from which considerable flight activity over young groundnuts may occur (Davies and Kasule, 1964; Davies, 1972). Experimental host-range tests have identified potential virus hosts among leguminous weed species (Okusanya and Watson, 1966; Adams, 1967) but there is no evidence that alternate hosts act as virus reservoirs in the field. In areas where volunteer groundnuts do not survive the dry season, initial infection may depend on the influx of viruliferous aphids on prevailing winds (Bunting, 1950; Adams, 1967; Rossel, 1977). While winged aphids are responsible for primary infection, secondary spread depends to a larger extent on both apterae and nymphs, and rosette incidence may be positively correlated with the number of aphid-infested plants (Booker, 1963; Davies, 1972). In Uganda, less than a third of rosette incidence has been ascribed to alatae. Large, but often well concealed, populations of apterae may build up, becoming a potentially important means of dissemination of rosette (Davies, 1972), perhaps by their movement over the soil surface (Storey and Ryland, 1955). The population dynamics of *A. craccivora* on groundnuts, therefore, are likely to influence the rate of secondary spread of rosette. Under certain conditions, parasitism by *Entomophthora* (Storey, 1935), or predation by syrphid larvae (Evans, 1954) and the coccinellid *Cheilomenes sulphurea* (Davies, 1972) may keep rosette incidence at a very low level (Bunting, 1950). However, it appears that plant spacing, in addition to its influence on initial infection as discussed above, may also affect the rate of secondary spread of rosette. At close spacing, plant growth is often depressed, and it has been observed that aphid fecundity may be reduced on less nutritious plants in a dense crop (Farrell, 1976a, b). Under conditions favouring the build-up of aphid populations, rosette development is rapid. In Tanzania, it was noted that early in the season about 10 per cent of the crop was aphid-infested a week before symptoms of rosette were seen.Weekly observations thereafter showed tenfold increases in rosette incidence in a fortnight; some six weeks later, rosette incidence had reached 65 per cent (Evans, 1954). It is known that the transmission efficiency of *A. craccivora* varies with biotype (Storey and Ryland, 1955; Watson and Okusanya, 1967; Bock *et al.,* 1976), distinct biotypes transmitting different strains of rosette in different regions (Watson and Okusanya, 1967). Within biotypes that can transmit, all larval forms as well as alate and apterous adults may transmit. Alates may be more efficient than wingless forms, or *vice versa* (Storey and Ryland, 1955). *A. craccivora* can acquire and transmit rosette before the appearance of symptoms, but with low efficiency. Since most transmissions occur after symptom development, the potential rate of secondary spread depends on the length of the incubation period of the virus(es) in groundnut, which in turn is influenced by plant spacing and maturity and their effects on the rate of leaf production (Farrell, 1976a, b).

Cultural control of rosette therefore depends on early sowing at dense plant populations (Booker, 1963; A'Brook, 1964; Farrell, 1976a). There may also be potential in the use of mixed cropping to control the spread of rosette (Guillemin, 1952; Farrell, 1976c); and foliar application of insecticides, notably the aphicide menazon, reduces rosette spread through vector control (Davies, 1975a, b; 1976). Sources of resistance have been identified and utilized in groundnut breeding, as will be shown later (Section 4.3).

4.2.5 Bud necrosis

Bud necrosis, which is caused by tomato spotted wilt virus (TSWV), is probably the most important virus disease of groundnuts in India where crop losses of up to 80 per cent have been reported. The disease is present in every groundnut-growing region of India, incidence ranging from 5–80 per cent (Ghanekar *et al.,* 1979). TSWV also causes important losses in groundnut crops in Queensland (Helms *et al.,* 1961) but elsewhere the virus appears not to cause severe losses, though it it reported from groundnut in the USA, Brazil, South Africa (Ghanekar *et al.,* 1979), Zaire (B. E. L. Lockhart, personal communication, 1981), and possibily also Nigeria (McDonald and Raheja, 1980). Symptoms under field conditions develop within about six weeks of sowing, appearing initially as chlorotic ringspots on the leaves, and subsequently as necrosis of petioles, stems, and the terminal bud. New leaves may be reduced in size and show distortion, mosaic mottling, or chlorosis. Seedling infection may lead to stunting. Although kernels from diseased plants may be shrivelled, and the testae discoloured, TSWV is not seedborne in groundnut. TSWV is transmitted by various species of thrips, including *Scirtothrips dorsalis* Hood (Ghanekar *et al.,* 1979) and *Frankliniella schultzei* (Trybom) (Ghanekar, 1980), the latter being the more important vector.

Preliminary experiments on the control of bud blight in India suggest that early sowing may decrease disease incidence and reduce subsequent losses. Planting at high density, and intercropping with pearl millet, may also reduce disease incidence. Screening for resistance to bud blight is under way at ICRISAT (Ghanekar, 1980).

4.2.6 Peanut mottle

Peanut mottle virus (PnMV) is probably the most widespread virus infecting groundnuts, as well as various other legumes. PnMV is known to occur in the USA, Venezuela, Europe, East Africa, Australia, Japan, Malaysia (Bock and Kuhn, 1975), India (Reddy *et al.,* 1978), and the Philippines (Ghanekar, 1980). In groundnut, PnMV induces a systemic mottle and necrosis; older leaves may become distorted and show mosaic symptoms. Yield losses of 25–30 per cent have been recorded (Kuhn, 1965; Ghanekar, 1980). The incidence of transmission through groundnut seed ranges from 0–8.5 per cent (Adams and Kuhn, 1977). PnMV is transmitted by aphids in the non-persistent manner.

While some measure of cultural control of peanut mottle may possibly be achieved through the production of seed in areas free of PnMV (Demski *et al.,* 1975), there appears to be scope for the selection of groundnut cultivars in which PnMV is not seed-transmitted (Ghanekar, 1980). Sources of tolerance have also have identified (Kuhn *et al.,* 1978).

4.3 DISEASE RESISTANCE AND GROUNDNUT IMPROVEMENT

Resistance, of various levels and types, has been found within *A. hypogaea* to the cercospora leaf spots (Hemingway, 1957; Abdou *et al.,* 1974; Sowell *et al.,* 1976; Nevill, 1980), rust (Bromfield and Cevario, 1970; Cook, 1972), *Aspergillus flavus* (La Prade and Bartz, 1972; Mixon and Rogers, 1973; Zambettakis *et al.,* 1977), crown rot caused by *A. niger* (Chohan *et al.,* 1969), sclerotinia blight (Porter *et al.,* 1975b), verticillium wilt (Smith, 1961; Frank and Krikun, 1969), cylindrocladium black rot (Wynne *et al.,* 1975), and pythium pod breakdown (Porter *et al.*, 1975a). Sources of resistance are available also against bacterial wilt (Jenkins *et al.,* 1966; Simbwa-Bunnya, 1972), rosette (Berchoux, 1958; 1960; Nutman *et al.,* 1964; Daniel and Berchoux, 1965; Dhéry and Gillier, 1971), root knot, *Meloidogyne hapla*

(Castillo *et al.,* 1973a), and perhaps against bud necrosis and peanut clump (Ghanekar, 1980; Reddy and McDonald, 1981). Degrees of lesser susceptibility may exist against peanut stunt virus (Culp and Troutman, 1968) and peanut mottle virus (Kuhn *et al.,* 1978). However, despite these investigations, surprisingly few 'good sources' of resistance have been identified against the principal diseases of groundnut. No immunity to the cercospora leaf spots, nor to rosette, have been found within *A. hypogaea.* Attempts to find sources of resistance to root knot nematode, *M. arenaria* (Minton and Hammons, 1975), and peanut mottle virus (Kuhn *et al.,* 1968) among about 500 accessions in each case were unsuccessful. Although induced mutation has been attempted (Cooper and Gregory, 1960), attention has chiefly been directed at wild species of *Arachis* as potential sources of increased disease resistance. Recent evaluation of *Arachis* spp. has shown promise and, despite sterility barriers, interspecific hybridization is receiving considerable attention (Moss, 1980). We shall return to this subject later.

Turning now to consider what is known of the nature of disease resistance and its inheritance in groundnuts, let us first discuss the cercospora leaf spots. Higgins (1935) found that resistance to the two species was inherited independently, plant selections resistant to one being often very susceptible to the other. Resistance is often associated with characters such as maturity, plant habit, and the colour of foliage, as was discussed in Section 4.2. Disease reaction in controlled environments may not closely correspond with ratings obtained from field evaluation (Hassan and Beute, 1977) and there is evidence that field assessment of resistance may be confounded by variations in host development (Nevill and Evans, 1980). However, satisfactory correspondence between laboratory and field evaluations can be achieved through careful manipulation. For example, Nevill (1980) found that the components of resistance to *Cercosporidium personatum*, when estimated in the laboratory, were significantly correlated with field scores of resistance. The selected parameters of lesion diameter, incubation period, lesion number, and defoliation together accounted for over 50 per cent of the varation in the field score when they were included in a multiple regression analysis. Using this model, it was possible to rank 90 per cent of the cultivars in an order similar to that obtained by field evaluation. Subsequent work (Nevill, 1981) has confirmed that longer latent periods, reduced sporulation and less defoliation, though not fully complementary, are components of resistance to *C. personatum* and to *C. arachidicola.*

Variations in lesion number, lesion size, and their rate of growth were noted by Hemingway (1957), who attributed differences in lesion number to variation in stomatal size. Lesion growth rate appeared to relate to differences in the thickness of the palisade tissue of leaves. Subsequent studies on the resistance of wild *Arachis* spp. confirmed that small stomatal apertures on upper leaf surfaces were associated with cercospora resistance (Gibbons and Bailey, 1967). However, Cook (1981) found that variations in stomatal density and stomatal length were not related to resistance to infection by *C.*

personatum. Stomatal counts were highest on the least susceptible leaves and, although stomatal apertures tended to be longer on the most susceptible leaves, variation among cultivars could not account for differences in susceptibility. Differences may occur in the extent to which germ-tube growth is directed toward stomata: immune accessions of wild *Arachis* spp. show no directed growth of germ tubes (Abdou *et al.,* 1974). Together, these studies suggest that a stomatal exclusion mechanism, and/or an absence of directed growth of germ tubes toward stomata, may contribute to a 'resistance to penetration'. A 'resistance to colonization' of the invaded host also occurs and is associated with the formation, around the site of infection, of a barrier composed of a pectic deposit on the cell wall which prevents further lesion development (Abdou *et al.,* 1974). Recent studies on the genetic control of leaf spot resistance suggest that several components of resistance within *A. hypogaea* are inherited in a quantitative manner. It was found that defoliation was controlled by genes with additive action, but for other parameters including lesion size dominance effects were important and resistance was recessive. It appeared that genes at 3 or 4 loci were controlling the disease reaction (Nevill, 1980).

The expression and underlying mechanisms of resistance to rust in groundnuts have been less thoroughly studied. Resistance has been identified both within *A. hypogaea* and in wild *Arachis* species (Mazzani and Hinojosa, 1961; Bromfield and Cevario, 1970; Cook, 1972; Subrahmanyam *et al.,* 1980). Certain genotypes of *A. hypogaea* apparently possess hypersensitive resistance (Bromfield and Cevario, 1970), perhaps under bigenic control (Bromfield and Bailey, 1972), while *A. glabrata* appears to be immune. However, histological studies suggest that *P. arachidis* is able to penetrate these 'immune' wild species although development ceases after the formation of a single hypha from the sub-stomatal vesicle (Nevill, 1980). There is also evidence that cultivars of groundnut differ in their susceptibility to infection, which is related to changing wettability of leaves with maturity. Spore retention on leaf surfaces is affected, and probably so are both urediospore germination and appressorium formation. On leaf surfaces with low wettability, germ tubes tend to pass over stomata without making successful contact (Cook, 1980b), close adhesion of the germ-tube tip to the surface probably governing appressorium formation (Wynn, 1976).

Studies on rosette have shown that resistance, but not immunity, is available amongst groundnut material from Upper Volta and the Ivory Coast. Rosette resistance is expressed as delayed and reduced incidence of infection. Symptomless infections, from which aphid vectors are unable to transmit, are also quite common. It has been suggested that these effects relate to differences in the rate of virus multiplication in rosette-resistant and susceptible cultivars (Nutman *et al.,* 1964), or perhaps to the presence, or production, of a virus inhibitor in the resistant plant (Daniel and Berchoux, 1965). Resistance in the West African genotypes is thought to be governed by two recessive genes (Gibbons and Mercer, 1972).

One feature that is striking about resistance to several of the major diseases in groundnuts is its apparent stability across environments. The known sources of rosette resistance are effective not only in Nigeria and Upper Volta but also in Kenya (C. Harkness, cited by McDonald and Raheja, 1980). Similarly, certain Valencia types from the Tarapoto district of Peru where they were first recognized as rust-resistant (Mazzani and Hinojosa, 1961) are resistant also to isolates of rust from Puerto Rico and Texas (Bromfield and Cevario, 1970), as well as being rust-resistant under field conditions in Nigeria (McDonald, 1978) and India (Subrahmanyam *et al.,* 1980). Some measure of variation in pathogenicity occurs within certain groundnut pathogens including the cercospora leaf spot fungi (Miller, 1949), *Cylindrocladium crotalariae* (Hadley *et al.,* 1979), a root knot nematode (Minton, 1963) and peanut mottle virus (Sun and Herbert, 1972; Paguio and Kuhn, 1973). In striking contrast to other rust fungi, no pathogenic variants of *P. arachidis* have so far been distinguished. It is tempting to suggest that, in groundnut, the paucity of 'good sources' of resistance, the lack of defined 'physiologic races', and the stability of certain resistances are in some way related.

Some genotypes of *A. hypogaea* possess combined resistance to two or more diseases. For instance, some Peruvian accessions from Tarapoto have resistance to rust and to both cercospora leaf spots (Mazzani and Hinojosa, 1961; Mazzani *et al.,* 1972; Sowell *et al.,* 1976; Subrahmanyam *et al.,* 1980), while a selection of cv. Schwarz 21 combines resistance to verticillium wilt, bacterial wilt, and pythium pod rot (Jenkins *et al*., 1966; Frank and Krikun, 1969). A few rosette-resistant cultivars also possess some resistance to late leaf spot (Nigam *et al.,* 1980). However, few commercial cultivars of groundnut have adequate resistance to the major diseases, and there remains considerable scope for incorporating the available resistance into agronomically acceptable backgrounds, using conventional breeding methods for self-pollinated crops (Norden, 1980). Natural outcrossing occurs in groundnut but is uncommon (Smartt, 1960a; Culp *et al.,* 1968). Referring to North American conditions, Hammons (1976) considers that the absence of oligogenic control of the meagre sources of resistance present within *A. hypogaea* is in part responsible for the slow progress made. Nevill and Evans (1980) ascribe the failure to produce heavy yielding, cercospora leafspot-resistant cultivars to inadequate methods of evaluation. Nevertheless, some significant progress has been made through conventional breeding methods. The group of cultivars from West Africa possessing rosette resistance belong to the Virginia group (ssp. *hypogaea*). No rosette resistance has been found in the early-maturing, sequentially branched cultivars belonging to ssp. *fastigiata.* Considerable success has been achieved in developing rosette resistance in oil-seed cultivars in West Africa (Dhéry and Gillier, 1971; Harkness, 1977, cited by McDonald and Raheja, 1980) as well as in larger seeded, confectionery types in Malawi (Gibbons and Mercer, 1972). Similarly, certain low-yielding Valencia types possessing resistance to seed infection by *Aspergillus flavus* have been used in crosses with agronomically superior material in the USA (Mixon and Rogers, 1973).

Much more progress may be expected. In 1976, a groundnut improvement programme was established at the International Crops Research Institute for the Semi-Arid Tropics (ICRISAT) in India. An extensive collection (currently standing at about 10,000 accessions) of *Arachis* germplasm is held at ICRISAT (Rao, 1980). Recent evaluation has confirmed the existence of many sources of disease resistance amongst this germplasm, and some progress has been made in the utilization of these resources in a programme of breeding for disease resistance on which much emphasis is currently placed (Nigam *et al.,* 1980). Considerable attention is also being devoted to the potential of wild species of *Arachis* in groundnut improvement, especially with respect to disease resistance. It has long been appreciated that certain species possess high levels of resistance, particularly to rust (Bromfield and Cevario, 1970) and to the cercospora leaf spots (Abdou *et al.,* 1974; Kolawole, 1976). Wild species may also be resistant to root knot nematodes (Castillo *et al.,* 1973a) and perhaps to tomato spotted wilt virus (Ghanekar, 1980; Reddy and McDonald, 1981). Some species, of which *A. chacoense* is an example, combine resistance to several different diseases (Subrahmanyam *et al.,* 1980; Reddy and McDonald, 1981).

There are two principal obstacles to the employment of wild species: differences in ploidy level and incompatibility between certain wild species and *A. hypogaea.* Genes have been successfully transferred to groundnut from compatible species, perhaps most frequently by chromosome doubling in a triploid hybrid resulting from a cross between a diploid species and the tetraploid groundnut. Other methods of gene transfer from compatible species are being studied, and sophisticated cytological investigation is under way at ICRISAT, in an attempt to make the whole gene pool within *Arachis* available to the groundnut breeder (Moss, 1980; Singh *et al*., 1980a). New variability derived from interspecific hybridization is thought by some to herald 'the beginning of a new era of groundnut evolution' (Gregory and Gregory, 1976).

Lest such optimism appears unduly euphoric, a word of caution may be warranted. Experience from the transfer of disease resistance between other species, though scanty, seems to indicate that pathogens have often developed host-specific pathogenicity to alien resistance. The extent to which such instability is due to the incomplete transfer of the resistance present in the parent species is hard to assess. Nor is it known whether the risk of 'breakdown' of resistance is confined to biotrophic parasites such as the rusts, or whether the same might apply also to typically less specialized, splash-dispersed pathogens, such as *Cercospora* species (Scott *et al.,* 1980). Until more information on the probable durability of resistance in *Arachis* becomes available, it would be prudent to concentrate also on the identification and synthesis of the components of resistance available within *A. hypogaea* to each of its major diseases. Refinement of screening technique, when coupled with the availability of an extensive germplasm collection and a multidisciplinary team working in an environment permitting two crop seasons each year, may also be expected to achieve rapid progress. The ultimate value of these two parallel approaches may be appraised by the use of standardized methods in international testing against a range of pathogen variants over a range of environments. Such an international testing network is currently being established (Nigam *et al*., 1980).

Chapter 5

Soyabean Diseases

5.1 EVOLUTION, ADAPTATION, AND PRODUCTION OF SOYABEANS

The genus *Glycine* consists of about nine species representing two sub-genera. *G. max*, the soyabean or soybean, and its putative ancestor *G. soja* (= *G. ussuriensis*), belong to the sub-genus *Soja* which has its centre of diversity in north-east China and adjacent areas of the USSR, Korea, Taiwan, and Japan. The weedy *G. gracilis*, which is now normally treated as conspecific with *G. max*, is thought to have evolved as a consequence of outcrossing between *G. max* and *G. soja* since it is found wherever these two species occur together. Members of the sub-genus *Soja* are fully interfertile, and both the cultigen *G. max*, which is unknown in the wild, and the wild *G. soja* are diploid ($2n = 40$) (Hymowitz, 1976; Hymowitz and Newell, 1980; 1981). *G. max* is also completely self-fertile, and the amount of natural outcrossing is less than 1 per cent (Garber and Odland, 1926) although, where genetic male sterility occurs, natural cross-pollination occurs at frequencies exceeding 11 per cent (Caviness *et al.*, 1970). The sub-genus *Glycine*, which occurs in Australia and the South Pacific as well as south-east Asia, contains seven perennial species none of which is interfertile with soyabean (Hymowitz and Newell, 1980; 1981). Formerly, a third sub-genus, *Bracteata*, was recognized to comprise the single species *G. wightii* (= *G. javanica*), a pasture legume endemic to Africa and south-east Asia. However, recent evidence suggests that *G. wightii* is atypical and its transfer to a new genus, *Neonotonia*, has been proposed (Lackey, 1977; Hymowitz and Newell, 1981). Such a move is supported also by the distinctive phytoalexin spectrum of this species (Ingham *et al.*, 1977).

It is probable that the soyabean was first domesticated around the eleventh century BC in the eastern half of northern China, from there spreading southwards during the expansion of the Chou dynasty. Later, Manchuria became a centre of secondary diversity. The soyabean was known in Europe by 1712, and the crop was first mentioned in American literature in 1804, although extensive cultivation in the USA was not to develop until the late 1940s. The earliest known date for the crop's introduction into Brazil is 1822 (Hymowitz,

1970; 1976), and into South Africa in 1903 (Auckland, 1970). The first attempt to establish soyabeans in East Africa was made in 1907 with material from America, and further trials were made in 1911 with seed imported from Japan, China, and South Africa (Greenway, 1945). The crop has been more recently established in West Africa (Yuwa, 1964).

The soyabean is now the most important grain legume crop in the world, in terms of total production and international trade. World production is approximately 75 million tonnes per annum, of which 60 per cent is produced in the USA at an average yield of about two tonnes per hectare. In 1973, soyabeans became the most important cash crop in the USA, overtaking wheat and maize. Other major producers are China and Brazil, together accounting for 33 per cent of world production. Total production has trebled over the last 20 years, a rate reflecting the threefold increase in US production over the same period. However, Brazilian production has increased at an even greater rate: from 0.35 million tonnes in 1960, 11.80 million tonnes in 1977 to about 15 million tonnes in 1980. Significant amounts of soyabeans are also produced in Indonesia, Korea, Japan, Romania, Canada, Mexico, Colombia, Argentina, and Paraguay (Hymowitz, 1976; Yang, 1980), and there is growing interest in the crop in many other countries.

The bulk of the soyabean crop is produced with high inputs on an intensive scale, chiefly outside the tropics (Brazilian production is mainly in the south, at a latitude of about 25–30 ° S). Most is used in industry for the extraction of oil and protein; the seeds contain 13–25 per cent of oil while having the largest protein content (up to 50 per cent in some cultivars) of all cultivated legumes (Cobley and Steele, 1976). It has long been appreciated that soyabeans have considerable potential also as a food crop in developing countries (Leng, 1968). In the Orient, soyabean seed is popular as a protein-rich human food, being utilized in liquid, powder, and curd forms, and immature green seeds and sprouts are considered highly nutritious. However, none of these uses has become popular outside eastern Asia, and soyabeans are often regarded as unpalatable to communities not accustomed to eating them, despite their great food value. Nevertheless, soyabeans are beginning to be accepted as a new component of complex cropping systems in subsistence agriculture, as for instance in parts of Zaire where soyabean meal is being used to enrich cassava flour (D. J. Allen, unpublished). Meal is being used in soups in Rwanda where soyabeans are now replacing *Phaseolus* beans in marginal areas. And in Nigeria locust beans (*Parkia clappertoniana*) are also starting to be replaced by soyabeans which are fermented for use as 'dadawa' (E. A. Kueneman, personal communication 1981). Earlier expectations that breeders would produce cultivars well adapted to the humid tropics are beginning to be realized.

Despite its temperate origins, the soyabean is adapted to a wide range of latitudes, from 0 ° to 52 ° N or S, and within the range of seasonal durations (90–180 days) these encompass shows substantial ability to compensate both by vegetative and reproductive growth. The means by which soyabeans accomplish such adaptabiltiy, a subject beyond the scope of this book, are discussed by Shibles (1980); only the essential points are summarized here. Soyabeans are short-day photoperiodic. Cultivars adapted to the long-day conditions of temperate regions tend to flower very early in the tropics, with the consequence that insufficient vegetative growth is made to be able to support heavy seed yields. However, there are many genotypes which do not flower so early under short tropical days. Although photoperiodic responses may be strongly influenced by day/night temperature regimes, genotypes adapted to short days, and so tropically adapted, appear to be less sensitive to small changes in day length than are most temperature cultivars (E. A. Kueneman, personal communication 1981). In North America, soyabeans are classified into various maturity groups based upon the effects of day length on the time of appearance of first flowers. Although seasonal changes in photoperiod may be pronounced in temperate climates, at the equator photoperiod is fairly uniform and other factors including temperature may be more important in determining adaptation than is photoperiod; therefore other systems of classification appear warranted for tropical soyabeans (Leakey and Rubaihayo, 1970; Shanmugasundaram *et al*., 1980). Various attempts have been made to evaluate introduced germplasm. In Africa, small programmes of breeding and cultivar selection have been conducted in Zimbabwe (Arnold, 1942), Zambia (Smartt, 1960b; Javaheri, unpublished 1977), Kenya (Weiss, 1967), Uganda (Rubaihayo and Leakey, 1970; Hawtin, 1974), and Nigeria (van Rheenen, 1972, cited by Rachie and Silvestre, 1977). One of the most successful soyabean breeding programmes carried out at low latitudes (< 10 ° N or S) was conducted in southern Tanzania in the late 1950s and the 1960s (Auckland, 1970). In sub-tropical India, a successful breeding programme has been carried out at the G.B. Pant University of Agriculture and Technology, Pantnagar. During the last decade, increased emphasis has been given to the development of tropically adapted soyabean cultivars at three international centres: the Asian Vegetable Research and Development Centre (AVRDC) in Taiwan, the International Soybean Program (INTSOY) at Illinois and Puerto Rico in the USA, and the International Institute of Tropical Agriculture (IITA) in Nigeria. Extensive germplasm collections are now available, and promising materials are emerging from improvement programmes.

A principal constraint confronting the improvement of soyabeans in lowland tropical environments is poor seed quality. Physiological factors and micro-organisms together account for the deterioration of soyabean seed, the relative importance of each varying with environment, and perhaps also with the investigator. The role that pathogens play in this problem will be discussed later. In Africa, the second major constraint is the incompatibility of the most productive (chiefly American) cultivars with indigenous *Rhizobium*, necessitating artificial inoculation. Locally adapted landraces and certain other genotypes are able to nodulate with rhizobia of the cowpea group, but their yield potential has been found inadequate. Thus the major emphasis of IITA's soyabean breeding programme is to combine high-yield potential with good seed quality and the ability to fix nitrogen in association with rhizobia indigenous to African soils (IITA, 1980). Whether or not there is a similar potential for the exploitation of mycorrhizal associations for the improvement of phosphorus nutrition of tropical soyabeans remains largely unknown (Ross and Harper, 1970; Sanders *et al*., 1975). With the exception of the pathogens associated with poor seed quality, diseases do not yet appear to be amongst the major production problems of the crop in Africa. Bacterial pustule, caused by *Xanthomonas campestris* pv. *glycines*, is probably the most widespread disease; with increased attention, coupled with an expansion of the crop, virus diseases are likely to increase in importance in Africa. The importance of virus diseases to soyabeans also in the American tropics and in the Indian sub-continent is appreciated by INTSOY, with its twin bases in Illinois and Puerto Rico. As part of a breeding programme committed to the development of tropically adapted soyabeans, considerable attention is being given both to soyabean mosaic virus and to whitefly-transmitted viruses in addition to studies on seedborne pathogens in relation to seed quality (Goodman *et al*., 1981). Rust, caused by *Phakopsora pachyrhizi*, is a major disease of soyabeans in the Orient, and screening for resistance to rust is an important part of AVRDC's breeding programme (Shanmugasundaram, 1979).

The relative importance of diseases as production problems, and other aspects of the pathology of soyabeans, will be discussed in the next section.

5.2 THE PATHOLOGY OF SOYABEANS

Despite the Asiatic origin of soyabeans, the bulk of the available information on the crop's diseases is American, a reflection of the main region of present production. While many of the diseases of soyabean in the USA appear to have been introduced from the Orient (Wolf and Lehman, 1926), others appear to be indigenous to America (Abel, 1977). With increasing interest in soyabeans as a food crop for developing countries, rather than as an industrial crop of temperate regions, it becomes evident that relatively little is known of the pathology of the crop in the tropics. Much more research is required, perhaps particularly in the Asian tropics and sub-tropics where the diversity of parasites of *G. max* and its relatives may be expected to be greatest. Work at AVRDC in Taiwan has confirmed the cardinal importance of rust (*Phakopsora pachyrhizi*) throughout the region, so that the control of rust is emphasized in that centre's programme. The potential importance of whitefly-transmitted virus diseases is beginning to become apparent from the recent work in Thailand (Iwaki *et al*., 1981), endorsing earlier attention devoted to their control in India (Singh *et al*., 1974). New Indian records of fungal diseases, such as drechslera blight (Narayanasamy and Durairaj, 1971), stemphylium leaf blight and leptosphaerulina leaf spot (Anahosur and Fazalnoor, 1972), possibly require monitoring; and in Australia, it remains to be seen whether pathogens of native *Glycine* species, such as *Pseudocercos-*

pora glycines (Deighton, 1976) and glycine mosaic virus (Bowyer *et al.*, 1980), have potential as soyabean pathogens. It has been suggested that pyrenochaeta leaf spot in Africa may have spread to soyabean from the indigenous *Neonotonia wightii* (Stewart, 1957). In lowland tropical Africa, bacterial pustule, web blight, anthracnose, and seedborne *Phomopsis* are common, and there is a growing awareness that cowpea mild mottle virus is also important. In cooler highland areas of Africa, downy mildew, bacterial blight, pyrenochaeta leaf spot and peanut mottle virus also naturally infect the soyabean crop, though their relative importance is unknown. Rust is not reliably reported from soyabean in Africa (Section 2.2.5), and old East African records of mosaic probably do not refer to soyabean mosaic virus (SMV). Other, rarely recorded, pathogens of soyabeans in Africa include *Pseudomonas syringae* pv. *tabaci*, the wildfire pathogen (Angus, 1962–66), *Dactuliophora glycines* (Leakey, 1964), *Calonectria crotalariae* (Delassus, 1978), and *Phaeoseptoria* sp. (Allen, 1979a).

The diseases of soyabeans in the Americas, particularly in the USA are well-known and extensively reviewed. The fungal diseases of the crop have been reviewed by Athow (1973), while the bacterial diseases are covered by Kennedy and Tachibana (1973), and the virus diseases by Dunleavy (1973) and Goodman and Nene (1976). Good (1973) has reviewed the nematodes which infest soyabeans. General reviews of the pathology of the crop include those made by Sinclair (1977), Yang (1980), and Goodman *et al.* (1981), the latter emphasizing the work of INTSOY. An annotated bibliography of soyabean diseases is also available (Sinclair and Dhingra, 1975). But the most important, and well-illustrated, review is the *Compendium of Soybean Diseases* (Sinclair and Shurtleff, 1975) on which the following account has depended heavily. No attempt is made here to emulate that review, but rather to focus on a few aspects of the pathology of tropical soyabeans, and the reader is referred to the *Compendium*, now revised (Sinclair, 1982), for a wider coverage of the crop's diseases.

Numerous fungi parasitize soyabeans, under both temperate and tropical conditions. The more important fungal diseases, their causal agents, distribution, and approximate economic importance are summarized in Table 16. Those of greatest concern to the crop under tropical and sub-tropical conditions include the seedborne fungi implicated in seed deterioration, rust and, to a lesser extent, anthracnose, charcoal rot, diaporthe stem blight and canker, rhizoctonia web blight and root rot, and fusarium wilt. Only the first two will be discussed further in a subsequent section.

Soyabeans are susceptible to a very wide range of viruses and, because of this, they are commonly used as test plants in virology. However, comparatively few of these viruses are of economic importance to the soyabean crop, the most widespread and important being SMV. Other well-known viruses, such as tobacco ringspot, bean yellow mosaic, peanut mottle, bean pod mottle, cowpea chlorotic mottle, alfalfa mosaic, and cowpea severe mosaic, are known to occur in natural infections of soyabean in the field but as yet all are

TABLE 16 Fungal Diseases of Soyabean

Disease	Causal fungi	Distribution	Importance/Crop loss	References
1. Seed and seedling diseases				
Seed decay	Associated principally with *Phomopsis* spp., and also with *Colletotrichum truncatum* (Schw.) Andrus and Moore, *Macrophomina phaseolina* (Tassi) Goid, *Fusarium* spp. and various others. Physiological factors may also be important	Widespread	Often the major factor limiting the spread of soyabean production in the lowland tropics	Kmetz *et al.*, 1978; Ndimande *et al.*, 1981
Purple stain	*Cercospora kikuchii* (Matsu. and Tomoy.) Gard., *C. sojina* Hara	Widespread	No yield loss, but seed viability may be impaired	Murakishi, 1951; Sherwin and Kreitlow, 1952; Hepperly and Sinclair, 1981
Seedling mortality	Associated with various fungi including *Colletotrichum truncatum* (Schw.) Andrus and Moore, *Pythium* spp. and *Rhizoctonia* spp.	Widespread	Of local, or seasonal, importance. Incidence >90% on calcium-deficient soils in Brazil	Meyer *et al.*, 1974; Sinclair and Shurtleff, 1975; Muchovej *et al.*, 1980
2. Stem and root rots, and wilts				
Anthracnose	*Colletotrichum truncatum* (Schw.) Andrus and Moore (= *C. dematium* (Pers. ex Fr.) var. *truncata* (Schw.) Arx) and *Glomerella glycines* (Hori) Lehman and Wolf	Widespread	Damaging in warm temperate and tropical regions	Sinclair and Shurtleff, 1975
Charcoal rot, ashy stem blight	*Macrophomina phaseolina* (Tassi) Goid (sclerotial state = *Rhizoctonia bataticola* (Taub.) Butl.)	Widespread	Potentially important under hot, dry conditions	Sinclair and Shurtleff, 1975

TABLE 16 *(Continued)*

Disease	Causal fungi	Distribution	Importance/Crop loss	References
Pod and stem blight, stem canker	*Diaporthe phaseolorum* (Cooke and Ellis) Sacc. var. *sojae* (Lehman) Wehm. (conidial state *Phomopsis*) and *D. phaseolorum* var. *caulivora* Athow and Caldwell	Widespread	Often destructive. Stem canker can cause losses of 20–50% in North America	Punithalingam and Holliday, 1972; Sinclair and Shurtleff, 1975; Kmetz *et al.*, 1978
Pythium root rot	*Pythium debaryanum* Hesse, *P. ultimum* Trow., *P. aphanidermatum* (Edson) Fitz., and other species	Widespread	Minor	Lehman and Wolf, 1926a; Sinclair and Shurtleff, 1975
Rhizoctonia root and stem rot	*Rhizoctonia solani* Kühn (perfect state = *Thanatephorus cucumeris* (Frank) Donk)	Widespread	Locally important on heavy, poorly drained soils	Sinclair and Shurtleff, 1975
Phytophthora rot	*Phytophthora megasperma* Drechs f.sp. *glycinea* Kuan and Erwin	USA, Canada	Very destructive in North America. Yield losses >50%	Kaufmann and Gerdemann, 1958; Hildebrand, 1959; Caviness and Walters, 1971; Kuan and Erwin, 1980
Sclerotinia stem rot, white mould	*Sclerotinia sclerotiorum* (Lib.) de Bary	USA, Canada, Hungary, India	Minor, though occasionally damaging under wet conditions	Chamberlain, 1951; Grau and Bissonnette, 1974; Sinclair and Shurtleff, 1975
Thielaviopsis root rot	*Thielaviopsis basicola* (Berk. and Br.) Ferr.	USA, Canada, Germany	Minor. Favoured by cool soil temperatures	Lockwood *et al.*, 1970; Sinclair and Shurtleff, 1975
Mycoleptodiscus root rot	*Mycoleptodiscus terrestris* (Gerd.) Ostazeski	USA	Minor, but possibly overlooked	Gerdemann, 1954; Sinclair and Shurtleff, 1975
Neocosmospora stem rot and wilt	*Neocosmospora vasinfecta* Smith	USA, Nigeria	Minor	Phillips, 1972; Oyekan, 1978
Cylindrocladium black rot	*Cylindrocladium crotalariae* (Loos) Bell and Sobers (perfect state = *Calonectria crotalariae* (Loos) Bell and Sobers)	USA, Cameroun and Japan (on groundnut)	Locally important; losses may exceed 50%	Sinclair and Shurtleff, 1975; Delassus, 1978; Beute, 1980

Sclerotium stem rot	*Sclerotium rolfsii* Sacc.	Widespread	Minor	Sinclair and Shurtleff, 1975
Brown stem rot	*Cephalosporium gregatum* All. and Chamb. (perfect state = *Phialophora gregata* (All. and Chamb.) Gams)	USA, Canada, Mexico, Egypt	Locally important; losses may exceed 25%. Favoured by cool conditions	Allington and Chamberlain, 1948; Chamberlain and McAlistair, 1954; Morgan and Dunleavy, 1966; Sinclair and Shurtleff, 1975; Abel, 1977
Fusarium root rot	*Fusarium oxysporum* Schlecht., and probably other spp.	Widespread	Minor	French and Kennedy, 1963
Fusarium wilt	*Fusarium oxysporum* Schlecht. f.sp. *tracheiphilum* (Sm.) Snyd. and Hans., f.sp. *vasinfectum* (Atk.) Snyd. and Hans., f.sp. *glycines* Armst. and Armst.	Widespread	Potentially important. Favoured by warm conditions, and exacerbated by nematode damage	Armstrong and Armstrong, 1950; 1958; 1965a
3. Foliar diseases				
Cercospora leaf spot, frogeye leaf spot	*Cercospora sojina* Hara, *C. kikuchii* (Matsu. and Tomoy.) Gard. and *C. canescens* Ell. and Mart.	Widespread	Potentially important. *C. sojina* can cause up to 15% crop loss	Sinclair and Shurtleff, 1975; D. J. Allen, unpubl.
Target spot	*Corynespora cassiicola* (Berk. and Curt.) Wei	Widespread	Potentially important, especially on late-maturing cvs. About 30% crop loss recorded	Hartwig, 1959
Septoria brown spot	*Septoria glycines* Hemmi	USA, Canada, Colombia, Brazil, Europe, USSR, China, Japan, Korea	Minor, though premature defoliation can lead to 17% yield loss	Punithalingam and Holliday, 1972c; Sinclair and Shurtleff, 1975; Williams and Nyvall, 1980
Ascochyta leaf spot	*Ascochyta sojicola* Abramoff, *A. phaseolorum* Sacc.	Eastern Africa	Minor	Riley, 1960; Angus, 1962–66; Stewart and Yirgou, 1967; Ondieki, 1973
Pyrenochaeta leaf spot	*Pyrenochaeta glycines* Stewart	Eastern Africa, Cameroun	Potentially of local importance, but crop loss unmeasured	Stewart, 1957; Javaid and Ashraf, 1978; D. J. Allen, unpubl.

TABLE 16 (*Continued*)

Disease	Causal fungi	Distribution	Importance/Crop loss	References
Phyllosticta leaf spot	*Phyllosticta glycinea* Tehon and Daniels (= *P. sojaecola* Massal.)	USA, Canada, Europe, Cameroun, Eastern Africa, USSR, Japan, China, Malaysia	Minor, though premature defoliation occurs	Olive and Walker, 1944; Sinclair and Shurtleff, 1975; Delassus, 1978; Ebbels and Allen, 1979
Rhizoctonia web blight	*Rhizoctonia solani* Kühn (perfect state = *Thanatephorus cucumeris* (Frank) Donk)	Widespread	Locally important under warm and wet conditions; 80% infection reported from China	Sinclair and Shurtleff, 1975
Scab	*Sphaceloma glycines* Kurata and Kuribayashi	Japan	Locally destructive	Jenkins, 1951
Downy mildew	*Peronospora manshurica* (Naoum.) Syd. ex Gaum.	Widespread; but in South America known only from Brazil, Colombia, and Argentina. In Africa, reported from Ethiopia, Zambia, Zimbabwe and S Africa	Generally minor; yield loss <8%	Sinclair and Shurtleff, 1975; CMI Distrib. Map 268 (1979); Allen, 1979a
Powdery mildew	*Microsphaera diffusa* Cke and Pk.	North America, Brazil, Peru, S Africa, Europe, China, India	Occasionally important; yield loss 10%	Paxton and Rogers, 1974; Sinclair and Shurtleff, 1975; McLaughlin *et al.*, 1977; Dunleavy, 1978
Rust	*Phakospora pachyrhizi* Syd.	Widespread, especially in Asia	Major. Complete crop loss may occur	Anahosur and Waller, 1978a

of comparatively minor importance, at least as far as tropical soyabeans are concerned. The distribution, means of transmission and estimated economic importance of the better characterized viruses naturally infecting soyabeans are shown in Table 17. Further investigation of the viruses infecting soyabeans in tropical regions may be expected to reveal new viruses. Recent studies have led to the partial characterization of a persistent, aphidborne virus, Indonesian soyabean dwarf virus (Iwaki *et al*., 1980), and research in Nigeria suggests that an apparently undescribed virus is present in the Abuja area of northern Nigeria, in the local landrace, Malayan (IITA, 1979; 1980). A further virus of potential importance to the soyabean crop is glycine mosaic virus, a comovirus from the Australian native species *Glycine clandestina* and *G. tabacina* (Bowyer *et al*., 1980).

In the next section of this chapter, we shall focus attention only on the viruses of probable importance to soyabeans in tropical and sub-tropical regions, dealing with SMV, the bud blights caused by tobacco ringspot and tobacco streak viruses, and various viruses known to be whitefly-transmitted. For more comprehensive coverage of the virus diseases of soyabeans, the reader is referred to the reviews by Dunleavy (1973), Sinclair and Shurtleff (1975), and Goodman and Nene (1976). A bibliography of viruses infecting the soyabean is also available (Kreitlow *et al*., 1957).

Symptoms in soyabeans referred to as witches' broom or phyllody occur widely, and are usually assumed to be associated with mycoplasma-like organisms (see Section 3.2). Witches' broom diseases have been reported from Peru (J. B. Sinclair, personal communication, 1981), and from Japan and Taiwan where they are transmitted by the leafhopper *Nesophrosyne orientalis* (Murayama, 1966; Lo and Han, 1969). An apparently similar disease affects soyabeans in Colombia where it is known as 'machismo' disease; the vector is the leafhopper *Scaphytopius fuliginosus* (Granada, 1979a). Soyabean phyllody occurs in Tanzania (Ebbels and Allen, 1979) and is also occasionally seen in Nigeria. Although too little is known of this group of pathogens to enable accurate assessment of their economic importance, it is probable that mycoplasmas are not amongst the major pathogens of tropical soyabeans.

Several bacterial diseases affect soyabeans wherever they are grown; both wilts and foliar diseases occur. *Bacillus subtilis* is involved in a soyabean seed decay (Ellis *et al*., 1977). A *Corynebacterium* sp. causes a vascular wilt of soyabean in the USA (Dunleavy, 1963) and *Pseudomonas solanacearum* occasionally also causes a wilt (Muras, 1963) but neither is important. Three pathovars of *Ps. syringae* (Section 3.3) infect soyabeans: pv. *syringae* is an occasional foliar pathogen of soyabean in highland areas of Kenya (Kaiser and Ramos, 1980); pv. *tabaci* causes wildfire of soyabean in the USA (Allington, 1945), Brazil (Hagedorn *et al*., 1969) and possibly Zambia (Angus 1962–66); and pv. *glycinea*, which is much the most important, causes bacterial blight. A strain of pv. *glycinea* causes distinct symptoms, called bacterial leaf crinkle (Tachibana and Shih, 1965). 'Chocolate spot' is caused by an unidentified bacterium (Dunleavy, 1964). Bacterial pustule, caused by

TABLE 17 Virus Diseases of Soyabean

Disease	Causal virus	Distribution	Transmission		Importance/Crop loss	References
			Vector	Seed (%)		
Mosaic	Soyabean mosaic virus (SMV)	Widespread	Aphid	<0.2–30	Major. Crop loss of 35% (natural infection), 45–86% (artificial inoculation)	Ross, 1977; Goodman and Oard, 1980
Yellow mosaic	Bean yellow mosaic virus (BYMV)	Widespread	Aphid	0	Minor. Occasional in mixed infections with SMV	Conover, 1948
Yellow mosaic	Mung bean yellow mosaic virus (MYMV)	Tropical Asia	Whitefly	0	Major	Suteri, 1974; Goodman and Nene, 1976; Iwaki *et al.*, 1981
Mild mottle	Cowpea mild mottle virus (CMMV)	Tropical Africa and Asia	Whitefly	<91.7	Probably major	Brunt and Kenten, 1973; Brunt and Phillips, 1978; Iwaki *et al.*, 1981; Thouvenel *et al.*, 1982
Mottle (top necrosis, in mixed infections + SMV	Bean pod mottle virus (BPMV)	USA	Beetle	0	Locally important, causing 10–15% yield loss alone, and >60% in mixed infections + SMV	Skotland, 1958; Ross, 1968a; Lee and Ross, 1972
Mottle	Peanut mottle virus (PnMV)	Widespread	Aphid	0	Minor	Kuhn *et al.*, 1972; Bock and Kuhn, 1975

Mottle	Cowpea chlorotic mottle virus (CCMV)	Temperate and tropical America	Beetle	0	Minor. Early infections can cause 20–30% crop loss	Kuhn, 1968; Harris and Kuhn, 1971
Mosaic	Cowpea severe mosaic virus (CSMV)	Tropical America	Beetle	?	Potentially important. Seedling infection (of 25% incidence), or infection at flowering (50% incidence) can cause significant yield loss	Thongmeearkom *et al.*, 1978; Lin *et al.*, 1981a
Mosaic	Alfalfa mosaic virus (AMV)	Widespread	Aphid	?	Minor	Allington *et al.*, 1960
Stunt	Peanut stunt virus (PSV)	USA, Japan	Aphid	? 0	Minor; often latent	Mink, 1972; Milbrath and Tolin, 1977
Dwarf	Soyabean dwarf virus (SDV)	Japan	Aphid	0	Locally important. 50% field incidence can cause 40% yield loss	Tamada and Kojima, 1977
Bud blight	Tobacco ringspot virus (TRSV)	USA, Canada, Australia, China	Nematode, thrips, grasshopper	<100	Locally important. Yield losses of 25–100% occur	Allington, 1946; Dunleavy, 1957; Athow and Bancroft, 1959; Messieha, 1969; Sinclair and Shurtleff, 1975
Brazilian bud blight	Tobacco streak virus (TSV)	Brazil, USA	?	2.6–30	Locally destructive in SE Brazil	Costa and Carvalho, 1961; Fagbenle and Ford, 1970; Ghanekar and Schwenk, 1974; Sinclair and Shurtleff, 1975

Xanthomonas campestris pv. *glycines*, is very widely distributed, and better adapted to the hot lowland tropics than bacterial blight. These two important bacterial diseases will be the subject of a further section. Aspects of the epidemiology of bacterial diseases are discussed in Section 3.3. The reader wishing a fuller account of the bacterial pathogens of soyabeans is referred to the reviews of Kennedy and Tachibana (1973) and Sinclair and Shurtleff (1975).

Nematodes may also be of major importance to the soyabean crop, yet their damage is generally unrecognized (Good, 1973). The soyabean cyst nematode (*Heterodera glycines*) can cause significant loss, but is essentially confined to temperate conditions. *H. glycines* was first reported in the USA in 1955, apparently having been introduced from China or Japan (Winstead *et al*., 1955). Much the most important group is the root knot nematodes (*Meloidogyne* spp.), although like the cyst nematode, the damage they cause is more readily recognized than that resulting from infestation by ectoparasitic species. Maximum damage from root knot nematodes occurs in warm climates on sandy soils, leading to losses in yield in susceptible cultivars of up to 90 per cent (Minton and Parker, 1974; Sinclair and Shurtleff, 1975). The reniform nematode (*Rotylenchulus reniformis*), which also infests soyabean roots, has been reported from Ghana (Peacock, 1956) as well as from the USA where severe losses are reported (Sinclair and Shurtleff, 1975). Some of the most destructive ectoparasitic nematodes are dagger, sting, lance, spiral, and stubby root (Good, 1973), some of which are discussed briefly in Section 3.4. Many other species are reported to attack soyabeans, but the damage they cause is thought to be insignificant. Dagger nematodes (*Xiphenema* spp.), which may be locally abundant in soyabean fields (Taylor *et al*., 1958), are vectors of tobacco ringspot virus although their efficiency in transmitting the virus to soyabean is low (Bergeson *et al*., 1964).

Unlike other of the food legumes, soyabeans appear not to be prone to attack by parasitic higher plants, discussed in general in Section 3.5, although *Alectra vogelii* has been recorded on soyabeans in South Africa (Wild, 1954) and dodder (*Cuscuta* spp.) is recognised as a potentially serious problem in China (R. M. Goodman, personal communication 1981).

5.2.1 Fungi and seed deterioration

The deterioration of seeds proceeds inexorably at a rate dependent on temperature and water content, starting as the seed matures and continuing through harvest and storage to sowing and germination. In soyabean, the process of deterioration precedes death by a longer period than in many other species (Roberts and Ellis, 1980). The production of good quality soyabean seed and the maintenance of its viability during storage have proved difficult in the humid tropics, so that seed deterioration is often seen as the major factor limiting the spread of soyabean production in tropical regions. Various pathogens have been implicated in this deterioration, both prior to harvest and during storage, but their cardinal importance in seed decay is, under certain circumstances open to question. Here, we shall discuss the apparent role that pathogens, particularly seedborne fungi, play in affecting soyabean seed quality.

Although bacteria (Nicholson and Sinclair, 1971; Tenne *et al.*, 1974; Ellis *et al.*, 1977) have been shown to reduce the potential rate of germination of soyabean seed and are thus implicated in seed deterioration, the fungi are the predominant seed pathogens of soyabean. Viruses have also been associated with decreased germination (Quinones *et al.*, 1971) but there is now evidence that this may be due to predisposition to *Phomopsis* rather than to a direct effect of virus on seed viability (Hepperly *et al.*, 1979). Early work tended to focus on deterioration in storage where fungi, especially species of *Aspergillus*, were isolated, their incidence being related to the moisture content of seed as well as to temperature. Species of *Alternaria, Fusarium*, and *Penicillium* were also often found (Tervet, 1945; Kennedy, 1964; Christensen and Dorworth, 1966; Dhingra *et al.*, 1973). Although invasion by such fungi has often been shown to be related to decreased seed viability, there is abundant evidence to suggest that seed deterioration during storage may also have non-pathological causes. Indeed, fungi often may be a consequence, not the principal cause, of seed deterioration during storage. Recent work has confirmed that seedborne fungi play a role subsidiary to physiological factors, enhanced by high temperature and moisture content, in soyabean seed deterioration during storage (Ndimande *et al.*, 1981).

Conversely, seedborne fungi are an important factor in soyabean seed decay prior to harvest. Although it had long been appreciated that species of *Phomopsis* (including the pycnidial state of *Diaporthe phaseolorum*) were seedborne in soyabean, research was initially preoccupied with the importance of these fungi as the causes of pod and stem blight, and stem canker. Only recently has it been appreciated that seed decay, with which *Phomopsis* spp. are most commonly associated, can occur independently of pod and stem blight and canker (Kmetz *et al.*, 1978). *Phomopsis* spp., which are internally seedborne, are known to impair both germination and emergence in soyabean, especially when their incidence in seed exceeds 25 per cent (Nicholson *et al.*, 1972; Chamberlain and Gray, 1974; Ellis *et al.*, 1974). Delayed harvesting tends to increase the prevalence of *Phomopsis* in seed (Ellis and Sinclair, 1976; Kmetz *et al.*, 1978; Paschal and Ellis, 1978; Ndimande *et al.*, 1981), and the relative prevalence of *Phomopsis* and other seedborne fungi may vary with season and locality (Hepperly and Sinclair, 1981), probably being regulated largely by the prevailing moisture conditions at harvest (Shortt *et al.*, 1981). Weed development can also affect the incidence of *Phomopsis* and other seedborne fungi in soyabean seed (Dhingra and da Silva, 1978). Other species of fungi which may have a role in seed decay prior to harvest include *Colletotrichum truncatum, Macrophomina phaseolina, Fusarium* spp., and *Cercospora kikuchii* (Hepperly and Sinclair, 1981; Ndimande *et al.*, 1981). *C. kikuchii* may be antagonistic to *Phomopsis* in both pods and seed (Roy and Abney, 1977), and this may explain why purple-stained seed (caused by *C. kikuchii*) may show higher germination than unstained seed, when the latter has a higher incidence of *Phomopsis* (Hepperly and Sinclair, 1981). Soyabean mosaic virus infection predisposes to seed infection by *Phomopsis* (Hepperly *et al*., 1979).

Some degree of protection against seed deterioration may be afforded by cultural measures such as prompt harvesting at maturity, the avoidance of mechanical damage, and the use of suitable storage conditions. The field application of fungicide sprays may reduce the incidence of seedborne fungi (Ellis and Sinclair, 1976) but efficacy may depend on environmental conditions (Ndimande *et al*., 1981). However, soyabean genotypes vary in seed quality characteristics, and it appears that certain tropical Asian cultivars have superior seed quality (IITA, 1976; Paschal and Ellis, 1978) which may be exploited in breeding. Methods have now been developed enabling the breeding for resistance to field weathering, which may logically include the identification of resistance to specific pathogens, especially *Phomopsis* (IITA, 1980; Ndimande *et al*., 1981).

5.2.2 Rust

Soyabean rust is widely distributed and of economic importance in tropical and sub-tropical Asia, and in many sub-tropical and maritime regions of the Pacific including eastern Australia. Although the causal fungus, *Phakopsora pachyrhizi*, is also widely distributed elsewhere, in these regions (central and south America; Africa) the pathogen is more commonly found on legume hosts other than soyabean. The extent to which *P. pachyrhizi* is a potential threat to soyabean production in the USA and Brazil, as well as in parts of tropical Africa, is a subject of current debate. Aspects of the epidemiology of soyabean rust have been discussed elsewhere (see Section 2.2.5); here, we are concerned with the symptoms and crop loss caused by rust, and its control in soyabeans.

The uredial stage is much the most common. After urediospore germination, infection occurs directly through the host cuticle (Bonde *et al*., 1976), leading to the appearance of chlorotic spots on the infected leaves. These lesions enlarge, becoming angular and tan to dark brown in colour. From 6–10 days after infection, uredial pustules develop on lower leaf surfaces (Figure 20). Pustules, which are not erumpent but sub-epidermal, arise as translucent blisters from the surface of a necrotic lesion, developing into a cone covered by the host epidermis. The pustule opens through a pore through which the hyaline urediospores are extruded in clumps (Bromfield, 1978; Vakili, 1978b).

Rust causes premature defoliation and increases the proportion of unfilled pods; both seed size and the number of seeds per plant are affected, and seed viability is impaired (Bromfield, 1976). Crop loss estimates are variable, but all confirm the potential economic importance of rust. Losses in individual

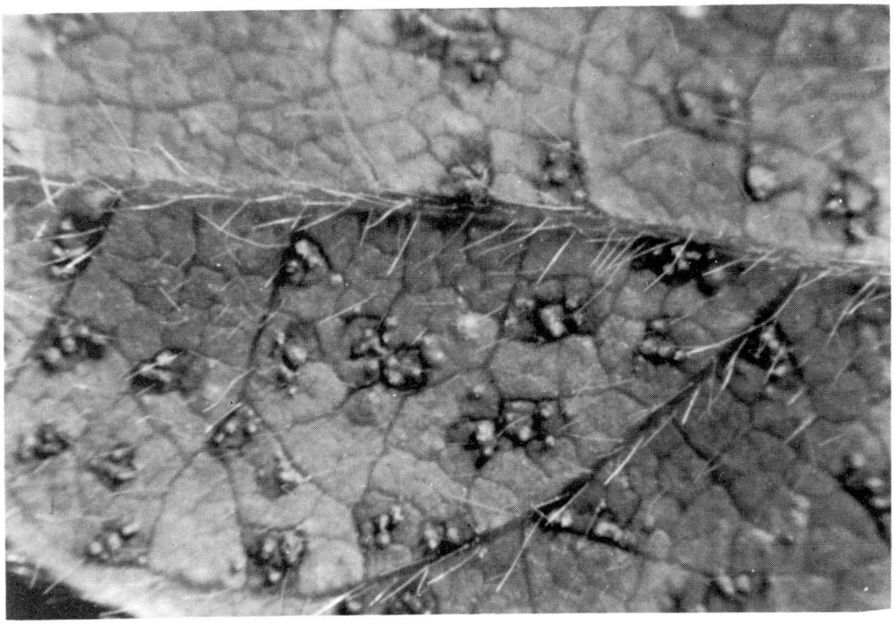

Figure 20 Uredia of the rust fungus, *Phakopsora pachyrhizi*, on the lower surface of soyabean leaves (photograph courtesy J. B. Sinclair)

fields are in the range of 15–80 per cent, depending on the locality, season, and cultivar. In Taiwan, annual rust losses may amount to 20–30 per cent of the total soyabean production (Liu, 1966, cited by Bromfield, 1976). Work at AVRDC has confirmed that yield losses of 50–70 per cent may occur in susceptible cultivars (Yang, 1978) and complete crop loss is occasionally reported in unadapted material.

No fully satisfactory measure of control has been developed. Both foliar application and seed dressing with fungicides have been tested, and chemical control may be economically justified where rust is severe. When more is understood of the biology of the pathogen, there may be potential for the development of cultural practices, possibly including the control of alternate weed hosts (Sinclair, 1978). The most promising approach to rust control is the breeding of resistant cultivars and their use, possibly in conjunction with cultural and chemical control measures. The expression and mechanisms of rust resistance and its use in soyabean improvement are discussed later (Section 5.3).

5.2.3 Mosaic

Mosaic, caused by soyabean mosaic virus (SMV), is very widely distributed, occurring in most parts of the world where viruses have been identified in soyabeans. The symptoms induced by SMV, described by Sinclair and Shurtleff (1975), are not especially characteristic. They vary with the source and time of infection, prevailing temperature, virus strain, and host genotype. Seedborne infection may produce seedlings whose primary leaves are distorted and prematurely chlorotic, while trifolilate leaves tend to be mottled and rugose. Early field infection of susceptible cultivars can cause stunting, and leaves are typically puckered. In certain cultivar-isolate combinations, a systemic necrosis and bud blight may develop (Ross, 1969; Cho *et al.*, 1977; Cho and Goodman, 1979). Symptoms may develop on pods which may become misshapen and contain fewer, smaller, and discoloured seed. Root nodulation may also be adversely affected. Soyabean plants may become doubly infected by SMV and other viruses. In the presence of bean pod mottle virus, SMV-infected plants may develop a top necrosis (Lee and Ross, 1972), and mixed infections of the two viruses can reduce nodulation more than single infections do (Tu *et al.*, 1970a). Natural infection with SMV can lead to yield losses of 20–30 per cent (Ross, 1977), while mechanical inoculation of seedlings can cause from 45–86 per cent loss (Goodman and Oard, 1980).

SMV has a relatively narrow host range and is thought to persist between seasons chiefly in seed (Hill *et al.*, 1980). The production and use of SMV-free seed is therefore a potentially important mosaic control measure, as discussed by Irwin and Goodman (1981). Other cultural methods may possibly include the manipulation of ground cover to control the field spread of SMV (Halbert and Irwin, 1981), in a manner similar to the control practices advocated for groundnut rosette (see Section 4.2.4). However, unlike rosette, SMV is transmitted in the non-persistent manner by numerous species of aphid, principally transient alatae (Halbert *et al.*, 1981). Since response to ground cover varies with aphid species (Halbert and Irwin, 1981), a knowledge of local vector prevalence would be an essential prerequisite for successful manipulation of ground cover.

There is great potential for the control of SMV through breeding for resistance. Sources of resistance have been identified amonst soyabean germplasm in Japan (Koshimizu and Iizuka, 1963), Taiwan (AVRDC, 1978), and the USA (Ross, 1969; Goodman *et al.*, 1981).

5.2.4 Bud blight

Three distinct viruses cause bud blight of soyabeans; these are tobacco ringspot virus (TRSV), tomato ringspot virus, and tobacco streak virus (TSV). All three occur in the USA but TRSV is the most important (Kahn and Latterell, 1955; Fagbenle and Ford, 1970). In south-eastern Brazil, TSV is of considerable local importance (Costa and Carvalho, 1961). The bud blight caused by TRSV, which we shall consider first, has been reported from Australia, China, and Canada as well as the USA. Tomato ringspot virus produces symptoms indistinguishable from those caused by TRSV (Sinclair and Shurtleff, 1975), but it is apparently a less important cause of bud blight than the other two viruses and will not be discussed further. The symptoms induced by TRSV, described by Allington (1946) and Kahn and Latterell (1955), include the curving of the terminal bud. Young leaves may be bronzed and some defoliation may occur. Subsequently, the growing point becomes necrotic and as bud blight develops the stem pith becomes discoloured. Stunting may occur, both vegetative and reproductive buds may proliferate and pods may abort. Early infection leads to reduced seed yield and impaired seed quality. Infection at flowering may cause pod damage, including a blotching. Seed transmission is important in the spread of TRSV, the rate depending on the time of infection. Up to 100 per cent transmission may occur (Athow and Bancroft, 1959), but at least 50 per cent seed infection is necessary significantly to affect soyabean seed yield in a subsequent season. Below this incidence, it seems that the remaining healthy plants can compensate by competing out severely diseased plants (Athow and Laviolette, 1961). Nevertheless, crop loss may exceed 50 per cent from natural infection, in addition to which TRSV impairs seed viability and tends to increase the incidence of purple-stained seed (Crittenden *et al.*, 1966). Horn *et al.* (1973) estimate that the economic injury threshold of TRSV infection in soyabeans is between 20 and 40 per cent field incidence. At 40, 80, and 100 per cent incidence, TRSV caused 33, 53, and 76 per cent crop loss respectively. Owing to the very wide host range of TRSV, control measures rely on cultural practices including isolation of soyabeans from other susceptible crops and the destruction of alternate weed hosts (Sinclair and Shurtleff, 1975). Despite the recognition of various vectors (see Table 17), none appears to be particularly efficient, and the field importance of the various modes of transmission remains largely unresolved. Significant variation in seed transmission rates between cultivars has not been found (Demski and Harris, 1974), although there may be scope for the development of soyabean cultivars with some resistance to bud blight (Orellana, 1981).

The symptoms of Brazilian bud blight are effectively indistinguishable from those induced by TRSV, although the disease is found only on more mature plants. The disease can be destructive in south-eastern Brazil. No vector of TSV has been identified, but the virus is seedborne up to about 30 per cent. Considerable differences occur between soyabean cultivars in seed transmission (2.6–30 per cent; Ghanekar and Schwenk, 1974), but resistance to seed transmission of TSV has not been sought.

5.2.5 Whitefly-transmitted virus diseases

There is growing evidence that various whitefly-transmitted viruses are the cause of damaging diseases of soyabeans, as well as other legumes, throughout the tropics. Brunt and Kenten (1973), who first described cowpea mild mottle virus (CMMV) from Ghana, showed that artificial inoculation of soyabean with CMMV led to severe disease, and that the virus was seedborne in soyabean. Later, CMMV was identified as the cause of a severe disease of soyabean in Thailand (Brunt and Phillips, 1978) and the Ivory Coast (Thouvenel *et al*., 1982). Recent work has shown that CMMV is transmitted, in a semi-persistent manner, by the whitefly, *Bemisia tabaci* (Iwaki *et al*., 1981; D. V. R. Reddy, personal communication 1981). In soyabean field trials, both in south-eastern Nigeria and central Ghana, it has been observed that stunted plants with cupped leaves may be associated with breeding populations of *B. tabaci* (D. J. Allen, unpublished) but whether CMMV was the cause is as yet uncertain. However, it is evident that CMMV, which is the first carlavirus shown to be whitefly-transmitted, is widespread in Africa and Asia; the virus has a relatively wide natural host range and is seed-transmitted. It seems probable that CMMV will soon be shown to be of considerable economic importance to tropical soyabeans.

There is also increasing evidence that mung bean yellow mosaic virus (MYMV), probably a geminivirus (Thongmeearkom *et al*., 1981), causes an important yellow mosaic of soyabeans in various parts of Asia, including India, Sri Lanka, and Thailand (Suteri, 1974; Goodman *et al*., 1981; Iwaki *et al*., 1981). MYMV is also transmitted by *B. tabaci* but unlike CMMV is not seedborne. Sources of resistance to MYMV have been identified in soyabean (Singh *et al*., 1974a).

Various other virus-like diseases are reported. Soyabean leaf crinkle is a new disease reported from Thailand, where it is transmitted by *B. tabaci* in the presistent manner. Affected plants are characterized by a yellow netting and enation of veins, especially on lower leaf surfaces (Iwaki *et al*., 1981). A 'crinkle mosaic' is caused by abutilon mosaic virus (AMV) in Brazil where it is of relatively minor importance to soyabean. It has been suggested that the disease may increase in significance in areas where AMV is abundant, and where there is a high density of the whitefly vector. Artificial inoculation with viruliferous whiteflies can cause losses in yield of 13–87 per cent, depending on the cultivar of soyabean (Costa, 1955; 1975). 'Dwarf mosaic' of soyabean in Brazil is caused by euphorbia mosaic virus, which is also whitefly-transmitted (Costa, 1975). *B. tabaii* is also thought to be the vector of an unidentified yellow mosaic in Venezuela (Debrot and Ordosgoitti, 1975). Virus-like leaf crinkle in soyabean can also be induced by soil chemical factors (Tollenaar and Martin, 1972; Delassus, 1978).

Until more is known about the aetiology of these whitefly-transmitted virus diseases, assessment of their distribution and economic importance is liable to be inaccurate. But there is sufficient evidence to suggest that this group of diseases is a considerable threat to tropical soyabeans.

5.2.6 Bacterial blight and pustule

Bacterial blight of soyabeans is caused by *Pseudomonas syringae* pv. *glycinea* which is distributed widely, though often limited to the cooler conditions of higher altitude regions within the tropics. Bacterial pustule, caused by *Xanthomonas campestris* pv. *glycines*, is less limited by high temperatures and is the more common disease under lowland tropical conditions, although both may occur together. The symptoms of the two diseases are not easy to distinguish. The first symptoms of bacterial pustule appear on the leaves as minute, pale green lesions with raised centres (Figure 21); bacterial blight lesions tend to be angular, translucent, and water-soaked, not raised in the centre. As the lesions develop and coalesce, leaves affected by either of these diseases often become torn and ragged (Figure 22), and fall early (Sinclair and Shurtleff, 1975). However, estimates of the crop loss caused by bacterial pustule do not exceed 15 per cent in susceptible cultivars (Hartwig and Johnson, 1953; Laviolette *et al*., 1970).

Factors influencing the survival and spread of these pathogens are discussed in Section 3.3. Since both pathogens are seedborne, the production and use of clean seed can be an important control measure. Chemical control of bacterial pustule has been investigated (Rao and Patel, 1973b), but the use of resistant cultivars is the best approach to control of both these diseases. Resistance to bacterial pustule has been identified in soyabean cultivars in the USA (Lehman and Woodside, 1929), and more recently in India (Patel *et al*., 1972) and Nigeria (Kueneman and Allen, 1980). Sources of bacterial blight resistance are also available (Kennedy and Cross, 1968; Tisselli *et al*., 1980). The development of resistant cultivars in breeding programmes is discussed in the following section.

Figure 21 Raised lesions of bacterial pustule, caused by *Xanthomonas campestris* pv. *glycines*, on lower (leaf) and upper surfaces of soyabean leaves (photograph courtesy J. B. Sinclair)

Figure 22 Coalesced lesions of bacterial blight, caused by *Pseudomonas syringae* pv. *glycinea*, on soyabean leaves which have become ragged (photograph courtesy J. B. Sinclair)

5.3 DISEASE RESISTANCE AND SOYABEAN IMPROVEMENT

Extensive screening for disease resistance has been carried out largely in the USA, and numerous sources of resistance have been identified. The available sources of resistance to 19 of the diseases of soyabean are summarized in Tisselli *et al*. (1980). Resistance, of various types, has been found against rust (Bromfield and Hartwig, 1980), powdery mildew (Demski and Phillips, 1974), downy mildew (Dunleavy and Hartwig, 1970; Bernard and Cremeens, 1971), phytophthora rot (Athow *et al*., 1974), pythium seed rot (Keeling, 1972), sclerotium blight (Agrawal and Kotasthane, 1971; Quebral and Pua, 1976), brown stem rot (Chamberlain and Bernard, 1968), cercospora leaf spot (Athow and Probst, 1952) and purple seed stain (Wilcox *et al*., 1975). Degrees of lesser susceptibility to septoria brown spot have also been recognized (Lim, 1979). Resistance is also available to bacterial blight (Kennedy and Cross, 1968), bacterial pustule (Lehman and Woodside, 1929; Patel *et al*., 1972), soyabean mosaic virus (Ross, 1969; Cho and Goodman, 1979), peanut mottle virus (Demski and Kuhn, 1975), cowpea chlorotic mottle virus (Boerma *et al*., 1975), bean yellow mosaic virus (Provvidenti, 1975), mung bean yellow mosaic virus (Singh *et al*., 1974a) and bud blight (Orellana, 1981). Sources of resistance to both the soyabean cyst nematode (Chambers and Epps, 1966) and root knot nematodes (Williams *et al.*, 1973) have also been identified. As yet, there is no adequate source of resistance to anthracnose, nor to the root rots caused by *Pythium, Rhizoctonia*, and *Fusarium* spp. (Yang, 1980).

In the USA the identification of resistance has, in many cases, led to the development of productive cultivars with high levels of resistance (Hartwig, 1975). In some instances, a single source of resistance has provided durable protection for many years; the cultivar Clemson Non-Shatter (CNS) shows no evidence of bacterial pustule development and has been used as a source of resistance in breeding programmes over the last 30 years (Hartwig and Lehman, 1951; Tisselli *et al.*, 1980; Jindal *et al.*, 1981). In other cases, resistance has been of short duration, and new races of the pathogen have been continuously reported. The latter is well exemplified by phytophthora rot (Laviolette and Athow, 1977), downy mildew (Dunleavy and Hartwig, 1970; Dunleavy, 1971), cercospora leaf spot (Ross, 1968b; Yorinori and Homechin, 1978), and both cyst and root knot nematodes (Epps and Duclos, 1970; Williams *et al.*, 1973). Let us now examine what is known of the nature of resistance to the principal pathogens of tropical regions before attempting to assess the progress and prospects for breeding for a resistance which will endure. We will consider rust resistance first.

Despite the availability of fairly extensive germplasm collections of *Glycine*, relatively few sources of rust resistance have been identified. One of the earlier sources (cv. Komata, PI 200492) to be used in the development of rust-resistant cultivars (Tainung 3 and Tainung 4) in Taiwan (Chan *et al.*, 1971) was subsequently found to have an inadequate level of resistance under field conditions, and crop losses of up to 25 per cent occurred in supposedly rust-resistant material (Yang, 1978). The accession PI 200492 has also been used in breeding programmes in the USA (Hartwig, 1978) and in Australia (McLean and Byth, 1976) but again there are indications that its resistance may not endure. What then is known of the nature and extent of pathogenic variation within *P. pachyrhizi*? It is evident that host specificity occurs within *P. pachyrhizi*, probably indicating that distinct *formae speciales* exist. Lin (1966) compared nine single spore isolates from soyabean in Taiwan. He found that, although they differed little in pathogenicity to soyabean, they could be divided into six distinct groups based on their virulence on three other hosts, catjang (*Vigna unguiculata* ssp. *cylindrica*), bean (*Phaseolus vulgaris*), and yam bean (*Pachyrrhizus erosus*). An isolate of *P. pachyrhizi* from Puerto Rico apparently lacks virulence on soyabean (Bromfield *et al.*, 1980) and cowpea (Vakili and Bromfield, 1976), while *P. pachyrhizi* is a common field pathogen of cowpea but not of soyabean in Nigeria (see Section 2.2.5). The pasture legume, *Neonotonia wightii*, is a host of *P. pachyrhizi* in Australia (Deverall *et al.*, 1977) but an Indian population lacks virulence on this host (Singh *et al.*, 1974). While it remains to be established whether or not certain host-adapted forms of *P. pachyrhizi* possess the capacity to become adapted to soyabean, it is evident that isolates do differ both in virulence and aggressiveness (*sensu* van der Plank, 1968) on soyabean. Isolates from Taiwan, India, Australia, and Puerto Rico were found to differ in virulence on a range of soyabean genotypes. The same isolates also differed in aggressiveness on a single soyabean cultivar (cv. Wayne) (Bromfield *et al*., 1980). From measurement of epidemiological parameters during a single pathogen life cycle, differences were found between isolates in infection ratio (the number of lesions per unit leaf area per unit inoculum) but not in latent period (the time required from penetration until the appearance of lesions). On this basis, the Indian isolate was the most aggressive and the Australian isolate the least aggressive. However, in terms of disease over the entire developmental cycle, which also takes into account the isolates' potential for secondary spore production, it was found that total sporulation in the Taiwanese and Indonesian isolates was greater than that of the Indian isolate. The Australian isolate produced substantially fewer spores than the other isolates (Melching *et al*., 1979).

While it is thus evident that pathogenic variation occurs within *P. pachyrhizi*, the implications for breeding of the differences in virulence between isolates is hard to assess, at least until more is known about the existence of *formae speciales*. Let us turn now to examine the available information on the expression and underlying mechanisms of resistance to rust in soyabean.

Three categories of response to infection by *P. pachyrhizi* are recognized in soyabean. These reactions have been referred to in a variety of ways, but essentially they comprise an immunity, a conspicuous hypersensitivity and susceptibility (Singh *et al*., 1974a; Thapliyal and Choudhary, 1976; McLean, 1979; Bromfield *et al*., 1980). All three infection types involve the development of cellular necrosis, and phytoalexin production is a feature both of susceptible and resistant reactions (Deverall *et al*., 1977). Histological comparisons of the three types of host-parasite interaction have been made in Australia (McLean, 1979). It was found that although 'immune' hosts showed no macroscopic symptoms, at the microscopic level they exhibited a hypersensitive response which was limited to a few cells. In the second category, the hypersensitive response was less limited, and fungal development and associated host necrosis were more extensive, so that symptoms of infection were visible macroscopically. In immune hosts, pre-penetration processes, penetration, the early spread of hyphae and development of haustoria, and necrosis of the penetrated epidermal cell were all similar to those in susceptible hosts. However, within 2–3 days of infection, both colony size and the number of haustoria were smaller in immune than susceptible hosts. Differences in the rate of development of cellular necrosis were detected between cultivars within the 'immune' category (McLean, 1979). Overtly hypersensitive cultivars may or may not bear uredia, apparently depending on the isolate (Bromfield *et al*., 1980; Keogh *et al*., 1980). Thus, these infection types, which are necessarily arbitrary, may differ only in degree; they do so not only in histological terms but also physiologically. Phytoalexins, which comprise at least two fractions probably containing related pterocarpanoid compounds (Deverall *et al.*, 1977), may form earlier in resistant cultivars relative to a susceptible one. However, even in a rust-susceptible host, there is a degree of incompatibility between the infection hyphae and the first host cells contacted. In view of the early formation of phytoalexins, this implies that phytoalexins may be produced by live soyabean cells around the infection site, and that the mycelium may be able to tolerate and grow in a phytoalexin-rich environment (Deverall *et al.*, 1977; Keogh *et al.*, 1980). These studies show that there are several different possible fates for infecting hyphae, suggesting there are several distinct defence mechanisms to rust in soyabeans.

Too little is known to be able to predict the potential stability of these resistance mechanisms. It has been suggested that the resistance of cv. Komata (PI 200492) has 'broken down' in Taiwan, and a Taiwanese isolate possesses virulence on cv. Komata under experimental conditions (Bromfield *et al.*, 1980). The same cultivar is immune to one Australian isolate and susceptible to another (McLean and Byth, 1976; McLean, 1979). Yet there is also evidence that cv. Komata possesses some resistance even when its apparently race-specific resistance is inoperable. Using a Taiwanese isolate, Marchetti *et al.* (1975) found that the latent period was longer and the sporulation period shorter in cv. Komata than in the rust-susceptible cv. Lee 68, suggesting that the former cultivar would tend to reduce the rate of rust spread. This appears to be borne out by Yang (1978) who found that the crop loss in cv. Tainung 4, a derivative of cv. Komata, was about half that obtained in a susceptible control cultivar. Other accessions (PI 230970, PI 230971) possess hypersensitive resistance to several diverse isolates (Bromfield *et al.*, 1980).

Such work as has been done on the inheritance of rust resistance in soyabean suggests that one or occasionally two dominant genes are involved. Breeding studies in Australia provided evidence that the resistance of cv. Komata was simply inherited, and the same was found in the case of cv. Ankur in India, as well as PI 230970 and PI 230971 under controlled environment conditions in the USA (Bromfield and Hartwig, 1980). Work in Indonesia has provided evidence that resistance in some genotypes may be conditioned by two dominant genes, both of which must be present for resistance to be expressed (Sumarno and Sudjadi, 1977). A selection in the F_8 generation from a cross between Nung-shih H-ll and cv. Komata was found to combine high-yield potential with rust resistance in Taiwan; after evaluation in regional trials, the selection was released to commerce in 1971 as cv. Tainung 4 (Chan *et al.*, 1971). However, the level of resistance cv. Tainung 4 possesses is now regarded as inadequate, perhaps because of the emergence of a new pathogenic race as discussed above, and new sources of resistance have been sought. Accessions PI 230970 and PI 230971 are currently among the most promising (AVRDC, 1979), although they each combine rust resistance with agronomically undesirable characters. Hartwig (1978) advocates using a modified backcrossing programme to incorporate their rust resistance, preferably screening early generation material under field conditions in south-eastern Asia where rust development would be comparatively consistent. Mutation breeding for rust resistance has also been attempted (Lu, 1970; Tsai *et al.*, 1974), and wild *Glycine* spp. have been evaluated for their rust reaction (Singh *et al.*, 1974a; Burdon and Marshall, 1981). Although certain accessions of *G. tabacina, G. tomentella*, and other species appear to possess high levels of rust resistance in India and Australia, their potential value as sources of resistance genes for soyabean breeding is limited by sterility barriers between the sub-genera of *Glycine*. Numerous attempts to cross *G. max* with species in the sub-genus *Glycine*, to which the rust-resistant species belong, have proved unsuccessful (Hymowitz and Newell, 1980). Until renewed attempts are made to overcome incompatibility within the genus, wild species will not play a significant part in contributing rust resistance to soyabeans, thus contrasting with the position in groundnut improvement (see Section 4.3). Perhaps the best approach would be to intercross lines possessing moderate levels of resistance, especially those known to rely on distinct mechanisms of rust resistance, in the hope that lines with higher levels of durable resistance could be selected from amongst the progeny.

Studies on the interaction between soyabean cultivars and soyabean mosaic virus (SMV) isolates suggest there are different genetic systems, possibly different genes, for SMV resistance (Ross, 1969; Cho and Goodman, 1979; Kiihl and Hartwig, 1979). Several distinct types, or components, of resistance (*sensu* Russell, 1978) to SMV can be recognized. Resistance to virus infection, often called ‘immunity’, has been identified in plants which remain symptomless on artificial inoculation; virus cannot be retrieved from the inoculated plant. In other soyabean-SMV reactions, infection occurs but the virus remains localized and does not spread, and is often expressed as hypersensitive necrosis (Ross, 1969). In some instances, SMV-resistant cultivars develop a severe systemic necrosis and bud blight when infected with certain pathogenic variants of SMV (Cho *et al.*, 1977), perhaps favoured by high temperatures (Cho and Goodman, 1979). Although the genetic basis of this reaction is not fully understood (Kiihl and Hartwig, 1979; Kwon and Oh, 1979), soyabean necrosis superficially resembles the ‘black root’ reaction in *Phaseolus vulgaris* (see Section 6.2.4). Nor are the breeding implications of systemic necrosis sufficiently clear. But it appears that cultivars resistant to certain necrosis-inducing strains of SMV are susceptible to others. While Ross (1969) found that cv. Ogden and PI 96983 were not infected by any isolate, including the most pathogenic (SMV-3) obtained from a plant with systemic necrosis, Kiihl and Hartwig (1979) found that cv. Ogden, but not PI 96983, was susceptible to a necrotic isolate. Cho and Goodman (1979) found that PI 96983 developed systemic necrosis on inoculation with their isolate SMV-G7, indicating that neither of these sources alone is likely to provide durable protection. There is also evidence that SMV tolerance occurs. Accessions of soyabean shown to be symptomless carriers of SMV have been identified (AVRDC, 1978), and a second kind of tolerance, in which certain cultivars show no significant loss in seed yield despite SMV infection, has also been noted in soyabean (Ross, 1977). Recent results have revealed that accessions which generally showed less yield reduction tended to be those with lower rates of seed transmission (Goodman and Oard, 1980). In view of the importance of seed transmission in the survival of SMV, emphasis has been given to the screening for 'resistance to seed transmission' at INTSOY. Although SMV was seedborne in all the accessions tested, five had transmission rates consistently below 0.25 per cent, relative to 2–20 per cent found in the other accessions (Goodman and Oard, 1980). There may also be potential for SMV control by use of resistance to aphid vectors. Resistance to *Aphis glycines*, one of the many vectors of SMV, is being sought (AVRDC, 1979), but the effects of aphid resistance on SMV spread have apparently not been investigated. There is evidence that resistance to SMV infection, which may be governed by single dominant genes (Koshimizu and Iizuka, 1963; Goodman *et al.*, 1981) and resistance to seed transmission are each virus isolate-specific so that neither may be stable under field conditions. Until more information is available, a breeding programme which aimed at combining the components of SMV resistance would seem a sound strategy. The identification of tropically adapted soyabean accessions which combine tolerance of SMV infection and resistance to seed transmission is a good start in this direction (Goodman and Oard, 1980; Bowers and Goodman, 1982), provided these two characters can be satisfactorily combined. Genotypes with resistance to SMV infection are free of seedborne infection, and the degree of resistance to seed transmission is therefore masked. Kueneman (personal communication 1981) suggests that if a recurrent parent with resistance to seed transmission was used, then resistance to infection could be incorporated by backcrossing from a donor, but combination of the two characters would be difficult to confirm.

Resistance to bacterial blight in soyabeans relies chiefly on the hypersensitive reaction (Kennedy and Cross, 1968) with which the phytoalexin hydroxyphaseollin and related isoflavonoids are associated (Keen and Kennedy, 1974). Such resistance is race-specific, and eight pathogenic races of *Pseudomonas syringae* pv. *glycinea* are now recognized (Cross *et al.*, 1966; Thomas and Leary, 1980). Resistance is governed by a single dominant gene, or by more than one gene, depending on the race involved (Mukherjee *et al.*, 1966). Conversely, resistance to bacterial pustule has proved to be durable. Resistance was recognized by Lehman and Woodside (1929) and later shown to be controlled by a single recessive gene (Hartwig and Lehman, 1951; Rao and Patel, 1973). The cv. Clemson Non-Shatter (CNS) has maintained its level of resistance, which is effective not only in the USA but also in India (Patel *et al.*, 1972). The stability of resistance apparently relates to the lack of significant pathogenic variation in *Xanthomonas campestris* pv. *glycines* (Jindal *et al.*, 1981), but whether the durability of resistance is the cause or consequence of the pathogen's stability is not known. Bacterial pustule resistance is generally sufficient to obviate the need for specific breeding. Selection should be made against such susceptibility as may be inadvertently introduced in breeding for other characters (Hartwig, 1975). For this purpose, it is important to ensure uniform disease pressure, perhaps through the use of a simple field inoculation technique (Jones and Hartwig, 1959). However, there is evidence that the relative precision of selection for bacterial pustule resistance in segregating populations of soyabean may rely more on the choice of an appropriate susceptible cultivar and its arrangement in breeding nurseries as a disease spreader than on artificial inoculation alone (Kueneman and Allen, 1980).

Sources of combined resistance to disease are available, both amongst named cultivars and in germplasm accessions. A Japanese cultivar, cv. Okushirome, is resistant to at least two fungal pathogens (*Sphaceloma glycines* and *Cercospora kikuchii*), two viruses (soyabean mosaic and soyabean stunt), and to a nematode (*Heterodera glycines*) (Matsumoto *et al*., 1974, cited by Russell, 1978). Soyabean cv. CNS possesses resistance to *Phytophthora* rot, downy mildew, cercospora leaf spot, purple seed stain, bacterial pustule, peanut mottle virus, and cowpea chlorotic mottle virus, while cv. Harosoy combines resistance to *Phytophthora* rot, *Sclerotium* and *Sclerotinia* stem rots, downy mildew, cercospora leaf spot and root knot. Many other genotypes possess combined resistance to two or more diseases (Tisselli *et al*., 1980). One of the principal objectives of AVRDC's soyabean breeding programme is to incorporate relevant disease resistance into high-yielding and tropically adapted backgrounds. Sources of resistance to bacterial pustule, downy mildew, and purple seed stain have been identified at AVRDC, and crosses have been made to combine them. Sources of resistance to soyabean mosaic and rust are also available (AVRDC, 1978; 1979). In some areas of Asia, resistance to mung bean yellow mosaic virus (MYMV) may be an important breeding objective. The soyabean accession PI 171443 has a high degree of MYMV resistance which is controlled by two recessive genes (Singh *et al*., 1974a; Singh and Malick, 1978). In lowland tropical Africa, immediate priorities for breeding are to identify and incorporate resistance to *Phomopsis* and other factors involved in seed deterioration into material possessing high-yield potential and susceptibility to indigenous rhizobia. Selection against susceptibility to bacterial pustule is desirable. Sources of resistance to cowpea mild mottle virus, as well as to peanut mottle virus, pyrenochaeta leaf spot and root knot nematode at medium elevations in eastern Africa, should be sought and utilized. The importance of SMV in Africa is uncertain: the virus is not reliably identified from East Africa although it occurs in Morocco (Lockhart and Fischer, 1976) and Nigeria (IITA, 1981). However, there may be merit in preventive breeding against SMV, and perhaps also rust. Preventive breeding for rust resistance is already well advanced in the USA (Hartwig, 1978), and breeding for resistance to SMV and for improved seed quality are amongst the priorities of INTSOY's programme (Goodman *et al*., 1981).

It appears that adequate resistance is available within *G. max*. The use of wild species of *Glycine*, even if sterility barriers could be overcome, appears not to be a high priority for disease-resistance breeding. The comparative wealth of 'good sources' of disease resistance in soyabeans is associated with the recognition of races of several pathogens, and with race-specific resistance, thus contrasting with the situation in groundnuts (see Section 4.3). Too little is known to be able to predict the epidemiological consequences of using race-specific resistance in breeding tropical soyabeans. But it is reasonable to assume that, whereas the extensive monoculture of genetically uniform cultivars in the USA enhances crop vulnerability (Anon, 1972), the smaller scale production of locally adapted cultivars of soyabeans for food in tropical regions might be expected to decrease vulnerability to epidemics. The most practicable strategy for breeding soyabeans for disease resistance in the tropics is to utilize all available sources of resistance and simultaneously to monitor the potential for host-specific pathogenicity within the pathogen population. In cases where the level of resistance is inadequate, or where the resistance is insufficiently durable, there may be scope for intercrossing and a search for transgressive resistance (Johnson, 1981b). Intercrossing to combine distinct components of resistance to a single disease, as discussed earlier with reference to rust and to SMV, would also appear a sound policy since they often may be expected to reinforce one another.

Chapter 6

Bean Diseases

6.1 EVOLUTION, ADAPTATION, AND PRODUCTION OF BEANS

The genus *Phaseolus* contains about 50 species of New World origin. Other taxa formerly placed in *Phaseolus* include some Asiatic species which are now accommodated in the genus *Vigna;* others of American origin have been transferred to *Macroptilium* (Verdcourt, 1970). There are four cultivated species of *Phaseolus* in the New World: *P. vulgaris,* the common, haricot, navy, French or snap bean; *P. coccineus,* the scarlet runner bean; *P. lunatus,* the lima, sieva or butter bean; and *P. acutifolius*, the tepary bean, all of which are diploid ($2n = 2x = 22$). The four species occupy a range of habitats in the wild, from the cool humid highlands of Guatemala at altitudes of about 1800 m where *P. coccineus* grows, to the arid semi-desert where *P. acutifolius* is found. *P. vulgaris* occurs in an ecological transition zone at 500–1800 m elevation, especially in Mexico and Guatemala, whereas *P. lunatus* is found under lowland tropical conditions (Evans, 1976; 1980a). Here we are concerned only with *P. vulgaris* to which all reference to 'beans' will relate, unless otherwise qualified. The lima, tepary, and scarlet runner beans are relatively minor crops and will not be discussed further, except in relation to interspecific hybridization.

Interspecific hybridization between *P. vulgaris* and the other cultivated species of *Phaseolus* is seldom successful (Smartt, 1970). In general, hybrids of *P. vulgaris* (♀) x *P. coccineus* (♂) can be produced easily while the reciprocal is obtained only with difficulty. Since these reciprocal differences in crossability have not been demonstrated with wild populations, it has been supposed that the unilateral incompatibility has developed under cultivation (Miranda and Evans, 1973). This is not uncommon, particularly when one species is cross-pollinated and the other self-pollinated, as is the case with *P. coccineus* and *P. vulgaris* respectively. *P. vulgaris* has also been hybridized with *P. lunatus* (Honma and Heeckt, 1958) as well as with *P. acutifolius* with the aid of embryo culture (Honma, 1956; Mok *et al.,* 1978). However, the authenticity of some of these claims has been questioned (Smartt, 1979). The use of *P. coccineus* as a bridging species between *P. acutifolius* and *P. vulgaris* allowed

Coyne (1964) to obtain tri-hybrid plants from the cross *P. acutifolius* × (*P. vulgaris* × (*P. vulgaris* × *P. coccineus*)) but the progeny were self-sterile. Very few successful crosses have been made between wild and cultivated species, with the exception of those made between cultivated species and their wild progenitors. However, a successful cross was made between *P. vulgaris* and *P. ritensis* by Braak and Kooistra (1975). Attempts at intergeneric hybridization between *P. vulgaris* and species of *Macroptilium* have produced pods but no hybrids have yet been developed (Evans, 1980a).

Radiocarbon dating of the remains of *P. vulgaris* from Callejon de Huaylas in Peru and from the Tehucan Valley in Mexico places the domestication of beans in the period 8000–5500 BC (Kaplan *et al.,* 1973). There is evidence that domestication also occurred in Brazil and northern Argentina as a wild form from which *P. aborigineus* is a modern survivor. Beans spread widely in post-Columbian times and reached Africa from Brazil with the slave trade. *P. vulgaris* reached Europe during the sixteenth century, and numerous introductions were made from Europe into the USA during the late nineteenth century. Wild populations of both annual and perennial forms are still available, primitive cultivars are found in Central and South America, landraces occur in the African tropics, and a range of more advanced cultivars exist in Europe and North America (Evans, 1976, 1980a). The most obvious effect of domestication in beans, in common with other grain legumes (see Section 1.2.1), is modification of the growth habit which ranges from indeterminate climbers to determinate bush types (Evans and Davis, 1978). The most important changes in the reproductive structure have been in pod and seed size, accompanied by an increase in flower size. Seed number per pod has decreased, while seed permeability has increased with domestication. A wide range of testa colours is found in *P. vulgaris* within which five main colour groups have been identified. Superimposed on these basic ground colours are various patterns of spots and flecks.

Beans are best adapted to cool tropical regions in which the crop is now widespread. They are not grown in hot regions as they are susceptible to drought and are also ill-adapted to areas with very high temperatures or year-around heavy rainfall and high humidity. *P. vulgaris* is also cultivated in temperate regions but the species is frost-sensitive. Sensitivity to photoperiod is common in beans which are then quantitative short-day plants (Wallace, 1980). In many temperate regions, beans are grown for their immature pods which are used as a fresh or frozen vegetable. Cultivars of either dwarf or climbing 'snap bean' types may be yellow-podded ('wax') or green-podded. In certain parts of the tropics, both the leaves and the immature seed are eaten but these are unimportant relative to the use of mature, dry seed which is important in the human nutrition of Latin America and, locally, of eastern and southern Africa. Strong preferences for testa colour and, occasionally, also seed size influence the local production of particular cultivars.

The annual world production of dry seed is about 14.6 million tonnes grown on about 26 million hectares (FAO, 1981). Approximately one-third of this total is produced in Latin America, 82 per cent of which is grown in Brazil and Mexico. Production has increased comparatively little in Brazil and Mexico over the last decade with the consequence that *per capita* consumption in Latin America is falling because production increases lag behind population growth rates. Increased production is therefore urgently needed, preferably obtained in a predominantly subsistence crop (Graham, 1978). Some 80 per cent of Latin America's bean crop is grown in mixed cropping systems (Francis, 1978), especially with maize with which beans initially co-evolved. However, in certain areas, such as the Cauca Valley in Colombia as well as much of Argentina, Chile, and Mexico, beans are produced in monoculture both for local consumption and export. Here, production is highly commercial: certified seed, fertilizer and pesticides are commonly utilized, and more than half the farms in the region rely on irrigation and mechanized land preparation. In general, subsistence farming predominates in the highlands where intercropping is common whereas, at intermediate and low elevations, more intensive monoculture is frequent (Sanders, 1977, H. F. Schwartz, personal communication 1982). Similar contrasts can be found in the production systems of eastern Africa (CIAT, 1980) where beans are the most important food legume at altitudes of about 1000 m; under hotter and often drier conditions at lower elevations, beans tend to give way to cowpeas and other legumes. Again the majority of the bean crop is grown on small farms for local consumption as dry seed; intercropping with maize is widespread. In Uganda, it has been estimated that about 75–90 per cent of all beans are grown in mixtures with coffee, bananas, or cereals on small farms, producing yields of about 400 kg/ha. Landraces of mixed seed type are common (Leakey, 1970a). Essentially similar bean-farming systems are to be found in Kenya (Fisher, 1974), Malawi (Mughogho *et al.,* 1972; Spurling, 1973), Tanzania, eastern Zaire, and north-western Cameroun (D. J. Allen, unpublished). In some areas, such as in the Southern Highlands of Tanzania and in Malawi, despite a unimodal rainfall pattern two successive crops of food beans are grown, the first using a climbing cultivar which is intersown with maize. Since the crop matures under wet conditions, support provided by an intercrop of maize may help to keep the pods from the ground. Bush types are then grown during the main season as a monocrop which matures into the dry season (Spurling, 1973; D. J. Allen, unpublished). In such a system, the time of sowing can strongly influence seed yield (Spurling, 1973). In areas where rainfall is more evenly distributed throughout the year, the sowing date appears to be less critical since almost continuous bean cultivation may be found, as in the Kivu district of Zaire (D. J. Allen, unpublished). Although mixed cropping of beans with maize tends to reduce bean seed yield, the association usually does not greatly decrease maize yields unless moisture is limiting. The combined seed yield of mixed stands, and the value of the produce, are almost invariably higher than those of the pure stand; the bean harvest represents a bonus (Spurling, 1973; Fisher, 1974, 1977; Edje *et al.,* 1976b). In the Arusha region of northern Tanzania, beans are also grown on contract to European seed firms. The cultivars grown are bush types selected for their acceptability in Europe as green ('snap') beans for canning or for the fresh vegetable market. All are produced in monoculture and receive more inputs, including aerial application of insecticide, than does the food bean crop (D. J. Allen, unpublished). Several attempts have been made in various parts of Africa to produce a class of white-seeded ('navy') bean which meets the stringent requirements of the canning trade. In Tanzania, the production of navy beans for export started in 1937 and expanded to more than 2500 tonnes in 1952. Rising interest in the crop attracted inexperienced producers with the result that quality declined rapidly at a time when canners were becoming increasingly demanding. In an effort to keep the industry alive, the cv. Michigan Pea bean, which was widely recognized as ideal for canning, was introduced into East Africa without any careful testing for adaptation. However it was found that, unlike the cv. Comptesse de Chambord which was the principal cultivar grown in the early years, Michigan Pea bean was especially susceptible to rust, and was almost totally destroyed by this disease (Leakey, 1970a). Subsequent work focused on screening a collection of white-seeded beans for suitability for East African production for canning. The cultivar Mexican 142 was among those selected and is now one of the most widely grown navy beans in the region (Macartney, 1966; Leakey, 1970a; Fisher, 1974). The lesson to be learnt here is that it can be dangerous to discount an apparently minor disease during the early stages of selection. The extent to which the low priority accorded powdery mildew at CIAT (CIAT, 1981) may prove to have been unwise is yet to be fully appreciated.

Although beans are notoriously low in yield when compared with the cereals, it is evident that there is considerable potential within *P. vulgaris* for increased yield. The principal restraints to higher yield in the subsistence agriculture of Latin America and eastern Africa include erratic rainfall, low soil fertility, pests and diseases. The importance of pathogens in the production of beans, and the potential for increased production of the crop through disease management are the subjects of the rest of this chapter.

6.2 THE PATHOLOGY OF BEANS

The literature on the diseases of *P. vulgaris* is extensive. An important monograph on bean diseases, with special reference to North American conditions, was published in 1957 (Zaumeyer and Thomas, 1957). Since then, a comprehensive review of recent advances in the pathology of beans in Latin America has become available (Schwartz and Gálvez, 1980), and this chapter relies heavily on that excellent review. A colour-illustrated manual for the field identification of bean diseases in Latin America is also available (Schwartz *et al.,* 1978a). The diseases of beans which prevail in Europe (Hubbeling, 1957; Evans and Davis, 1978) and Australia (Ballantyne, 1971a) have also been reviewed.

As might be expected, the prevalence of bean diseases varies considerably with environment (both locality and season), as well as with the cultivar. In Latin America, beans produced under cool conditions at intermediate to high altitude are often affected by some combination of bean common mosaic virus (BCMV), rust (*Uromyces appendiculatus*) anthracnose (*Colletotrichum lindemuthianum*) angular leaf spot (*Phaeoisariopsis griseola*), halo blight (*Pseudomonas syringae* pv. *phaseolicola*), various root rots, *Ascochyta* leaf spot and white leaf spot (*Pseudocercosporella albida*). Beans produced under hot, dry conditions are liable to infection by BCMV, bean golden mosaic virus (BGMV), rust, angular leaf spot, common blight (*Xanthomonas campestris* pv. *phaseoli*) and root rots; and those grown in a hot, moist climate may be affected by web blight (*Rhizoctonia solani*), common blight, and various rots including those caused by *Sclerotium rolfsii* (Shands, *et al.,* 1964; Echandi, 1976; Schwartz and Gálvez, 1980; Schwartz *et al.,* 1981a). The relative importance of diseases as constraints to production in Latin America has been assessed from surveys by Pinstrup-Andersen *et al.* (1976) and Sanders (1977). A crop loss profile showing the relative importance of different factors to bean production in the Cauca Valley of Colombia revealed that unfavourable rainfall conditions caused the greatest yield loss (22.3 per cent of the total loss), closely followed by the effect of rust (22.0 per cent), common blight (17.5 per cent), leafhopper damage (14 per cent), angular leaf spot (10.3 per cent) and the use of uncertified seed. The estimated economic value for these losses ranged from US$552,000 to US$1.2m for angular leaf spot and rust, respectively. Outside the Cauca Valley where large farms produce beans in monoculture for export, under subsistence farming conditions there were more soil fertility problems, as well as more pests and diseases. Control of diseases, notably BCMV, anthracnose, a stem rot, and angular leaf spot, would have increased the local value of bean production by US$1.1m in one season (Sanders, 1977). Such analyses emphasize the importance of pathogens amongst other production constraints in Latin America where, overall, the principal diseases are considered to be BCMV, BGMV, rust, anthracnose, angular leaf spot, powdery mildew, and common blight (Graham, 1978).

The relative importance of bean diseases under eastern African conditions, first reviewed by Wallace (1939), is less well quantified although again pathogen prevalence reflects ecological conditions. Thus, halo blight and ascochyta leaf spot may be locally important under cool conditions and at high altitude, while common blight is the more important disease at higher temperatures. Under wet conditions, sclerotium blight can be damaging (Leakey, 1970a). Rhizoctonia web blight can also cause significant losses under humid conditions, while floury leaf spot (*Mycovellosiella phaseoli*) and dry root rot (*Fusarium solani* f.sp. *phaseoli*) may also be important under certain conditions (Leakey, 1970a; Mughogho, 1970; Mukunya, 1974). However, throughout eastern Africa, anthracnose, rust, and angular leaf spot are important and these are probably the principal diseases of the region. Anthracnose has been estimated to cause over 92 per cent crop loss in Malawi

(Peregrine, 1971a); rust can cause complete crop loss, especially in white-seeded types for canning (Macartney, 1966); angular leaf spot can cause extensive premature defoliation under wet conditions in Tanzania and Malawi (Hocking, 1967; Peregrine, 1971a), but the extent of crop loss has not been estimated under African conditions. BCMV is also common throughout the region, but it is uncertain how important a production constraint mosaic may be (Kulkarni, 1973). In many parts of eastern Africa, there are two arable rainfed cropping seasons per year. Since farmers usually save their own seed, seedborne diseases, such as anthracnose and the bacterial blights, can become epidemic.

The aetiology, geographical distribution, and relative economic importance of the fungal diseases which affect beans are summarized in Table 18. Considerably more attention has been given to those diseases which prevail under temperate North American conditions than to those elsewhere, but recent work in tropical Latin America, especially at the Centro Internacional de Agricultura Tropical (CIAT) in Colombia, has done much to rectify this imbalance. Furthermore, the major fungal diseases of beans, rust, anthracnose, and angular leaf spot occur in virtually all the bean-producing regions of the world, and research on these is likely to be of wide relevance. These three fungal diseases are discussed in more detail in Sections 6.2.1, 6.2.2, and 6.2.3. Amongst the various root rots, *Fusarium solani* f.sp. *phaseoli* is probably the most important root rot pathogen, at least within the tropics. Aspects of the epidemiology of the dry root rot which this species causes have been reviewed in Section 2.2.2. Amongst the fungal pathogens with restricted geographical distribution, several (e.g. *Cercospora vanderysti, Chaetoseptoria wellmanii, Pseudocercosporella albida*) are not known outside Latin America. Conversely, the bean scab pathogen (*Elsinoë phaseoli*) is known from *P. vulgaris* only in Africa; elsewhere, other legume hosts are involved (see discussion in Section 2.2.4). Numerous other fungi have been recorded from beans (see Zaumeyer and Thomas, 1957; Schwartz and Gálvez, 1980) but none is of importance to bean production in the tropics, with one possible exception. *Aspergillus flavus* has been reported in the Sudan where market samples of beans were found to contain high levels of aflatoxin which is produced by this fungus (Habish, 1972). This subject is discussed further with reference to the groundnut crop, in Section 4.2.3.

There are at least 16 well-characterized and formally described viruses which naturally infect *Phaseolus vulgaris;* these are summarized in Table 19. Only four of these viruses are known to be seedborne in beans, and most are of only sporadic or local importance as bean pathogens. Bean common mosaic virus (BCMV) is much the most important, probably occurring wherever beans are grown. In Latin America, bean golden mosaic virus (BGMV) can also cause extensive losses, especially in hot lowland areas of Brazil and the coastal plains of El Salvador (Gálvez and Cárdenas, 1980). These two virus diseases will be discussed more fully later in this chapter (Sections 6.2.4 and 6.2.5). Curly top is a major virus disease in western USA

TABLE 18 Fungal Diseases of Beans

Disease	Causal fungi	Distribution	Importance/Crop loss	References
1. Seedling diseases				
Damping-off	*Pythium ultimum* Trow., *P. myriotylum* Dresch. and other spp; *Rhizoctonia* sp. and *Sclerotium rolfsii* Sacc.	Widespread	A major limiting factor to production in some areas of USA. Of minor importance elsewhere	Schroth and Cook, 1964; Gay, 1969a; York *et al.*, 1977; Bolkan, 1980
2. Stem and root rots, and wilts				
Charcoal rot (ashy stem blight)	*Macrophomina phaseolina* (Tassi) Goid.	Widespread	Locally and seasonally damaging, especially under warm conditions. Losses of 60–65% reported from USA	Zaumeyer and Thomas, 1957; Schwartz and Gálvez, 1980
Sclerotium blight (southern blight)	*Sclerotium rolfsii* Sacc. (perfect state = *Corticum rolfsii* Curzi)	Widespread in warm temperate and tropical regions	Locally important in Latin America	Bolkan, 1980
Pythium blight (root rot, wilt)	*Pythium aphanidermatum* (Edson) Fitz.; *P debaryanum* Hesse, and other spp.	USA, Latin America	Minor	Harter and Zaumeyer, 1931; Adegbola and Hagedorn, 1969; Hoch *et al.*, 1975; Bolkan, 1980
Rhizoctonia root rot	*Rhizoctonia solani* Kühn (perfect state = *Thanatephorus cucumeris* (Frank) Donk)	Widespread	Locally important. Losses of 10% reported from USA	Zaumeyer and Thomas, 1957; Bolkan, 1980
Texas root rot	*Phymatotrichum omnivorum* (Shear) Dugg.	USA, Mexico	Minor	Bolkan, 1980
Black root rot	*Thielaviopsis basicola* (Berk. and Br.) Ferr.	USA, Europe	Minor	Zaumeyer and Thomas, 1957
Fusarium dry root rot	*Fusarium solani* f.sp. *phaseoli* (Burk.) Snyder and Hansen	Widespread (see Table 6)	Locally, or seasonally, important. Up to 84% crop loss estimated in USA	Harter and Zaumeyer, 1931; Adegbola and Hagedorn, 1969: Hoch *et al.*, 1975; Bolkan, 1980
Fusarium wilt (yellows)	*Fusarium oxysporum* f. sp. *phaseoli* Kendrick and Snyder	Temperate and tropical America	Minor	Bolkan, 1980

3. Foliar and pod diseases				
Rust	*Uromyces appendiculatus* (Pers.) Ung. (= *U. phaseoli* (Pers.) Wint.)	Widespread in all bean-growing regions	Major. Complete crop loss may occur	Zaumeyer and Thomas, 1957; Howland and Macartney, 1966; Ballantyne, 1974b
Anthracnose	*Colletotrichum lindemuthianum* (Sacc. and Magn.) Bri. and Cav.	Worldwide	Major, almost complete crop loss may occur. Over 92% loss estimated in Malawi	Leakey, 1970a; Peregrine, 1971a; Chaves, 1980
Angular leaf spot	*Phaeoisariopsis griseola* (Sacc.) Ferraris (= *Isariopsis griseola* Sacc.)	Widespread	Major. Losses may be as high as 60–80%	Barros *et al.*, 1957; Ferraz, 1980; Schwartz *et al.*, 1981a
Web blight	*Rhizoctonia solani* Kühn	Widespread in warm temperate and tropical regions	Locally important. Occasionally very destructive under wet conditions	Weber, 1939; Gálvez *et al.*, 1980
White mould	*Sclerotinia sclerotiorum* (Lib.) de Bary	Temperate and tropical America, E Africa, Australia	Locally important, especially in temperate N America where average losses may reach 30%, with losses in individual fields as high as 92%. Occasionally also damaging in the tropics	Riley, 1960; Ballantyne, 1971a; Kerr *et al.*, 1978; Schwartz and Steadman, 1980
Floury leaf spot	*Mycovellosiella phaseoli* (Drummond) Deighton (= *Ramularia phaseoli* (Drummond) Deighton)	Tropical America; Eastern Africa, NW India, Malaysia, Papua New Guinea	Potentially important at high altitudes in the tropics where premature defoliation may occur	Cardona-Alvarez and Skiles, 1958; Sohi *et al.*, 1965; Deighton, 1967a, 1974
Alternaria leaf and pod spot	*Alternaria alternata* (Fr.) Keissler (= *A. tenuis* C. G. Nees), and other species	Widespread, reported from the USA, Latin America, Europe and Eastern Africa	Minor, though premature defoliation may occur	Saad and Hagedorn, 1969; Abawi *et al.*, 1977; Russell and Brown, 1977; Ebbels and Allen, 1979; Schwartz and Gálvez, 1980
Ascochyta leaf spot	*Ascochyta phaseolorum* Sacc. (= *Phoma exigua* Desm.) and *A. boltshauseri* Sacc.	Widespread, including the USA, Latin America, Europe and Eastern Africa	Locally important. Occasionally damaging under cool conditions at high altitude. 43% crop loss reported in Colombia	Sprague, 1935; Hendrickx, 1939; Sneep, 1945; Zaumeyer and Thomas, 1957; Leakey, 1970a; Boerema, 1972; Schwartz *et al.*, 1981a

TABLE 18 (*Continued*)

Disease	Causal fungi	Distribution	Importance/Crop loss	References
Phyllosticta leaf spot	*Phyllosticta phaseolina* Sacc.	USA, Latin America, Europe, southern Africa, Japan	Minor	Zaumeyer and Thomas, 1957; Goth and Zaumeyer, 1963; Peregrine and Siddiqi, 1972; Schwartz and Gálvez, 1980
Chaetoseptoria leaf spot	*Chaetoseptoria wellmanii* Stev.	Tropical America	Minor	Muller, 1953; Yerkes, 1956
Scab	*Elsinoë phaseoli* Jenkins	Kenya, Tanzania, Malawi, Zambia, Zimbabwe	Local and minor	Bates, 1957; Angus, 1962–66; Peregrine and Siddiqi, 1972; Mutitu and Mukunya, 1979; D. J. Allen (IMI 237697)
Grey leaf spot	*Cercospora vanderysti* P. Henn.	Latin America	Potentially important, especially at high altitudes	Skiles and Cardona-Alvarez, 1959; Schwartz and Gálvez, 1980
Cercospora leaf spot	*Cercospora canescens* Ell. and Mart., *Pseudocercospora cruenta* Sacc. (Deighton), and other species	Widespread	Minor	Schwartz and Gálvez, 1980
White leaf spot	*Pseudocercosporella albida* (Matta and Bell.) Yoshii and Aamodt	Guatemala, Colombia	Locally important in highland areas where up to 47% crop loss may occur	Yoshii and Aamodt, 1978; Schwartz *et al.*, 1981b
Leaf smut	*Entyloma* sp.	Tropical America	Minor	Schieber and Zentmyer, 1971; Vakili, 1978a
Powdery mildew	*Erysiphe polygoni* DC	Worldwide	Locally important; seed yield losses up to 69%	Zaumeyer and Thomas, 1957; Schwartz *et al.*, 1981b
Phytophthora pod rot ('downy mildew')	*Phytophthora nicotianae* var. *parasitica* (Dast.) Waterh., and *P. phaseoli* Thaxt.	USA, Zaire, India	Minor	Hendrickx, 1939; Venkatakrishnaiya, 1950; Schwartz and Gálvez, 1980
Grey mould	*Botrytis cinerea* Pers. ex Fries	Widespread	Of local and sporadic importance under wet, cool conditions	Campbell, 1949; Dixon and Doodson, 1975; Polach and Abawi, 1975

(Zaumeyer and Thomas, 1957) but it has not been found in tropical regions. Bean yellow mosaic virus (BYMV) is of uncertain status in Latin America (Schwartz and Gálvez, 1980) although its incidence in Chile may be considerable (CIAT, 1980). Several beetle-transmitted viruses are also of potential importance although, as yet, the extent of crop loss they cause under tropical field conditions is largely unknown. In addition to the viruses listed in Table 19, bean rugose mosaic virus, an incompletely characterized comovirus (Fulton and Scott, 1977; Gámez, 1980), may be of importance in Central America. There are also several incompletely known geminiviruses which occur in natural infections of bean. In Australia, bean summer death (Ballantyne, 1971b) is now known to be caused by a geminivirus related to beet curly top, transmitted by a leafhopper (Thomas and Bowyer, 1980). In Latin America, abutilon mosaic and euphorbia mosaic viruses and perhaps other whitefly-transmitted viruses (in addition to BGMV) infect beans (Costa, 1955, 1965; Gálvez and Cárdenas, 1980); and in India mung bean yellow mosaic virus also infects *P. vulgaris* (Singh, 1979).

Beans in eastern Africa remain free of many of these viruses. Although BCMV is common and widespread, BYMV is apparently rare. Despite an early report of a yellow mosaic (Wallace, 1941), presumed to have been BYMV (Zaumeyer and Thomas, 1957), the virus has not been found in extensive surveys of beans throughout the region (Kulkarni, 1973). An outbreak of BYMV at Kitale in Kenya is thought to have been eradicated (Leakey, pers. comm. 1982). The only other virus reliably identified from *P. vulgaris* in eastern Africa is a strain of tobacco mosaic virus (Hollings *et al.*, 1981). Cucumber mosaic virus is not known from beans (Bock *et al.*, 1975), nor have any whitefly or beetle-transmitted viruses been found in *P. vulgaris* within that region. This contrasts with the situation in cowpea, an African native, as we shall see in Chapter 7.

Mycoplasma-like organisms are occasionally implicated in bean diseases as, for example, in Japan (Murayama, 1966) and Colombia (Granada, 1980), but they are seldom of economic importance.

Bacterial diseases are arguably of more importance in beans than in any of the other legumes. The bacterial pathogens of beans, summarized in Table 11 in Section 3.3, have representatives in three genera. *Corynebacterium flaccumfaciens* causes a vascular wilt which is locally important in southern Africa, and in the USA where two colour-pigmented isolates have also been recognized (Schuster and Christiansen, 1957; Zaumeyer and Thomas, 1957; Schuster *et al.*, 1968). *Pseudomonas solanacearum* is also recorded from beans in which it induces a brown rot and wilt (Smith and McCulloch, 1919) but this disease is not of economic importance. Three pathovars of *Pseudomonas syringae* naturally infect beans. *Ps. syringae* pv. *syringae* is the cause of brown spot which is reported from beans in Kenya (Kaiser and Ramos, 1980), Brazil and the USA (Schwartz and Gálvez, 1980), while *Ps. syringae* pv. *tabaci* is known from Brazil where it causes bean wildfire (Ribeiro *et al.*, 1979). But the most important pathogen of this genus is *Ps.*

TABLE 19 Virus Diseases of Beans

Disease	Causal virus	Distribution	Transmission		Importance/Crop loss	References
			Vector	Seed (%)		
Common mosaic, black root	Bean common mosaic virus (BCMV)	Worldwide	Aphid	<83	Major. Yield loss estimated 35–98%	Bos, 1971; Schwartz and Gálvez, 1980
Yellow mosaic	Bean yellow mosaic virus (BYMV)	Widespread, but local	Aphid	0	Potentially important. May cause 33% loss of pods and 41% loss of seed	Bos, 1970; Kulkarni, 1973; Hampton, 1975
Cucumber mosaic	Cucumber mosaic virus (CMV)	Worldwide, but not known from beans in E Africa	Aphid	<30	Minor	Bos and Maat, 1974; Bock *et al.*, 1975; Schwartz and Gálvez, 1980
Peanut stunt	Peanut stunt virus (PSV)	USA, Japan, Morocco	Aphid	0	Locally important. Pod production reduced by 90% in artificially infected plants	Echandi and Hebert, 1971; Tsuchizaki, 1973; Fischer and Lockhart, 1978
Yellow dot, yellow mosaic, vein necrosis, spot mosaic	Alfalfa mosaic virus (AMV)	USA, S Africa	Aphid	?	Minor	Zaumeyer, 1953, 1963; Neveling, 1956
Pod mottle	Bean pod mottle virus (BPMV)	USA	Beetle	0	Locally important	Zaumeyer and Thomas, 1948; Semancik, 1972

Curly dwarf mosaic	Quail pea mosaic virus (QpMV)	El Salvador, Costa Rica	Beetle	0	Minor. 1–15% field incidence reported	Moore, 1973; Meiners *et al.*, 1977; Hobbs, 1981; Moore and Scott, 1981
Southern bean mosaic	Southern bean mosaic virus (SBMV)	Warm temperate and tropical America; W Africa	Beetle	+	Potentially important. Losses of 83–94% obtained	Yerkes and Patiño, 1960; Shepherd, 1971; Schwartz and Gálvez, 1980
Yellow stipple	Cowpea chlorotic mottle virus (CCMV)	Southern USA, Central America	Beetle	0	Minor	Zaumeyer and Thomas, 1950; Fulton *et al.*, 1975
Mild mosaic	Bean mild mosaic virus (BMMV)	El Salvador, Colombia	Beetle	?	Minor	Waterworth *et al.*, 1977; Hobbs, 1981; Waterworth, 1981
Golden mosaic	Bean golden mosaic virus (BGMV)	Latin America	Whitefly	0	Major. 40–100% crop loss	Goodman and Bird, 1978; Gálvez and Cárdenas, 1980
Curly top	Beet curly top virus (BCTV)	North America	Leafhopper	0	Locally important	Thomas and Mink, 1979
Red node	Tobacco streak virus (TSV)	Widespread ?	?	+	Minor	Thomas and Zaumeyer, 1950a; Mink *et al.*, 1966
Stipple streak	Tobacco necrosis virus (TNV)	Holland	Fungus	0	Minor	Bawden and Van der Want, 1949; Kassanis, 1970
Tobacco mosaic	Tobacco mosaic virus (TMV)	Kenya	?	0?	?	Hollings *et al.*, 1981
Spotted wilt	Tomato spotted wilt virus (TSWV)	Brazil, Canada	Thrips	0	Minor	Schwartz and Gálvez, 1980

syringae pv. *phaseolicola* which causes halo blight, and this disease is discussed further in Section 6.2.6. Several xanthomonads also occur in natural infections of bean. *Xanthomonas campestris* pv. *vignicola* causes a foliar blight of beans (as well as of cowpeas) in the USA and Puerto Rico (Burkholder, 1944; Vakili *et al*., 1975), but *X. campestris* pv. *phaseoli* and its brown pigmented variety *fuscans* are of considerable economic importance throughout the tropics, as the causes of common and fuscous blight, respectively. These, too, are the subject of a subsequent section of this chapter (6.2.7).

Numerous species of nematode are associated with bean crops, as will be seen from Table 12 in Section 3.4, but the most important are the root knot nematodes (*Meloidogyne* spp.) and the root lesion nematodes (*Pratylenchus* spp.). Estimates of the potential yield loss from root knot infestation may be as high as 50–90 per cent (Agudelo, 1980), and 10–80 per cent crop loss may result from root lesion infestation (Robbins *et al.,* 1972). In Kenya, *Meloidogyne* spp. can cause up to 60 per cent loss in beans (Ngundo and Taylor, 1974). Species of *Meloidogyne* and *Pratylenchus* may interact with *Fusarium solani* f.sp. *phaseoli* in such a way as to influence the initiation of fungal infection but not its subsequent colonization of the bean plant (Hutton *et al.,* 1973). Other groups of nematode which may infest beans include the reniform nematode, the sting, ring and spiral nematodes, a bud and leaf nematode and several species of dagger nematode (Agudelo, 1980). Amongst the parasitic higher plants, dodder (*Cuscuta* spp.), broomrape (*Orobanche minor*), and *Alectra vogelii* are occasionally reported from beans (See Table 13).

We shall now consider the major diseases of beans in greater detail, paying attention to their field symptoms, potential crop loss, and control.

6.2.1 Rust

Although the bean rust fungus is widely referred to as *Uromyces phaseoli* (Pers.) Wint., in accordance with the International Rules of Botanical Nomenclature, the prior name *U. appendiculatus* (Pers.) Ung. should be accepted as more correct (Laundon and Waterston, 1965b; Wilson and Henderson, 1966), and it is the latter which is used here. *U. appendiculatus* is an autoecious macrocyclic rust. De Bary (1863, cited by Wilson and Henderson, 1966) proved the genetic connection of the aecia (the terminology of the various spore stages follows that advocated by Laundon, 1967) with the urediospores and teliospores, and the complete life cycle of the bean rust fungus was induced under greenhouse conditions by Andrus (1931) and, more recently, by Groth and Mogen (1978). The aecial stage is rare under field conditions. Isolates of *U. appendiculatus* may differ in their ability to produce teliospores (Harter *et al.,* 1935; Allen, 1975c). Urediospores may vary in the position of their germ pores, a basis on which varieties have been erected (Arthur, 1934; Almeida, 1977), as well as in the degree of urediospore pigmentation (Allen, 1975c). But much the most well-known and important type of variation within the bean rust fungus is in its pathogenicity, which will be discussed later.

The selective retention of bean rust urediospores on bean leaves may depend on the release of film-forming substances from the spores, the amount produced depending on host cuticle wettability (Yarwood, 1968; Woodbury and Stahmann, 1970); the extent to which urediospore deposition may vary with cultivar has not been examined with bean rust. The optimum temperature for urediospore germination of *U. appendiculatus* is about 14.5 °C and germ tubes are longer at temperatures where the most germination takes place (Harter *et al.,* 1935). Bean rust spores germinate poorly if their concentration is too dense (Yarwood, 1956a) because of the presence of an auto-inhibitor (Macko *et al.,* 1970). There is evidence that light may inhibit rust urediospore germination and Maheshwari *et al.* (1967) have shown that infection structures of *U. appendiculatus* do develop in darkness. Wynn (1976) demonstrated that appressoria form over stomata in response to a surface contact stimulus, and the fine structure of the subsequent infection process has been investigated by Locci and Bisiach (1970), Hardwick *et al.* (1971), and Mendgen (1973). In a susceptible reaction symptoms of bean rust first develop about 5–6 days after infection, appearing as minute, chlorotic, raised spots both on surfaces of leaves and on pods and petioles. These spots then enlarge and, on the eighth or ninth day after infection, the epidermis is ruptured to form reddish-brown coloured pustules about 2 mm in diameter. These uredia reach maximum development on the fourteenth day when they are about 5 mm in diameter (Rey and Lozano, 1961) but there is much variation in size. The uredia may be surrounded by chlorotic halos and eventually by rings of secondary sori. As infection progresses the leaf becomes debilitated and the chlorotic areas surrounding the pustules become confluent, while tissue ramified by the fungus remains green as a result of starch accumulation (Wang, 1961). Urediospores are released passively and are disseminated principally by wind (see Section 2.2.5). Toward the end of the growing season, especially in temperate regions, the pustules become darker as pigmented, thick-walled teliospores (Figure 23) are produced. In severe infections leaves may abscise after about 25 days.

The extent of crop loss caused by rust depends on the plant growth stage at which infection occurs and the susceptibility of the cultivar. Early infection of white-seeded canning types can lead to almost complete crop loss in some seasons (Howland and Macartney, 1966); rust is usually most severe in tropical and sub-tropical regions (Ballantyne, 1974b). Fromme and Wingard (1921) reported 81–95 per cent loss of seed yield in rust-susceptible cultivars in the USA while, in the Cauca Valley of Colombia, rust was held responsible for 22 per cent of the total crop loss from all causes (Pinstrup-Andersen *et al.*, 1976).

Numerous chemical control measures for bean rust have been reported. Of the older fungicides, sulphur dusts have often given good control (Zaumeyer, 1946) but in general protectant fungicides tend to fail in areas where there is

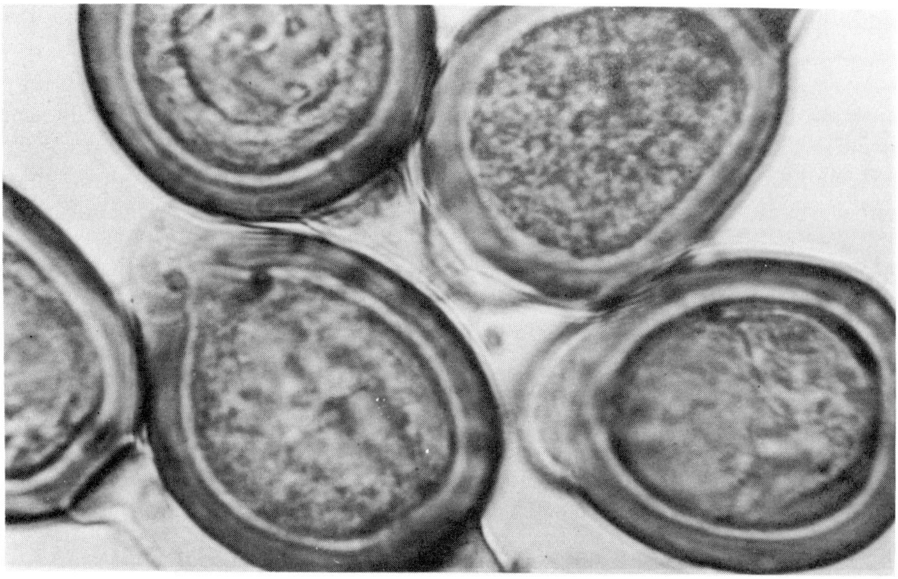

Figure 23 The teliospores of *Uromyces appendiculatus* on bean

frequent rainfall because the deposit is washed off too soon. Effective systemic activity in beans has been displayed by a range of substances the most important of which is oxycarboxin (Bates and Tweedy, 1971; Vargas, 1980). However, while certain fungicides may be effective, the overriding factor that should regulate their use is their estimated cost-effectiveness.In tropical regions heavy infections may necessitate several fungicide applications and this is often considered impracticable and uneconomic, even when technically feasible (Wallace, 1939; Issa and Arruda, 1964 cited by Vargas, 1980; Howland and Macartney, 1966). But in the USA, where spray can be applied from aircraft, the use of fungicides can be economic. Bissonnette and Johnson (1972) have given a guide, based on plant maturity and rust severity, to aid the management decision on when to spray. They also recommend cultural practices, such as rotation and the burying of debris, which may be important in reducing initial infection. However, cultural practices are best regarded as ancillary control measures. Since chemical control is usually too costly in tropical subsistence agriculture, the breeding of resistant cultivars potentially provides the most attractive long-term means of control of bean rust.

Differences in the field susceptibility of bean cultivars to rust infection were first reported in 1909 by Duggar in the USA and by Gassner in Uruguay (cited by Fromme and Wingard, 1921). Since then, extensive selection and breeding for rust resistance has been made in all the main bean-growing areas of the world. Resistance to rust in *P. vulgaris* is typically based on the hypersensitive reaction (Wingard, 1933, 1935) shown to be under oligogenic control (Zaumeyer and Harter, 1941; Augustin *et al.,* 1972) and associated with the accumulation of phytoalexins (Bailey and Ingham, 1971; Sempio *et al.*, 1975). Cultivars possessing hypersensitive resistance to rust usually have short life expectancy in commercial agriculture because their resistance has proved to be race-specific. This reliance upon race-specific resistance has led inexorably to the now classic battle between the plant breeders' ability to identify new sources of resistance and the pathogen's ability to produce new races. This 'boom and bust' cycle, which is characteristic of many other biotrophic pathogens against which race-specific resistance has been used, usually relates to the existence of a gene-for-gene relationship in which a gene for resistance in the host is matched, and overcome, by a gene for virulence in the pathogen; such a system does occur in the bean-bean rust relationship (Christ and Groth, 1979). Meanwhile, plant pathologists have tended to be preoccupied with rust race definition. Physiologic races, each with a distinct pattern of virulence as determined by inoculation of a set of differential cultivars, have long been recognized in the USA (Harter *et al.*, 1935; Dundas and Scott, 1939; Harter, 1939; Harter and Zaumeyer, 1941; Fisher, 1952; Zaumeyer, 1960; Goode, 1961; Hikida, 1961; McMillan, 1972), Mexico (Crispin and Dongo, 1962), Costa Rica (Christen and Echandi, 1967), Nicaragua, Honduras, El Salvador, Guatemala (Vargas, 1980), Brazil (Netto *et al.*, 1967; Dias and Costa, 1968; Augustin and Antunes, 1973), Colombia (Rey and Lozano, 1961; Zúñiga de Rodríguez and Victoria, 1975), Peru (Guerra and Dongo, 1973), Hawaii (Frazier *et al.*, 1948), Australia (Waterhouse, 1953; Ogle and Johnson, 1974; Ballantyne and McIntosh, 1976), New Zealand (Yen and Brien, 1960), Europe (Hubbeling, 1957; Allen, 1975c), and East Africa (Howland and Macartney, 1966; Allen, 1975c). However, comparison of results between race surveys and regions is essentially impossible due to the absence of a standardized inoculation technique (Davison and Vaughan, 1964), as well as a standardized rating system and an internationally accepted set of differential cultivars (Harter and Zaumeyer, 1941; Crispin and Dongo, 1962; Davison and Vaughan, 1963; Wilkinson and Vieira, 1972; Ballantyne, 1974c; Pereira and Chaves, 1977). The implications of pathogenic variation to the breeding of resistance cultivars, and the strategies to stabilize resistance to rust in beans are discussed in Section 6.3.

6.2.2 Anthracnose

Bean anthracnose is caused by *Colletotrichum lindemuthianum* (Sacc. and Magn.) Bri. and Cav. which is widely distributed in all bean-producing regions of the world. There is a very extensive literature on bean anthracnose, which has been studied principally in Europe and the USA where it has ranked amongst the most important diseases of beans. However, in recent years the economic importance of anthracnose has declined in the USA and in Britain, through the effective use of clean seed (Goth and Zaumeyer, 1965; Dixon and Doodson, 1973), while the widespread use of resistant cultivars has been responsible for the decline in its importance in continental Europe (Kooistra,

1965 cited by Leakey and Simbwa-Bunnya, 1972; Fouilloux, 1979). However, in tropical countries anthracnose remains a major problem: yield losses of 95 per cent have been recorded in the highlands of Colombia (Chaves, 1980), and potential losses in Malawi have been estimated to exceed 92 per cent (Peregrine, 1971a). Throughout Latin America and eastern Africa anthracnose is considered to be one of the principal diseases of beans.

Anthracnose symptoms may develop on all above-ground parts of *P. vulgaris,* depending on the time of infection and the source of inoculum. Symptoms on seed consist of cankers of variable size and colour. Infection of the seedling's hypocotyl leads to the development of small pink or brown lesions which enlarge and extend along the hypocotyl. Such lesions have a sunken appearance and contain the fungal fruiting bodies (acervuli) which produce large quantities of pink-to-beige coloured spores. Foliar symptoms are most conspicuous on the lower surface of leaves where they appear as a darkening of veins (Figure 13). Characteristic symptoms commonly occur on pods (Figure 24) on which pink-to-brown lesions develop into sunken cankers delimited by a black ring surrounded by a reddish margin. During wet weather, the pinkish spore masses ooze from acervuli in the centre of these lesions (Wheeler, 1969; Dixon and Doodson, 1973; Chaves, 1980). Spores germinate within nine hours and form appressoria in close contact with the

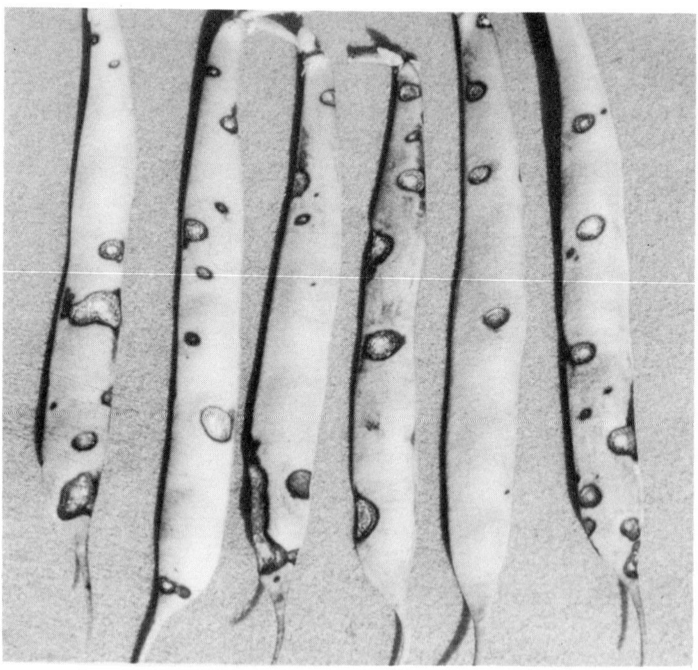

Figure 24 Anthracnose lesions of bean pods (photograph courtesy H. F. Schwartz)

host surface. From each appressorium, an infection peg penetrates the host cuticle by mechanical means (Dey, 1919; Mercer *et al*., 1975) and the fungus enters an epidermal cell and ramifies throughout susceptible host tissue. Mycelium then aggregates to form an acervulus which ruptures the host cuticle (Chaves, 1980).

The production of pathogen-free seed in semi-arid areas of western USA has considerably reduced the importance of anthracnose in North America (Goth and Zaumeyer, 1965; Zaumeyer and Meiners, 1975), and the use of clean seed is seen as a potentially powerful control measure in areas where strict standards of seed health can be maintained (Dixon and Doodson, 1973). In tropical regions, the production of pathogen-free seed is usually difficult and alternative control strategies are sought (Yerkes and Ortiz, 1956; Schwartz and Sanders, 1979). Although fungicides may have potential in seed multiplication (Peregrine, 1971a), in general their high cost precludes their use in the control of anthracnose in the tropics.

That bean cultivars may differ in their susceptibility to anthracnose was first demonstrated by Barrus (1911). However, he also notice that isolates of *C. lindemuthianum* differed in their pathogenicity (Barrus, 1918). Barrus named his two physiologic races of the anthracnose fungus 'alpha' and 'beta', and resistance to each was shown to be governed by a single dominant gene (Burkholder, 1918; McRostie, 1921) but then a new race, named gamma, was identified (Burkholder, 1923). Work by Schreiber (1932) in Germany confirmed that *C. lindemuthianum* also varied in Europe; he identified 34 distinct races which could be divided into three groups, roughly corresponding to the alpha, beta, and gamma races in the USA. When resistance to the three race-groups was combined in a single cultivar, again a new race ('delta') able to overcome this combined resistance was recognized (Andrus and Wade, 1942). This situation, which has been nicely traced by Leakey and Simbwa-Bunnya (1972), has continued to the present, and studies have shown that an ever-increasing number of physological races of *C. lindemuthianum* can be differentiated by using an increasing number of differential host cultivars. Most of these investigations have been done either in the USA or in Europe, where the most recently recognized races have been found (Charrier and Bannerot, 1970; Schnock *et al.,* 1975; Hubbeling, 1977; Kruger *et al.,* 1977). But there is no reason to suppose that the pathogen is any less variable in tropical regions; indeed, it has been suggested (Yerkes and Ortiz, 1956; Yerkes, 1958) that the centre of origin of the host (Mexico–Guatemala; see Section 6.1) is likely to contain a genetically diverse parasite population, as suggested by their demonstration of 13 new races of *C. lindemuthianum* in Mexico. Similarly, races not conforming to those previously recognized in Europe and the USA have been reported from Brazil (Augustin *et al.,* 1973; Oliari *et al.,* 1973), Uganda (Leakey and Simbwa-Bunnya, 1972) and Malawi (Ayonoadu, 1974), although the race-groups alpha, beta, gamma, and delta have also been found in tropical countries. Since there is evidence that species of *Colletotrichum* from other hosts may become adapted to beans in response to artificial selection (Muller, cited by Zaumeyer and Thomas, 1957), it has been suggested (Leakey and Simbwa-Bunnya, 1972) that a wide range of *Colletotrichum* species may share a common gene pool. This would increase the wealth of pathogenicity genes available to the fungus. Furthermore, the perfect state of *C. lindemuthianum* is occasionally reported (Zaumeyer and Thomas, 1957; Chaves, 1980), permitting sexual recombination. Leakey and Simbwa-Bunnya (1972) have pointed out that, even if the *Glomerella* state is rare in nature, its development could nevertheless be very important in terms of variability.

Although Bannerot and Richter (1968) have studied the evolution of races, the genetics of virulence in *C. lindemuthianum* has received little attention; the genetic control of resistance in beans has been investigated more fully. Burkholder (1918) and McRostie (1919) were the first to show that single dominant genes governed resistance to alpha and beta races. However, in certain host-parasite combinations, the genetic system is more complicated. Cardenas *et al.* (1964) found that, while resistance to the alpha race in the nine cultivars they studied was controlled by dominant alleles at either of two independent loci, several complementary alleles appeared to govern resistance to the beta race in some crosses. Susceptibility was sometimes dominant to resistance. Distinct systems of complementary factors were involved in resistance to the beta and gamma races. A broad-spectrum resistance found in the cv. Cornell 49–242 is controlled by a single dominent gene (Mastenbroek, 1960). The underlying mechanism of anthracnose resistance is a hypersensitivity (Mercer *et al*., 1974), like the reaction of beans resistant to rust. In an incompatible reaction (between an avirulent race and a resistant cultivar), the fungus spore germinates, produces an appressorium and penetrates the host's epidermal cells which rapidly become necrotic. Subsequent hyphal growth is at first slower than in a compatible reaction, and then ceases. A dense deposit of 'reaction material' may form at the site of penetration in some cells and this, too, may prevent further hyphal development (Skipp and Deverall, 1972). The phytoalexin, phaseollin, which is present in necrotic tissue in both resistant and susceptible reactions, is found earlier but not in larger amounts, in an incompatible reaction (Bailey and Deverall, 1971; Rahe, 1973). In comparing the ability of different races of *C. lindemuthianum* to metabolize phaseollin, Van den Heuvel and Vollaard (1976) found that four races metabolized the phytoalexin in the same way, although there was some difference in the rate of conversion. They concluded that such differences in the rate of conversion and sensitivity to phaseollin or its metabolites were insufficient to explain the differential pathogenicity of these races to different bean cultivars. Infection of the bean plant with an incompatible race of *C. lindemuthianum,* or with alien non-pathogens, may protect the bean plant against disease caused by a compatible race of the anthracnose fungus (Rahe *et al.,* 1969; Elliston *et al.,* 1971; Skipp and Deverall, 1973). But it is not known whether such induced resistance to anthracnose may operate under field conditions, as has been suggested may be the case with rust (Johnson and Allen, 1975; Johnson and Taylor, 1976).

The implications of race-specificity to breeding for anthracnose resistance, and possible strategies to increase the durability of that resistance, will be discussed in Section 6.3

6.2.3 Angular leaf spot

Angular leaf spot, which is caused by the fungus *Phaeoisariopsis griseola* (Sacc.) Ferr., has received much less attention than either bean rust or anthracnose but the available literature has recently been reviewed by Ferraz (1980). Like the other two diseases, angular leaf spot is widely distributed and occurs in most major bean-producing areas. *P. griseola* has been known from *P. vulgaris* in East Africa since 1900 (as *Cercospora stuhlmanni* P. Henn., Ebbels and Allen, 1979), and in India since 1953 (Srinivasan, 1953). In Latin America, angular leaf spot was estimated to have caused about 30 per cent loss of crop in a new cultivar in the Cauca Valley of Colombia in 1955 (Barros *et al.,* 1957), while a recent assessment in Colombia places losses as high as 80 per cent (Schwartz *et al*., 1981a). Yield losses of 50 and 80 per cent have been reported from the USA and Mexico respectively (Ferraz, 1980). Angular leaf spot can cause considerable losses to seed bean crops in Australia (Brock, 1951), and it is one of the principal diseases of beans in eastern Africa, as well as in the North-West Province of Cameroun (D. J. Allen, unpublished).

Angular leaf spot symptoms usually appear first on the lower leaf surface as grey lesions which are typically delimited by the veins. These angular lesions turn brown and become covered with small black columnar structures (synnemata) on which conidia are borne. Brown angular lesions are also apparent on the upper leaf surface (Figure 25) but the synnemata are effectively confined to the lower surface. In severe infections, leaf spots may coalesce causing extensive necrosis, chlorosis, and premature defoliation; lesions may develop also on petioles, stems and pods. Pod infection can lead to poor quality seed (Barros *et al.,* 1958a). The pathogen is seedborne (Orozco-Sarria and Cardona-Alvarez, 1959) and survives the intercrop period on infected plant debris (Cardona Alvarez and Walker, 1956). Other aspects of the epidemiology of angular leaf spot have been discussed in Section 2.2.3.

There is some evidence that epidemics of angular leaf spot are often attributable to poor cultural practices. Rotation, good land preparation and, presumably, the use of good quality seed are likely to decrease disease severity. Amongst the fungicides tested for efficacy, zineb has given good control (Barros *et al.,* 1958b). Other chemical control measures are reviewed by Ferraz (1980). Sources of host-plant resistance have been identified (Brock, 1951; Barros *et al.,* 1957; Silvera, 1967; Sohi and Dass, 1972; Singh and Sharma, 1975; Santos-Filho *et al*., 1976; Schwartz *et al*., 1982). The extent to which populations of *P. griseola* may vary in pathogenicity is uncertain. A highly aggressive form of *P. griseola* from Tanzania caused circular, not angular, lesions on bean leaves, and bore longer conidia on robust synnemata. The new form was infective at a lower inoculum concentration than a 'typical'

Figure 25 Angular leaf spot (*Phaeoisariopsis griseola*) lesions on bean leaves, NW Province, Cameroun

isolate, and the lesions it caused grew faster and attained a larger diameter (Hocking, 1967). Preliminary investigations of the pathogenic variability of *P. griseola* in Colombia have suggested that resistance identified in cv. Caraota 260 in Brazil is not effective against certain isolates from Colombia (Alvarez-Ayala and Schwartz, 1979), but further studies appear warranted to confirm whether differences in virulence do exist, and to determine the probable stability of the available resistance to angular leaf spot.

6.2.4 Common mosaic

Common mosaic is caused by the bean common mosaic virus (BCMV) which is a seedborne pathogen transmitted by aphids in the non-persistent manner. The virus occurs in all regions where beans are grown and is economically important in many areas. The host range of BCMV is narrow. Although potential wild host species have been identified (Provvidenti and Braverman, 1976; Meiners *et al.,* 1978), carry over occurs principally as seedborne infection which, in some regions, is probably encouraged by farmers' practice of saving seed from a previous harvest. In Morocco, it appears that seedborne infection rather than the secondary spread of BCMV is the principal cause of yield loss in beans (Lockhart and Fischer, 1974). Crop loss estimates are variable: severe infection may cause up to 64 per cent loss in the number of pods and 68 per cent loss of seed yield (Hampton, 1975), although almost complete crop failure may occur in Latin America (Gálvez, 1980). The extent of damage depends on the susceptibility of the cultivar, the aggressiveness of the virus strain, the time and source of infection, and the environment (Burke and Silbernagel, 1974b). Symptoms caused by BCMV may be expressed either as 'common mosaic' or as 'black root'. Common mosaic symptoms typically are associated with a mottle and mosaic of leaves which are often malformed and rolled; affected plants are often stunted, and flowers and pods may be misshapen. Black root is characterized by a vascular necrosis, appearing first as an apical die-back which is followed by plant death (Jenkins, 1941; Grogan and Walker, 1948; Bos, 1971; Schwartz *et al.,* 1978a).

Some degree of control of the disease may possibly be achieved through the chemical control of the aphid vector population (Schwartz *et al.,* 1978a), by sowing to avoid periods of peak vector activity (Burke, 1964), or by the use of virus-free seed. However, the implementation of seed certification schemes is difficult in tropical regions, and the development of resistant cultivars is much the most satisfactory means of controlling BCMV. Indeed, in the USA resistant cultivars have now made BCMV of little economic importance (Zaumeyer and Meiners, 1975).

Sources of resistance to BCMV have been known for many years (Reddick and Stewart, 1918; McRostie, 1921; Rands and Brotherton, 1925). The cultivars Robust and Corbett Refugee, in particular, have been widely used to breed mosaic-resistant cultivars of both dry and snap beans (Zaumeyer and Meiners, 1975). Early studies on the genetics of resistance to BCMV revealed that, while resistance derived from cv. Robust was controlled by a single recessive 'a' gene, resistance in cv. Corbett Refugee was governed by a single dominant 'I' gene (Pierce, 1935; Wade and Andrus, 1941; Ali, 1950). Parker 1936) obtained evidence of a cytoplasmic influence on resistance in certain reciprocal crosses. The expression of these two types of resistance is also different. While cultivars in which resistance has been derived from Robust are effectively immune to the type strain of BCMV, Corbett Refugee and its derivatives possess a hypersensitive resistance conferred by the dominant 'I' gene. However, the 'I' gene is temperature-sensitive, since hypersensitivity is localized at temperatures less than 30 °C, above this temperature a lethal systemic necrosis (black root) can occur (Grogan and Walker, 1948) in cultivars with this type of resistance. Ali (1950) has suggested that the dominant gene 'A' is required for virus infection, thereby rendering the host susceptible. The dominant 'I' gene, in the presence of the 'A' gene, inhibits mosaic symptom expression and conditions systemic necrosis. In the presence of gene 'a', acting as a recessive epistatic, the plant becomes resistant to both mosaic and top necrosis. On this basis, the genotypes of a fully susceptible cultivar might be AAii, while those of cvs Corbett Refugee and Robust might be AAII and aaii respectively.

The dominant 'I' gene is reported to give resistance against all known strains in the USA (Silbernagel, 1969) and the resistance of cv. Corbett Refugee has remained effective for more than 45 years (Zaumeyer and Meiners, 1975). However, some BCMV strains isolated in Europe have been found to overcome the resistance conferred by the dominant 'I' gene and can induce necrotic symptoms at lower temperatures (20 °C; Hubbeling, 1969). Many dry bean cultivars derive their BCMV resistance from the recessive 'a' gene present in cv. Robust (Zaumeyer and Meiners, 1975). This resistance is strain-specific; it is effective against the type strain of BCMV but is overcome by the 'New York 15' strain (Richards and Burkholder, 1943). Other BCMV strains have been differentiated subsequently (Zaumeyer and Goth, 1964; Silbernagel, 1969; Drijfhout and Bos, 1977), these usually being discovered in newly bred cultivars resistant to previous strains in countries with long-established resistance breeding programmes (Drijfhout and Bos, 1977). A comprehensive investigation of BCMV strains has now been made in an attempt to promote international standardization of test conditions, procedures, and criteria for strain differentiation (Drijfhout *et al.*, 1978). Seven strain groups (pathotypes) were recognized and a standard set of differential cultivars recommended. Study of the genetic interaction between BCMV and *P. vulgaris* has shown there is a gene-for-gene relationship between the pathogen and its host (Drijfhout, 1978). Four strain-specific resistance genes (bc-1, bc-1^2, bc-2, bc-2^2) are overcome by 'matching' pathogenicity genes present in the virus. For a fifth strain-specific resistance gene (bc-3), a corresponding pathogenicity gene has not yet been found. A strain non-specific gene (bc-n), which confers no resistance on its own, was found to be complementary to the strain-specific genes. Innes and Walkey (1980) found that all the cultivars they tested appeared to carry the strain non-specific gene bc-n in addition to certain strain-specific genes. However, segregation patterns in F_2 populations of one of their crosses indicated that, in addition to a major recessive gene, polygenes were present; the distribution patterns of resistance in F_3s of these populations suggested that a high degree of resistance may be achieved by accumulating polygenes only (Innes and Walkey, 1980).

6.2.5 Golden mosaic

Golden mosaic is caused by bean golden mosaic virus (BGMV) which is a whitefly-transmitted geminivirus. The disease is widely distributed in tropical and sub-tropical America (Goodman and Bird, 1978) where it is of greatest importance under hot conditions at low altitudes. Field incidence frequently reaches 100 per cent. Golden mosaic was first reported in 1961 from southern Brazil and has subsequently spread to all major bean producing regions of Brazil where it now ranks amongst the most serious diseases of the crop (Costa, 1976). Gámez (1971) considers that golden mosaic is the principal bean disease in the Pacific coastal plains of El Salvador, and Pierce (1975) states that BGMV is the most important limiting factor on bean production in the Jamaican lowlands. Estimates of the crop loss caused by BGMV range from 40–100 per cent (Gálvez and Cárdenas, 1980), varying with the time and intensity of infection and the susceptibility of the cultivar. The extent to which isolates of BGMV may vary in pathogenicity is not yet known.

The field symptoms of golden mosaic (Gálvez and Cárdenas, 1980) may be distinctive. Affected plants become a brilliant golden yellow colour, such symptoms appearing within two weeks of sowing if vector populations are dense. Young leaves of diseased plants usually become rolled and cupped, and severely affected plants may be stunted. Pods often exhibit blotching and seed may be discoloured, as well as reduced in size and number (Costa and Cupertino, 1976). The concentration of virus increases during symptom development and reaches a peak 1–2 weeks after infection. Thereafter, virus titre decreases rapidly and is very low in plants with well-developed symptoms (Shock and Goodman, 1981). BGMV has a narrow host range which, in addition to *P. vulgaris,* includes the lima bean (*P. lunatus*) and the related leguminous weed *Macroptilium lathyroides* (Alconero and Santiago, 1973; Goodman and Bird, 1978); since BGMV is not seedborne, these natural hosts presumably act as reservoirs for the infection of beans. The vector, *Bemisia tabaci,* which is probably of Asian origin, has a very broad host range (Mound, 1981). It has been suggested that the increased importance of golden mosaic to the Brazilian bean crop is attributable to the expansion in production of soyabeans from which populations of *B. tabaci* migrate to young bean plants (Costa, 1975). Other aspects of the epidemiology of golden mosaic are discussed in Section 3.1.5.

Elimination of plant species liable to act as BGMV reservoirs and the isolation of beans from crops known to favour whitefly population build-up have each been suggested as measures for the cultural control of golden mosaic. Similarly, the adjustment of the time of sowing to avoid periods of peak vector activity may also provide some protection (Gálvez and Cárdenas, 1980). The spread of golden mosaic can also be reduced by controlling the vector by chemical or, perhaps, by biological means. Various insecticides are effective against whiteflies, and the combination of a systemic insecticide with mineral oil, which can immobilize *B. tabaci* within three minutes of its alighting on an oil-sprayed leaf, has been advocated (Nene, 1973). Since it has been claimed that the acquisition and inoculation of BGMV by *B. tabaci* can be effected in a total time of less than six minutes (Goodman and Bird, 1978), an effective chemical control measure would need to be quick to immobilize the vector. However, it is not known whether a latent period exists in the BGMV-B. *tabaci* relationship, as it does in the case of apparently similar diseases (Section 3.1.5). If a latent period does occur, then it is reasonable to assume that the potential for chemical control of golden mosaic is greater than if there were none. Certain parasitic fungi including *Paecilomyces farinosus* may possibly also play some role in the natural regulation of *B. tabaci* populations (Nene, 1972). The relative potential for the manipulation of ground cover so as to interfere with the alighting response of whiteflies (Mound, 1962) appears not to have been investigated, as it has been with aphidborne virus diseases (Sections 4.2.4 and 5.2.3).

The search for host-plant resistance amongst bean germplasm has led to the identification of a few promising accessions. Gámez (1971) screened some

4000 accessions in Costa Rica, and some resistance was identified in cvs Turrialba 1 and Porrillo 1. Extensive screening has subsequently been conducted in Colombia, but in no case has immunity been found. Attempts have also been made to produce BGMV-resistant material by irradiation and chemical mutagenesis (Neto *et al*., 1977) but with rather limited success. The available partially resistant ('tolerant') cultivars apparently show both delayed and milder expressions of disease symptoms which were associated with relatively low virus concentrations (Gálvez and Cárdenas, 1980), suggesting that a 'resistance to virus multiplication' (*sensu* Russell, 1978) may be involved. Under heavy infection pressure, yield loss in such partially resistant cultivars as Turrialba 1 and ICA Pijao may be as high as 74 per cent and 55 per cent respectively, in comparison with a 78 per cent loss in the golden-mosaic-susceptible cv. Pecho Amarillo (Yoshii *et al.,* 1980). Intercrossing among resistant material at CIAT has produced lines with golden mosaic resistance superior to any of the three parents used, indicating that transgressive segregation had occurred (Yoshii *et al.,* 1980).

6.2.6 Halo blight

Halo blight is caused by the bacterium *Pseudomonas syringae* pv. *phaseolicola* (see Section 3.3) which occurs worldwide. Halo blight is of greatest economic importance in temperate regions, and under cool conditions at medium altitude within the tropics. It occurs in various parts of Latin America, including southern Chile and Brazil (Schwartz and Gálvez, 1980). In eastern Africa the disease can be of economic importance in Malawi and Kenya (Mughogho, 1970; Mukunya, 1974) but rarely in Uganda except at the highest altitudes of cultivation (Leakey, personal communication 1982). Yield losses of 23–43 per cent have been reported under experimental conditions in Michigan (Saettler and Potter, 1970, cited by Schwartz and Gálvez, 1980). Initial symptoms appear 3–5 days after infection as small, water-soaked spots on the leaves. A halo of greenish-yellow tissue then appears around these lesions (Figure 26). Under epidemic conditions, stems and pods may also become infected, and systemic infection leads to water-soaking and discoloration of tissue bordering veins. A whitish exudate is sometimes found. Seed contamination can cause a stem girdling and joint rot at nodes of seedlings (Schwartz *et al.,* 1978a).

Halo blight can, to some extent, be controlled by chemical means, using various formulations of copper or the antibiotic streptomycin, applied as foliar sprays (Taylor and Dudley, 1977) or as a seed dressing (Ralph, 1976). However, chemical control may be neither sufficiently effective nor practicable for adoption in tropical regions. And there is the added risk of the development of antibiotic-resistant mutants (Russell, 1975). In the USA, the seriousness of halo blight has been considerably lessened by the production of clean seed under arid conditions, notably in Idaho. While field inspection and roguing of diseased plants may help to ensure that seed is pathogen-free,

Figure 26 Lesions of halo blight (*Pseudomonas syringae* pv. *phaseolicola*) on bean leaves, Southern Highlands, Tanzania

contamination of seed may still occur even where halo blight incidence is negligible, including certain resistant cultivars (Stadt and Saettler, 1981). Since chemical seed treatments are generally ineffective in eradicating internal seed infections, clean seed production cannot rely solely on seed dressings, field inspection, or seed multiplication under arid conditions for a single season (Grogan and Kimble, 1967). Various methods of seed testing have been developed and used to detect contaminated seed lots; these include the use of a specific bacteriophage (Katznelson, 1950), serology (Guthrie *et al.*, 1965), inoculation of excised cotyledons (Parker, 1966) and ultra-violet fluorescence (Wharton, 1967). However, in view of the low admissible level for halo blight (1 seed/5 kg may be sufficient to initiate an epidemic; Wharton, 1967), a more precise, quantitative technique may be preferable (Taylor, 1970). Other aspects of the cultural and chemical control of halo blight are reviewed by Zaumeyer and Thomas (1957) and Schwartz and Gálvez (1980). The epidemics of halo blight which occurred in the seed bean crop in Idaho during 1963–67 (Butcher *et al*, 1969), are thought by Zaumeyer and Meiners (1975) to have stimulated interest in research, especially in the breeding of resistant cultivars. During the last decade, a considerable amount of work has been devoted to the identification of sources of resistance to halo blight, and to other bacterial diseases of beans, most notably in Nebraska, France, and Bulgaria.

Both qualitative (Patel and Walker, 1965) and quantitative (Schroth *et al.*, 1971; Hubbeling, 1973; Szarka and Velich, 1980) variation in the pathogenicity of *Ps. syringae* pv. *phaseolicola* to beans have been reported. Qualitative differences (virulence) have conventionally depended on the reaction of the bean cultivar Red Mexican U.I.3 to artificial inoculation, two physiological races being so defined (Patel and Walker, 1965). Isolates referable to each of these two races are apparently widespread (Epton and Deverall, 1965) and existence of a third race has been suggested (Coyne *et al.*, 1979). Quantitative differences (aggressiveness) apparently relate to varation in toxin production. 'Halo-less' isolates have long been recognized (Jensen and Livingston, 1944; Johnson, 1969), and their level of pathogenicity is often low (Hubbeling, 1973). Both the development of halos and toxin production are temperature-dependent (Goss, 1940; Jensen and Livingston, 1944; Skoog, 1952), and the area of the halo is positively correlated with the concentration of a culture filtrate and presumably also of the toxin (Hoitink *et al.*, 1966). Further evidence that toxin production may in part govern aggressiveness has been obtained by Patil *et al.* (1974) who found that the pathogenicity of an induced non-toxigenic mutant of *Ps. syringae* pv. *phaseolicola* had been altered relative to the parent wild type.

Considerable variation in the resistance of beans to halo blight can also be recognized. The resistance of cv. Red Mexican U.I.3 is based on the hypersensitive reaction, is race-specific, and is governed by a single dominant gene (Walker and Patel, 1964b; Patel and Walker, 1965; Taylor *et al.*, 1978). 'Race 2' of the halo blight bacterium is able to multiply as rapidly in cv. Red Mexican U.I.3 as it does in a susceptible cultivar (Omer and Wood, 1969), suggesting that its resistance is completely overcome by the matching race. This race-specific, hypersensitive resistance is associated with the accumulation of phytoalexins, including phaseollin, phaseollinisoflavan, coumestrol, and kievitone (Stholasuta *et al.*, 1971; Lyon and Wood, 1975; Gnanamanickam and Patil, 1977); the ultrastructure of the incompatible reaction has been investigated by Sigee and Epton (1975) and Roebuck *et al*., (1978). Such ultrastructural studies indicate that bacteria in intercellular spaces bind to plant cells in resistant but not in susceptible reactions (Sing and Schroth, 1977), and there is now evidence that the different responses of cv. Red Mexican to races 1 and 2 of *Ps. syringae* pv. *phaseolicola* relate to differential binding (Slusarenko and Wood, 1981). The toxin produced by the bacterium, phaseotoxin, also plays a role in the hypersensitive reaction. Although the toxin is the principal cause of chlorosis in infected tissue, as we have seen, phaseotoxin does not exhibit the same host specificity as the pathogen which produces it so that the toxin constitutes but one of several determinants in host specificity. There is evidence that phaseotoxin production may be suppressed in infected bean plants possessing hypersensitive resistance, but not in infected susceptible ones. But there is also evidence that pre-treatment of resistant cultivars with the toxin can suppress the hypersensitive response and phytoalexin accumulation. Thus, it seems that the factor which determines hypersensitive resistance to halo blight is associated with suppression of phaseotoxin (Patil and Gnanamanickam, 1976). No sources of hypersensitive resistance to race 2 of *Ps. syringae* pv. *phaseolicola* were found by Patel and Walker (1965), but Russell (1976) has identified fifteen accessions with hypersensitive resistance to race 2 and two with resistance to both races 1 and 2. Other accessions, most notably PI 150414 from El Salvador, possess resistance effective against a wide range of halo blight isolates (Patel and Walker, 1965, 1966; Dickson and Natti, 1966; Taylor *et al.,* 1978; Coyne *et al.,* 1979), but a mechanism other than hypersensitivity is involved. The resistance of PI 150414 has often been called 'tolerance' since it confers a less extreme resistance than that conferred by hypersensitivity. As stressed by Taylor *et al.,* (1978), such usage of the term 'tolerance' is incorrect and, despite its wide acceptance, is better avoided. Patel and Walker (1966) showed that, in crosses with cv. Tenderwhite, the resistance present in PI 150414 was controlled by a single recessive gene. However, from crosses between PI 150414 and cv. Gallatin 50, Hill *et al.* (1972) found that the resistance of different plant parts of PI 150414 was controlled by different dominant genes. Subsequent work reveals that the resistance factor in PI 150414 may behave as a recessive in some crosses but may be partially dominant in crosses with cv. Gallatin 50, suggesting that genetic background can influence the expression of the resistance of PI 150414 (Taylor *et al.*, 1978). The resistance present in PI 150414 is, in some crosses, qualitative, but it is apparently race non-specific (*sensu* Scott *et al.,* 1980).

6.2.7 Common and fuscous blight

Three xanthomonads cause blight of beans. Common blight is caused by *Xanthomonas campestris* pv. *phaseoli* and fuscous blight is caused by a brown-pigmented variant, *fuscans*. Both of these are widespread and have been recognized in eastern Africa (Leakey, 1970a) and India (Patel and Jindal, 1972c) as well as in the Americas. In Puerto Rico, *X. campestris* pv. *vignicola* also infects *P. vulgaris* (Vakili *et al.,* 1975). For present purposes, common and fuscous blights are treated together; they often occur together, their field symptoms are similar (Yoshii, 1980) though greater tissue collapse is associated with fuscous blight (Leakey, personal communication, 1982). Unlike halo blight, common blight is important in regions with moderate to high day temperatures. Estimates in Colombia have shown that natural and artificial infection can cause up to 22 and 45 per cent yield loss respectively (Yoshii *et al.,* 1976). Common blight is responsible for about 17 per cent loss of crop in farms in the Cauca Valley of Colombia (Pinstrup-Andersen *et al.,* 1976). In Canada losses as high as 38 per cent have been recorded (Wallen and Jackson, 1975). Initial infection appears as water-soaked lesions on the undersurface of leaves. These spots then enlarge and adjacent lesions coalesce to produce irregular areas of necrosis. The blighted areas are typically bordered by a narrow zone of yellow tissue (Plate 1). A yellow exudate is often found on these lesions. Stems and pods are also often infected, and pod infection can lead to seed discolouration (Figure 27) as well as to subse-

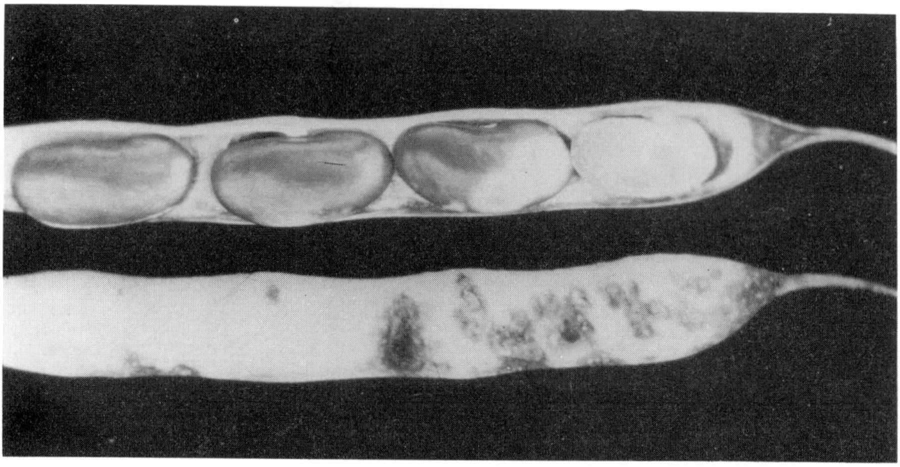

Figure 27 Bean pod and seeds infected with *X. campestris* pv. *phaseoli* (photograph courtesy H. F. Schwartz)

quent seed transmission of the bacterium (Schwartz *et al.,* 1978). Selected aspects of the epidemiology of these bacterial blights are discussed in Section 3.3.

Chemical control of common blight is generally unsatisfactory (Zaumeyer and Thomas, 1957; Saettler, 1971a). Though antibiotics such as streptomycin (Mitchell *et al.,* 1954) and copper sprays may afford some control of foliar infection in certain areas (Oshima and Dickens, 1971), pod infection may not be significantly reduced or yield increased (Weller and Saettler, 1976). Seed treatment has also been tried but with limited success. As with halo blight, the use of clean seed is an important means of controlling common blight in the USA (Zaumeyer and Thomas, 1957) and in Australia (Wilson, 1947). While the production of disease-free seed in the USA relies largely on the selection of suitably arid areas, such as Idaho, in New South Wales there are no sufficiently low rainfall areas and seed is raised for certification under high rainfall where it is easier to detect a very low incidence of blight. However, seed transmission of *X. campestris* pv. *phaseoli* may occur in symptomless seed harvested from symptomless pods of either resistant or susceptible cultivars (Cafati and Saettler, 1980), thus emphasizing the need for sensitive detection methods in seed production programmes (Katznelson *et al.,* 1954). Since both cultural and chemical control practices have limited application to subsistence agriculture in the tropics, the use of resistant cultivars may yet prove to be the best alternative control strategy.

Variation in the degree of susceptibility to *Xanthomonas* blight was first noted amongst accessions of *P. vulgaris* by Rands and Brotherton (1925), but they found no very high level of resistance. The absence of an extreme resistance, confirmed by later investigation, led subsequently to the search for resistance in other species of *Phaseolus*. *P. coccineus* was shown to have some resistance (Burkholder and Bullard, 1946), and a rather high level was recognized in *P. acutifolius* (Schuster, 1955). Extensive screening of *Phaseolus* germplasm has led to the identification of further sources of moderate resistance ('tolerance') (Coyne and Schuster, 1973), including the cultivar Great Northern Nebraska No. 1 selection 27 which is derived from an interspecific cross between *P. vulgaris* and *P. acutifolius* (Coyne *et al.,* 1965b). This resistance, which apparently does not involve hypersensitivity, is quantitatively inherited (Honma, 1956; Coyne *et al.,* 1965b) and transgressive segregation for blight resistance occurs in progenies derived from cultivars possessing it (Valladares-Sanchez *et al.,* 1979). Certain cultivars of *P. vulgaris* and *P. coccineus* give differential reactions of leaves and pods on artificial inoculation, suggesting that different genes control the response of different plant parts (Coyne and Schuster, 1974b; Valladares-Sanchez *et al.,* 1979).

Variation in the degree of pathogenicity of *X. campestris* pv. *phaseoli* has also been recognized. There is a tendency for isolates of var. *fuscans* to be relatively more pathogenic to beans than common blight isolates (Leakey, 1973; Ekpo and Saettler, 1976). Although it is commonly stated that such isolates differ in virulence, a term which often connotes specificity (van der Plank, 1968; Scott *et al*., 1980), the degree of host specificity amongst xanthomonads from beans seems uncertain. Although the genetic basis of pathogenicity has not been studied, it seems likely that pathogenicity varies quantitatively. Available data on the *Xanthomonas–Phaseolus* interaction (Schuster and Coyne, 1971; Schuster *et al.,* 1973; Schuster and Coyne, 1975; Ekpo and Saettler, 1976; Yoshii *et al.,* 1978) do not clearly reveal a reversed ranking order of cultivars in resistance, nor of isolates in pathogenicity; a formal test of differential interation might help to clarify the degree of specificity to be expected (Scott *et al.,* 1980). It is evident that blight resistance, effective in one region, may afford inadequate protection in another: this was the case with cv. Great Northern Nebraska No. 1 selection 27 in Colombia. Although pathogenic variation can account for this (Schuster and Coyne, 1971), it is not certain that host-specific pathogenicity was involved. Environmental factors are also involved (Schwartz, personal communication, 1982). The important point to be made here is that host-specific and host-non-specific variation in pathogenicity have different implications for the plant breeder.

6.3 DISEASE RESISTANCE AND THE IMPROVEMENT OF *PHASEOLUS* BEANS

The breeding of beans for disease resistance has a long history; Burkholder (1918), McRostie (1921) and Rands and Brotherton (1925) were amongst the pioneers. The literature, which is perhaps more extensive than that of any other food legume, has been reviewed recently (Silbernagel and Zaumeyer, 1973; Zaumeyer and Meiners, 1975; Graham, 1978; Meiners and Elden, 1980). Despite the wealth of recognised sources of resistance, the greatest progress in breeding for disease resistance has probably been the development of common mosaic resistant cultivars of both dry and snap beans in the USA. Combined resistance remains relatively uncommon among commercial cultivars (Meiners and Elden, 1980). Disease resistance breeding has also received considerable attention in Europe (Hubbeling, 1957; Innes *et al.*, 1977; Fouilloux, 1979) and, since the establishment of the Programa Cooperativo Centroamericano para el Mejoramiento de Cultivos Alimentios (PCCMCA) and the Bean Program at the Centro Internacional de Agricultura Tropical (CIAT), significant progress has also been made more recently in Latin America. CIAT's programme, which gives high priority to disease resistance as a component of integrated control (Graham, 1978; Schwartz and Sanders, 1979), focuses on the introduction of resistance into tropically adapted genotypes to which relatively little attention has been previously devoted. The high priority accorded to pathogens reflects the diversity which is to be expected in the parasites of a crop whose major region of production is also its centre of origin.

Sources of resistance have been identified against the following fungal diseases of beans: rust (Ballantyne, 1974b; Coyne and Schuster, 1975), anthracnose (Mastenbroek, 1960; Fouilloux, 1976; Schwartz *et al*., 1982) angular leaf spot (Santos-Filho *et al.,* 1976; Schwartz *et al.,* 1982), fusarium dry root rot (Boomstra *et al.,* 1977; Beebe *et al.,* 1981), fusarium wilt (Ribeiro and Hagedorn, 1979a, b), black root rot (Hassan *et al*., 1971), *Rhizoctonia* diseases (Deakin and Dukes, 1975), pythium blight (Adegbola and Hagedorn, 1970), grey mould (Dixon and Doodson, 1975), white mould (Schwartz *et al.,* 1977; Hunter *et al.,* 1981), white leaf spot (Yoshii and Aamodt, 1978; Schwartz *et al.,* 1981a) and powdery mildew (Schwartz *et al.,* 1981b). Amongst the virus diseases of beans, resistance of various types has been found against common mosaic (Drijfhout, 1978), yellow mosaic (Baggett and Frazier, 1957), golden mosaic (Yoshii *et al.,* 1980), curly top (Silbernagel and Jafri, 1974), summer death (Ballantyne, 1970), pod mottle (Thomas and Zaumeyer, 1950), rugose mosaic (Machado and Pinchinat, 1975), and peanut stunt (Meiners and Gillaspie, 1980). Resistance is also available to halo blight (Patel and Walker, 1965; Russell, 1976; Taylor *et al.,* 1978), common and fuscous blight (Leakey, 1973; Coyne and Schuster, 1974a), brown spot (Coyne and Schuster, 1969; Daub and Hagedorn, 1979), bacterial wilt (Coyne *et al.,* 1965a), and root knot (Fassuliotis *et al.,* 1970; Ngundo, 1977).

There is, of course, considerable variation in the expression of resistance as well as in its mechanism of operation and genetic control. Some types of resistance provide complete control of the disease while others afford only partial protection. Some resistance has proved durable. For instance, resistance to curly top derived from a Red Mexican type has remained effective in agriculture for over 45 years, despite strain variation in the virus, it includes both a tendency to escape infection by vector non-preference and a post-infection resistance (Silbernagel and Jafri, 1974). In other instances, as we have seen, resistance has proved transient, as a result of reliance being placed on a race-specific mechanism. Hypersensitive resistance, recognized in beans against rust, anthracnose, and halo blight is, in each of these cases, known to be race-specific and has repeatedly proved to be of short life expectancy in agriculture. Hypersensitivity is associated with a high degree of resistance which tends to exert considerable selection pressure in favour of resistance-breaking variants of the pathogen. Hypersensitivity is also associated with the accumulation of phytoalexins. It has long been recognized that legumes produce these post-infectional antibiotics in response to invasion by pathogens, but their role is not entirely understood (Smith and Ingham, 1980). In certain bean-parasite combinations, the localized accumualtion of phytoalexins is temporally associated with cytological changes and the inhibition of the pathogen. For instance, in incompatible reactions with anthracnose, the accumulation of phytoalexins, principally phaseollin and phaseollinisoflavan, is associated with lesion limitation (Bailey, 1974). Similarly, phaseollin accumulation is associated with the local necrosis which typifies hypersensitive resistance to rust (Bailey and Ingham, 1971), while phaseollin, phaseollinisoflavan, coumestrol and kievitone are associated with halo blight hypersensitivity (Section 6.2.6). Phytoalexins have also been detected in bean plants with resistance to *Fusarium solani* f.sp. *phaseoli*, *Thielaviopsis basicola* and *Rhizoctonia solani* (Pierce, 1971; Smith *et al*., 1975) but in none of these is resistance known to be race-specific. However, *F. solani* f.sp. *phaseoli* can tolerate and detoxify several phytoalexins (Kuhn and Smith, 1979; Smith *et al*., 1980; Kistler and Van Etten, 1981) which suggests that host-specificity occurs in the pathogen, but it is not known whether differences in this ability occur between isolates within f.sp. *phaseoli.*

Race-specific resistance is often a qualitative character in which a gene-for-gene relationship between the host and its parasite occurs. Such a relationship has been demonstrated in some combinations of bean genotype and rust (Christ and Groth, 1979) and common mosaic (Drijfhout, 1978). Race non-specific resistance can be recognized only by the absence of specificity; it cannot be proven (Scott *et al*., 1980). In beans, resistance to the root rots, common blight, and golden mosaic is not known to be race-specific; all are under polygenic control (Coyne *et al.,* 1965b; Bravo *et al.,* 1969; Hassan *et al.,* 1971; Boomstra and Bliss, 1977; Dickson and Boettger, 1977; Yoshii *et al*., 1980). Yet the evidently durable resistance to curly top is governed by only two factors, in dominant and recessive epistasis (Schultz and Dean, 1947). There are instances in which single genes confer resistance to a wide spectrum of pathogenicity, apparently race non-specific until the emergence of a matching race. The 'Are' gene for anthracnose resistance (Mastenbroek, 1960; Schnock *et al.,* 1975) is one such example, the resistance of cv. NEP 2 to rust is another (Ballantyne and McIntosh, 1977; Schwartz, 1980).

These inconsistencies confirm that there are no simple indices of durable resistance (Johnson, 1981a, b). But it is evident from experience in bean breeding that new strategies are warranted, especially where reliance on race-specific resistance has led to the breakdown of that resistance. Both rust and anthracnose resistance have tended to have a short life expectancy in agriculture. We shall now consider some of the strategies available to stablize race-specific resistance and possible means of identifying race non-specific resistance.

While some have expressed concern over the genetic uniformity and vulnerability of both dry and snap bean production in the USA (Adams, 1972; Zaumeyer, 1972a, b), others have remarked on the great genetic diversity and stability of bean crops in tropical regions (Yerkes and Ortiz, 1956; Leakey, 1970a; Vieira and Wilkinson, 1972). The potential of multiline cultivars and cultivar mixtures in the control of disease in self-pollinated crops has been appreciated (Browning and Frey, 1969; Wolfe *et al.,* 1981) and their effects on the evolution of pathogens discussed (Section 1.3.2). Van der Plank (1968) asserts that a stabilizing selection operates against races which possess superfluous genes for host-specific pathogenicity (virulence), virulence being negatively correlated with aggressiveness (host non-specific pathogenicity). That the latter is not invariably so has subsequently been demonstrated in cereal rusts (Brown and Sharp, 1970; Ogle and Brown, 1970) and bean rust (Allen, 1975d), showing that there are cases in which the race with the largest number of genes for virulence is also the most aggressive one. Parlevliet (1981) failed to find clear evidence for the existence of stabilizing selection. This has important implications for the development of multilines, especially those whose components possess unique single resistance genes, since the operation of stabilizing selection is fundamental to their presumed disease-resistance stability. Marshall and Pryor (1978) concluded that such 'dirty crop' multiline cultivars provide stable disease control only in limited and relatively rare circumstances. Relative to multilines whose components carried single genes for resistance, Marshall and Prior (1979) found that components with two or more genes were just as, or more, effective in stabilizing the racial composition of the pathogen population only when the components possessed disjoint sets of genes (of the type AB, CD, EF); components carrying overlapping sets of genes (of the type AB, BC, AC) were considered less likely to lead to stability. The extent to which the development of a multiline whose components possess resistance to all prevalent races of a pathogen may confer stability is unclear, but such a 'clean crop approach' is being investigated (Marshall and Pryor, 1978). Since multiline cultivars of bean have been advocated (Vieira and Wilkinson, 1972; Coyne and Schuster, 1975) and developed (Voysest *et al.,* 1976) for the control of bean rust, it is clearly important that their effects on populations of *U. appendiculatus* should be better understood.

Certain sources of monogenic resistance, though race-specific, are effective against a wide spectrum of pathogenicity. While the overuse of such genes invites their breakdown, careful deployment may possibly prolong their efficacy. The almost exclusive reliance in Europe on the 'Are' gene for anthracnose resistance has apparently produced a precarious situation: resistance-breaking races of *C. lindemuthianum* now occur in Europe, and alternative resistance genes are hurriedly being sought (Fouilloux, 1979). Yet elsewhere, in the absence of the matching race, the 'Are' gene may yet continue to be effective, especially if its overuse is avoided. Similarly, certain single genes may confer resistance to all races of rust recognized within a specific region (Ballantyne and McIntosh, 1977). Such genes are of considerable potential value and may be expected to provide lasting protection until the evolution, or accidental introduction, of a matching race. Some cultivars possess resistance which has been shown to be effective against all local races but the number of genes involved is apparently unknown. Examples include the cv. Westralia in Australia (Cass Smith *et al.,* 1951; Ballatyne, 1974b), cv. Ecuador 66 in East Africa (Howland and Macartney, 1966; Allen, 1975c) and cvs Costa Rica 1031 and Negro Jalpatagua in the USA (Meiners, 1979). In general, single genes for rust resistance have tended to provide transient protection and it is often argued that combinations of single genes, though race-specific, are likely to give more durable resistance. The greater the number of resistance genes, the less likely is the occurrence of a mutation to virulence to match it, provided that the component genes are not used singly in commercial cultivars. Otherwise, they might permit single, sequential mutations to virulence, recombination and thus the accumulation of multiple virulence. Such gene pyramiding has been advocated for the control of bean rust by Ballantyne (1974a), but the extent to which this approach may provide lasting protection from rust remains in doubt. Yet it is pertinent to note that the resistance of cv. NEP 2, known to be effective across a wide spectrum of pathogenicity (Allen, 1975c; Ballantyne and McIntosh, 1977; Meiners, 1979a), is race-specific (Allen, 1975c) and controlled by at least four genes (Ballantyne and McIntosh, 1977).

Differences in the field susceptibility of bean genotypes have been reported and considered evidence for the existence of race non-specific, or horizontal, resistance to rust (Leakey, 1970a; Leakey *et al.,* 1972; Vieira, 1972; Atkins, 1973; Ballantyne, 1974b). Atkins (1973) attempted to quantify these differences by measuring the rate of rust development using the parameters of frequency (number of rusted plants per plot), intensity (number of rusted leaflets per plant), and severity (per centage rusted leaf area). From this, she suggested that certain cultivars including cv. Mexican 142 are 'slow-rusters' in Uganda. Other slow-rusting cultivars have subsequently been identified in the USA (Meiners *et al.,* 1975; Meiners, 1979b) and Australia (Ballantyne and McIntosh, 1977). Although slow-rusters are usually found susceptible in seedling tests, few attempts have been made to determine the nature of slow-rusting in beans. Some 60 years ago, Fromme and Wingard (1921) detected quantitative differences in rust response and showed that there were several components of resistance to rust in beans. These components included a reduction in the number of infections, a smaller pustule size, and a longer incubation period. More recent studies have confirmed that cultivars may differ in infection ratio (Ogle and Johnson, 1974; Allen, 1975c; Groth and

Urs, 1982), latent period (van Breukelen, 1979), and pustule size (Pastor-Corrales and Schwartz, unpublished). Since such differences may be found between cultivars apparently lacking vertical (Section 10.2.1) resistance genes, it seems probable that these are components of race non-specific resistance to rust in beans (Allen, 1975c). Transgressive segregants for rust resistance have been found among advanced inbred populations derived from the slow-rusting cvs Apollo and California White Kidney (Ballantyne and McIntosh, 1977), suggesting a polygenic inheritance of slow-rusting. Despite the attention given to bean rust, race non-specific resistance is seldom sought, and even less frequently are attempts made to utilize it in breeding.

In general, it seems that where a high level of race-specific resistance is evident it has received attention almost to the complete exclusion of race non-specific resistance. This appears to be the case with rust, as discussed above, as well as with anthracnose. The pioneering work of Fromme and Wingard (1921) on rust has received little recognition, and the work of Bredemann and Koolman (1927) on anthracnose, to which Leakey (1979) has drawn attention, has similarly been overlooked. From their studies on the pathogenic variation within *C. lindemuthianum*, they warned that race-specific resistance seemed an undesirable breeding objective, suggesting instead that a major alteration in plant type might constitute an effective mechanism of resistance in the field. However, in cases where high levels of resistance are not available, quantitative methods have often been developed to detect degrees of partial resistance which more often appear to be race non-specific. A case in point would seem to be the work on canopy structure in relation to the susceptibility of cultivars to white mould (Schwartz *et al.*, 1978). It seems that where resistance has proved transient, more attention should be given to quantitative rather than qualitative methods of assessment, and to devising new breeding strategies designed to increase durability.

Conventional methods of breeding self-pollinated crops have been widely used in bean improvement and have led to the successful incorporation of resistance, especially that under oligogenic control, into agronomically superior backgrounds. Numerous genotypes possess combined resistance and these are of considerable value as sources of resistance in the further improvement of the crop. Combining resistance to several diseases implies the combination of a fairly large number of genes governing resistance. Hubbeling (1957) estimated that about 20 genes had to be combined in order to attain resistance to the six principal diseases in Europe, and Frazier *et al.* (1961) considered that a minimum of 8 genes, and/or many modifiers, were involved in combining resistance to dry root rot, rust, common mosaic, and yellow mosaic. In screening for combined resistance, a continuously heavy selection pressure is required and reliance is usually placed on the artificial induction of epidemics. Screening procedures, discussed more fully in Section 10.2.3, must reliably distinguish resistance in segregating populations, and those conducted under laboratory or greenhouse conditions must accurately predict field reaction (Webster *et al.*, 1980). Certain non-destructive screening methods are available for root pathogens (Wyatt and Fassuliotis, 1979).

Amongst the many cultivars of beans which have been developed, there are some which have combined resistance to four fungal root rots (Beebe *et al.,* 1981), to three bacterial diseases, common mosaic and anthracnose (Hagedorn and Rand, 1977), to common mosaic, yellow mosaic, halo blight and anthracnose (Fouilloux and Bannerot, 1977), and to common mosaic, golden mosaic and rust (Yoshii *et al.,* 1980). As was discussed in Section 2.1, disease prevalence varies markedly with environment so that the particular combination of resistance must be appropriate to the region for which the breeding programme is designed.

Conventional breeding methods, including bulk and pedigree selection and backcross procedures, have been successfully used in developing disease-resistant cultivars of beans, especially when resistance is controlled by relatively few genes. However, such methods impose severe restrictions on genetic progress; population improvement and recurrent selection methods, in which such restrictions are considerably reduced, would appear to have potential, particularly where quantitatively inherited resistance is involved (Bliss, 1977). Recurrent selection procedures should employ mechanisms which enhance natural outcrossing. Although genetic male sterility is available within *P. vulgaris* (Van Rheenan *et al*., 1979; Singh *et al*., 1980b), it has apparently not yet been utilized. Transgressive segregation for improved resistance has been detected from work on dry root rot (Boomstra and Bliss, 1977), common blight (Valladares-Sanchez *et al.,* 1979), golden mosaic (Yoshii *et al.,* 1980) and rust (Ballantyne and McIntosh, 1977), suggesting that in these cases a recurrent selection scheme might profitably be used. Recurrent selection procedures are now being utilized in breeding for resistance to dry root rot at Wisconsin (Bliss, 1977), and common blight and golden mosaic at CIAT (Temple and Song, 1980). A somewhat similar strategy has been applied to bean breeding in Ethiopia by Stoetzer (in Robinson, 1976). Stoetzer's programme, which was designed to produce new cultivars of white-seeded navy bean with horizontal resistance to all locally important pathogens, involved a series of random polycrosses relying on seed-coat colour to indicate cross-pollination.

It is obvious that the potential for the improvement of a crop rests heavily upon the diversity and availability of its genetic resources. There is considerable variation within *P. vulgaris* and extensive collections of germplasm are maintained; the world collection held at CIAT currently has about 27,000 accessions (Evans, 1980a, b; Temple and Song, 1980 CIAT, 1981). Nevertheless, in several cases other species of *Phaseolus* have been utilized as sources of disease resistance in interspecific hybridization. *P. coccineus* has been used as a donor of resistance to yellow mosaic (Rudorf, 1955; Dickson and Natti, 1968), brown spot (Hagedorn and Rand, 1976), dry root rot, black root rot (Baggett and Frazier 1959; Wallace and Wilkinson, 1965; Hassan *et al.,* 1971), anthracnose (Richardson and Evans, 1972), and white mould (Abawi *et al.,* 1978; Hunter *et al.,* 1981). Common blight resistance has been found in *P. acutifolius* in which it is under quantitative genetic control (Schuster, 1955; Honma, 1956). The bean cultivar Great Northern Nebraska No. 1

originated from an interspecific cross between *P. vulgaris* and *P. acutifolius* (Coyne *et al.*, 1965b), and a selection from it has been used in breeding the improved cvs Jules and Tara in which the inheritance of common blight resistance is again quantitative. Rather greater use has been made of interspecific hybridization to improve disease resistance in beans than of that in other food legumes. Further use may be expected from improved success in species hybridization both between the cultivated species and between cultivated and wild species, perhaps achieved through use of a greater number of genotypes in species hybridization (Evans, 1980a). Induced mutation has seldom been exploited for increased disease resistance in *P. vulgaris,* although Neto *et al.* (1977) found a mutant which possessed some golden mosaic resistance.

Considerable advances in the improvement of beans may be expected within the next decade. The Programa Cooperativo Centroamericano para el Mejoramiento de Cultivos Alimenticios (PCCMCA) and, later, the establishment of the bean programme at CIAT have stimulated international co-operation in research. The operation of the International Bean Yield and Adaptation Nursery (IBYAN) and the International Bean Rust Nursery (IBRN) are two examples. An important purpose of multilocational testing is the search for site non-specific resistance as an index of race non-specificity. In the case of the IBRN, no entry has yet remained resistant across all sites and seasons (Schwartz, 1980). This may largely reflect the difficulty of identifying race non-specific resistance relative to the ease of recognizing race-specific resistance to rust. The value of multilocational testing is likely to be limited as much by the degree of standardization and precision of disease evaluation as by the bean genotype on trial. While it is important to appreciate that in agriculture the economic contribution of race-specific resistance, though transient, can often be very considerable (Schwartz and Sanders, 1979), it seems that more emphasis should be given to the development of screening methods able to detect small quantitative differences in a breeding programme designed to accumulate polygenes. Meanwhile, sources of broad spectrum, yet race-specific oligogenic resistance should be utilized where matching races are absent; and the two types of resistance might perhaps be usefully combined.

Mention has already been made of the possible merits of gene pyramiding to stabilize rust resistance. An analogous approach would be the pyramiding of genes governing distinct components, or mechanisms, of resistance to bean diseases. Such a strategy might be expected not only to stabilize race-specific resistance, with which the above comments were concerned, but perhaps also to reinforce race non-specific resistance. Others have advocated the synthesis of resistance components: Coyne and Schuster (1974c) have suggested that genes controlling the reaction of different plant parts to bean bacterial pathogens be recombined; and Schwartz *et al.* (1978b) have recognized the potential for combining partial resistance to white mould with plant architectural characters such as upright growth habit and an open canopy, which constitute a disease escape mechanism. Similarly, leafy cultivars may be able to compensate for premature defoliation (Edje *et al.,* 1976a) and it seems possible that this may constitute a mechanism of escape from the consequences of premature defoliation. A compensatory mechanism is known also to operate against dry rot (*Fusarium solani*) in that plants are able to produce adventitious roots above the diseased area (Wheeler, 1969); but whether cultivar differences exist seems uncertain. Other plant characters may also be associated with disease resistance in beans. For instance, dry root rot resistance may be correlated with seed size (Wallace and Wilkinson, 1973), or with seed-coat colour (Burke and Silbernagel, 1980); resistance to both pythium and rhizoctonia root rots with seed-coat colour (Deakin and Dukes, 1975; Prasad and Weigle, 1976; Dickson and Boettger, 1977; Burke and Silbernagel, 1980); and common blight resistance with later maturity and indeterminate growth habit (Valladares-Sanchez *et al.,* 1979). There is some evidence that extracts from coloured but not from white-seeded cultivars may be fungitoxic, suggesting that the epistatic effect is associated with phenolic metabolism (Deakin and Dukes, 1975; Prasad and Weigle, 1976) but the linkage between seed-coat colour and resistance has now been broken (Dickson and Boettger, 1977). Certainly, some of these association can be overcome simply by expanding the genetic base. There are other opportunities for combining resistance components. For example, in breeding against common mosaic, the dominant 'I' gene and the recessive 'a' gene (Section 6.2.4) have been combined. But there may also be scope for the use of 'field resistance' which slows the rate of vector spread of common mosaic (Grogan and Walker, 1948): whether or not such a mechanism entails a host resistance to the aphid vector (Johnson, 1953) appears unknown. Nor is it known whether resistance to the beetle, *Epilachna* (Campbell and Brett, 1966) could be used to supplement hypersensitive resistance to pod mottle (Thomas and Zaumeyer, 1950b) which the beetle transmits. Vector non-preference does appear to play a role in curly top resistance, which depends also on a resistance to virus multiplication (Silbernagel and Jafri, 1974).

The best strategy for the successful management of pathogens of beans in tropical subsistence agriculture is likely to be an integrated one, in which the production of clean seed, the use of good husbandry including crop rotation, and the exploitation of genetic diversity will each play a role. More studies are required on the effects of cropping system on disease development, and much remains to be learnt of the nature of pathosystems balance in traditional agriculture. The development of disease-resistant cultivars will play a central role, especially if emphasis is given to the combination of distinct mechanisms of resistance where its durability or level of protection given by a single mechanism is inadequate. The measurement of disease priorities across environments is fundamental to breeding for combined resistance, with the objective of developing cultivars adapted to tropical environments with comprehensive, durable resistance appropriate for each ecological zone.

Chapter 7

Cowpea Diseases

7.1 EVOLUTION, ADAPTATION, AND PRODUCTION OF COWPEAS

The genus *Vigna* comprises some 160 species, most of them African (Steele and Mehra, 1980). Only seven species are cultivated and five of these are Asian. Here, we are concerned only with *V. unguiculata* (= *V. sinensis*), the cowpea. Five sub-species of *V. unguiculata* are recognized: ssp. *unguiculata*, the cowpea; ssp. *cylindrica*, the catjang; ssp. *sesquipedalis*, the yard-long or asparagus bean; ssp. *dekindtiana*, and ssp. *mensensis*, the progenitors of the cultivars. All five sub-species are fully interfertile (Steele and Mehra, 1980). Maréchal *et al.* (1978) have emphasized the unity of the cultivated forms by erecting 'cultigroups' within sub-species *unguiculata* in which they include the catjang (cultigroup *biflora*) and the yard-long bean (cultigroup *sesquipedalis*). Various outcrossing mechanisms exist (Rachie *et al.*, 1975; Lush, 1979) and gene exchange occurs between the weed forms and cultivars in West Africa (Rawal, 1975). It has been suggested that the wild sub-species *dekindtiana* is more likely to have been the progenitor of modern cowpea than the other wild sub-species, *mensensis* (Lush and Evans, 1981). Domestication has been associated with changes in pod structure and seed coat as well as with an increase in the rate of inbreeding (Lush, 1979; Lush and Evans, 1981). As a species, *V. unguiculata* is reproductively isolated. Attempts to hybridize cowpeas with other *Vigna* species have not yet been successful, although there is preliminary evidence that *V. vexillata* may have potential as a bridging host between cowpeas and the Asiatic species of *Vigna* (Evans, 1975).

Both Africa and India are centres of genetic diversity of *V. unguiculata*. Sub-species *unguiculata* is thought to be a West African neolithic domesticate (Faris, 1965; Rawal, 1975; Steele, 1976), whose progenitors were the wild and weed sub-species *dekindtiana* and *mensensis*. Rawal (1975) considers that an annual weed form which is adapted to disturbed habitats has evolved from the perennial wild cowpea which is now confined to humid and sub-humid regions of West Africa. Cultivars were perhaps then domesticated from the weed form. Modern photosensitive cultivars appear so well adapted to the

West African cereal farming system as to suggest that they have evolved with it, or even been domesticated into it (Steele, 1972; Steele and Mehra, 1980). Photoinsensitive forms of ssp. *unguiculata* evolved with adaptation to the yam-farming system of the more humid environments in West Africa. Limited archaeological evidence from stone implements from the Niger river and from Ghana suggests that these events took place around 3000 BC (Steele and Mehra, 1980).

The current view is that cowpeas were introduced from Africa to the Indian sub-continent between 2000 and 3500 years ago, probably at about the same time as the introduction of sorghum and millet. In India, sub-species *cylindrica* evolved from ssp. *unguiculata*, and ssp. *sesquipedalis* evolved in south-east Asia from vegetable types selected from ssp. *unguiculata*. Cowpeas reached Europe from Asia and perhaps from North Africa before 300 BC, and the Spanish took the crop to the West Indies in the seventeenth century AD. More cultivars reached the New World from West Africa with the slave trade, reaching the southern USA early in the eighteenth century (Steele and Mehra, 1980).

The cowpea is now grown throughout the tropics and sub-tropics, principally for its dry seed but also as a vegetable, for fodder and as a cover crop. The dried pulse may be left whole, or ground into a meal or paste, before preparation in a variety of ways (Dovlo *et al*., 1976). In Africa, the young leaves are eaten as a spinach or dried for use in soups, while the haulms are fed to livestock. Some cultivars provide a fibre obtained from the peduncles. In the humid tropics, especially in south-east Asia, cowpeas are grown as a leaf and pod vegetable and in north-east Brazil cowpeas are used both as a pulse and as fresh immature seed. In advanced agriculture, cowpeas are used mostly for fodder and as a cover crop, though Blackeye types are grown for dry seed on a large scale in California.

Cowpeas can be cultivated under a wide range of conditions. While the vegetable types (ssp. *sesquipedalis*) can be cultivated under relatively high rainfall conditions because mature seed is not the primary product, grain types (ssp. *unguiculata*) are adapted to hot, dry lowland tropical conditions. Although cowpeas are not particularly sensitive to drought during vegetative growth, large yield reductions may result if drought stress occurs during grain development. Similarly, responses to temperature vary according to crop growth stage. While, in the vegetative stage, high night temperature induces early flowering and low night temperatures lengthen the vegetative phase, in the reproductive phase it is day temperature which is of cardinal importance. High day temperatures decrease pod-filling time, and grain yield decreases. Conversely, low day temperatures lengthen the pod-filling period; the number of seed per pod and seed size increase, with a concomitant increase in yield (Summerfield and Wien, 1980). There are genetic differences between cowpea cultivars in their response to day and night temperatures, and in the rate of pod development. Such variation in response in part governs the adaptation of traditional cultivars to their environment (Wien and Summerfield, 1980) and to the system of farming of which they are part (Steele and Mehra, 1980; see Section 1.2.3).

West Africa, Brazil, and India are the main centres of production of dry seed. There are no accurate estimates of world production, but the annual total probably exceeds two million tonnes (Steele and Mehra, 1980). Nigeria annually produces some 0.75 million tonnes; other African countries which produce significant amounts include Niger and Upper Volta. Brazilian production, concentrated in the North-eastern region, is estimated at 0.5 million tonnes; annual consumption in the rural areas is 30–40 kg/caput (Rios *et al.*, 1980). Significant amounts of cowpeas are also produced in Central America and the Caribbean, most notably in Venezuela, Panama, El Salvador, Haiti, Trinidad (including green pod production), Jamaica, Colombia, and Guyana (Woolley, 1980). In India cowpeas are most important as pod vegetables and for fodder; they are a relatively minor grain legume (Steele and Mehra, 1980). In Africa the bulk of the crop is grown in traditional systems of subsistence farming, typically as a subordinate companion in complex crop mixtures which receive few inputs. Here grain yields are low, seldom exceeding 400 kg/ha, for which the ravages of diseases and especially pests are principally responsible (Singh and Allen, 1980). Although soil moisture may be limiting, it seems that traditional late sowing forfeits some yield to escape insect damage. In north-east Brazil cowpeas are locally a major crop, often in association with maize, in the medium-to-low rainfall zones of Pernambuco, Ceara, and Piaui states. Except on the most fertile soils, maize is considered the subordinate component of the mixture on many farms, a reversal relative to tropical Africa. Cowpeas are also extensively grown as a sole crop from which yields of 1000–2000 kg/ha are not uncommon. The principal constraints to production in north-east Brazil are low soil fertility and drought, especially irregularity of rainfall distribution; here pests and diseases are not the chief determinants of crop loss, as they are in Africa (Allen *et al.*, 1980). Compared with Africa, cowpeas in India are grown where human populations are dense and in diverse farming systems which range from intensive, modern, irrigated agriculture to subsistence traditional farming (Steele and Mehra, 1980). But the relative importance of the principal restraints to higher yields of cowpeas in India has not been clarified.

Like *Phaseolus* beans, cowpeas are principally a subsistence crop and as such they are very low yielding. However, average experimental yields of improved genotypes in Africa are usually about 1500 kg/ha (Smithson *et al.*, 1980), without the addition of heavy inputs of pesticide. Much higher yields may be obtained under optimum conditions. In California, commercial yields from Blackeye types may reach about 2500 kg/ha (Purseglove, 1968). This suggests that there is considerable opportunity for the improvement of cowpea yields from their current level of less than 400 kg/ha. The extent to which pathogens are responsible for the present low yields, and the gains to be expected from their control through developing disease-resistant cultivars, will be discussed in the remainder of this chapter.

7.2 THE PATHOLOGY OF COWPEAS

Cowpeas are susceptible to a very wide range of pests and pathogens which attack the crop at all stages of growth. In tropical Africa, which is both the chief region of production and the principal centre of genetic diversity of cowpeas and, presumably, also of its parasites, pests and diseases are the major restraints to increased production: without some protection, grain yields are negligible (Singh and Allen, 1980). In north-eastern Brazil, the second most important region of production, pests and diseases are subordinate to agronomic constraints (Allen *et al*., 1980). As with *Phaseolus* beans (Section 6.2), the economic importance of cowpea pathogens varies considerably with ecological zone. Thus, web blight, cercospora leaf spot, anthracnose, rust, cowpea (yellow) mosaic and bacterial pustule are considered the major diseases of the West African forest belt (Williams, 1975b; Oyekan, 1979), while scab (*Sphaceloma* sp.), septoria leaf spot, brown blotch (*Colletotrichum* spp.), cowpea aphidborne mosaic, and bacterial blight are the principal diseases of the African savannahs (Allen *et al*., 1981b). *Asochyta phaseolorum* causes a blight which is destructive at elevations above 1000 m in Africa and tropical America. In dry land areas of north-eastern Brazil the most important diseases are cowpea severe mosaic, cowpea aphidborne mosaic and its relatives, scab, leaf smut (*Entyloma vignae*), cercospora leaf spot, and powdery mildew (Allen *et al*., 1980). In California and the southern states of the USA fusarium wilt and root knot nematode have long been important, especially in sandy areas (Webber and Orton, 1902; Mackie, 1934; Hawthorne, 1943; Toler *et al*., 1963). Rust, *Septoria* leaf spot, powdery mildew, cercospora leaf spot, bacterial blight, and cowpea banding mosaic are among the major diseases of cowpea in India (Verma and Patel, 1969; Patel and Jindal, 1970; Sharma and Varma, 1975b; Sokhi and Sohi, 1976). In general, however, crop loss expectation from disease has been inadequately quantified. More precise studies, such as those made by Schneider *et al*. (1976), are required particularly where a pathogen causes inconspicuous, and thus possibly insidious, damage. Despite a relatively extensive literature, few reviews of the diseases of cowpea are available. Williams (1975b) has described the diseases present in Nigeria, and Singh and Allen (1980) have reviewed the subject worldwide. A colour-illustrated manual covering the most important diseases is also available (Singh and Allen, 1979).

Some 40 species of fungi are cowpea pathogens; the most important are shown in Table 20 where an attempt is made to categorize them by the symptoms they cause. However, any such grouping is necessarily somewhat arbitrary, and there is often overlap. For instance, *Sclerotium rolfsii* and *Macrophomina phaseolina* can cause leaf spots as well as the stem diseases under which they are given (Weimer and Luttrell, 1945; Singh and Allen, 1979); and scab, listed in Table 20 as a pod disease, involves the development of symptoms on all aerial parts of affected plants (see Section 7.2.2). Various other species of fungi, omitted from the table, are also recorded as cowpea

TABLE 20 Fungal Diseases of Cowpea

Disease	Causal fungi	Distribution	Importance/Crop loss	References
1. Seed and seedling diseases				
Seed decay and seedling mortality	Caused principally by *Pythium aphanidermatum* (Edson) Fitz. and *Rhizoctonia solani* Kühn though *R. bataticola* (Taub.) Butl. and *Phytophthora* spp. cause seedling mortality in some areas	Wisespread	Locally and seasonally damaging. Up to 75% incidence recorded in Nigeria	Tompkins and Gardner, 1935; Weimer, 1949; Nirwan and Upadhyaya, 1972; Onuorah, 1973; Williams, 1975a
2. Stem, collar and root rots, and wilts				
Anthracnose	*Colletotrichum lindemuthianum* (Sacc. and Magn.) Bri. and Cav.	Widespread but local, being recorded from E and W Africa and Brazil	Locally important. Losses up to 50% recorded in Nigeria	Angus, 1962–66; Onesirosan and Barker, 1971; Williams, 1975b; Rios *et al.*, 1980
Phytophthora stem rot	*Phytophthora vignae* Purss and *P. cactorum* (Leb. and Cohn) Schroet.	Widespread but local. Reported from USA (*P. cactorum*), Australia and India (*P. vignae*). *P. vignae* is known from adzuki bean (*Vigna angularis*) in Japan; an unidentified *Phytophthora* causes a blight of cowpea in Tanzania	Locally important; especially damaging in Queensland, Australia	Weimer, 1949; Purss, 1953, 1957; Nirwan and Upadhyaya, 1972; Kitazawa *et al.*, 1978; P. N. Patel and C. Kuwite, pers. comm., 1979
Pythium stem rot	*Pythium aphanidermatum* (Edson) Fitz.	The pathogen occurs worldwide but the disease is reported only from Nigeria, Tanzania, and Brazil	Minor; field incidence normally ranges between 0.5–10% but occasionally up to 30%	Onuorah, 1973; Williams and Ayanaba, 1975; Ebbels and Allen, 1979; Rios *et al.*, 1980
Sclerotium stem rot	*Sclerotium rolfsii* Sacc. (perfect state = *Corticium rolfsii* Curzi)	Widespread	Minor	Mordue, 1974a; Maduewesi, 1975; Williams, 1975b

Ashy stem blight	*Macrophomina phaseolina* (Tassi) Goid (sclerotial state = *Rhizoctonia bataticola* (Taub.) Butl.)	Widespread	Minor	Luttrell and Weimer, 1952; Williams, 1975b
Fusarium collar and dry root rot	*Fusarium solani* (Mart.) Sacc. (perfect state = *Nectria haematococca* Berk. and Br.)	Widespread but local. Recorded from Brazil, Puerto Rico, Nigeria, Uganda, Malaysia, and the Philippines	Minor; perhaps locally important	Bailey, 1966; da Ponte and da Silva, 1977; Liu, 1977; R. M. Abilay, J. Mukiibi and N. G. Vakili, pers. comm., 1977
Fusarium wilt	*Fusarium oxysporum* f.sp. *tracheiphilum* (E. F. Smith) Snyd. and Hans.	Widespread. Reported from the USA, Brazil, Nigeria, Uganda, India, Malaysia, and Australia	Locally and seasonally important. Over 50% mortality recorded in Nigeria, and field incidence up to 75% obtained in India	Kendrick, 1931; Armstrong and Armstrong, 1950, 1980; Simmonds, 1955; Singh and Sinha, 1955; Holliday, 1970; Oyekan, 1975; Williams and Liu, 1976; J. Mukiibi, pers. comm., 1977; Rios *et al.*, 1980
Verticillium wilt	*Verticillium* sp.	USA and Australia	Minor	Simmonds, 1955; McNeill, 1976
3. Foliar diseases				
Cercospora leaf spot	*Pseudocercospora cruenta* (Sacc.) Deighton (perfect state = *Mycosphaerella cruenta* Latham) and *Cercospora canescens* Ell. and Mart.	Widespread	Major. Defoliation can lead to yield losses of up to 20% (*C. canescens*) and over 40% (*P. cruenta*)	Latham, 1934; Verma and Patel, 1969; Schneider *et al.*, 1976; Fery *et al.*, 1977
Target spot	*Corynespora cassiicola* (Berk. and Curt.) Wei	Worldwide	Generally minor though considered important in the Yangtze valley of China and in southern USA. Often in mixed infections with *C. canescens*	Wei, 1950; Ellis and Holliday, 1971; Singh and Allen, 1979

TABLE 20 (*continued*)

Disease	Causal fungi	Distribution	Importance/Crop loss	References
Septoria leaf spot	*Septoria vignae* P. Henn., *S. vignicola* Vasant Rao, *S. kozo-polzanskii* Nikolajeva and *S. vignae-sinensis* Sawada	Widespread. *S. vignae* reported from E and W Africa and Brazil; *S. vignicola* occurs in eastern Africa and India. *S. kozo-polzanskii* is known only from the USSR and Zambia, and *S. vignae-sinensis* is reported from Taiwan	Probably important in savannahs of Africa though extent of crop loss unknown. Relative importance of *S. vignae* and other species is also uncertain	Sokhi and Sohi, 1976; Allen, 1979a; Singh and Allen, 1979; Allen, unpubl. (IMI 248316a, leg. 1980); Rawal and Sohi, 1981
Chaetoseptoria leaf spot	*Chaetoseptoria wellmanii* Stevenson	USA and C America	Minor	Tehon, 1937; Muller, 1953
Aristastoma leaf spot	*Aristastoma guttulosum* Sutton and *A. oecono-micum* (Ellis and Tracy) Tehon	Nigeria and the USA	Minor	Tehon, 1933; Lefebvre and Stevenson, 1945; Bailey, 1966
Leptosphaerulina leaf spot, pepper spot	*Leptosphaerulina trifolii* (Rost.) Petr. = *L. vignae* Tehon and Stout	Widespread. Recorded from USA, Brazil, Malawi, Ethiopia, and India	Minor	Tehon and Stout, 1928; Booth and Pirozynski, 1967; Peregrine and Siddiqi, 1972; Anahosur *et al.*, 1974; Allen, unpubl. (IMI 207118, leg. 1976; IMI 236789, IMI 236792b, leg. J. B. Smithson 1979)
Ascochyta blight, zonate leaf spot	*Ascochyta phaseolorum Sacc.* (= ?*Phoma exigua* Desm.) and *A. boltshauseri* Sacc.	Widespread in Africa and Central America. Reported recently from India	Major. Causes severe losses under cooler conditions at elevations above 1000 m	Sutton and Waterston, 1966; Stewart and Yirgou, 1967; Boerema, 1972; Moreno, 1975; Singh *et al.*, 1978; Singh and Allen, 1979
Web blight	*Rhizoctonia solani* Kühn (perfect state = *Thanatephorus cucumeris* (Frank) Donk.)	Worldwide	Major. Devastating under hot, wet conditions in the low tropics	Mordue, 1974a; Williams, 1975b; Oyekan, 1979; Singh and Allen, 1979
Dactuliophora leaf spot	*Dactuliophora tarrii* Leakey	Widespread in tropical Africa	Minor	Leakey, 1964; Mukiibi, 1969a, 1976a
False rust, yellow blister	*Synchytrium dolichi* (Cooke) Gaüm.	Widespread in tropical Africa	Minor, though occasionally damaging	Robinson, 1960; Mukiibi, 1969b; Singh and Allen, [illegible]

Pink rust	*Phakopsora pachyrhizi* Syd.	Widespread in tropical Africa. *P. pachyrhizi* occurs also in Asia and Puerto Rico but some isolates are apparently not pathogenic to cowpea	Minor; perhaps seasonally of economic importance in southern Nigeria but crop loss estimates not made	Singh and Allen, 1979; see also Section 2.2.5
Brown rust	*Uromyces appendiculatus* (Pers.) Ung. = *U. vignae* Barcl. = *Aecidium caulicola* P. Henn.	Widespread	Locally and seasonally damaging	Fromme, 1924; Williams, 1975b; Sokhi and Sohi, 1976
Leaf smut	*Entyloma vignae* Batista = ?*Protomycopsis phaseoli* Ramak. and Subram., but more comparative studies warranted	Tropical America (*E. vignae*), Africa and India (*P. phaseoli*)	Locally important in Ceara, Brazil where yield losses of 30–40% are reported	Batista *et al.*, 1966; da Ponte, 1966; Vakili, 1978a; Prabhu and Albuquerque (1980). See discussion in Section 2.2
Powdery mildew	*Erysiphe polygoni* DC	Widespread	Locally important in tropical America and India, but of minor importance in Africa	Fennell, 1948; Sohi and Rawal, 1975; Singh and Allen, 1979
4. Pod diseases				
Scab	*Sphaceloma* sp. (presumed to be the conidial state of *Elsinoë phaseoli* Jenk.)	Widespread in tropical Africa; also in Surinam, NE Brazil and Bangalore, India	Major. Probably the most damaging fungal pathogen in savannah areas though often not recognized as such	Sivanesan and Holliday, 1971b; Emechebe, 1980; Rios *et al.*, 1980; J. B. Smithson, pers. comm., 1981; see also Section 2.2.4
Cladosporium pod spot, scab	*Cladosporium vignae* Gardner	Widespread but local. Recorded from the USA, Brazil, Uganda, Zimbabwe and Australia	Minor	Gardner, 1925; Hansford, 1937; Weber, 1941; Simmonds, 1955; Whiteside, 1959 (cited by Smartt, 1976b); da Ponte, 1976
Brown blotch	*Colletotrichum capsici* (Syd.) Butl. and Bisb. and *C. truncatum* (Schw.) And. and Moore	Though the pathogens are pantropical, the disease is recognized only from Nigeria, Upper Volta, Cameroun, Kenya, Zambia, and Japan	Locally important in the African savannahs	Hemmi, 1955; Singh and Allen, 1979; Emechebe, in press; D. J. Allen, unpubl.
Choanephora pod rot, lamb's tail pod rot	*Choanephora cucurbitarum* (Berk. and Rav.) Thaxt. and *C. infundibulifera* (Currey) Sacc.	Widespread. Reported from the USA, Brazil, Nigeria, and India	Minor. Of occasional importance under humid conditions, exacerbated by insect damage. 1–2% loss recorded in Georgia, USA	Lefebvre and Weimer, 1939; Toler and Dukes, 1965; Wilson and Jose, 1968; Fery and Cuthbert, 1972; Williams, 1975b; da Ponte, 1976

pathogens: these include *Alternaria atrans* (Sohi and Rawal, 1974; da Ponte, 1976), *Cephalosporium zonatum* Sawada (Turner, 1971), *Myrothecium roridum* Tode ex Fr. (Williams and Liu, 1976), *Phaeoseptoria* sp. (leg. J. B. Smithson, IMI 236790), *Phoma* spp. (Kirby and Archer, 1927; leg. D. J. Allen, IMI 207113, 207119-21), and *Phyllosticta* spp. (da Ponte, 1976; Williams and Liu, 1976), all from leaves. *Botrytis cinerea* Pers. ex Fr. causes a pod rot under irrigation in Brazil (da Ponte, 1976), and *Aspergillus flavus* Link. ex Fr. occurs as a seed contaminant (Esuruoso, 1975; Sinha and Khare, 1977) though its importance as a source of aflatoxin in cowpea is uncertain (see Section 4.2.3).

Certain fungi, not known to be seedborne in cowpea, have restricted distributions though they may be locally important. Examples of the latter include: *Phytophthora vignae*, which causes a damaging blight in Queensland, Australia (Purss, 1957; Section 2.2.1); *Entyloma vignae*, the cause of leaf smut which is important in Ceara, Brazil (da Ponte, 1966; Section 2.2); and *Synchytrium dolichi*, which causes a false rust (Figure 28) in Africa where it can be important in Uganda (Mukiibi, 1969b). However, many fungal pathogens are seedborne (Singh and Chohan, 1974; Esuruoso, 1975; Sinha and Khare, 1977; Emechebe and McDonald, 1979) and are distributed wherever the crop is cultivated, although the diseases they cause may be important less widely. Overall, the most important fungal diseases of cowpea are considered to be anthracnose, scab, cercospora leaf spot, web blight, and ascochyta blight, and it is these that are discussed in detail later in this chap-

Figure 28 False rust of cowpea caused by *Synchytrium dolichi*, Ibadan, Nigeria

ter. When more studies have been made, it may well emerge that others, given little emphasis here, are in fact of greater economic importance than hitherto recognized. Indeed, the recognition of the importance of scab is recent (Emechebe, 1980), and more attention should be given to the septoria leaf spots and to brown blotch (*Colletotrichum capsici* and *C. truncatum*).

Virus diseases can cause very severe damage to cowpea crops and, accordingly, viruses have perhaps received more attention than other pathogens. Natural infections of cowpea often involve virus mixtures (Kuhn, 1964; Bozarth and Shoyinka, 1979) which may lead to synergism (Harrison and Gudauskas, 1968; Fischer and Lockhart, 1976b; Pio-Ribeiro *et al.*, 1978) and probably also to erroneous records when these are based only on symptomatology (see Section 3.1). Nomenclatural confusion tends to compound the problem; of the many virus diseases reported in the literature, only the more reliably identified are given in Table 21, in which details of the geographical distribution, transmission, and approximate economic importance are summarized. Other, less fully described, viruses which have been reported from natural infections of cowpea include pea leafroll, transmitted by aphids in a persistent manner (Kaiser, 1972b), and cowpea ringspot which is a cucumovirus (Phatak *et al.*, 1976); both occur in Iran. Cowpea banding mosaic virus is a cucumovirus known in India (Chenulu *et al.*, 1968; Sharma and Varma, 1975a, b; Prakash and Joshi, 1980). More work is required to compare the cucumoviruses of cowpea. Similarly, various potyviruses have been described from this host but it remains uncertain whether they are clearly distinct from cowpea aphidborne mosaic virus (CAMV; Bock and Conti, 1974). In 1955, Anderson reported a virus from Florida which he referred to as Blackeye cowpea mosaic virus (BlCMV) which is thought to be a potyvirus closely related to, but distinct from others from legumes (Pio-Ribeiro *et al.*, 1978; Lima *et al.*, 1979; Taiwo and Gonsalves, 1982). A virus identified as BlCMV has been reported subsequently from Brazil (Lin *et al.*, 1981b), and from India (Mali and Kulthe, 1980) but, as Conti (1981) has pointed out, in the latter case no reference was made to CAMV. Potyvirus isolates from cowpea have also been referred to bean yellow mosaic virus (Brierly and Smith, 1962; Kuhn *et al.*, 1965; Harrison and Gudauskas, 1968) but there is evidence that at least one such isolate is, in fact, BlCMV (Pio-Ribeiro *et al.*, 1978). Bean common mosaic virus has also been reported from cowpeas (Uyemoto *et al.*, 1973; Sachchinanda *et al.*, 1973; Behncken and Maleevsky, 1977) and in Brazil an apparently distinct potyvirus named cowpea green vein-banding virus has been isolated (Lin, 1979; Lin *et al.*, 1981b). In view of the known strain variation within CAMV (Bock, 1973d), it appears that further comparative studies of cowpea potyvirus isolates are warranted. There is also a need for further research on whitefly-transmitted virus diseases. While cowpea mild mottle virus, which is apparently of little economic importance to cowpeas, has been satisfactorialy characterized, little is known of the possible geminivirus aetiology of cowpea diseases. In Nigeria, a geminivirus is associated with a golden mosaic (Figure 29; Vetten *et al.*,

TABLE 21 Virus Diseases of Cowpeas

Disease	Causal virus	Distribution	Transmission		Importance/Crop loss	References
			Vector	Seed (%)		
Mosaic	Cowpea aphidborne mosaic virus (CAMV) (including blackeye cowpea mosaic virus, B1CMV)	Widespread	Aphid	0–40	Major. Losses of up to 87% reported	Bock and Conti, 1974; Kaiser and Mossahebi, 1975; Ladipo, 1977; Taiwo and Gonsalves, 1982
Mosaic (stunt, in mixed infections with potyvirus)	Cucumber mosaic virus (CMV)	Worldwide, but not known from legumes in E Africa	Aphid	4–26	Minor, except in mixed infections	Brantley *et al.*, 1965; Bock *et al.*, 1975; Fischer and Lockhart, 1976b; Pio-Ribeiro *et al.*, 1978
Yellow mosiac	Alfalfa mosaic virus (AMV)	Worldwide	Aphid	0?	Locally important; early infection can reduce seed yield by 50%	Kaiser, 1979; Jaspars and Bos, 1980
Mosaic (chlorotic spot)	Sunnhemp mosaic virus (ShMV)	India, Nigeria, and USA	?	4–20	Potentially important. 56% crop loss recorded	Lister and Thresh, 1955; Kassanis and Varma, 1975; Sharma and Varma, 1975b

Mosaic	Broad bean wilt virus (BBWV)	Europe, Japan	Aphid	0	Unknown; probably negligible	Tsuchizaki *et al.*, 1970; Selassie-Gebre *et al.*, 1976
Yellow mosaic	Cowpea (yellow) mosaic virus (CYMV)	Africa; occasionally reported from America	Beetle	1–5	Major. Yield losses up to 95% reported	Chant, 1959, 1960; Bock, 1971; Gilmer *et al.*, 1974; van Kammen and de Jager, 1978
Severe mosaic	Cowpea severe mosaic virus (CSMV)	Warm temperate and tropical America	Beetle	0–10	Major. Losses of up to 80% recorded	Shepherd, 1964; Lima and Nelson, 1977; de Jager, 1979
Mosaic	Southern bean mosaic virus (SBMV)	Warm temperate and tropical America; W Africa	Beetle	3–44	Potentially important. Yield loss of 59% reported	Shepherd, 1971; Lamptey and Hamilton, 1974; Shoyinka and Okunsanya, 1976; Givord, 1981
Mottle	Cowpea mottle virus (CMeV)	Nigeria	Beetle	0–10	Locally important. Yield loss may exceed 75%	Robertson, 1966; Shoyinka *et al.*, 1978; Bozarth and Shoyinka, 1979; Allen *et al.*, 1981a; Allen *et al.*, 1982
Chlorotic mottle	Cowpea chlorotic mottle virus (CCMV)	Warm temperate and tropical America	Beetle	0	Apparently negligible	Bancroft, 1971; Hobbs and Fulton, 1979
Mild mottle	Cowpea mild mottle virus (CMMV)	Africa, Asia	Whitefly	90	Unknown; probably minor	Brunt and Kenten, 1973; 1974; Iwaki *et al.*, 1981
Golden mosaic	Cowpea golden mosaic virus (CGMV)	Nigeria; similar diseases occur elsewhere in Africa, and in Asia	Whitefly	0	Locally important; almost complete crop loss may occur	Singh and Allen, 1979; Vetten and Allen, in press; Vetten *et al.*, unpublished

Figure 29 Cowpea golden mosaic which is spread by whiteflies, Onne, Port Harcourt, Nigeria (photograph courtesy H. J. Vetten)

unpublished), the symptoms of which closely resemble similar diseases of cowpea in Niger, Kenya, Tanzania (Singh and Allen, 1979), and Pakistan (Ahmad, 1978). A whitefly-transmitted yellow fleck disease occurs in India (Sharma and Varma, 1976), and uncharacterized virus-like diseases occur elsewhere (Gathuru and Buruchara, 1979).

Despite the number and diversity of the virus diseases of cowpea, on present evidence only three are of significant economic importance internationally. These are: cowpea aphidborne mosaic, which is the most widely distributed virus; cowpea severe mosaic virus (CSMV), which is amongst the principal pathogens throughout tropical America; and cowpea (yellow) mosaic virus (CYMV), which can cause major losses to the crop in Africa. These three virus diseases will be discussed more fully in subsequent sections of this chapter.

Witches' broom, phyllody, and 'little leaf' diseases have been reported to affect cowpeas in Indonesia (Semangoen, 1958), Australia (Bowyer and Atherton, 1970), India (Ramiah and Narayanasamy, 1974; Varma *et al.*, 1978), Nigeria (D. J. Allen, unpublished), and Brazil (Allen *et al.*, 1980). The presence of mycoplasma-like bodies in diseased plants has been confirmed in certain cases (Bowyer and Atherton, 1970; Varma *et al.*, 1978) but the relationship between these diseases remains unknown (see Section 3.2). An unidentified pathogen thought to be a mycoplasma induces an intense leaf yellowing of cowpea in Kenya (Bock *et al.*, 1971).

Few bacterial pathogens of cowpea have been described or studied in detail. Pseudomonads are seldom reported from cowpea crops, perhaps because the genus *Pseudomonas* appears better adapted to relatively cool conditions whereas cowpeas are essentially a lowland tropical crop. Nevertheless, *Ps. syringae* pv. *syringae* is known as a cowpea pathogen in the USA (Gardner and Kendrick, 1923, 1925), Tanzania (Riley, 1960), and Australia (Wilson, 1936), and a bacterial spot is caused by an unidentified fluorescent group 1b pseudomonad in Ethiopia (Allen, 1979a). *Ps. syringae* pv. *tabaci* occasionally causes wildfire (Tisdale, 1924), and *Ps. solanacearum* sometimes causes bacterial wilt in cowpea (da Ponte and Santos, 1973). Two important diseases are caused by xanthomonads: *X. campestris* pv. *vignicola* causes bacterial blight; and certain isolates apparently referable to the same pathovar cause bacterial pustule. These two diseases are discussed in Section 7.2.8.

At least 17 species belonging to 9 genera of nematodes have been found associated with cowpea roots (Caveness, 1963; Bridge, 1973, 1981). The root knot nematodes (*Meloidogyne* spp.) are much the most important; they are discussed in Section 7.2.9. Angiosperm parasites of cowpea include *Striga gesnerioides*, with pink or mauve flowers, and *Alectra vogelii*, with yellow flowers (Rattray, 1932; Okonkwo and Nwoke, 1975; Singh and Allen, 1979); both belong to the family Scrophulariaceae. *S. gesnerioides* is the more important species, particularly in the Sudan savannah belt of West Africa. Dodder (*Cuscuta* sp.) has also been found on cowpeas in Nigeria (P. Matteson, personal communication, 1980) and Brazil (Allen *et al.*, 1980).

The following sections will focus in greater detail on the diseases which are considered to be of principal importance to cowpeas in the tropics, paying particular attention to their control through host-plant resistance.

7.2.1 Anthracnose

Cowpea anthracnose is caused by the fungus *Colletotrichum lindemuthianum* which is also the cause of bean anthracnose (see Section 6.2.2). However, unlike the latter, anthracnose is principally a stem disease of cowpeas (Onesirosan and Barker, 1971). Individual lesions are lenticular to circular, sunken and tan to brown in colour. Highly susceptible cultivars develop large spreading lesions which rapidly coalesce to girdle the stems, and often also the branches, peduncles, and petioles. Lesions may also develop on leaves and pods but the symptoms on the stem and branches are more severe (Williams, 1975b). Grain yield losses of 35–50 per cent have been recorded in experimental plots of the highly susceptible cv. TVu 91 in southern Nigeria (Williams, 1974a) from where an anthracnose epidemic was first reported. *C. lindemuthianum* is able to survive in infected cowpea debris (Onesirosan and Sagay, 1975) as well as in seed (Onesirosan and Barker, 1971; Esuruoso, 1975; Emechebe and McDonald, 1979). Despite the wide geographical distribution of *C. lindemuthianum* (Mordue, 1971a), the distribution of cowpea anthracnose is not well known, compounded perhaps by the frequent confusion of scab with anthracnose; 'brown blotch', caused by *C. capsici* and *C. truncatum*, is a distinct disease (Singh and Allen, 1979). In Nigeria, anthracnose can become devastating, particularly under experimental conditions in the more humid regions of southern Nigeria; it is effectively absent from the drier savannahs in the north (Section 2.2.4). The disease is also reported from Brazil (Rios *et al*., 1980) and from Zambia (Angus, 1962–66); early records from the USA are reviewed by Onesirosan and Barker (1971).

Anthracnose can be partially controlled by fungicides (Williams, 1975c) and presumably by the use of disease-free seed but neither is likely to be practicable in subsistence farming; the development of anthracnose-resistant cultivars remains the most promising control measure. Sources of anthracnose resistance have been identified (Williams, 1977a), and a detached petiole test has been used to screen cowpea germplasm (Williams, 1974b). In collaborative work between the International Institute of Tropical Agriculture (IITA) in Nigeria and Wye College, London University, it was found that several distinct mechanisms, operating at different points in the infection process, appeared to restrict fungal growth. These mechanisms included failed penetration, penetration leading to encapsulation of hyphae within 'reaction material, failure of the primary vesicle to develop (associated with the appearance of a red halo in the cell wall around the appressorium), and cell death which leads to the cessation of fungal development (hypersensitivity). Hypersensitive resistance, which is associated with the accumulation of phytoalexins, is thought to be governed by a single dominant gene in cv. New Era (TVu 57; Skipp, 1975; IITA, 1976, 1977) whose resistance is race-specific: a resistance-breaking race (referred to as I57) was found in Ibadan, Nigeria in 1972 (IITA, 1973). There is some evidence (Fakorede, 1976) that the greater pathogenicity of I57 is associated with a faster mycelial growth rate and more abundant sporulation at higher temperatures, relative to the 'original' race (I47), first isolated at Ife, Nigeria in 1969. Comparison of the two isolates has shown that I57 is also very slightly more tolerant of certain phytoalexins (Preston, 1975a).

In the wake of this demonstration of the potential transience of hypersensitive resistance, interest was aroused by the observation that certain cultivars, including cvs. Prima and VITA 3 (TVu 76 and 1190), though susceptible in detached petiole tests in the laboratory, remained largely anthracnose-free in the field. Tissue age was found to be important in such 'field-resistant' cultivars: only the youngest petioles developed symptoms whereas mature petioles of field susceptible cultivars showed obvious damage. Examination of symptomless but inoculated tissue from cultivars with such adult plant resistance revealed that penetration had occurred almost as frequently as in younger tissue but that cells containing expanded primary vesicles were reddish brown and granular, this intense response being analogous to hypersensitivity, which it apparently replaces (Skipp, 1975). Essentially, the mechanism of 'field resistance' differs from both hypersensitive resistance and susceptibility in that when the primary vesicle eventually does kill the first penetrated cell, the latter still has the capacity to produce a sufficiently anti-fungal environment to prevent the development of secondary hyphae. By contrast, the hypersensitive reaction is a very rapid response to infection by an incompatible race and, in a susceptible reaction, the initial growth of the primary vesicle goes undetected by the host cell (IITA, 1977).

Inoculation of cowpeas with *C. lindemuthianum* leads to the production of at least eight anti-fungal compounds, the most important of which are vignafuran and methyl phaseollidin isoflavan (Lampard, 1974; Preston, 1975a, 1975b; Preston *et al*., 1975). Although there is little doubt that phytoalexins are present collectively in sufficient concentrations to control anthracnose in resistant seedlings (Preston, 1975a), their precise role in resistance remains uncertain. While vignafuran is the major phytoalexin of anthracnose-infected leaves, this substance is present only in trace quantites in the stem, perhaps reflecting the particular susceptibility of stem tissue. Several anthracnose-resistant cultivars, including TVu 57 and 317, produce methyl phaseolliniso-flavan in infected stem tissue but this is replaced by phaseollidin in the anthracnose-susceptible cultivar TVu 32 and the field resistant TVu 76 (Preston, 1975a, b). On current evidence, it appears that in cowpeas there are several distinct mechanisms of anthracnose resistance, some mediated by the synthesis of phytoalexins in response to infection. While it appears that resistance based on the hypersensitive reaction may be race-specific and under monogenic control, nothing is known of the specificity of other mechanisms, and information on the genetic control of anthracnose resistance (IITA, 1974; Fakorede *et al*., in press) is generally scanty. While more work is required to elucidate the nature of the interaction between cultivars and races in cowpea anthracnose, salutory lessons may be learnt from experience of bean anthracnose resistance (Section 6.2.2).

7.2.2 Scab

Cowpea scab is caused by a species of *Sphaceloma* often presumed to be the conidial state of *Elsinoë phaseoli* (Emechebe, 1980; see Section 2.2.4). Owing to the comparative difficulty of isolating the causal fungus, coupled with its confusion with anthracnose, the geographical distribution of scab is uncertain. Cowpea scab is widespread in tropical Africa where it is known in northern Nigeria (Emechebe, 1980), Upper Volta, Kenya, Tanzania, Ethiopia (D. J. Allen, unpublished), Uganda (Mukiibi, 1979), Zambia (Angus, 1962–66), and Zimbabwe (Bates, 1955). The disease also occurs in Surinam (Sivanesan and Holliday, 1971b), north-eastern Brazil (Allen *et al*., 1980; Rios *et al*., 1980), and southern India (J. B. Smithson, personal communication, 1981). *Cladosporium vignae* can apparently also cause a scab of cowpea whose symptoms, illustrated by Gardner (1925), appear very similar to those induced by *Sphaceloma*, but whether or not the two diseases have at times been confused is not known. Cladosporium scab is considered important in the USA (Weber, 1941) and is also reported from Brazil (da Ponte, 1976), Australia (Simmonds, 1955), Uganda (Hansford, 1937), and Zimbabwe (Whiteside, 1959). There appears to be a need for comparative studies

Figure 30 Scab of cowpea caused by *Sphaceloma* sp., Utinga, Bahia, NE Brazil

between the cladosporium and sphaceloma scab diseases as well as an investigation of the relationship between *Sphaceloma* and *Elsinoë* in cowpeas.

The symptoms of sphaceloma scab are characterized by the development of silvery grey, circular to oval lesions first on stems, then on leaves and their petioles, peduncles and, ultimately, the pods (Figure 30). Leaves of diseased plants are often cupped and bear numerous, small whitish scab lesions along the veins, and 'shot-hole' lesions occur. In severe infections, peduncle lesions coalesce, causing distortion and flower bud abortion. Those pods which do develop are usually heavily spotted, curled, and transformed to 'mummies', containing virtually no seed (Singh and Allen, 1979; Emechebe, 1980). Crop loss estimates have not been made, but it is evident that under experimental conditions a severe epidemic could cause almost complete loss. Although scab has usually been regarded as a relatively minor disease, recent evidence suggests this is not the case, rather that scab is one of the most destructive diseases of cowpea, not only throughout the savannahs of Africa but also in north-eastern Brazil (Allen *et al.*, 1980; Emechebe, 1980; Allen *et al.*, 1981b).

Sphaceloma scab survives on infected debris and probably also on, or in, seed so that rotation, sanitation, and seed treatment are potential control measures (Sivanesan and Holliday, 1971b); Emechebe, 1980). The fungicide, Mancozeb, also effectively controls scab (Mukiibi, 1979). However, it has recently been shown that cowpea cultivars differ in their degree of scab susceptibility, and partial resistance has been identified (Rios *et al.*, 1980; Allen *et al.*, 1981b). Nothing is known of the variation in pathogenicity between isolates of *Sphaceloma* from cowpea, nor has the mechanism and inheritance of scab resistance been investigated. Yet the recent independent identification of cv. VITA 4 as scab-resistant in Nigeria and Brazil suggests that scab resistance may, under some circumstances, provide useful protection against populations of *Sphaceloma* across diverse environments (Allen *et al.*, 1981b).

7.2.3 Cercospora leaf spot

Cercospora leaf spot is caused by *Pseudocercospora cruenta*, which is the imperfect state of *Mycosphaerella cruenta*, and *Cercospora canescens*; both species are widespread in warmer regions (see Section 2.2.3). *P. cruenta*, the more important of the two pathogens, is reported to be of economic importance on cowpeas in India (Verma and Patel, 1969), Egypt (Fahim *et al.*, 1969), Nigeria (Schneider *et al.*, 1976), Puerto Rico (Vakili, 1977), and the USA (Fery *et al.*, 1977). Since infection occurs late, it has been widely held that cercospora leaf spot does not cause significant yield loss. This is now known not to be the case. Both *C. canescens* and *P. cruenta* can cause considerable leaf spotting after flowering, when defoliation can lead to yield losses of up to 20 and over 40 per cent respectively, at least in southern Nigeria (Schneider *et al.*, 1976). Fery *et al.* (1977) have shown that *P. cruenta* can cause up to 36 per cent loss in seed yield under North American conditions;

all components of seed yield were adversely affected. *C. canescens* induces circular to irregular cherry red to reddish brown lesions on both leaf surfaces, whereas leaf spots caused by *P. cruenta* appear first as a chlorosis on the upper surface of leaves which become dotted with necrotic lesions; these enlarge until the entire affected area becomes necrotic. The lower leaf surfaces infected by *P. cruenta* bear areas of profuse sporulation in which the masses of conidiophores appear as downy greyish black mats (Williams, 1975b; Singh and Allen, 1979).

Both pathogens are seedborne in cowpea (Williams, 1975b), although Emechebe and McDonald (1979) were unable to demonstrate seed-to-plant transmission of *C. canescens*. Both fungi are controlled by the use of clean seed as well as by the foliar application of fungicides (Williams, 1975c; Oyekan, 1979). Cowpea germplasm has been evaluated for resistance to cercospora leaf spot and numerous cultivars with resistance to *P. cruenta* have been identified (Verma and Patel, 1969; Fery *et al*., 1976; Vakili, 1977; Williams, 1977a). Some of these possess resistance against both leaf spot pathogens, and some, but not all, sources of resistance are effective at several widely different locations (Williams, 1975d, 1977a; Raj and Patel, 1977; Vakili, 1977). No studies have yet been made of the pathogenic variation within either of the species so that the relative race-specificity of leaf spot resistance remains unknown. Similarly, little is known of the underlying mechanisms of cercospora leaf spot resistance, although there is some evidence that pre-formed fungal toxins may contribute to resistance to *C. canescens*. Noting that there appeared to be an effect of leaf age on disease development in the field, Schneider and Sinclair (1975) found that conidial germination and germ tube growth were each inhibited on the surface of young, but not old, leaves of the leafspot-susceptible cv. Prima. Similar inhibition was caused by diffusates from apical but not basal leaves of cv. Prima, and by diffusates from both apical and basal leaves of the resistant cv. Lalita. Somewhat similar results were obtained with *P. cruenta* by Ekpo and Esuruoso (1977), who found that leaf extracts from old plants of cv. New Era tended to enhance conidial germination while extracts from young plants did not. Each investigation indicates that susceptibility in part depends on a pre-penetration phenomenon influenced by host plant maturity. The role that phytoalexins may play in post-penetration resistance of cowpea to cercospora leaf spot remains unknown. Resistance to *P. cruenta* is governed by at least two separate genes each of which conditions a high level of resistance. The cultivar CR 17-1-34 possesses a single dominant gene the expression of which is not significantly affected either by modifying factors or by environment. The cultivar Ala. 963.8 possesses a single incompletely dominant gene the expression of which is strongly influenced by environment when in the heterozygous condition. The two genes are neither allelic not linked (Fery *et al*., 1976; Fery and Dukes, 1977). The extent to which cercospora leaf spot resistance has been utilized in cowpea improvement is discussed in Section 7.3.

7.2.4 Web blight

Web blight is caused by *Rhizoctonia solani*, the sclerotial state of *Thanatephorus cucumeris*, which is an ubiquitous soil fungus. *R. solani* is an aggregate species on which there is a substantial literature, reviewed most notably in Parmeter (1970). The extremely wide host range of the species, and the diversity of the diseases which it causes, together account for the attention which *R. solani* has received as a plant pathogen. Yet there is specialization among isolates which are referable essentially to one of three ecological types, based on their vertical distribution on hosts (Durbin, 1959). Thus aerial, surface, and subterranean types of *R. solani* are recognized. Isolates adapted to aerial habitats are typically fast-growing, carbon dioxide-intolerant types which produce sclerotia readily. Subterranean isolates, which tend to have slower growth rates, tolerate higher carbon dioxide concentrations and produce sclerotia less readily; surface types are intermediate (Durbin, 1959). In cowpea, diseases of the stem base are caused by subterranean types while both seedling mortality and web blight are associated with isolates adapted to aerial habitats (Onesirosan, 1977). Since Onesirosan found that none of his isolates from the soil were aerial types, he has suggested that web blight may have only a cursory association with the soil. Carry-over depends on the availability of infected debris (Onesirosan and Sagay, 1975), weed hosts and seedborne inoculum (Onesirosan, 1975). Secondary spread, which is favoured by relatively high temperature and moisture, relies on the dissemination of sclerotia (Weber, 1939; Onesirosan, 1975), and basidiospores (Echandi, 1965; Saksena and Dwivedi, 1973), as well as on direct mycelial growth through plant contact.

Web blight is a serious disease of cowpea throughout the humid tropical lowlands. It is among the major diseases in the forest belt of West Africa (Oyekan, 1979) though it can also be damaging under locally waterlogged conditions in the drier savannahs. Oyekan (1979) estimates that cowpea seed yields may be reduced by 28–40 per cent by web blight in Nigeria, but field observations suggest that almost complete crop loss may occur under epidemic conditions. The disease is reported from cowpeas also in India (Dwivedi, 1977) and Brazil (Rios *et al*., 1980). Initial symptoms on the leaves appear as small, circular, reddish brown spots which enlarge, becoming surrounded by irregular shaped water-soaked areas. Under humid conditions, the lesions develop rapidly and coalesce, leading to extensive blighting and defoliation. All aerial parts of the plant are affected, becoming covered with sclerotia which resemble a coarse brown dust (Weber, 1939; Williams, 1975b; Singh and Allen, 1979).

Measures for the cultural control of web blight include the use of clean seed, field sanitation, and the avoidance of dense monocropping during periods of peak rainfall (Onesirosan and Sagay, 1975; Williams, 1975b). Certain foliar fungicides may also be effective (Oyekan, 1979). Although there is evidence that some variation exists in the degree of susceptibility between cowpea genotypes, no high levels of resistance have been found (Oyekan *et al*., 1976; IITA, 1980). Cultivars such as VITA 1 possess partial resistance (IITA, 1980) but under heavy disease pressure such protection as this affords is inadequate, and the integration of several control measures appears warranted. Detection of partial resistance depends heavily upon the appropriate choice of a screening technique. Consistent small differences were found between genotypes in Nigeria through the choice of a site, season, and plant spacing which favoured the uniform development of a web blight epidemic (IITA, 1980). Resistance tends to increase with host-plant age (Christou, 1962; IITA, 1976), and to decrease with increasing inoculum concentration (Warren, 1975). No studies have been made of the mechanism of web blight resistance in cowpea, nor is any information available on the extent of pathogenic variation among aerial types of *R. solani* from this host. Work on other hosts with a range of isolates has shown that a degree of host specificity exists within the fungus, and that several distinct resistance mechanisms may operate at different stages in the infection process. A pre-penetration resistance, controlled by an inhibitory host exudate, operates in certain cases (Kerr and Flentje, 1957; Flentje *et al*., 1963), while in others a resistance to penetration, or a hypersensitive resistance, may occur (Flentje, 1957), the latter being associated with the production of phytoalexins (Smith *et al*., 1975). However, it remains uncertain whether each of these resistance mechanisms may operate within a single host species, nor is the relative race-specificity of these mechanisms known. Attempts have been made to define physiologic races of *R. solani* (Le Clerg, 1939; Kernkamp *et al*., 1952) on the basis of variation in morphology, growth rate, and pathogenicity. An isolate's habitat is often consistently associated with a particular degree of host specificity as well as with a particular mode of penetration (Houston, 1945; Flentje and Saksena, 1957; Dodman *et al*., 1968). There is evidence that ecological factors provide a selective mechanism which has permitted divergent development within *R. solani*, resulting in a degree of genetic isolation. Thus, distinct anastomosis (hyphal fusion) groups have been recognized. Anastomosis behaviour appears to be a good measure of the natural genetic relatedness of isolates: morphology, pathogenicity, growth rate, mode of penetration, habitat, and anastomosis behaviour are often strongly associated (Dodman *et al*., 1968; Parmeter *et al*., 1969; Sherwood, 1969). Aerial isolates from web blight are referable to anastomosis groups (AG)1 or 2, while AG3 and 4 isolates are commonly associated with diseases at or below the soil surface. Serological methods have been developed for determining the anastomosis grouping of isolates (Adams and Butler, 1979). In cognizance of these studies, work is now required on the genetics of the interaction between web blight isolates of *R. solani* and cultivars of cowpea so as to elucidate the degree of specificity as an aid to resistance breeding.

7.2.5 Ascochyta blight

Ascochyta blight, which is sometimes also called ascochyta leaf spot, zonate leaf spot or 'target spot' (the latter is preferably avoided because of confusion with corynespora leaf spot; see Table 20), is caused by the fungus *Ascochyta phaseolorum*. *A. phaseolorum* occurs worldwide as a pathogen of various crops, including beans as well as cowpeas (Sutton and Waterston, 1966). *A. boltshauseri* is also known to infect cowpeas (Stewart and Yirgou, 1967). Species of the related form genus, *Phoma*, notably *P. bakeriana* and *P. exigua*, are also commonly associated with leaf and pod lesions of cowpea (Kirby and Archer, 1927; D. J. Allen, unpublished), relevant since Boerema (1972) considers that *P. exigua* is a synonym of *A. phaseolorum*. Ascochyta blight is of economic importance to cowpeas principally under cool humid conditions, as found at elevations above 1000 m in southern and eastern Africa, extending westward to the highlands of northwestern Cameroun (Allen, 1979a). The disease is also reported from Costa Rica (Moreno, 1975), Brazil (Rios *et al*., 1980), India (Singh *et al*., 1978), and Australia (Alcorn, 1968). Isolates referable to *P. exigua*, as well as to *P. bakeriana*, have been obtained from cowpea outside the ecological range of ascochyta blight but it remains unclear whether such isolates can cause blight under suitable environmental conditions. *A. phaseolorum* is considered to be a weak parasite and the disease is seldom destructive unless the crop has been predisposed to infection by unfavourable growing conditions, by wind or insect damage, or by prior infection with another pathogen (Sutton and Waterston, 1966; Pegg and Alcorn, 1967). Leaf lesions are at first circular, with grey to brown centres surrounded by a yellow halo. Under favourable conditions, such lesions become zonate (Figure 31) and spread rapidly, causing extensive blighting of leaves leading to defoliation. Lesions develop also on stems and pods, and epidemics can lead to complete destruction of the crop (Angus, 1962–66; Sutton and Waterston, 1966; Singh and Allen, 1979).

A. phaseolorum is seedborne in cowpea (Noble and Richardson, 1968). The pathogen may also survive in infected debris and in certain perennial hosts (Sutton and Waterston, 1966; Pegg and Alcorn, 1967). The use of clean seed, field sanitation, isolation from infected reservoirs, and the use of *Pennisetum* windbreaks have been suggested as cultural control measures (Pegg and Alcorn, 1967). The efficacy of the latter is confirmed by Moreno (1975) who has shown that ascochyta blight is less severe, and spreads more slowly, in crops of cowpea associated with maize than when in pure stands of cowpea. Foliar application of the fungicide Mancozeb can also afford some protection (Mukiibi, 1979). Small differences exist between cultivars in their susceptibility, but so far no satisfactory degree of resistance has been identified. The cvs Floricream and V-44 possess partial resistance to *Ascochyta* in Costa Rica (Moreno, 1975) and other genotypes were possibly slightly less susceptible in Zambia (IITA, 1978). Susceptibility appears to increase with age (Angus, 1962–66) and it has been suggested that early maturity may enable escape

Figure 31 Zonate lesions caused by *Ascochyta phaseolorum* on cowpea leaves, Mufulira, Zambia

from total destruction (Sutton and Waterston, 1966). Crossan (1958) has shown that isolates of *A. phaseolorum* may differ in pathogenicity, but there is probably little host-specificity (Alcorn, 1968). For the moment, it appears that reliance must be placed on cultural practices for the control of ascochyta blight, at least until attempts have been made to increase the level of resistance, perhaps through intercrossing sources of partial resistance.

7.2.6 Aphidborne mosaic and Blackeye cowpea mosaic

Cowpea aphidborne mosaic virus (CAMV) and blackeye cowpea mosaic virus (B1CMV) are two closely related but apparently distinct potyviruses. Recent evidence indicates that there is much confusion between the two. Isolates identified as CAMV from Kenya and Nigeria are now thought to be more correctly assigned to B1CMV, to which isolates from the USA are also referred (Taiwo and Gonsalves, 1982; Taiwo *et al*., 1982). For present purposes, the two viruses are treated together as variants of cowpea aphidborne mosaic, but the reader should bear in mind that new information may in time prove this to be unsatisfactory.

Cowpea aphidborne mosiac virus (CAMV) is the most widespread virus infecting cowpeas. It is reported from Europe, Asia, Africa (Bock and Conti, 1974), and Australia (Behncken and Maleevsky, 1977); similar or identical viruses occur in India (Mali and Kulthe, 1980; Conti, 1981), the USA (Anderson, 1955; Lima *et al.*, 1979), Brazil (Lin, 1979), and the south-west Pacific (Bock and Conti, 1974). Within Africa, CAMV has been positively identified from Kenya, Tanzania, Uganda (Bock, 1973a), Egypt (Phatak, 1974), Morocco (Fischer and Lockhart, 1976a), and Nigeria (Ladipo, 1976). The seed transmissibility of CAMV, reflected in its wide geographical distribution, probably also governs virus survival in dry areas. CAMV is present in the dry savannah of northern Nigeria (Rossel, 1977), where the development of dry season irrigation schemes leads to the outbreak of epidemics through encouraging the massive build-up of populations of aphids by which the virus is spread (Raheja and Leleji, 1974). The nature and severity of the symptoms induced by CAMV vary with host cultivar, virus isolate, and the time of infection. Diseased plants show varying amounts of green vein-banding (Figure 32) or interveinal chlorosis, leaf distortion, blistering and stunting (Bock and Conti, 1974). Crop loss from natural field infection has been estimated at 87 per cent in Iran (Kaiser and Mossahebi, 1975).

The production of disease-free seed is a potential measure for the cultural control of CAMV, particularly if certified seed can be produced in areas where the virus is not known to occur (Zettler and Evans, 1972). Field inspection and roguing of diseased plants may help to eliminate seedborne inoculum but, since there is evidence that CAMV may also occasionally be seed-transmitted in symptomless plants (Aboul Ata *et al.*, 1982), there is

Figure 32 Green vein-banding induced by cowpea aphid-borne mosaic virus in cowpea, Ilonga, Tanzania

potential value in implementing a rapid indexing procedure for the detection of CAMV in seed-lots. It is well established that the level of seed transmission varies with cultivar (Bock and Conti, 1974; Kaiser and Mossahebi, 1975; Ladipo, 1977; Aboul Ata *et al*., 1982), and Ladipo (1977) has identified certain cowpea cultivars in which apparently no seed transmission of CAMV occurs. While detection of virus in seed depends partly on the sensitivity of the assay (Goodman and Oard, 1980), the incidence of seed transmission in some cultivars might be sufficiently low to control the disease significantly; but the minimum level of seedborne inoculum at which an epidemic may be initiated is unknown. It seems there may be potential for the exploitation of 'resistance to seed transmission' in cowpea breeding (Aboul Ata *et al*., 1982).

At least six species of aphid are vectors of CAMV (Bock and Conti, 1974). Of these, *Aphis craccivora* is a widespread and common pest of cowpea in Africa and India (Singh and Allen, 1979) but the relative importance of this and other aphid species as virus vectors is not known. Nor has any attention been given to the effect of vector control on the spread of mosaic. Since CAMV is transmitted in the non-persistent manner, it is unlikely that insecticides would provide adequate protection from mosaic. Cowpea cultivars possessing resistance and tolerance to *A. craccivora* have been identified (Chari *et al*., 1976; Singh and Allen, 1979, 1980), but it is not known whether their use would limit the rate of CAMV spread; this is currently under investigation in Nigeria (G. Atiri, personal communication, 1980). Sources of CAMV resistance have been identified (Ladipo and Allen, 1979b) and are being used in cowpea improvement in Nigeria. Three types of resistance are recognized: an 'immunity', in which inoculated plants remain symptomless and virus cannot be recovered; a hypersensitive resistance shown by the development of necrotic local lesions without systemic spread of virus; and virus tolerance, in which systemic infection occurs without the appearance of symptoms. Immunity is governed by a single recessive gene (Taiwo *et al*., 1981; Walker and Chambliss, 1981). Comparison of the reactions of certain cultivars to isolates from Kenya, Nigeria, and the USA shows that pathogenic variation occurs in CAMV (Taiwo, 1978; Ladipo and Allen, 1979b) but the implications this may have for breeding are not yet clear. Subsequent study of a 'resistance-breaking' isolate of CAMV from south-eastern Nigeria suggests that cultivars with immunity to one isolate may possess considerable tolerance to those isolates which can cause systemic infection (IITA, 1980), implying that resistance-breaking isolates need not necessarily constitute a threat to the durability of resistance. Although this requires further investigation, it is evident that certain cowpea cultivars including TVu 2480 possess resistance effective against diverse isolates of CAMV. However, these studies were limited to isolates now apparently referable to BlCMV. Although preliminary investigations have failed to identify a common source of resistance to BlCMV and 'true' CAMV (Taiwo *et al*., 1982), it remains to be seen whether differences in the geographical distribution of the two viruses, and the degree of pathogenic variation between them, will necessitate an altered resistance breeding strategy, in a manner resembling the recent distinction between cowpea severe mosaic and cowpea (yellow) mosaic viruses which we discuss next.

7.2.7 Cowpea (yellow) mosaic and severe mosaic

Beetle-transmitted viruses which cause mosaic symptoms in cowpea have long been recognized (Smith, 1924; Dale, 1949; Chant, 1959). Isolates from Nigeria, Surinam, Trinidad, and Arkansas have been compared and shown to belong to two distinct strains, the yellow and severe strains of cowpea mosaic virus (Chant, 1962; Agrawal and Maat, 1964; Shepherd, 1964; van Kammen, 1971). More recent studies have demonstrated that these strains are better regarded as distinct viruses, called cowpea mosaic virus (van Kammen and de Jager, 1978) and cowpea severe mosaic virus (CSMV; de Jager, 1979), respectively. Because the SB isolate, from Surinam, is the type member of the comovirus group, the name cowpea mosaic virus is correctly maintained for the yellow strain. For present purposes, however, 'yellow' is inserted in parenthesis so as to minimize possible confusion between what we shall thus call cowpea (yellow) mosaic virus (CYMV), CSMV and the other 'cowpea mosaic viruses' which abound in the literature.

CYMV is essentially an African virus. It is widespread and common in the humid and sub-humid regions of southern Nigeria (Chant, 1959, 1960; Gilmer *et al.*, 1974) and it occurs also in Togo (IITA, 1978), Tanzania (Patel, 1982), and coastal districts of Kenya (Bock, 1971). On current evidence, CYMV appears better adapted to relatively humid conditions than to the dry savannahs of Africa. Isolates referable to CYMV have been obtained also from the western hemisphere, including Surinam (Agrawal and Maat, 1964) and the USA (McLaughlin *et al.*, 1977), though the latter is apparently now in doubt (Fulton and Scott, 1979). An isolate from Cuba (Kvíčala *et al.*, 1973) has also been assigned to CYMV because of its ability to infect *Chenopodium amaranticolor* systemically (de Jager, 1979). This, too, seems dubious, especially since certain serotypes of CSMV are now known to cause systemic infection in *C. amaranticolor* (Lin *et al.*, 1981a). Surveys of cowpeas in warm temperate and tropical America suggest that CSMV is much the most common comovirus and that CYMV is not a common cause of mosaic in the western hemisphere (Fulton and Scott, 1979; Fulton and Allen, 1982). CSMV is not known to occur in Africa or Asia.

Both CYMV and CSMV induce a range of symptoms including various types of mosaic, leaf mottling, distortion and blistering, apical necrosis, and death of the plant. Symptomatology is not diagnostic and the severity of symptoms varies with the virus isolate, cowpea cultivar, time of infection and, perhaps, environmental conditions. CYMV can cause losses in seed yield of 40–100 per cent. Although infections at flowering time can cause significant decreases in growth and yield, early infection causes the greatest crop loss (Chant, 1960; Gilmer *et al.*, 1974). CYMV is said to be seed-transmitted at a level of 1–5 per cent (Gilmer *et al*., 1974) but the rather restricted geographical and ecological distributions of the virus suggest that more studies of the seed transmissibility of CYMV may be warranted. Little is known of the relative importance of seedborne inoculum and alternative hosts in the carry-over of CYMV. In view of its narrow natural host range (Robertson, 1963), it seems possible that wild and weedy cowpeas (Rawal, 1975), or volunteer seedlings of cultivars, may be sources of primary infection for newly sown crops at the start of the rainy season in southern Nigeria. The principal vector of CYMV is the chrysomelid beetle, *Ootheca mutabilis*, the larval development of which occurs in the soil. A period of quiescence, or diapause, apparently ensures the beetle's survival of the dry season, synchronizing the emergence of adults with the onset of the rains and with the emergence of the cowpea crop (Ochieng, 1978). The early establishment of *O. mutabilis* on cowpea seedlings is thus ensured, presumably accounting for the conspicuous spread of CYMV in the field. CYMV is transmitted by other species of beetle, and apparently also by grasshoppers (Whitney and Gilmer, 1974) and a nematode (Caveness *et al*., 1975), but the claim of transmission by thrips (Whitney and Gilmer, 1974) has now been refuted (Allen and van Damme, 1981).

In tropical America the field incidence of CSMV may reach 100 per cent and severe infections can result in 50 per cent reduction in fresh plant weight and in number and weight of pods (Debrot and de Rojas, 1967), and up to 80 per cent loss in seed yield (Lima and Nelson, 1977). The level of seed transmission of CSMV varies from 0–10 per cent (Dale, 1949; Shepherd, 1964; Haque and Persad, 1975), apparently depending on the cultivar. The natural host range of CSMV is wider than that of CYMV and varioui s virus reservoirs have been identified (Dale, 1949; Alconero and Santiago, 1973; Diaz, 1974; Lima and Nelson, 1977; Lin *et al*., 1982). About ten species of beetle are CSMV vectors (Smith, 1924; Walters and Barnett, 1964; Jansen and Staples, 1970b; Valverde *et al*., 1978); *Ceratoma ruficornis* and *C. trifurcata* are amongst the most important (de Jager, 1979). Neither CSMV nor CYMV exhibits specificity for beetle species within the sub-family Galerucinae of the Chrysomelidae (Jansen and Staples, 1971), despite the geographically distinct ranges of the two principal virus-vector combinations. Like the viruses they transmit, *Ceratoma* and *Ootheca* are exclusively American and African genera, respectively.

Since both viruses are thought to be seedborne, there may be potential for the cultural control of these mosaic diseases, through the production of clean seed and the enforcement of quarantine legislation. Whether or not a 'resistance to seed transmission', as discussed with cowpea aphidborne mosaic (Section 7.2.6), could be exploited is not known. The incidence of both CYMV and CSMV may be less in cowpeas grown in association with maize, so there may be potential for the development of cropping systems which provide a barrier to vector spread (Section 1.3.2). The efficacy of insecticidal sprays in controlling the spread of beetleborne mosaic diseases has not been examined, nor are there reports of beetle vector resistance in cowpeas. This neglect of potential cultural and chemical control measures probably reflects the considerable success achieved through breeding for resistance to both these diseases.

Sources of resistance to CYMV have been identified in Nigeria (Wells and Deba, 1961; Robertson, 1965; Williams, 1977b) and in the USA (Beier *et al.*, 1977). Immunity, hypersensitive resistance, and tolerance are each recognized. Hypersensitive resistance is controlled by a single dominant gene in the cv. Dixielee; three additive loci, each independent of the locus governing the hypersensitive reaction, are responsible for the CYMV tolerance shown by cv. Alabunch (Bliss and Robertson, 1971). Little is known of the underlying mechanisms of resistance. It appears that immunity, which is apparently not merely an exaggerated form of hypersensitivity, may operate in several ways. At the cellular level, protoplasts from immune cultivars are usually susceptible to inoculation with CYMV but cv. Arlington protoplasts are apparently able to prevent the replication of CYMV. In other cultivars, such as PI 339709, Black, Louisiana Purchase, Victor K798, Groit and Brabham K892, there may be a delay in CYMV increase in protoplasts. Protoplasts of PI 339609 tend to die after inoculation with CYMV, suggesting that something other than delayed virus replication is responsible for the observed low virus titre in this cultivar (Beier *et al.*, 1977, 1979).

Resistance to CSMV appears to be rare. Immunity to CSMV has been identified in cv. Macaibo (Lima and Nelson, 1977) and in the lines TVu 612, TVu 1460-2, TVu 1948, and TVu 2480. Other genotypes appear to possess tolerance (Fulton and Allen, 1982). TVu 3901 (PRCMV-10R-61) has been found resistant in some experiments (N. G. Vakili, personal communication, 1976; H. A. Hobbs, personal communication, 1979; Klas, 1979; Rios *et al.*, 1980) but not in others (J. P. Fulton, personal communication, 1981). There is a conspicuous lack of correspondence between resistance to CSMV and to CYMV: cultivars with resistance to one are almost invariably susceptible to the other, thus further supporting the separation of these two viruses. For instance, sources of resistance to CYMV isolates from Nigeria or from Surinam (the SB isolate) are almost all susceptible to isolates of CSMV from Puerto Rico (Allen, 1977b; Beier *et al.*, 1977); and TVu 612 and 2480 which possess resistance to a range of CSMV isolates (Fulton and Allen, 1982) are both highly susceptible to CYMV under field conditions in Nigeria (D. J. Allen, unpublished). Strong differential interactions such as these apparently do not occur, however, between cowpea cultivars and virus isolates within either CYMV or CSMV. Although pathogenic variation exists within each virus, present evidence suggests that isolates vary in their degree of aggressiveness rather than in virulence (*sensu* van der Plank, 1968). Serologically distinguishable isolates of CSMV may consistently induce symptoms of varying severity in susceptible cowpea cultivars, the most aggressive isolate causing the severest disease. A constant ranking of virus isolates and of cowpea cultivars is generally obtained, and in no case have the less aggressive isolates been able to infect a cultivar immune to the most aggressive isolate (Fulton and Allen, 1982). The same seems to apply to CYMV. Robertson (1965) obtained differential reactions in some cultivars in response to two Nigerian isolates; those immune to one isolate often expressed hypersensitive resistance to the other. However, in no case did a cultivar with resistance to one isolate become systemically infected by the other. Williams (1977b), who recognized that pathogenic variants occurred, was nonetheless able to identify genotypes with resistance effective against a range of CYMV isolates, and Ladipo (personal communication, 1981) found remarkably little difference in host reaction to isolates of CYMV from cowpea and *Crotalaria* among more than 150 cowpea accessions. Nevertheless, 'resistance-breaking' mutants of CYMV have been isolated (de Jager, 1979). Although such variants are able to overcome the hypersensitive resistance of cv. Early Red, suggesting they possess increased virulence, they normally induce milder symptoms and occur at lower virus titres than the parent strain, suggesting a decreased aggressiveness. It remains to be seen whether such mutants are able to survive under natural conditions. Meanwhile, it is pertinent to note that a variant of CYMV occurs under field conditions in Tanzania where it causes a severe top necrosis in cowpea cultivars resistant to 'normal' isolates of CYMV (Patel, 1982).

7.2.8 Bacterial blight and pustule

Bacterial blight, caused by *Xanthomonas campestris* pv. *vignicola*, is widely distributed in all major cowpea-growing regions. The disease was first reported in the USA from Oklahoma in 1931. The disease became important in that state during the 1940s, losses being estimated at nearly 10 per cent (Larsh, 1943; Hoffmaster, 1944). Simultaneously, bacterial blight was reported, then fully described, from Texas where its incidence reached 60 per cent in some plantings (Dunlap, 1943; Watkins, 1943; Burkholder, 1944). Later reports of the disease came from Florida (Lefebvre and Sherwin, 1945) and Illinois (Boewe, 1948). Now bacterial blight is known to occur in Puerto Rico (Vakili *et al*., 1975), Brazil (Rios *et al*., 1980), and India (Patel and Jindal, 1970), as well as throughout Africa. The first reliable record of bacterial blight in Africa appears to have come from Tanzania in 1964 (Ebbels and Allen, 1979). The disease occurs also in Kenya (Kaiser and Ramos, 1979), Ethiopia, Zambia, Cameroun, Upper Volta, Niger (D. J. Allen, unpublished), and Nigeria (Williams, 1975b). A claim that the pathogen may have been introduced recently into East Africa on seed from Nigeria appears insubstantial since records from East Africa antedate those from West Africa (Kaiser and Ramos, 1979; Allen, 1981). Ekpo (1979) has estimated that the potential loss in seed yield in cv. Ife Brown may exceed 26 per cent. However, since it appears that cv. Ife Brown is only moderately susceptible to bacterial blight (Allen *et al*., 1981c), the extent of crop loss might be expected to be considerably greater than 26 per cent in a highly susceptible cultivar.

There are distinct syndromes of bacterial blight: stem canker, foliar blight, and seedling mortality. Stem canker is the symptom most commonly referred to in the early North American literature (Burkholder, 1944; Hoffmaster, 1944) but foliar blight is the most common symptom under African field conditions. The initial symptoms are tiny water-soaked dots on leaves. These dots remain small and the surrounding tissue dies, developing a tan to orange coloration with a yellow halo. On heavily infected leaves the necrosis coalesces so that large areas of leaf are affected. The pathogen also infects the stem, causing cracking (canker), and causes water-soaking of pods from where the pathogen enters the seed (Williams, 1975b). Such symptoms are easily distinguished from those associated with bacterial pustule caused by a xanthomonad, for the moment probably best regarded as a strain of *X. campestris* pv. *vignicola*.

Although isolates from bacterial pustule tend to make slower growth in culture than do those from bacterial blight (J. F. Bradbury, personal communication, 1980), on other bacteriological grounds the two pathogens are effectively indistinguishable. There is some evidence that the bacterial blight pathogen is better adapted to the dry savannahs than bacterial pustule, which is known only from Nigeria (Williams, 1975b), Tanzania (Patel, 1981), and Kenya (Kaiser and Ramos, 1979). The suggestion that bacterial pustule also occurs in Brazil (Singh and Allen, 1979) has not yet been substantiated. The first collections of the disease were made by D. R. W. Watson at two sites in Tanzania in 1964–66 (IMI B2281 and B2962; Ebbels and Allen, 1979). A 'bacterial leaf spot', attributed to *X. phaseoli* (*X. campestris* pv. *phaseoli*), was recorded at about the same time in Nigeria (Bailey, 1966) where both cultivated and wild cowpeas are affected by bacterial pustule (Williams, 1975b). The symptoms first appear on the undersurface of leaves as tiny, dark, raised, water-soaked lesions which enlarge to about 3 mm diameter. Dark necrotic spots occur on the upper surface of infected leaves. Pustules in older infections are usually dry and sunken, and heavily infected leaves become chlorotic and fall prematurely (Williams, 1975b). The potential crop loss in a resistant (cv. VITA 3) and a susceptible cultivar (cv. Prima) is estimated at 1.8 and 26.6 per cent respectively (Ekpo, in press).

Both pathogens are seedborne in cowpea (Shekhawat and Patel, 1977; Emechebe and McDonald, 1979; Kaiser and Ramos, 1979); the production of clean seed is a potential control measure in some areas (Boettinger and Bowers, 1975). However, resistance to both diseases has been recognized so that breeding for resistance is the principal means of control. Differences in the susceptibility of cultivars to bacterial blight were first detected by Hoffmaster (1944) who identified cvs Buff, Iron, and Victor as resistant. The high degree of resistance of these three cultivars was confirmed by Sherwin and Lefebvre (1951) who identified many other sources of resistance in an extensive screening of germplasm. Two distinct mechanisms of resistance are recognized: hypersensitivity, and a partial resistance expressed as reduced and delayed disease development (Patel and Jindal, 1970; Allen *et al*., 1981c).

Although the histology of a susceptible reaction has been investigated (Shekhawat *et al*., 1977), the histology of resistance appears not to have been examined. Hypersensitive resistance appears to be controlled either by a single dominant or a single recessive gene, depending on the cross. There is evidence that modifying factors may also be involved in cv. Iron whose resistance is governed by a recessive gene (Lefebvre and Sherwin, 1950; Singh and Patel, 1977). Too little information is available to assess the probable durability of each resistance mechanism. Pathogenic variation has been demonstrated among isolates of *X. campestris* pv. *vignicola* and attempts have been made to define distinct races (Sherwin and Lefebvre, 1951; Jindal *et al*., 1981). Pathogenicity may be host-specific, revealed by the differential interactions between cultivars and isolates shown by Jindal *et al*. (1981); yet they obtained no evidence that sources of known monogenic resistance, as possessed by the cvs P309, P426, P910, and Iron (Singh and Patel, 1977), are race-specific. Furthermore, data presented by Jindal *et al*. (1981) suggest that variation in pathogenicity may also be quantitative; the most aggressive isolates were also the more virulent. This also suggests that increased pathogenicity may not necessarily be associated with decreased survival ability. While the extent to which pathogenic variants pose a potential threat to the durability of resistance remains unknown, it is interesting to note, as Jindal *et al*. (1981) have done, that certain cowpea cultivars including cvs Iron and Victor have been recognized as resistant to bacterial blight over at least 25 years (Hoffmaster, 1944; Sherwin and Lefebvre, 1951; Patel and Jindal, 1970).

Sources of resistance to bacterial pustule have also been identified (Williams, 1977a) and thought to be independent of resistance to bacterial blight. Again, there is evidence of pathogenic variation which appears to be host-specific. Three races are recognized (Patel, 1981). Amongst the cultivars found resistant in Nigeria (Williams, 1977a), those possessing hypersensitive resistance governed by a single dominant gene have proved to be susceptible to bacterial pustule in Tanzania where more virulent races have been isolated. However, in cultivars such as TVu 43, resistance does not depend on hypersensitivity but on a partial resistance controlled by a single recessive gene. So far, there is no evidence that such resistance is race-specific (Patel, 1978, 1981) but, until more widely used in breeding, it is premature to suggest that such sources would provide durable protection.

7.2.9 Root knot

Three species of nematode cause root knot in cowpeas; these are *Meloidgyne incognita, M. javanica*, and *M. arenaria*. Each is widespread and common, and they often occur in mixed populations. Their larvae have been recovered from cowpea roots within 48 hours of sowing (Caveness, 1963), indicating their rapid colonization of susceptible hosts. The root systems of invaded plants become heavily galled, leading to premature senescence of the whole plant. *M. incognita* is thought to be responsible for annual yield losses of 5–10 per cent in Georgia, USA (Toler *et al*., 1963), and nematicide treatment has led to yield increases of up to 69 per cent in the USA (Dukes *et al*., 1979) and up to 64 per cent in Nigeria (F. E. Caveness, personal communication, 1978). There is evidence that damage may in part result from an antagonistic interaction between *Meloidogyne* and nitrogen-fixing bacteria (Sharma and Sethi, 1976b; Ali *et al*., 1981).

Control measures include the use of trap crops, rotation, soil flooding, and soil fumigation, though the latter is unlikely to be economic, as well as the use of root knot-resistance cultivars (Caveness, 1963). Resistance to root knot was first found in cv. Iron (Webber and Orton, 1902). Subsequent studies have led to the identification of other sources of resistance in the USA (Hare, 1959; Thomason and McKinney, 1960), Nigeria (Caveness, 1963, 1975; Amosu, 1974; Ogbuji, 1978), India (Sharma and Sethi, 1976a; Patel *et al*., 1977), and Brazil (Lemos and da Ponte, 1978). There is little known of the nature of resistance. Roots of resistant plants often show slight galling and support some *Meloidogyne* egg production (Fery and Dukes, 1980). There is evidence that a resistance to larval penetration occurs (Sethi and Sharma, 1976). The genetics of root knot resistance has been investigated by Hare (1959), Amosu and Franckowiak (1974) and Fery and Dukes (1980), all of whom have found that resistance is governed by single dominant genes. The same gene, designated R_k, is present in cvs Colossus, Iron, and Mississippi Silver in which it determines resistance to all three species of *Meloidogyne* (Fery and Dukes, 1980). Host-specific races of both *M. incognita* and *M. arenaria* are recognized (Sasser, 1980), but whether or not specificity occurs within, as well as between, host species appears uncertain.

7.3 DISEASE RESISTANCE AND COWPEA IMPROVEMENT

The identification of resistance to disease in cowpeas began more than 80 years ago when, in South Carolina, USA, it was noticed that cv. Iron was resistant to wilt (*Fusarium oxysporum* f.sp. *vasinfectum*) and to root knot (*Meloidogyne* sp.) (Orton, 1902; Webber and Orton, 1902). Combined resistance to the two diseases was considered especially important because of the observation that they frequently occurred together and that, where they did so, much greater injury was done than where either disease occurred alone (Webber and Orton, 1902), an observation confirmed much later (Thomason *et al*., 1959). Orton (1902) foresaw the need to combine this resistance with the higher yielding potential of other cultivars by breeding work. Mackie (1934, 1939) utilized cv. Iron in crosses with susceptible cv. California Blackeye to produce wilt- and root knot-resistant Blackeye cultivars for California, and resistant crowder types were produced in Louisiana also by using cv. Iron as a parent (Hawthorne, 1943). Subsequent work elsewhere has revealed many other sources of resistance both to fusarium wilt (Kendrick, 1936; Oyekan, 1977) and to root knot (Kendrick, 1936; Hare, 1959; Thomason and McKinney, 1960; Caveness, 1963, 1975; Patel *et al*., 1977; Lemos and da Ponte, 1978; Ogbuji, 1978).

Resistance of various types has also been identified in cowpeas against many other fungal diseases including anthracnose (Williams, 1974b, 1977a), sphaceloma scab (Allen *et al*., 1981b), cercospora leaf spot (Verma and Patel, 1969; Fery *et al*., 1976; Vakili, 1977; Williams, 1977a), rust (*Uromyces appendiculatus*; Heath and Heath, 1971; Williams, 1977a), smut (*Entyloma vignae*; da Ponte *et al*., 1976), powdery mildew (Fennell, 1948; Sohi and Rawal, 1975), target spot (*Corynespora cassiicola*; Williams, 1977a), septoria leaf spot, brown blotch (*Colletotrichum* spp.; Allen *et al*., 1981b), ascochyta blight (Moreno, 1975; IITA, 1978), web blight (Oyekan *et al*., 1976; IITA, 1980), ashy stem blight (*Macrophomina phaseolina*; Mackie, 1934; Luttrell and Weimer, 1952), phytophthora stem rot (Purss, 1958), dry root rot (*Fusarium solani*; da Ponte and da Silva, 1977), and verticillium wilt (McNeill, 1976). Genotypes are also available with resistance to virus diseases, including cowpea aphidborne mosaic and its relatives (Kuhn *et al*., 1965; Ladipo and Allen, 1979b; Walker and Chambliss, 1981), cowpea (yellow) mosaic (Wells and Deba, 1961; Robertson, 1965; Williams, 1977b), cowpea severe mosaic (Fulton and Allen, 1982), cowpea mottle (Bliss and Robertson, 1971; Allen, 1980; Allen *et al*., 1982), cowpea chlorotic mottle (Sowell *et al*., 1965), cowpea golden mosaic (IITA, 1980; Vetten and Allen, in press), southern bean mosaic (Kuhn and Brantley, 1963; Lamptey and Hamilton, 1974; Ladipo and Allen, 1979a; O'Hair *et al*., 1981), cucumber mosaic (Sinclair and Walker, 1955), and the related cowpea banding mosaic (Sharma and Varma, 1975a; Raj and Patel, 1977). Resistance to bacterial blight (Sherwin and Lefebvre, 1951; Patel and Jindal, 1970; Kishun *et al*., 1980; Allen *et al*., 1981c) and bacterial pustule (Williams, 1977a; Patel, 1981) has also been identified.

Thus, resistance is available to most of the major diseases. The considerable variabiltiy within *V. unguiculata* has largely obviated the need to search for sources of resistance by other means. Induced mutations apparently have not been attempted, and sterility barriers currently prevent interspecific hybridization (see Section 7.1). Should wide crossing become possible, perhaps with the aid of embryo culture, an opportunity should be taken to examine the potential value of pubescence, available in many *Vigna* species denser than is present within *V. unguiculata*. Intersub-specific hybridization has been used in breeding for resistance to powdery mildew (Fennell, 1948). While most resistance appears to afford sufficient protection to dispense with other control measures, the available levels of resistance against web blight and ascochyta blight are inadequate alone. Owing to their very wide host ranges, it seems improbable that interspecific hybridization could provide resistance to the two pathogens concerned: intercrossing of partially resistant genotypes would seem worth trying first.

Evaluations of germplasm have sometimes suggested that resistance is associated with collections from particular regions (Ahmad *et al*., 1972; W. M. Steele, personal communication, 1980), presumably reflecting their co-evolution (see Section 1.3.3). While the incidence of resistance clearly depends partly on the genetic diversity of the available germplasm and on the efficacy of screening procedure, it is evident that disease resistance is relatively common in cowpea. On screening a germplasm collection of 5000 accessions in Nigeria (IITA, 1974a), Williams (1977a) found that some 720 accessions (14.4 per cent) possessed resistance to at least one disease and 208 (4.2 per cent) were resistant to four diseases. Similarly, 44 out of 315 accessions (14 per cent), selected for their resistance elsewhere, were found resistant to one or more diseases in India (Raj and Patel, 1977). Several cultivars of cowpea are now known to combine resistance to as many as ten diseases. For instance, cv. Iron (= TVu 990) has been shown to possess resistance to fusarium wilt, root knot (Webber and Orton, 1902), anthracnose, rust, cercospora leaf spot, bacterial pustule (Williams, 1977a), bacterial blight, cowpea banding mosaic (Raj and Patel, 1977), cowpea (yellow) mosaic (Williams, 1977b), and southern bean mosaic (Ladipo and Allen, 1979a). Others, including cvs VITA 1 (= TVu 201) and VITA 3 (= TVu 1190), combine resistance to at least nine diseases, chiefly those of economic importance, in humid lowland Africa (Williams, 1977a, b; Ladipo and Allen, 1979a, b; Allen, 1980; IITA, 1980), whereas cv. VITA 4 (= TVu 1977-OD; selection from PI 225922) possesses resistance to four of the major diseases of the African savannahs (Allen *et al.*, 1981b). Combined resistance to up to five virus diseases has also been identified (Gay, 1976; Allen, 1977b, 1978, 1980), although earlier studies (Kuhn *et al.*, 1966) had shown that no cultivar, plant introduction, or breeding line examined possessed resistance to more than one of the five viruses present in Georgia, USA. Strategies for the incorporation and ecological deployment of combined resistance will be discussed after some consideration has been given to the nature and inheritance of the resistance available within *V. unguiculata*.

In 1974 the International Institute of Tropical Agriculture (IITA) launched an international testing programme for germplasm shown to possess combined resistance in Nigeria. The objectives of the International Cowpea Disease Nursery (ICDN) were then, and continue to be, the identification of broad spectrum, stable resistance to distinct populations of pathogens over a wide range of environments, while providing a means by which resistant germplasm could be exchanged between co-operators. Results from the first nurseries revealed that, although in some cases resistance to certain diseases was effective at all locations, in others there was evidence of site-differential interactions (Williams, 1975a). As the ICDN programme progressed, it became clear that site, or season, effects could be attributed to one or more of the following:

(i) Disease escape, through insufficient infection pressure, with the result that susceptible entries are liable to be scored as resistant.
(ii) Masking of resistance, through use of an inappropriate screening technique. Under field conditions, excessive infection pressure can lead to a 'breakdown' of partial resistance, as reported in cv. Cristaudo against phytophthora stem rot (Purss, 1958), and in cv. VITA 1

against rhizoctonia web blight (IITA, 1980). Under greenhouse conditions, 'field resistance' to anthracnose may be masked (Williams, 1974b); and the hypersensitive resistance of cv. Clay to southern bean mosaic virus may be overcome at high temperatures (Kuhn and Adams, 1976).

(iii) Genetic impurity of entries, resulting in mixed reactions within a seed lot, or in differential reactions between accessions of the same name. An example is the variation in response to bacterial blight found between six accessions of cv. Iron (Singh and Patel, 1977).

(iv) Misidentification of the pathogen. Strong location effects were noted in reactions of ICDN entries to mosaic (Williams, 1975a; Allen, 1977b), now known to be attributable to reactions to two distinct viruses (see Section 7.2.7). Apparent 'breakdown' of anthracnose resistance across locations is now thought to have been due to misidentification of scab for anthracnose (Section 7.2.2).

(v) Mixed infections altering reactions. It has long been recognized that fusarium wilt resistance can 'break down' in the presence of root knot nematodes, as shown in cv. Grant (Thomason *et al*., 1959). Mixed infections of viruses can also cause differential reactions which are absent when either component virus occurs alone (Pio-Ribeiro and Kuhn, 1980).

(vi) The existence of host-specific pathogenicity and race-specific resistance, which we shall now discuss.

The race-specificity of phytophthora stem rot resistance in cv. Caloona was clearly demonstrated in Queensland, Australia, in 1962. At that time, cv. Caloona had just been selected from among the progeny of a cross between cv. Poona and the stem rot-resistant cv. Blackeye 5. A seed multiplication plot was then found to contain a few diseased plants of cv. Caloona, suggesting a new race of *Phytophthora vignae* had appeared. This was subsequently confirmed (Purss, 1963, 1972). Similar cases of breakdown are known to have occurred in Nigeria with anthracnose-resistant cv. New Era (Skipp, 1975; Section 7.2.2), and with rust resistance in cvs VITA 1 and VITA 3 (IITA, 1979). That the latter was due to a race change in *U. appendiculatus* was confirmed by Dr. Michèle Heath in Toronto. Pathogenic variation is recognized also within the fusarium wilt pathogen (Armstrong and Armstrong, 1950; Hare, 1953); race 1 has been identified from Nigeria (Oyekan, 1975; Armstrong and Armstrong, 1980) but the geographical distribution of races remains largely unknown. Host-specific pathogenicity probably also occurs in the powdery mildew fungus (Fennell, 1948), though more than one species may possibly be involved (Jhooty and Munshi, 1980). Resistance-breaking variants of *Xanthomonas campestris* pv. *vignicola* have been reported from both bacterial blight and bacterial pustule (Jindal *et al*., 1981; Patel, 1981; see Section 7.2.8), as well as from cowpea aphidborne mosaic and cowpea (yellow) mosaic. However, in neither of the latter two cases is there clear evidence of cultivar specificity (Sections 7.2.6 and 7.2.7). Certain mechanisms of resistance appear more commonly associated with race-specificity than are others. The resistance of cv. VITA 3 to anthracnose, and the resistance of cv. TVu 43 to bacterial pustule, are in neither case known to be race-specific, although in other cultivars, whose resistance is based on the hypersensitive reaction, race-specificity is recognized.

Histological study of the mechanisms of resistance to rust (Heath, 1971; Heath and Heath, 1971; Heath, 1974), and to anthracnose (Skipp, 1975; IITA, 1977) suggests that, in each case, there are 'switching points' in the infection process beyond which further fungal development may, or may not, continue, depending on the host cultivar concerned. Ultrastructural examination of the interaction of various cowpea cultivars and rust (*U. appendiculatus*) has revealed that there are at least four histologically distinct resistance mechanisms, including three 'immune' responses and a necrotic fleck reaction (Heath, 1974). One of the 'immune' responses involves the development of a haustorial sheath without hypersensitive death of the invaded host cells, suggesting that hypersensitivity is not the only effective means of limiting the growth of *U. appendiculatus* in infected tissue (Heath, 1971). It is not yet known whether distinct histological mechanisms are under independent genetic control, nor whether they are liable to differ in their race-specificity, though it is known that rust resistance in cv. TVu 4552 relies on the necrotic fleck response which is controlled by a single dominant gene (IITA, 1976; M. C. Heath, personal communication, 1977). A similar range of mechanisms is involved in restricting the development of *Colletotrichum lindemuthianum* in host tissue, in which phytoalexins occur (see Section 7.2.1). Biochemical studies have shown that phytoalexins probably also play a role in the resistance of cowpeas to other fungal diseases. Partridge and Keen (1976) have provided evidence that kievitone is the basis for the expression of monogenic resistance in cv. Caloona to phytophthora stem rot, and McNeill (1976) has suggested that differential synthesis of chlorogenic acid may account for verticillium wilt resistance, which is also under the control of a single dominant gene (Moore, 1974).

Virus disease resistance in cowpea may be expressed in various ways and distinct components can be recognized. For instance, resistance to southern bean mosaic includes an immunity, in PI 186454 (= TVu 1948; Ladipo and Allen, 1979a), as well as hypersensitive resistance and tolerance (Kuhn and Brantley, 1963; Kuhn *et al*., 1966); since there is evidence that some cultivars do not transmit SBMV through seed (O'Hair *et al*., 1981), a 'resistance to seed transmission' may possibly also occur. Immunity, hypersensitivity, tolerance (Ladipo and Allen, 1979b), and resistance to seed transmission (Ladipo, 1977; Aboul Ata *et al*., 1982) are each also components of resistance to cowpea aphidborne mosaic, though whether or not resistance to the aphid vector *Aphis craccivora* (Singh and Allen, 1980) confers disease resistance is not yet known. There is evidence that vector non-preference of the whitefly, *Bemisia tabaci*, is involved in field resistance to cowpea golden mosaic (Vetten and Allen, in press). Cultivars possessing tolerance to infection by cowpea mottle virus under greenhouse conditions have been shown to exhibit both tolerance and resistance to infection by beetle vectors under field conditions, resistance to infection being exemplified by delayed and reduced incidence of infection resulting from natural infection pressure, apparently unrelated to vector feeding preference (Allen *et al.*, 1982). Very little is known about the histological and biochemical bases of such resistance. Hypersensitive necrosis induced by virus inoculation of cowpea is associated with the accumulation of anti-fungal isoflavonoids (Bailey, 1973) but the nature of anti-viral factors in resistance remains largely unknown. Immunity, which has been shown to be mechanistically distinct from hypersensitivity to cowpea (yellow) mosaic virus, operates either through prevention or by delaying the onset of virus replication (Beier *et al.*, 1977; Section 7.2.7). Resistance to cowpea chlorotic mottle virus, which is expressed not as in immunity but as symptomless local infection without systemic spread, depends on a greatly reduced rate of replication as well as restricted movement of virus (Sowell *et al.*, 1965; Wyatt and Kuhn, 1979).

The genetics of disease resistance in cowpea, reviewed recently by Fery (1980) and Meiners (1981), has been studied in some 20 diseases. In almost all cases, monogenic control has been demonstrated, in the majority of cases resistance being dominant to susceptibility (Dundas, 1939; Sinclair and Walker, 1955; Brantley and Kuhn, 1970; Purss, 1972; Moore, 1974; IITA, 1976; Fery and Dukes, 1980); single recessive genes have been identified in some instances (Reeder *et al.*, 1972; Rogers *et al.*, 1973; Raj and Patel, 1978; Walker and Chambliss, 1980). Resistance to a single disease may be conditioned by different genetic factors. For instance, resistance to cercospora leaf spot is controlled by a single dominant gene in one breeding line but an incompletely dominant gene, strongly influenced by environment, governs resistance in another (Fery *et al.*, 1976; Fery and Dukes, 1977). Similarly, either a single dominant gene (Dundas, 1939) or two or three recessive genes (Fennell, 1948) control powdery mildew resistance. Distinct mechanisms of resistance may be governed independently. Patel (1978) has suggested that hypersensitive resistance to bacterial pustule appears to be simply inherited as a dominant trait, while resistance not involving hypersensitivity may be controlled by a single recessive gene. Three additive loci are responsible for tolerance to cowpea (yellow) mosaic in cv. Alabunch but hypersensitive resistance of a Dixielee selection is conferred by a single dominant gene (Bliss and Robertson, 1971). Several separate major genes may each govern the same resistance mechanism, as is the case with phytophthora stem rot (Purss, 1972). Investigation of the genetics of combined resistance has confirmed that distinct mechanisms of inheritance determine a host's reaction to separate diseases (Raj and Patel, 1978).

Having considered the specificity of host-pathogen interactions, the mechanism of operation and genetic control of resistance in cowpeas, what can be concluded with regard to the probable durability of resistance in agriculture? It is evident that there are instances in which resistance has proved transient, but these appear to be confined to intensive agricultural systems in developed countries (e.g. Queensland, Australia; Purss, 1972) or to experimental conditions in developing countries (e.g. Nigeria; IITA, 1973, 1979). While the extension of improved cultivars into commerical production has so far made little headway in tropical Africa, so that incorporated resistance may not yet have faced an adequate trial, it seems probable that the very nature of the farming systems of which cowpeas are a component (see Section 1.2.3) confers a protection against epidemic development: the intrinsic vulnerability of an individual plant's resistance is thereby masked. The more intensive monocrop production of cowpeas in certain parts of north-eastern Brazil may provide a more stringent test of the potential durability of resistance. If there is a lesson to be learnt from current knowledge, it may be that it is the mechanism of resistance, rather than its genetic control, that provides the more reliable index of durability. It is certainly premature to conclude that the preponderance of publications on oligogenic resistance reflects a natural prevalence over polygenic resistance, rather than on its greater ease of recognition. Cowpea is a relatively unimproved crop in which, it is reasonable to assume, the degrees of horizontal resistance and the diversity of genetic systems governing all types of resistance are each comparatively large. It remains to be seen to what extent polygenes are involved. It is also premature to suggest, as Meiners (1981) has done, that oligogenic inheritance of resistance necessarily infers vulnerability. Certainly, resistance governed by single dominant genes has proved transient, as discussed earlier in this chapter., but they are also associated with apparently more stable resistance. Fery and Dukes (1980) have shown that the gene 'R_k', which confers a high level of resistance to three species of root knot nematode, is present in cv. Iron whose resistance has endured for some 80 years (Webber and Orton, 1902). The oligogenic resistance of the same cultivar to fusarium wilt has also apparently remained effective over a long period, despite the recognition of distinct races.

Disease resistance has long been an important objective in cowpea breeding programmes in the USA, where the main commercial cultivars are classified by horticultural group, based on seed characters (Brittingham, 1946). In California, where Blackeye types predominate, early work focused on the incorporation of fusarium wilt and root knot resistance, from cvs Iron and Virginia Blackeye, into locally acceptable Blackeye cultivars (Mackie, 1934; 1939; Kendrick, 1936). In the southern states, emphasis was placed in the improvement of brown crowder types, culminating in the release of cv. Mississippi Silver which was found to possess resistance to all three races of the fusarium wilt fungus, combined with resistance to three species of root knot nematode and tolerance to several viruses (Hawthorne, 1943; Hare, 1957b, 1967). Mississippi Silver was also to become a leading cultivar in Georgia where, however, it was found to be virus-susceptible. Its hybridization with cv. Pinkeye Purplehull led to the selection of cv. Worthmore which combines resistance to at least three viruses (Gay, 1976). In Queensland, Australia, where cowpeas are grown chiefly as a green manure or fodder crop, the cv. Poona was widely used until the appearance of phytophthora stem rot to which it was highly susceptible. California Blackeye 5 was used as the resistant parent in crosses with cv. Poona leading to the development of cv. Caloona (Purss, 1963). In tropical developing countries, cowpea improvement has most often been confined to selection within germplasm collections. This is generally true of India; in the few cases where new cultivars have been bred, emphasis has been placed on agronomic attributes at the expense of disease resistance. But collections have been evaluated for resistance at various centres, including New Delhi, Ludhiana, Coimbatore, and Bangalore (Patel and Jindal, 1970; Ahmad *et al.*, 1972; Sohi and Rawal, 1975; Raj and Patel, 1977; Kishun *et al.*, 1980), and a programme to incorporate combined disease resistance into the popular cultivar Pusa Dofasli has been initiated (Raj and Patel, 1977, 1978). Germplasm has also been evaluated for resistance to a whitefly-transmitted mosaic disease at Faisalabad in Pakistan (Ahmad, 1978; M. Ahmad, personal communication, 1980). Within the tropics, most research has been conducted in West Africa where the crop is thought to have originated and which is now the main region of production (Section 7.1). Breeding programmes, in both Nigeria and Senegal, were established in the 1960s and a number of improved cultivars, including cvs Westbred, Ife Brown, and N'Diambour, have been released (Smithson *et al.*, 1980). However, disease resistance was again accorded low priority in these programmes with the result that improved material emerging from them has tended to be highly susceptible. On the establishment of the International Institute of Tropical Agriculture (IITA) in Nigeria, breeding objectives began to change, and resistance was recognized as a key component in the crop's improvement. Recent strategy has been the incorporation of resistance into a range of plant types and photoperiod reactions suited to different cropping systems and environments (Smithson *et al.*, 1980).

The development of the cowpea breeding programme of IITA has been reviewed recently (Smithson *et al.*, 1980). Initial emphasis was on the assembly of germplasm and its evaluation, including disease reaction. The collection now stands at well over 10,000 accessions (IITA, 1980) and almost 6000 of these have been evaluated for up to 46 plant characters, including resistance to eight diseases (IITA, 1974). Seven lines combining good yield performance and resistance to a range of pests and diseases plus an eighth, comprising an F_2 population segregating for male sterility, have been described as VITA lines. Both conventional breeding and population improvement methods have been employed. Though conventional breeding methods have led to significant crop improvement, especially with regard to characters under oligogenic control, genetic progress is restricted by continued selfing and the relatively small numbers of parents which can be handled in crossing so that the probability of novel recombinants is reduced. These restrictions may be lessened by the adoption of population improvement and recurrent selection methods, as successfully used in cross-pollinated crops, by utilizing genetic male sterility. At IITA, a system has been developed in which sub-populations selected for specific characters, including disease resistance, feed into a back-up and thence into a main population, represented by a recurrent selection system operating with conventional crossing rather than male sterility. Some evidence of changes in the populations has been obtained, including increased frequencies of genes for resistance to cercospora leaf spot and anthracnose (Smithson *et al.*, 1980). But whether or not population breeding has, indeed, favoured the accumulation of polygenes and avoided eroding the horizontal resistance which is presumed to exist is not yet known. Effective selection for resistance depends on the disease pressure under which selection is made (Asafo-Adjei, 1982), emphasizing the importance of the ability to regulate epidemics in screening programmes. No attempt has been made at IITA, in either conventional breeding or in population improvement, to select against the expression of mechanisms likely to confer race-specific resistance; rather, selection has been against high susceptibility, thereby intending to retain a diversity in the components of resistance. Time alone will show whether such a strategy is satisfactory, or whether it is indeed necessary to select initially for susceptibility, as advocated by Robinson (1976).

We have seen that the distribution and importance of cowpea pathogens varies markedly with ecological zone (Section 7.2), accounting for some of the large interactions that occur between cultivars and environments. In view of such environmental effects, a set of testing locations has been selected to take advantage of the ecological diversity within Nigeria, while limiting the need for seed movement and attendant quarantine risks. Three sites (Onne, near Port Harcourt, Ibadan, and Funtua) were chosen for disease-resistance screening purposes; respectively, these represent the very humid lowland region, where 2400 mm rainfall is received over 9 months, the forest-savannah transition zone (1250 mm rainfall with a bimodal distribution), and the northern Guinea savannah, with a mean annual rainfall of 1100 mm received in five months (Allen, 1979b; Smithson *et al.*, 1980; Allen and Goldsworthy, 1981). At Onne, early sowing at the start of the rains encourages the uniform development of golden mosaic, while close spacing (10 × 50 cm) enhances web blight, both from natural infection. Moderate levels of combined resistance to the two diseases have thus been identified among advanced breeding lines (Allen and Goldsworthy, 1981). At IITA headquarters in Ibadan, field inoculation enables selection for resistance to anthracnose, cercospora leaf spot, bacterial pustule, and cowpea (yellow) mosaic (Williams, 1977a, b); aphidborne mosaic and cowpea mottle have been screened against under greenhouse conditions (Ladipo and Allen, 1979b; Allen, 1980; IITA, 1980). At Funtua, combined resistance has been sought against scab, septoria leaf spot, brown blotch (*Colletotrichum* spp.) and bacterial blight (Allen *et al.*, 1981b). Thus early generation material and more advanced breeding lines are screened against a total of six fungal, two bacterial, and four virus diseases each year, using environments which favour uniform disease pressure. Greatest progress has been made with the incorporation of resistance to anthracnose, cercospora leaf spot and cowpea (yellow) mosaic at Ibadan, where selection has continued virtually uninterrupted for more than 20 seasons in 10 years. Improved materials now emerging from IITA's cowpea breeding programme possess sufficient combined disease resistance to obviate the need for chemical protection under most conditions; exceptions include web blight and ascochyta blight against which available levels of resistance are insufficient.

In common with several other crop improvement programmes within the international centres network (CGIAR, 1976), IITA has virtually pantropical responsibility for the improvement of cowpeas, a situation never before faced by plant breeders. The prospect of 'breeding a crop in one place for adaptation elsewhere to a vast range of climates, latitudes, soils, diseases, agricultural systems and consumer preferences' (Simmonds, 1979) is formidable indeed. Clearly, considerable progress can be made through the availability of a large and diverse germplasm collection, a multidiscipline team of scientists, the ability to advance several (with cowpea at IITA, three or four) generations a year, coupled with multilocation testing of promising material. Multilocation testing is a powerful tool in breeding for disease resistance (Dubin and Rajaram, 1981), and one year's trials at a number of sites can be as useful as trials at a single location for a number of years, thereby greatly accelerating the evaluation of material (Day, 1973). But some caution is called for: as we have seen, site non-specific reactions to disease are not reliable indices of race non-specificity, and the use of cultivars selected for their wide adaptation, often due largely to their photoperiod insensitivity, can ultimately lead to genetic vulnerability (Day, 1973). Recent work on cowpea at IITA has shifted emphasis from the development of lines which perform well across environments to a strategy of environment specificity, suggested both on physiological (Wien and Summerfield, 1980) and pathological grounds. Thus, disease-resistance breeding becomes stratified, based on the ecological deployment of specific combinations of resistances, with the ultimate objective of developing durable comprehensive (*sensu* Robinson, 1976) resistance in each environment. The validity and application of such a strategy is demonstrable within West Africa, where environments are fairly clearly defined, and where the logistics of serving national improvement programmes from IITA are relatively well established (Smithson *et al.*, 1980). Disease resistance identified at sites in northern Nigeria (Allen *et al.*, 1981b) is directly useful to the project in Upper Volta which has now evolved into a regional programme for the semi-arid areas of West Africa. The extension of the strategy to other regions, with widely different ecological conditions, is more difficult. The nuclei of regional programmes have been established in East Africa and in Brazil, and in each case a start has been made on the evaluation of genetically diverse breeding material, including sources of combined disease resistance, for resistance to local populations of the major pathogens.

Plate 1 Common blight *(Xanthomonas campestris* pv. *phaseoli)* lesions showing yellow margin (photograph courtesy H. F. Schwartz)

Plate 2 Wilt of pigeon pea caused by *Fusarium udum;* Hyderabad, India

Plate 3 Cowpea golden mosaic resistance was identified by sowing to synchronize seedling emergence with peak populations of the whitefly vector; Onne, Port Harcourt, Nigeria

Plate 4 'Sick-plots' have been established to screen against fusarium wilt in chickpea; Hyderabad, India (photograph courtesy Y. L. Nene)

Chapter 8

Pigeon Pea Diseases

8.1 EVOLUTION, ADAPTATION AND PRODUCTION OF PIGEON PEAS

The genus *Cajanus* to which pigeon pea belongs is often considered to be monospecific, though it contains at least two validly published species: *C. kerstingii* Harms, which is endemic to the West African savannah, and *C. cajan* (L.) Millsp. (= *C. indicus* Spreng.), the pigeon pea. Other species of *Cajanus* have been erected, though most have been reduced to synonymy with *Atylosia* species (van der Maesen, 1975, 1979; Lackey, 1981). Some authors recognize two sub-species or, more commonly, two botanical varieties within *C. cajan*. Variety *flavus* encompasses the 'tur' cultivars of peninsular India; these are shorter and earlier, have predominantly yellow flowers and green pods containing three lightly coloured seeds. Variety *bicolor*, the 'arhar' cultivars of northern India, includes the large bush types which are late-maturing perennials with red, purple or darkly veined flowers and pubescent purple pods containing four or five dark, or speckled, seeds (Krauss, 1932; Purseglove, 1968). However, most combinations of these characters occur so that the maintenance of these varieties may no longer be useful, and other methods of classifying cultivars may be preferable (Akinola and Whiteman, 1972). Most cultivars are diploid ($2n = 22$) but tetraploids and hexaploids have been found. *C. cajan* can be crossed readily with diploid species of the closely related genus *Atylosia* having the same chromosome number. It seems that the pigeon pea is nothing more than a cultivated form of *Atylosia,* and a merger of the two genera has been proposed (van der Maesen, 1981). *A. lineata* Wight and Arn. is close to *C. cajan* but whether or not this or another species of *Atylosia* is the progenitor of pigeon pea remains uncertain (De, 1974; Royes, 1976; ICRISAT, 1980; Lackey, 1981). Although *C. cajan* is predominantly self-pollinated, a higher degree of outcrossing occurs than in most other grain legumes. The extent of natural outcrossing is variable (1–48 per cent) though 20 per cent is average (Wilsie and Takahashi, 1934; Abrams, 1967; Khan, 1973; Venkateswarlu and Singh, 1980). Bees and thrips have both been implicated in cross-pollination (Abrams, 1967; Rachie and Roberts, 1974). Higher levels of outcrossing can be obtained through the use of genetic male sterility (Reddy *et al*., 1978b). In view of the ease with which *C. cajan* may be crossed with *Atylosia* species, it seems possible that natural introgression may occur in certain regions.

Owing to the inadequacy of archaeological evidence, the origin of pigeon pea is disputed, and opinion is divided as to whether *C. cajan* is an African or an Indian native. Although *C. kerstingii* is an African species (van der Maesen, 1979), *C. cajan* does not show the diversity in Africa which might be expected if it had originated there. Furthermore, if we are to accept that *C. cajan* was domesticated from certain species of *Atylosia,* the rarity of that genus in Africa suggests the crop has its origins elsewhere. The current view (De, 1974; Royes, 1976; Van der Maesen, 1980; Lackey, 1981) is that pigeon pea is an Asian domesticate. De (1974) points out that the greatest differentiation and abundance of *Cajanus* and *Atylosia* is found in the broad-leaf evergreen forest area of the Western Ghats and the Malabar coast of India, concluding that it is here that the pigeon pea originated. From this centre the whole Indian sub-continent has become the area of diversity. Smartt (1980) suggests that the cultigen, *C. cajan,* has morphologically been changed little by domestication; according to De (1974), it can escape from cultivation and survive in the wild. Such wild populations obviously do not represent any stage in the evolution of the cultigen, and probably possess some of its relict characters. If we accept an Indian origin, it is probable that the crop spread eastwards to Malaysia about 2000 years ago, later being taken to China, reaching Australia through Indonesia. *Cajanus* was carried to East Africa between 1000 and 1500 years ago, moving northwards and westwards into West Africa. The crop was probably taken to the New World from Zaire or Angola by the Portuguese in early post-Columbian times (De, 1974; Royes, 1976). The crop is now well established throughout the tropics.

Pigeon pea is widely adapted to lowland tropical conditions. It is notably tolerant of heat, drought, and poor soil, with a deep root system which permits good growth in semi-arid regions. The crop is less suited to the very humid tropics: it will not tolerate waterlogging nor soils markedly deficient in lime. Its frost sensitivity limits its adaptation in highland areas. Most cultivars, especially the tall, late-maturing types, are strongly photosensitive, and the interactions between photoperiod, temperature, and the genotype's maturation rate govern environmental adaptation (Gooding, 1962; Purseglove, 1968; ICRISAT, 1976). In India, medium- to late-maturing landraces are typically sown in June or July when the monsoon (kharif) season begins. These develop into woody shrubs which flower and set pods into the post-monsoon (winter, or rabi) season, relying on residual soil moisture (Narayanan and Sheldrake, 1979). Sorghum, pearl millet, cotton, or another legume crop are common companions (Sharma and Green, 1975). In northern Uganda, pigeon pea is usually grown in mixtures with finger millet (*Eleusine coracana*) after a crop of cotton. The pigeon pea, which is sown in March or April with the 'long rains', is some 2 m tall by June or July when the millet is harvested. The stubble is then dug over, and either sorghum or sesame (*Sesamum indicum*) is sown in August when the 'short rains' begin. Both the sorghum, or sesame, and the pigeon pea are harvested between November and January (Khan and Ashley, 1975), though average yields of dry grain of *Cajanus* may be as low as 168 kg/ha (Khan and Rachie, 1972). Rather higher average yields, of 716 kg/ha, are recorded from India, which produces 93 per cent of the world's crop, estimated at about 2 million tonnes from 2.8 million hectares. The principal use of pigeon pea in India is as split seed in 'dhal'. Outside India, the crop is grown for its dry seed throughout the semi-arid and sub-humid regions of Africa, though Uganda, Malawi, and Tanzania are the only recorded producers (FAO, 1976). In the Caribbean, the crop is grown for its immature green seed for fresh use or for canning (Gooding, 1962), while in Hawaii and in Queensland, Australia, pigeon pea is used as a forage (Krauss, 1932; Royes, 1976). It has also been used as a green manure in rotation with pineapple, as a windbreak, and as a source of fuel (Krauss, 1932).

In recent years, increased attention has been given to the improvement of pigeon pea. There seems little doubt that sufficient variability exists within *Cajanus* to transform it into a highly developed crop (Royes, 1976). In India, which is both the centre of diversity and the major area of production, various breeding programmes have developed, including that of the International Crops Research Institute for the Semi-Arid Tropics (ICRISAT) at Hyderabad where 8775 accessions of *Cajanus* were held in 1979 (ICRISAT, 1980). Some 5531 accessions are also held at the International Institute of Tropical Agriculture (IITA) in Nigeria (IITA, 1978), consisting of material which was first assembled in Uganda in 1969 (Rachie and Wurster, 1971; Khan and Rachie, 1972). Work in India has shown that cultivars may be classified into different maturity groups, based on their duration after sowing at the start of the monsoon rains in June or July at Hyderabad. Early, medium, and later cultivars mature in 4–5, 5–6, and 6–9 months, respectively. Considerable attention has been devoted to the short-duration cultivars which make it possible to grow two sequential crops a year (Khan and Rachie, 1972), or to grow a crop of pigeon pea before a winter- ('rabi') sown crop of wheat. Early-maturing, compact types are also suited to cropping as a companion to a cereal crop, in each case making it possible to increase production substantially (Ramanujam, 1974; Willey *et al.,* 1981). Furthermore, it appears that medium-to-late-maturing cultivars, sown in October at the start of the 'rabi' season at relatively high plant populations, could have considerable potential as a winter crop in peninsular India, where yields up to 1700 kg/ha within 4½ months have been obtained. Under these conditions, the short-duration cultivars, which have been selected under the warm conditions prevailing in the monsoon and immediately post-monsoon periods, may be less well adapted to the cooler temperatures of the 'rabi' season in which medium-to-late-maturing cultivars normally produce their flowers and pods (Green, 1979; Narayanan and Sheldrake, 1979). Similarly, in Trinidad, there is potential for the development of a completely new cropping system. By exploiting their photosensitivity, cultivars sown under short-day conditions in December, instead of the usual May–June sowing, produce small plants which can be grown at high population densities to maximize yields (Spence and Williams, 1972). Day-neutrality is seen as an important objective in the development of more widely adapted, early-maturing genotypes (Ariyanayagam and Spence, 1978a; Saxena *et al*., 1980). Although it has long been appreciated that considerable outcrossing occurs in *Cajanus,* early breeding work treated the crop as self-pollinated. However, crosses between inbred lines show heterosis; attempts have now been made to exploit this by encouraging outcrossing, enhanced by genetic male sterility, both in the formation of composites in population breeding and in hybrid seed production (Solomon *et al.,* 1957; Khan, 1973; ICRISAT, 1978; Reddy *et al.,* 1978b; Saxena *et al.,* 1980).

Pigeon pea is affected by several important diseases but pests are thought to cause more significant yield loss. Few attempts have been made to quantify the extent of crop loss attributable to pathogens under experimental conditions, and even less is known of their importance in traditional agriculture. In the face of changing patterns of pigeon pea production, it is evident that more information is required about the effects of cropping practice on disease development. The winter-sown crop is apparently more liable to yellow mosaic (Misra and Gurha, 1980) and to root rot (*Macrophomina phaseolina;* Nene, 1980a) than the traditional 'kharif' crop. Yet traditional cultivars are often of ten months duration, followed by a ratoon crop, thus providing a virtually continuous source of host tissue. The extent to which the adoptation of short-duration types in a rotation could provide a break sufficient to affect the survival of pathogens, especially those with a narrow host range (Koshy and Swarup, 1971b), remains uncertain. However, there is evidence that early-maturing dwarf types are less able to tolerate leaf loss, suggesting a greater susceptibility to foliar pathogens, relative to long-duration perennial cultivars (Hammerton, 1975; Tayo, 1982). There is also renewed interest in the potential of ratooning (Sharma *et al.,* 1977), a practice known to exacerbate fusarium wilt. The extent to which pathogens threaten the increased production of pigeon pea is considered in the next section of this chapter.

8.2 THE PATHOLOGY OF PIGEON PEAS

The literature on pigeon pea pathology is rather less extensive than that of the crops discussed in previous chapters. The bulk of the work on *Cajanus* has been done in India. Although a world list of pigeon pea pathogens is available (Nene, 1978, now revised), there is no review of the crop's diseases worldwide. The diseases of pigeon peas in India have been covered by Williams *et al.,* (1968), Nene and Rathi (1972), Singh and Pavgi (1972), and Sen Gupta (1974); those occurring in Africa and Puerto Rico have been reviewed by Kaiser (1976, 1981a), and Brathwaite (1981) has reviewed the crops' diseases in the Caribbean. An attempt is made, in Table 22, to summarize the major fungal diseases, their aetiology, geographical distribution, and approximate economic importance. It is evident that comparative studies are required in several cases. For instance, it seems uncertain whether the collar and stem cankers in the Caribbean and India have a common aetiology. In Trinidad, Leach and Wright (1930) concluded that a species of *Physalospora,* with both *Phoma* and *Macrophoma* pycnidal forms, was the principal pathogen, although a *Diplodia* produced local lesions at the collar. Subsequent work in Puerto Rico (Alvarez, 1960; López Rosa, 1969) has apparently confirmed that *Phoma* is the cause of canker, though no mention is made of a perfect state. Recently, *Botryodiplodia theobromae,* whose many synonyms include species of *Diplodia* and *Macrophoma* and whose perfect state is a *Physalospora* (Punithalingam, 1976), has been obtained from pigeon pea seed in Puerto Rico (Ellis and Paschal, 1979). In India, canker has been attributed to *Diplodia cajani* (Raychaudhuri, 1942) but *Macrophomina phaseolina,* of which *Macrophoma cajani* is a synonym (Holliday and Punithalingam, 1970), also causes a stem canker (ICRISAT, 1980). Thus there appears to be a need to determine whether or not there are one or more pathogens involved. Similarly, pigeon pea rust (*Uredo cajani*) has been frequently confused with *Uromyces dolicholi,* which is confined to *Rhynchosia,* a close relative of *Cajanus.* A distinction between the two rust fungi is only possible where telia are found and these have not been recorded for *U. cajani* (Anahosur and Waller, 1978b). Certain other fungi, not given in Table 22, are reported from pigeon pea, though their status as pathogens is uncertain; examples include species of *Cladosporium* (Hansford, 1937; Tarr, 1955; Rizwi, 1978) and *Nematospora* (Riley, 1960). In the semi-arid areas of India *Phytophthora* blight and wilt (*Fusarium udum*) are the most important fungal diseases; these are discussed fully in subsequent sections of this chapter. In eastern Africa wilt and leaf spot (*Mycovellosiella cajani* var. *cajani*) are the two most important diseases (Khan and Rachie, 1972; Onim, 1980); rust (*Uredo cajani*), canker (*Phoma* sp.), and *Phytophthora* blight are the major fungal diseases in the Caribbean (Abrams *et al.,* 1978; Kaiser, 1981a).

Various virus-like diseases have been reported from pigeon pea but in virtually no case has the virus been satisfactorily characterized. Cowpea severe mosaic virus occurs in natural infections of *Cajanus* in Trinidad (Dale, 1949); since CSMV is widespread throughout the American tropics (Section 7.2.7), it seems likely that some of the unidentified mosaics of pigeon pea in this region (e.g. ICRISAT, 1978) may be attributable to CSMV. However, cowpea (yellow) mosaic virus has been identified from *Cajanus* in El Salvador (R. F. Bozarth, personal communication 1977). A whitefly-transmitted mosaic of *Rhynchosia* sometimes naturally infects pigeon pea in Puerto Rico (Bird *et al.* 1975), while the yellow mosaic present in India is caused by mung bean yellow mosaic virus (Nene and Rathi, 1972). A tobamovirus causes 'arhar mosaic' in India (Singh and Mall, 1976; Singh and Pandey, 1979), considered distinct from the miscellaneous virus-like diseases reported from

TABLE 22 Fungal Diseases of Pigeon Pea

Disease	Causal fungi	Distribution	Importance/Crop loss	References
1. Seed and seedling diseases				
Seed decay, seedling rot	Associated principally with *Fusarium semitectum* Berk. and Rav., *Botryodiplodia theobromae* Pat., *Phomopsis* sp. and *Aspergillus* spp., incl. *A. flavus* Link ex Fr.	Puerto Rico, India	Potentially important	Sinha and Prasad, 1977; Ellis and Smith, 1978; Shukla and Bhargava, 1978; Ellis and Paschal, 1979
Seedling mortality, collar rot	*Sclerotium rolfsii* Sacc. (perfect state = *Corticium rolfsii* Curzi)	India, Trinidad, Puerto Rico	Perhaps locally important	Jehle and Wood, 1926; Spence, 1975; ICRISAT, 1978; Nene, 1980a; Kaiser, 1981a
2. Stem and root rots, and wilts				
Collar rot	*Pythium aphanidermatum* (Edson) Fitz.	India	Minor	Malik, 1945
Collar and stem canker	*Phoma* sp. and *Diplodia cajani* Raychaudhuri	Caribbean, India	Locally damaging	Nowell, 1924; Leach and Wright, 1930; Raychaudhuri, 1942; Alvarez, 1960; López Rosa, 1969
Stem blight	*Phytophthora drechsleri* Tuck. f.sp. *cajani* Kannaiyan and *P. nicotianae* Breda de Haan var. *parasitica* (Dast.) Waterh.	India (*P. drechsleri* f.sp. *cajani*) and Puerto Rico (*P. nicotianae* var. *parasitica*)	Major; especially damaging under wet conditions in northern India	Pal *et al.*, 1970; Amin *et al.*, 1978a; Kaiser and Melendez, 1978; Kannaiyan *et al.*, 1980
Stem anthracnose	*Colletotrichum truncatum* (Schw.) Andrus and Moore and *C. capsici* (Syd.) Butl. and Bisby.	India	Minor, potentially destructive?	Khan and Singh, 1974; Somani *et al.*, 1975; Hasan and Khan, 1979
Ashy stem blight, root rot	*Macrophomina phaseolina* (Tassi) Goid. (= *Macrophoma cajani* Syd. and Butl., sclerotial state = *Rhizoctonia bataticola* (Taub.) Butl.)	India, Uganda, Nigeria	Minor, perhaps locally important; but a winter-sown crop is liable to serious damage in C and S India	Hansford, 1937; Malik, 1945; Bailey, 1966; Rathi and Lal, 1977; ICRISAT, 1980; Nene, 1980

Root rot	*Armillaria mellea* (Vahl ex Fr.) Karst	Eastern Africa	Minor	Riley, 1960; Peregrine and Siddiqi, 1972
Root rot	*Fomes lignosus* (Kl.) Bres. and *F. noxius*	Nigeria	Minor?	Bailey, 1966
Root rot and wilt	*Neocosmospora vasinfecta* Smith	India	Minor	Mitra, 1934; Sarojini, 1954
Wilt	*Fusarium udum* Butl.	Widespread; perhaps pan-tropical	Major; the principal disease	See Section 8.2.1
3. Foliar and pod diseases				
Leaf spot	*Mycovellosiella cajani* (P. Henn.) Rangel ex Trotter (= *Cercospora cajani* P. Henn.)	Widespread; the three varieties have different distributions	Important, especially in East Africa where 85% crop loss has been reported	Singh, 1934; Khan and Rachie, 1972; Deighton, 1974; Onim and Rubaihayo, 1976; Onim, 1980
	var. *cajani*	Tropical America, Africa		
	var. *indica* (U. B. Singh) Deighton	India, Pakistan, Bangladesh, Burma		
	var. *trichophila* (Curzi) Deighton	Kenya, Somalia, Taiwan		
Brown leaf spot	*Cercoseptoria cajanicola* Pavgi and Singh	India, Kenya, Tanzania, Sierra Leone	Minor, often in mixed infections with *M. cajani*	Pavgi and Singh, 1965; Deighton, 1974; Ebbels and Allen, 1979; Waller and Sutton, 1979
Chaetoseptoria leaf spot	*Chaetoseptoria wellmanii* Stev.	Guatemala	Minor?	Muller, 1953
Phyllosticta leaf spot	*Phyllosticta cajani* Rangel	India, Tanzania, Brazil, ? Puerto Rico	Minor	Jehle and Wood, 1926; Batista and Vital, 1952; Saksena and Kumar, 1972; Nene, 1978; Ebbels and Allen, 1979
Pyrenochaeta leaf spot	*Pyrenochaeta cajani* Singh and Pavgi	India	Minor	Pavgi and Singh, 1965
Alternaria leaf spot	*Alternaria tenuissima* (Kunze ex Pers.) Wilts.	India	Minor, though important in the winter-sown crop in NE India	Kannaiyan and Nene, 1977
Leptosphaerulina leaf spot	*Leptosphaerulina trifolii* (Rost.) Petr. (= *L. briosiana* (Poll.) Grah. and Lutt.)	India	Minor	Pavgi and Singh, 1965; Booth and Pirozynski, 1967; Nene, 1979

TABLE 22 (*continued*)

Disease	Causal fungi	Distribution	Importance/Crop loss	References
Anthracnose	*Colletotrichum cajani* Rangel (perfect state = *Glomerella cingulata* (Stonem.) Spauld. and Schrenk)	India, Puerto Rico, Brazil	Minor, though perhaps occasionally important. Incidence of pod infection up to 86%, with 36% unmarketable seed, reported from Puerto Rico	Tucker, 1927; McRae, 1932
White mould	*Sclerotinia sclerotiorum* (Lib.) de Bary	Bermuda, Australia	Minor?	Waterston, 1944; Akinola and Whiteman, 1972
Web blight	*Rhizoctonia solani* Kühn (= *Thanatephorus cucumeris* (Frank) Donk)	India, Malaysia	Locally, or seasonally, important?	Turner, 1971; Dwivedi and Saksena, 1975; ICRISAT, 1978
Leaf rot	*Choanephora cucurbitarum* (Berk. and Rav.) Thaxter	India	Minor	Misra and Mehra, 1969; Singh and Pavgi, 1972
Powdery mildew	*Leveillula taurica* (Lév.) Arn. (conidial state = *Oidiopsis taurica* Lév.) Salmon)	India, Malawi, Sudan, Tanzania, Uganda	Minor, though can cause a shoot blight	Kamat and Patel, 1948; Tarr, 1955; Riley, 1960; Rachie and Wurster, 1971; Peregrine and Siddiqi, 1972; Kannaiyan and Nene, 1979a
Rust	*Uredo cajani* Syd.	Apparently pan-tropical but confusion with *Uromyces dolicholi* Arth. has occurred	Generally minor though considered important in the Caribbean where severe defoliation occurs	Hammerton, 1975; Spence, 1975; Anahosur and Waller, 1978b
False rust	*Synchytrium* spp.	India	Minor	Gupta and Sinha, 1951; Ramakrishnan and Sundaram, 1954; Singh and Pavgi, 1972

India (Bisht and Banerjee, 1965) and Sri Lanka (Newton and Peiris, 1953). Virus-like symptoms have also been noted in Malaysia (ICRISAT, 1978), Kenya, (Gathuru, 1979), Ethiopia and Nigeria (D. J. Allen, unpublished). Much the most important virus-like disease of pigeon pea is sterility mosaic (Capoor, 1952) which is discussed in Section 8.2.3. Rhabdovirus particles have been found alone, associated with mosaic symptoms (Licha-Baquero, 1980), and in dual infections with mycoplasma-like organisms associated with a witches' broom disease in Puerto Rico and the Dominican Republic (Hirumi *et al.,* 1973; Maramorosch *et al.,* 1974). There is some evidence that a leafhopper (*Empoasca fabae*) is the vector of witches' broom (Vakili and Maramorosch, 1974; Licha-Baquero, 1980), a disease which occurs also in Haiti, Jamaica (Vakili and Maramorosch, 1974), and Panama (ICRISAT, 1978), occasionally becoming severe. Mycoplasma-like organisms are also associated with a pigeon pea rosette in India (Maramorosch *et al.,* 1976), as well as with legume 'little leaf' disease in Australia, where the leafhopper, *Orosius argentatus,* is the vector (Hutton and Grylls, 1956; Bowyer, 1974). A witches' broom is also reported from Papua New Guinea (Anon, 1963). Nothing is known of the relationships between these diseases.

A bacterial leaf spot and stem canker occurs in India and the Sudan, where it is caused by *Xanthomonas campestris* pv. *cajani* (Kulkarni *et al.,* 1950; Sabet *et al.,* 1969). An unidentified bacterial leaf spot also occurs in Uganda, where it can become severe during the 'long rains' (Rachie and Wurster, 1971), and a bacterial blight has been observed in Panama (ICRISAT, 1978). Halo blight (*Pseudomonas syringae* pv. *phaseolicola*) has recently been found for the first time on *Cajanus* in Ethiopia (Allen, 1979a) and Queensland, Australia (A. C. Hayward, personal communication 1981). However, neither of these two bacterial pathogens is considered of economic importance to pigeon pea. The crop is also susceptible to a wide range of nematodes, including species from 12 genera, but only a few have been shown to cause crop loss. The root knot nematodes (*Meloidogyne* spp.) are commonly found infesting pigeon pea on which yield losses are estimated at 8–35 per cent (Bridge, 1981). *M. incognita* occurs on *Cajanus* in the Caribbean, Australia, India, West Africa, and the USA; *M. javanica* in Australia, Brazil, India, Puerto Rico, and the USA; *M. hapla* in East Africa and the USA; *M. arenaria* in the USA and *M. acronea* in Malawi (Bridge, 1981). The cyst nematode, *Heterodera cajani*, is a parasite of pigeon pea in India (Koshy *et al*., 1970; Koshy and Swarup, 1971a, b) and *Rotylenchulus reniformis* is known from the crop in Hawaii, Jamaica, Puerto Rico, and Trinidad (Ayala and Ramirez, 1964; Singh and Farrell, 1972; Hutton and Hammerton, 1975; Bridge, 1981). Although Hutton and Hammerton (1975) found no evidence in Jamaica of a relationship between population size of *R. reniformis* in the rhizosphere of pigeon pea and the incidence of decline symptoms (see also Section 3.4), they found that growth was more rapid, and green pod yield was increased, when the nematode populations were artifically suppressed. In Upper Volta, stunting and chlorosis of pigeon pea is related to populations of

Aphasmatylenchus straturatus which apparently can cause complete loss in yield (Germani and Dhery, 1973). Dodder (*Cuscuta hyalina*) is the only angiosperm parasite reported from *Cajanus* (Kotasthane *et al*., 1980).

The following three sections of this chapter are devoted to discussion of the field symptoms, crop loss potential, and control of wilt, phytophthora blight, and sterility mosaic which are considered the major diseases of pigeon pea in India, where more than 90 per cent of the world's crop is produced.

8.2.1 Fusarium wilt

Wilt is caused by the fungus *Fusarium udum* (Butler, 1910; Booth, 1978); certain other fusaria are also associated with the disease but they are of little importance (Kannaiyan and Nene, 1979b). Although wilt has been reported from the Caribbean, East Africa, and south-east Asia (Table 6), it is of greater importance in India where it is widespread, effectively occurring wherever the crop is extensively cultivated (Butler, 1906). Of the states surveyed recently, its incidence was found highest in Maharashtra (ICRISAT, 1978; Nene, 1980a). Affected plants appear chlorotic and start to wilt, exacerbated by periods of drought. Wilting may often be initally confined to individual branches. On peeling off the bark, black streaks are found to extend in the vascular tissue from the dead branches downward to one or more infected lateral roots. Later, other roots become affected. Infection of the tap root usually causes withering of the entire plant which remains standing with the dried leaves still attached (Plate 2; Butler, 1906; Nene, 1980). Discoloured portions of the stem and roots contain the pathogen, concentrated within the vessels. Systemic infection may lead to the development of spore masses externally on the bark; aerial infection apparently plays little part in disease development (Mohanty, 1946). The fungus can be retrieved from the roots of a susceptible cultivar within 15–30 days of sowing in 'wilt-sick' soil, and symptoms develop two weeks later (Nene, 1980). However, wilting is more usual during the crop's reproductive phase (Mundkur, 1935). Sheldrake *et al.* (1978a) have shown that disease development is influenced by the physiological state of the plant: the plant was less liable to wilt if pod development was prevented by flower removal, whereas wilting was exacerbated by defoliation during the reproductive phase. At this stage, developing pods compete with vegetative organs for assimilates and nutrients, and there is a mobilization of starch reserves from stems and roots. Sheldrake and his colleagues have suggested that the depletion of the assimilates in the roots facilitates the entry of the pathogen and/or its subsequent development. Defoliation during the reproductive phase might be expected to amplify this effect. The time of wilting strongly influences the extent of crop loss. When wilt occurs before pod development, grain yield loss is total; but when wilting does not occur until harvest time, losses are only estimated at about 30 per cent (Kannaiyan and Nene, 1981).

F. udum is not internally seedborne though it may carry-over as a contaminant of pigeon pea seed (Mohanty, 1946; Nene, 1980). The fungus occurs as a saprophyte in soil where it survives for at least three years, its persistence being governed by soil type and nutrient status (Sarojini, 1951; Nene, 1980), and presumably also by its propensity to form chlamydospores (Padwick, 1940b). Chlamydospores are produced in the vascular tissue of old infected plants (Butler, 1906) as well as in the hyphae and reproductive structures of other soil-inhabiting fungi (Upadhyay *et al.,* 1981). *F. udum* spreads through the soil at the rate of about 3 m during the growing season, probably depending in part on soil composition and organic matter content (McRae and Shaw, 1933; Butler and Jones, 1949; Shukla, 1975). Its growth is influenced by soil antagonists, especially the bacterium *Bacillus subtilis* which produces the antibiotic bulbiformin (Vasudeva and Roy, 1950; Vasudeva *et al.,* 1952, 1962), inhibition resulting in a reduced incidence of wilt. Soil amendment with organic matter has long been recognized as a potential measure for the cultural control of wilt (McRae and Shaw, 1933) and recent evidence suggests that amendment can increase populations of soil bacteria, including *B. subtilis,* thus suppressing the activity of *F. udum* through lysis of its mycelium (Singh and Singh, 1981). The treatment of pigeon pea seed with *B. subtilis* may also afford some protection against wilt at the seedling stage (Singh *et al.*, 1965). Certain soil fungi are also known to suppress the development of wilt in some cases, apparently through induction of resistance (Vasudeva and Govindaswamy, 1953; Kaiser and Sen Gupta, 1977); sterility mosaic infection is also reported to protect against wilt (Chadha and Raychaudhuri, 1966) but more recent work refutes this claim (ICRISAT, 1977). There is also a claim that the intercropping of pigeon pea with sorghum reduces the incidence of wilt by encouraging a deeper root system which then escapes infection (Gupta, 1954; cited by Deshpande *et al.*, 1963). More work is warranted on the effects of cropping systems on wilt development, including the efficacy of rotation in disease management. Farmers in India have long appreciated that the continuous cultivation of pigeon pea on the same land leads to crop failure, and this has been borne out by experiment (Butler, 1906). Certain rotations have seemed promising (Bose, 1938) though, in view of the persistence of the pathogen in soil, suitable cultural practices should probably be combined with the use of wilt-resistant cultivars.

Work begun in 1905 at Poona, in Maharashtra state, led to the selection of cultivars with wilt resistance (Butler, 1908). Subsequent screening has been conducted at many centres, including Delhi (Deshpande *et al.,* 1963), Pusa in Bihar (McRae and Shaw, 1933), West Bengal (Mukherjee *et al.*, 1971), Varanasi in Uttar Pradesh (Venkateswarlu *et al.,* 1980), in Madhya Pradesh (Sharma *et al.,* 1977), at Nagpur in Maharashtra (Patil and Sable, 1973), and at Hyderabad in Andhra Pradesh (Vaheeduddin, 1958; ICRISAT, 1978). Selections which have repeatedly been found resistant include cvs NP–80, BDN–1, and C–11 (McRae and Shaw, 1933; Vaheeduddin, 1958; Sharma *et al.,* 1977; ICRISAT, 1980), and some 33 lines have proved consistently resistant in 'wilt-sick' plots at ICRISAT (Nene, 1980a; Nene and Kannaiyan

1982). However, inconsistencies in reaction are common, and it is evident that resistance is environmentally unstable. For instance, cultivars possessing resistance when grown as annuals may develop severe wilt if they are grown on as a ratoon crop (Nene, 1980a); and it has long been appreciated that resistance may become ineffective in cultivars grown on the same land for three or more years (McRae and Shaw, 1933; Bose, 1938). While the first presumably relates to a physiological change in the host (Sheldrake *et al.*, 1978), the latter probably reflects a build-up of the pathogen population in the soil. It is possible that some sources of wilt resistance are inoculum-density-dependent; in other cases, such effects are consequences of disease escape through an initially insufficient infection pressure. The high degree of outcrossing in *Cajanus* contributes to the heterogeneity in wilt reaction, and soil microbiological influences on disease development are liable to compound the problem. Multilocation testing of promising material in the ICAR–ICRISAT uniform trials for pigeon pea wilt resistance in collaboration with the All India Coordinated Pulse Improvement Project (AICPIP) has confirmed that site differential reactions (see Section 7.3) are common, but the extent to which pathogenic variation, rather than the other factors we have discussed, are responsible for apparent 'breakdown' of wilt resistance remains unclear. It is recognized that *F. udum* is a highly variable species, particularly with regard to its ability to produce chlamydospores (Padwick, 1940b), but variation also occurs in morphological and colony characteristics (Sarojini, 1951; Subramanian, 1955) and pathogenicity. It is evident that certain isolates, which may cause pre-emergence seedling rot, are essentially soil saprophytes while others are specialized wilt pathogens (Sarojini, 1951). Isolates capable of profuse mycelial growth are often less pathogenic than those producing scanty mycelium, but no correlation has yet been shown between conidial production and pathogenicity (Shit and Sen Gupta, 1978). Host specialization within *F. udum* is recognized at the generic and specific levels; f.sp. *crotalariae* is the cause of wilt in sunnhemp (Mitra, 1934; Booth, 1978) and three races of this *forma specialis* have been distinguished on the basis of differential pathogenicity to separate species of *Crotalaria* (Armstrong and Armstrong, 1951). However, there is some evidence that pathogenicity may also be cultivar-specific. Certain pigeon pea isolates are apparently capable of overcoming the resistance of cv. C-11 (Baldev and Amin, 1974; Shit and Sen Gupta, 1978), which would therefore appear to be a useful race differential. The possible implications this may have for breeding are discussed in Section 8.3.

8.2.2 Phytophthora blight

A stem blight, first reported in India in 1966 (Williams *et al.*, 1968), was shown to be caused by a species of *Phytophthora* by Pal *et al.* (1970). Amin *et al.* (1978a) described the pathogen as a new species of *Phytophthora, P. cajani,* whereas isolates studied by Kannaiyan *et al.* (1980) have been classified as *P. drechsleri* f.sp. *cajani,* which is the nomenclature adopted here. A similar disease caused by *P. nicotianae* var. *parasitica* has been reported from Puerto Rico (Kaiser and Melendez, 1978) and apparently also from the Dominican Republic and Trinidad (Nene, 1978).

In India, the disease is known to occur in Delhi, West Bengal, Madhya Pradesh, Maharastra, Uttar Pradesh, and Andhra Pradesh where it is considered of economic importance, especially under wet conditions. Symptoms, which are confined to the aerial parts, first appear as dark-brown lesions at the stem base, or on the stem and branches up to 1 m above ground. These lesions, which often develop from a leaf scar, coalesce to girdle the stem (Figure 6), thus leading to a rapid wilting and dessication of foliage above. Petioles may become withered and leaflets, which may bear water-soaked lesions, often become upward-rolled. Longitudinal cuts into newly formed lesions reveal brown discoloration of the bark and cambium but not of the xylem, though in older infections the xylem may become discoloured and the stem may break at the lesion (Pal *et al.,* 1970; Williams *et al.,* 1975; Nene, 1980). The gross symptoms of phytophthora stem blight resemble those of fusarium wilt (Amin *et al.,* 1978b) and it seems probable that the two diseases have often been confused. Crop loss estimates have not been made, but there is no doubt that the disease has the potential to cause devastation to susceptible cultivars, especially under poorly drained conditions (Williams *et al*., 1975). There is no evidence that *P. drechsleri* f.sp. *cajani* is seedborne. The pathogen can survive in soil for at least four years (Nene, 1980; see Section 2.2.2). Too little is known of the epidemiology of stem blight to formulate satisfactory measures for its cultural control, and emphasis has been given to its management through host resistance. Pigeon pea lines possessing moderate resistance have been identified by Pal *et al*. (1970), and a pot-culture technique has been used at ICRISAT to screen 2835 accessions of *Cajanus* as well as seven species of *Atylosia*; the stem blight resistance of 75 of these was confirmed under field conditions (Kannaiyan *et al*., 1981). However, certain isolates of *P. drechsleri* f.sp. *cajani* from northern India are able to overcome resistance effective against an isolate from Hyberabad, indicating that pathogenic variation may complicate blight-resistance screening. The cultivar ICP-7065, amongst others, possesses race-specific resistance (Kannaiyan *et al*., 1980; Sharma *et al*., 1982).

8.2.3 Sterility mosaic

Sterility mosaic was first reported in India from Bihar in 1931. The disease is now known to be widespread throughout India; apparently similar diseases are known in Burma and Thailand. Capoor (1952) first named the disease and demonstrated its infectious nature, but its cause remains unknown. A virus aetiology has been generally assumed but the pathogen has not been characterized. Both sap (Capoor, 1952) and nematode (Narayanasamy *et al.,* 1963; Narayanasamy and Ramakrishnan, 1965) transmission has been claimed but neither has been confirmed (Nene and Rathi, 1972). However, Seth (1962, 1965) has demonstrated that the agent is transmitted by the eriophyid mite, *Aceria cajani;* this has subsequently been substantiated by Janarthanan *et al*. (1972) and Nene and Rathi (1972). It remains to be seen whether the aetiology of sterility mosaic is in any way related to other mite-transmitted diseases of woody plants, such as fig mosaic and peach mosaic whose causes remain unknown (Smith, 1972).

Symptoms of sterility mosaic include a stunting and a cessation of reproductive growth, although affected plants often bear both sterile and flowering branches. The leaves of affected plants are smaller, paler and usually show mosaic (Figure 33) without malformation (Capoor, 1952; Nene and Rathi, 1972). An epidemic of sterility mosaic occurred in 1959 in Coimbatore (Kandaswamy and Ramakrishnan, 1960), and subsequent surveys of the states of Tamil Nadu (Janarthanan *et al.,* 1973; ICRISAT, 1977), Karnataka, Maharashtra (ICRISAT, 1977), Madhya Pradesh (ICRISAT, 1978) and Uttar Pradesh (Nene and Rathi, 1972) have confirmed that the disease is widespread and prevalent in India. Crop loss estimates have not been made but, since diseased plants are either partially or completely sterile, local and seasonal outbreaks in which disease incidence has been reported to exceed 75 per cent are likely to result in considerable yield loss. Little is known of the epidemiology of sterility mosaic. The pathogen is not seedborne (Nene and

Figure 33 Symptoms of sterility mosaic, a disease of pigeon pea spread by mites, India (photograph courtesy Y. L. Nene)

Rathi, 1972). Various species of *Atylosia* are known to be susceptible to the agent as well as to its vector (ICRISAT, 1978) but their role in survival is unknown. There is some evidence that mite mortality may be high during the summer (ICRISAT, 1977). Individual mites remain buried in between the hairs on the undersurface of leaves (Seth, 1962) from where they may be windborne. Seth (1962) noticed that the incidence of sterility mosaic related to the proximity of an infection source, and recent work at ICRISAT has confirmed that vector spread of the disease occurs up to a distance of 35 m from an inoculum source (ICRISAT, 1980). Both nymphs and adults are effective vectors; a single mite can transmit, but more than 10 mites per seedling are required for fully efficient transmission (Janarthanan *et al.*, 1972; ICRISAT, 1977). The minimum inoculation access period apparently may be as brief as 5 min (ICRISAT, 1980), while the agent's incubation period in the host varies from one to six weeks, depending on mite numbers (Janarthanan *et al.*, 1972), host-plant age (Nene and Reddy, 1976a) and its intrinsic susceptibility (Nene and Reddy, 1976b). It has been claimed that sterility mosaic infection protects against wilt (Chadha and Raychaudhuri, 1966) but predisposes the plant to infection by powdery mildew (ICRISAT, 1980).

Little is known of the cultural control potential of destroying infection sources; nor has attention been given to vector control, although Rathi (1979) has shown that treatment of pigeon pea seed with a systemic pesticide can afford lasting protection against sterility mosaic. Most attention has been directed to the identification of host resistance, which appears a promising approach to the disease's management. Various resistance-screening techniques have been used, including grafting, the tying of mite-infested, diseased material to the test plant (Nene and Rathi, 1972; Subramanian *et al.*, 1973), the 'leaf-stapling' method (Nene and Reddy, 1976c), and the use of field infector rows (ICRISAT, 1980). The leaf-stapling inoculation technique has been used at ICRISAT to screen 2804 accessions of *Cajanus, Atylosia,* and progenies from intergeneric crosses. Five sources of immunity were found, confirmed by back-inoculation to a susceptible indicator (cv. Sharda). Some lines appeared to possess other mechanisms of sterility mosaic resistance, expressed as reduced or delayed development of mild symptoms (Nene and Reddy, 1976b). A ringspot reaction, typified by ICP-2376, appears to be a localized response to infection in which symptoms are transient and do not lead to sterility (Reddy and Nene, 1979). It has recently been shown (Reddy and Nene, 1980) that sterility mosaic-resistance, in some genotypes, is associated with a resistance to the mite vector exemplified by its decreased multiplication. Resistance to the agent when combined with such a vector resistance is expected to provide an increased stability of resistance (Reddy and Nene, 1980). The extent to which these mechanisms are indeed distinct components of resistance to sterility mosaic would be clarified by the development of a technique for the mechanical transmission of the agent, its purification and characterization. The progress made with the development of sterility mosaic-resistant cultivars is discussed in the next section.

8.3 DISEASE RESISTANCE AND PIGEON PEA IMPROVEMENT

Resistance within *C. cajan,* reviewed by Nene *et al.* (1981) has been found against rust (Hammerton, 1975), anthracnose (Hasan and Khan, 1979) and the leafspots caused by *Mycovellosiella cajani* (Onim and Rubaihayo, 1976), and *Phyllosticta cajani* (Saksena and Kumar, 1972), as well as against wilt, phytophthora blight, and sterility mosaic as we have already seen. Searches for resistance have often been made amongst very limited collections of pigeon pea germplasm, and this may in part account for certain failures to identify material possessing adequate resistance, especially to wilt. In view of this, Kumar *et al.* (1958) attempted to exploit the wilt resistance available within *Atylosia lineata* through 'intergeneric' hybridization, and Sheriff *et al.* (1977) utilized ethylmethyl sulphonate to induce wilt-resistant mutants of the susceptible cv. Co-1. Subsequent evaluation of more extensive collections of pigeon pea germplasm at ICRISAT has confirmed that adequate wilt resistance is available within *C. cajan* (Nene and Kannaiyan, 1982). There is evidence that resistance to wilt, on which most information is available, has a physiological basis. Subramanian (1963) showed that a vascular diffusate from a resistant (cv. NP-15) but not from a susceptible (cv. NP-24) plant was partially inhibitory to conidial germination in *F. udum*, especially when the diffusate was obtained from an infected plant; and Murthy and Bagyaraj (1980) have demonstrated that flavanols, as well as alkaloids, occur at higher concentrations at all stages of growth of a wilt-resistant cultivar (cv. C-11-6) than in the susceptible cv. TS-136-1. Ingham (1976a) has identified various anti-fungal isoflavanoids, including cajanol, from infected stems and Preston (1977) has shown that the phytoalexin, cajanone, which is the major phenolic compound in the roots of pigeon pea, has the ability to inhibit completely the growth of *F. udum*. Pinostrobin chalcone and two stilbene carboxylic acids have been shown to be phytoalexins in pigeon pea leaves (Dahiya *et al.*, 1982).

Little is known of the genetic control of disease resistance in *Cajanus.* Shaw (1936a, b) suggested that wilt resistance in cv. NP-80 was controlled by two complementary genes, and Joshi (1957) indicated that wilt resistance was thought to be governed by a single dominant gene pair in other material. A more complex inheritance is suggested by recent investigation (ICRISAT, 1980). Work on phytophthora blight has shown that a single dominant gene, designated 'Pd_1', governs race-specific resistance in all the seven resistant parents examined (Sharma *et al.,* 1982). Four allelic genes control the response to sterility mosaic: two control immunity, one conditions the ringspot reaction (Section 8.2.3), and one governs susceptibility, which is dominant to the three resistance genes (ICRISAT, 1980).

Disease resistance has been an objective of breeding programmes in India (Deshpande *et al.*, 1963) as well as in Uganda (Khan and Rachie, 1972), Puerto Rico (Abrams *et al.,* 1978) and Trinidad (Ariyanayagan and Spence, 1978b). In the Caribbean, the incorporation of rust resistance into green pod types is an important goal, but progress has been hampered by the 'break-down' of resistance, as occurred in the cultivar U.W.I.17 in Jamaica in 1972 (Hammerton, 1975). Most of the early work in India focused solely on wilt, and the development of wilt-resistant cultivars relied on selection, though some crosses were made to combine wilt resistance with heavy yield potential and other desirable agronomic characters in a programme at the Indian Agricultural Research Institute in Delhi (Deshpande *et al.,* 1963). Pigeon pea breeding at ICRISAT, recently reviewed by Byth *et al*., 1981 and Green *et al*., 1981, has given higher priority to disease resistance than it has hitherto been accorded elsewhere. Wilt-sick plots, a pot-culture technique, and a leaf-stapling inocualtion method have been successfully used to screen germplasm for resistance to wilt, phytophthora blight, and sterility mosaic, respectively. Certain lines have been found to possess combined resistance. For instance, ICP-7273 carries resistance to wilt, blight, and sterility mosaic, ICP-3753 and ICP-7182 are resistant to wilt and blight, and various other lines possess resistance to wilt and sterility mosaic or blight and sterility mosaic (ICRISAT, 1978; Kannaiyan *et al.,* 1981; Nene *et al.,* 1981). Such sources have been used as parents in crosses to incorporate disease resistance into agronomically superior material, using conventional breeding methods. Population-breeding procedures have been used in pigeon pea improvement (Khan, 1973) but, unlike work on cowpea (see Section 7.3), they have not yet been employed in the development of disease resistance in pigeon pea. Segregating progenies at ICRISAT are now screened in a 'multiple disease nursery' (ICRISAT, 1980). Promising material from this programme is tested across a range of environments in multilocational trials, some as entries in the All India Co-ordinated Pulse Improvement Project. Resistance to each of the three major diseases has shown location-specificity. It is not known whether pathogenic variation occurs in the agent causing sterility mosaic (ICRISAT, 1977), though such variation is recognized in *P. drechsleri* f.sp. *cajani* (Kannaiyan *et al.* 1980) and *F. udum* (Baldev and Amin, 1974; Shit and Sen Gupta, 1978; IARI, 1979). However, in no case is it yet firmly established that cultivar-specific pathogenicity does jeopardize the durability of resistance in pigeon pea. Experience with vascular wilt fusaria from other crops suggests that the existence of races need not necessarily terminate the commerical life of cultivars with race-specific resistance. For instance, race 3 of *F. oxysporum* f.sp. *lycopersici* can attack tomato cultivars that carry the two genes which protect against races 1 and 2; but race 1 remains the most widely distributed and economically important, and cultivars possessing the gene effective against race 1 are still making an important contribution to the control of fusarium wilt (Russell, 1978). Similarly, experience with Panama disease of banana, caused by *F. oxysporum* f.sp. *cubense* in which only two races have been distinguished, suggests that cultivars highly resistant to race 1 show no evidence of break-down despite being extensively grown where that race occurs (Holliday, 1980). Such breakdown as does occur appears to be attributable to adverse environment rather than to the emergence of new races (Stover and Malo,

1972). Time alone will show whether lessions from these two intensively studied fusaria have application to the wilt disease of pigeon pea. Experience with species of *Phytophthora,* on the other hand, does indicate that resistance is often transient, and cultivar-specific races have led to breakdown within a single environment, as shown by Purss (1972) with *P. vignae*. Current evidence (Sharma *et al*., 1982) suggests the same is true of phytophthora blight in pigeon pea. Identified sources of resistance to *P. drechsleri* f.sp. *cajani* has been shown to be race-specific, operative against the Hyderabad race but not against the Kanpur race, and governed by a single gene, 'Pd_1'. A search for genes conferring resistance to the Kanpur race is an obvious next step, but means of identifying race non-specific resistance should also be sought. There might be potential for the regional deployment of race-specific resistance genes, depending on the geographical distribution of races. In this context, it is of interest that ecotypes ('environmental races' *sensu* Green, 1973) are recognized within *P. drechsleri* (Shepherd and Pratt, 1973). The nature of the variation within *P. drechsleri* f.sp. *cajani* requires further investigation if guidance is to be given in the formulation of an appropriate breeding strategy.

Although sterility barriers seem unlikely to prevent the use of resistance present in *Atylosia,* there appears to be little urgency to utilize such sources. Good progress has been made recently at ICRISAT in the identification of resistance in *Cajanus* germplasm and these are being used in the development of superior genotypes. The extent to which resistance alone can be relied upon to provide adequate protection from disease in Indian agriculture remains uncertain, but it may be prudent to work towards an integrated system of disease management in which appropriate cultural practices would be employed to supplement host-plant resistance.

Chapter 9

Chickpea Diseases

9.1 EVOLUTION, ADAPTATION, AND PRODUCTION OF CHICKPEAS

The Old World genus *Cicer*, to which the chickpea or gram belongs, contains some 40 species of annual or perennial herb, or small perennial shrub; one species (*C. cuneatum* Hochst. ex Rich.) bears tendrils and is a semi-climber. Nine annual species are recognized: *C. chorassanicum* (Bge) M. Pop. and *C. yamashitae* Kitam. are native to Afghanistan; *C. judaicum* Boiss., *C. pinnatifidum* Jaub. & Spach, and *C. bijugum* K. H. Rech. are wild in the Middle East; *C. cuneatum* is endemic to Ethiopia and the Sudan; and the two species *C. echinospermum* Davis and *C. reticulatum* Lad. have recently been described from south-eastern Turkey. Investigation of the relationships between the latter six species and the cultivated *C. arietinum* L. suggests that three cross-compatibility groups can be recognized (Ladizinsky and Adler, 1976b). Group I comprises *C. arietinum, C. reticulatum,* and *C. echinospermum*; Group II species are *C. judaicum, C. pinnatifidum*, and *C. bijugum*; and Group III is represented by *C. cuneatum*. All have the same somatic chromosome number (2n = 16) but crosses between members of different groups yield no viable seed. Within a group interspecific hybridization is possible but relative fertility varies with the cross. Gene transfer from the wild species to the cultivated *C. arietinum*, is likely to be confined to Group I members. *C. echinospermum* is within the secondary gene pool (*sensu* Harlan and de Wet, 1971) of *C. arietinum*, while *C. reticulatum*, which crosses readily with the cultigen, is considered the wild progenitor of chickpea (Ladizinsky and Adler, 1976a). It has subsequently been suggested that the cultigen and its progenitor be treated merely as separate sub-species of *C. arietinum*, the wild chickpea as spp. *reticulatum* (Lad.) Moreno & Cubero, and the cultivated kinds as ssp. *arietinum* (Moreno and Cubero, 1978).

The centre of diversity of the genus is in western Asia. The two wild taxa, *C. echinospermum* and *C. arietinum* ssp. *reticulatum,* with which the cultigen has close phylogenetic affinity, are known only from a restricted region of south-eastern Turkey, suggesting that it is here that the chickpea has its origin. Indeed, the earliest record of chickpea is from the Hacilar site in Turkey, radiocarbon-dated to 5450 BC (van der Maesen, 1972; Ramanujam, 1976), but such archaeological evidence is scanty, perhaps partly a consequence of misidentification of carbonized remains, for the chickpea is characterized by a beak which can easily be destroyed by fire, thus becoming effectively indistinguishable from carbonized pea (*Pisum*). Nevertheless, seed has been unearthed at various archaeological sites throughout the Middle East, indicating the use of chickpea in the early Neolithic, together with wheat, barley, pea, and lentil (Ladizinsky and Adler, 1976a). It seems reasonably certain that this ancient crop was first domesticated in the 'Fertile Crescent' of the Near East from where it spread throughout the Mediterranean area in the third and fourth millenia BC (Ramanujam, 1976; Hawtin *et al.,* 1980). The earliest record of chickpea in India, at Atranji Khera in Uttar Pradesh, is dated at 2000 BC and presumably represents an overland introduction. Early records from peninsular India are more recent; whether or not the crop also reached southern India independently via maritime trade with the Middle East remains uncertain. Ethiopia, which is the third centre of diversity, probably also received the crop from the Mediterranean. The chickpea has been introduced into a few other areas in more recent times. In the sixteenth century, Spanish and Portuguese travellers took it to the New World, most notably Mexico (Ramanujam, 1976).

It has long been appreciated that considerable genetic variation exists within the cultigen and various attempts at intraspecific classification have been made (Howard *et al.,* 1915; Soomro and Baluch, 1968; van der Maesen, 1972). Various taxa including sub-species and varieties have been erected but, with the exception of the distinction of wild from cultivated forms of *C. arietinum* as ssp. *reticulatum* and ssp. *arietinum* respectively (Moreno and Cubero, 1978), there is apparently no sound taxonomic basis on which to maintain them. Nevertheless, a recent evaluation of chickpea germplasm collections has revealed that there is discrete clustering of characters between which there is considerable genetic divergence (Narayan and Macefield, 1976; Martinez *et al*., 1979; Moreno and Cubero, 1979).

Two races (Harlan and de Wet, 1971) are recognized within the cultivated chickpea, referred to as *microsperma* and *macrosperma* groups, which are roughly equivalent to the 'desi' and 'kabuli' types widely recognized by plant breeders (Moreno and Cubero, 1978; Hawtin and Singh, 1980). Kabuli types produce large seeds (exceeding 26 g/100) which are rounded ('sheep-headed') and pale cream in colour. Plants are relatively tall, bear white flowers, and have no anthocyanin pigmentation. The kabuli forms are characteristic of the Mediterranean and the Near East, but they are also cultivated on a small scale in the Indian sub-continent where they are a recent (1700 AD) introduction. The desi types produce smaller, irregularly shaped seed of various colours. Plants are relatively short and have smaller leaflets, they are sometimes prostrate and most often have anthocyanin pigmentation in flowers and stems. Cultivars of this group are characteristic of East Asia, Ethiopia, parts of Iran, Afghanistan, and India (Moreno and Cubero, 1978; Hawtin *et al.,* 1980). Thus the two types differ substantially not only in morphology but also in physiological and adaptive characteristics. The kabuli types are thought to have originated by selection from the more primitive desi types, the divergence having occurred in comparatively recent times. It is often held that kabulis have a narrower genetic base than the desis, but recent evidence suggests that the diversity within kabuli types may have been underestimated, perhaps owing to the greater emphasis on desi types in India. Whatever the extent of their respective gene pools, it is clear that each group has certain characters potentially useful to the other, and introgression is potentially an important strategy in the improvement of the crop (Hawtin and Singh, 1980). Study of the genetic systems governing certain quantitative characters in the two types has shown that no isolation barriers exist, although there is usually a dominance of 'primitive' over 'selected' characters (Martinez *et al.,* 1979). Indeed, there is evidence (Niknejad and Khosh-Khui, 1972) that natural introgression between the two types may occur, albeit at a low level (1.5 per cent).

Chickpeas are widespread in seasonally arid regions as a rainfed crop on poor soil. They are notably tolerant of drought and low management and have apparently not become adapted to conditions very different from their original habitat (van der Maesen, 1972). Yet, within the crop's range, there is considerable local adaptation among landraces of which the crop principally consists. Although these adaptive responses are yet to be fully quantified, it is probable that different gene constellations have been fixed by selection in different environments (Murty, 1975; Summerfield *et al.,* 1980). In the western part of the crop's range in the Mediterranean kabuli types are traditionally sown in spring, whereas in the eastern and southern parts of its distribution, including India, desi types are grown as a winter crop (Ramanujam, 1976). Narayan and Macefield (1976) concluded that the most important morphological characters operating in the genetic divergence between these two groups were plant type, seed colour, and flower colour. The kabuli group as a whole has a greater range in seed size, and tends to have more primary branches, greater cold tolerance, and a more upright habit. Desi types tend to have a bushier habit, more seeds per pod, more pods per plant, and greater tolerance to drought and heat (Hawtin and Singh, 1980). Cultivars adapted to peninsular India typically mature within 110 days, while those adapted to northern India are of longer duration. However, duration is environmental rather than genetic under field conditions, maturity being governed primarily by the onset of moisture stress with increased temperature and depleted soil water; in North India, this occurs much later than in peninsular India (J. B. Smithson, personal communication 1982). Although chickpeas are quantitatively long-day plants, it is known that cultivars vary in their photosensitivity; both temperature and day length strongly influence crop duration which is a powerful determinant of seed yield (Minchin *et al*., 1980; Roberts *et al*., 1980; Summerfield *et al.,* 1980).

Some 7.5 million tonnes of seed are produced from nearly 10.8 million hectares at an average yield of 692 kg/ha. India is much the largest producer, accounting for 78 per cent of the world's crop; other countries producing significant quantities are, in order of importance, Pakistan, Mexico, Turkey, and Ethiopia. Kabuli types contribute some 10–15 per cent of the total production (Hawtin *et al.,* 1980). There has been a slight fall in chickpea production in recent years (FAO, 1976); this trend probably reflects adverse competition from high-yielding cultivars of cereals arising from the extension of irrigation in northern India (Swaminathan and Jain, 1973). Low management inputs are the rule and, as a result of reliance on conserved soil moisture, production is erratic and low. In India chickpeas are most commonly either intercropped or grown in a rotation with wheat, as winter ('rabi') crops following millet or sorghum (van der Maesen, 1972). In Ethiopia, as in India, chickpeas are the most important grain legume; the crop is sown at the end of September after the main rains, usually as a sole crop in rotation with wheat or tef (*Eragrostis tef*), but occasionally also in association with sorghum or safflower (*Carthamus tinctorius*) (Westphal, 1974). In western Asia and the Mediterranean chickpeas are grown in monoculture, traditionally as a spring-sown crop. During the past two decades there has been an intensification of research into chickpea improvement, beginning with the establishment of the Regional Pulse Improvement Project (RPIP) in India and Iran in 1962. In 1973 work on the crop began at the International Crops Research Institute for the Semi-Arid Tropics (ICRISAT), based near Hyderabad in India. Meanwhile, the Arid Lands Agricultural Development programme (ALAD) in the Middle East and North Africa had set up a food legume project which embraced chickpea, and this was absorbed by the International Centre for Agricultural Research in the Dry Areas (ICARDA) in 1977. Collaborative work has subsequently revealed that there are several means by which increased production of chickpeas may be brought about, both through a changed system of farming and the crop's spread into new areas. Despite a reputation for low yields, it is evident that the crop has potentially high productivity; 4800 kg/ha is a commonly quoted maximum value (Summerfield *et al.,* 1980). Chickpeas are grown under irrigation in many regions, as for instance in Egypt (Mann, 1947), northern Sudan (Ageeb and Ayoub, 1977), and north-western Mexico (Arias, 1980). The use of supplementary irrigation may encourage the spread of the crop into new areas, like Zambia (Hill, 1981), and there is potential for the exploitation of short-duration cultivars responsive to inputs including fertilization and irrigation in cereal-farming areas of northern India. Here, too, cultivars adapted to late (November) sowing could have a place in rotation with crops sown in the monsoon season. Crop adaptation studies are also required to assess the potential of the early (September) sowing of chickpea in peninsular India (Byth *et al.,* 1980). However, scope for increased productivity through altered cropping systems is not confined to Indian agriculture; there exists the possibility of a fundamental change in chickpea production in the Mediterranean region. Joint work of ICRISAT and ICARDA has revealed that winter sowing can substantially increased seed yield, apparently through extending the growing season so that more vegetative growth may be made, and by advancing development so that reproductive growth is made under more favourable conditions, before moisture stress and high temperatures become severely limiting. Winter sowing may possibly also permit the cultivation of chickpeas in yet drier areas, and allow its inclusion in rotations with irrigated summer crops. Traditionally, however, *Cicer* is sown in spring in the Mediterranean, perhaps partly because local cultivars sometimes possess insufficient cold tolerance, but principally because blight, caused by *Ascochyta rabiei,* can devastate the winter crop; spring sowing sometimes enables a measure of escape from its ravages. Thus, the effective control of ascochyta blight is the key to the increased production of kabuli types in the Mediterranean region. This and other diseases of chickpea are reviewed in the next section.

9.2 THE PATHOLOGY OF CHICKPEAS

Although *C. arietinum* is susceptible to a relatively large number of pathogens (50 are listed by Nene, 1978), few diseases are currently recognized as economic restraints to chickpea production. The literature on the crop's diseases is not extensive; general reviews have been published by van der Maesen (1972) and Nene (1980b), and reviews of the diseases present in India (Sen Gupta, 1974) and the USA (Kaiser, 1981a) are also available. Predominant among the pathogens of *Cicer* are certain soilborne fungi which, in India, have long been associated with diseases of the root system and collar region of the plant. While still little is known of the interactions that may occur between the various pathogens during disease development, it has been established that the 'wilt complex' comprises a number of effectively distinct diseases, the separate diagnosis of which is nicely illustrated in colour by Nene *et al.* (1979). A vascular wilt, caused by *Fusarium oxysporum* f.sp. *ciceris* (widely and incorrectly referred to as f.sp. *ciceri*), and an aerial blight, caused by *Ascochyta rabiei,* are the most important chickpea diseases, and these are each discussed more fully in subsequent sections of this chapter. An attempt is made in Table 23 to summarize information on the other fungal diseases of the crop. Those omitted because of insufficient information include a root rot caused by *Ozonium texanum* (Mishra, 1955), a white root rot of uncertain aetiology (Haware and Nene, 1976), downy mildew (Nene, 1980b) and various diseases associated with species of *Phytophthora* (Frezzi, 1950; Bhelwa, 1962; Suryanarayana and Pathak, 1968).

Six viruses have been identified from natural infections of chickpea (Table 24), the most important of which is pea leafroll virus, the cause of stunt; this disease is the subject of Section 9.2.3. A phyllody disease has been reported from Coimbatore, Tamil Nadu in southern India (Venkataraman, 1959; Kandaswamy and Natarajan, 1974) but its cause has not been elucidated. A bacterial disease of seedlings has been shown to be caused by *Xanthomonas campestris* pv. *cassiae* (Rangaswami and Prasad, 1959). Various nematodes are pathogenic to *Cicer* but none is of great importance. The root knot

TABLE 23 Fungal Diseases of Chickpea

Disease	Causal fungi	Distribution	Importance/Crop loss	References
1. Seed and seedling diseases				
Pre-emergence damping-off	*Macrophomina phaseolina* (Tassi) Goid. and *Pythium ultimum* Trow	India, NW USA	Potentially important, local	Singh and Mehrotra, 1980a; Kaiser and Hannan, 1981
Seedling mortality	*Sclerotium rolfsii* Sacc. (perfect state = *Corticium rolfsii* Curzi)	India	Potentially important	Nene, 1980b
2. Stem and root rots, and wilts				
Stem anthracnose	*Colletotrichum capsici* (Syd.) Butl. and Bisby	India	Occasional, minor	Ramakrishnan, 1947; Nene, 1980b
Stem rot, white mould	*Sclerotinia sclerotiorum* (Lib.) de Bary	Iran, India, Pakistan, Australia, Chile	Locally important, especially under cool wet conditions	Kaiser, 1972a; Nene, 1978; Khan, 1980
Foot rot	*Operculella padwickii* Kheswalla	Northern India	Locally important	Kheswalla, 1941; Singh and Bedi, 1975
Collar rot	*Sclerotium rolfsii* Sacc.	Ethiopia, India, Syria	Locally important, favoured by high soil moisture and warm temperature	Stewart and Yirgou, 1967; Nene, 1978
Rhizoctonia root rot	*Rhizoctonia solani* Kühn (perfect state = *Thanatephorus cucumeris* (Frank) Donk)	Widespread?	Minor	Nene, 1978
Dry root rot, ashy stem blight	*Macrophomina phaseolina* (Tassi) Goid.	India, Pakistan, Iran, Lebanon, Syria, Turkey, Ethiopia, USA, Australia	Minor	Dastur, 1935; Westerlund *et al.*, 1974; Khan, 1980; Nene, 1980b
Fusarium root rot	*Fusarium solani* f.sp. *pisi* (Jones) Snyd. and Hans.	USA, ? India	Potentially important, local	Kraft, 1969; Grewal *et al.*, 1974; Westerlund *et al.*, 1974
Neocosmospora root rot	*Neocosmospora vasinfecta* Smith	India	Minor	Haware and Nene, 1976

Acrophialophora wilt	*Acrophialophora fusispora* (Saksena) Samson	India	Minor	Purkayastha and Chakraborty, 1981
Verticillium wilt	*Verticillium dahliae* Kleb. = microsclerotial form of *V albo-atrum* Reinke and Berth.	USA	Minor	Erwin, 1958a; Kaiser, 1981a
Fusarium wilt	*Fusarium oxysporum* f.sp. *ciceris* (Padw.) Matuo and Sato (= *F. lateritium* (Nees) Snyd. and Hans. f. *ciceri* (Padw.) Erwin)	Widespread. Reported from Burma, India, Pakistan, Ethiopia, Malawi, Tunisia, USA, Mexico, Peru	Major. Wilt at flowering to early pod-fill can cause 77–94% loss in seed yield	McKerral, 1923; Prasad and Padwick, 1939; Erwin, 1958b; Stewart and Yirgou, 1967; Echandi, 1970; Nene, 1978; Haware and Nene, 1980a; Reddy *et al.*, 1980; Kannaiyan, 1981
3. Foliar diseases				
Ascochyta blight	*Ascochyta rabiei* (Pass.) Lab. (= *Phyllosticta rabiei* (Pass.) Trott. = *Phoma rabiei* (Pass.) Khune and Kajoor)	Widespread	Major. Annual losses of 20–50% common in some regions	see Section 9.2.2
Stemphylium blight	*Stemphylium sarciniforme* (Cav.) Wilts.	India, Iran, Syria	Minor?	Das and Sen Gupta, 1961; Sen Gupta and Das, 1964; Prasad *et al.*, 1969; Kaiser, 1972a; Nene, 1980b
Alternaria blight	*Alternaria alternata* (Fr.) Keissler and *Alternaria* state of *Pleospora infectoria* Fuckel	India	Minor	Vishwakarma and Basu Chaudhary, 1974; Haware and Nene, 1976
Grey mould	*Botrytis cinerea* Pers. ex Pers. (perfect state = *Sclerotinia fuckeliana* (de Bary) Fuckel)	India, Australia, Argentina, Colombia	Locally damaging, as in N India	Joshi and Singh, 1969; Cother, 1977a; Nene, 1978
Powdery mildew	*Leveillula taurica* (Lév.) Arn. (conidial state = *Oidiopsis taurica* (Lév.) Salmon)	India, Ethiopia, Sudan	Minor	Nour, 1957; Stewart and Yirgou, 1967; Khune *et al.*, 1979
Rust	*Uromyces ciceris-arietini* Jacz. apud Boy. and Jacz.	Widespread in Mediterranean region, SE Europe and Asia. Present also in E Africa and Mexico	Locally important; regarded as the major factor of chickpea production in C Mexico	Mehta and Mundkur, 1946; Saksena and Prasad, 1955; Dalela, 1962; Payak, 1962; Gallegos *et al.*, 1965; Punithalingam, 1968; Bahadur and Sinha, 1970a, b

TABLE 24 Virus Diseases of Chickpea

Disease	Causal virus	Distribution	Transmission		Importance/Crop loss	References
			Vector	Seed (%)		
Yellow mosaic, yellowing	Bean yellow mosaic virus (BYMV)	USA, Iran	Aphid	0	Locally important. Seed yield loss estimated at 77–92% in Iran	Snyder *et al.*, 1956; Erwin and Snyder, 1958; Kaiser *et al.*, 1968; Kaiser and Danesh, 1971a, b
Mosaic, bud necrosis, wilt	Alfalfa mosaic virus (AMV)	USA, Iran, India	Aphid	0	Locally important in Iran where losses of 72–96% have been recorded	Erwin and Snyder, 1958; Kaiser *et al.*, 1968; Kaiser and Danesh, 1971a,b; Nene *et al.*, 1979
Mosaic	Cucumber mosaic virus (CMV)	Iran	Aphid	0	Locally important in Iran where a yield loss of 52% has been recorded	Kaiser *et al.*, 1968; Kaiser and Danesh, 1971a, b
Enation mosaic	Pea enation mosaic virus (PEMV)	USA	Aphid	0	Minor?	Erwin and Snyder, 1958
Stunt	Pea leafroll virus (PLRV)	Iran, India, and probably elsewhere	Aphid	0	Major. Causes the greatest yield loss (up to 100%) of all virus diseases	Kaiser and Danesh, 1971a, b; Kaiser, 1972b; Nene and Reddy, 1976a; Reddy *et al.*, 1979
Necrotic yellows	Lettuce necrotic yellows virus (LNYV)	Australia	Aphid	0?	Potentially important in northern New South Wales and southern Queensland	Corbin, 1975

nematodes, *Meloidogyne incognita* and *M. javanica,* are capable of causing significant yield loss, especially at high population densities favoured by irrigation (Srivastava *et al*., 1974; Reddy, 1975; Jamal, 1976). Chickpea is also occasionally infested by other nematode genera, including *Tylenchorhynchus* (Gill and Swarup, 1977), *Pratylenchus* and *Rotylenchulus* (Nene, 1978). Angiosperm parasites include dodder (Vyas and Joshi, 1975) and broomrape (J. B. Smithson, personal communication 1980).

9.2.1 *Fusarium* wilt

The history of chickpea wilt, traced by Nene (1980b), has its origins in India where the disease was first mentioned in 1918. Since that time, an often bewildering array of agents have been claimed as the cause of 'wilt', including various soil fungi of the genera *Fusarium* (Padwick, 1940a; Grewal *et al.,* 1974), *Verticillium* (Erwin, 1958a), *Rhizoctonia* (Dastur, 1935), *Sclerotium* (Mathur, 1963), *Ozonium* (Mishra, 1955), and *Acrophialophora* (Purkayastha and Chakraborty, 1981) as well as certain viruses (Kaiser and Danesh, 1971a) and physiological factors (Dastur, 1935). Confusion in the literature and, particularly, in the field was rife, and it became accepted in India that a 'wilt complex' existed (Nene, 1980b). Although mixed infections are common, it has subsequently become apparent that wilt (*sensu stricto,* see Section 2.2.2) and other wilt-like disorders are in fact separate, diagnosable diseases (Nene *et al.,* 1979). Chickpea wilt is caused by *Fusarium oxysporum* f.sp. *ciceris.*

Fusarium wilt is reported, outside India, from Burma (McKerral, 1923), Pakistan (Khan, 1980), Russia (Kuz'mina, 1966), Ethiopia (Stewart and Yirgou, 1967), Malawi (Kannaiyan, 1981), Tunisia (Reddy *et al.,* 1980), Spain (Benlloch, 1949), the USA (Erwin, 1958b), Mexico (Arias, 1980), and Peru (Echandi, 1970). In India, wilt is responsible for substantial crop loss; an annual loss of 10 per cent is often quoted (Mathur *et al.,* 1960; Nene, 1980b). Disease incidence, which is generally higher on light, sandy and alkaline soils than on heavier acid soils (Chauhan, 1962b; Echandi, 1970), is influenced by sowing date which reflects changes in soil moisture and temperature. In northern India, a delay in sowing from late September to mid-October can lead to decreased wilt incidence (Padwick and Bhagawagar, 1943), presumably because of declining soil moisture and falling temperature (Chauhan, 1963a, b). 'Early' and 'late' wilts are distinguished according to the time symptoms appear relative to crop phenology (Raheja and Das, 1957). Early wilt develops at the seedling stage and late wilt is not expressed until after flowering, there being little disease development during the intervening period because low temperatures inhibit wilting, at least in northern India. In peninsular India, where winter temperatures are not so low, the development of wilt is apparently unimpeded (Nene, 1980a), The time at which wilting occurs influences the extent of crop loss. Wilting during the early pod-filling period has been estimated to cause from 77.5–94.2 per cent loss on a single-plant basis, while wilt developing later, at pod maturity, may account for 57.7–82.6 per cent loss; wilting at harvest time may be responsible for losses in the range 23.6–65.4 per cent (Haware and Nene, 1980a). Typical symptoms of wilt are first seen as sudden drooping of leaves which become straw-brown in colour without being shed prematurely. Affected plants show no external signs of root rot but, on splitting open the stem at the collar, an internal discoloration of the xylem and pith is conspicuous. Sometimes only a few branches of a single plant are affected so that wilting is partial (Nene *et al.,* 1979).

Although it has long been appreciated that the wilt fungus is soilborne (McKerral, 1923), its longevity as chlamydospores in soil and infected plant debris remains uncertain, and crop rotation alone has not been successful in controlling the disease. There is evidence that *F. oxysporum* f.sp. *ciceris* is able to survive for at least 3 years in soil (ICRISAT, 1980), as well as in latent infections of roots of various crops, suggesting that alternate hosts may play some role in the pathogen's carry-over (Armstrong and Armstrong, 1948; Haware and Nene, 1979b); this has obvious implications for the choice of suitable rotations involving chickpea. The wilt fungus is also seedborne, and certain seed dressings have shown promise in the eradication of the pathogen from seed (Kuz'mina, 1966; Haware *et al.,* 1978). However, the identification and use of host-plant resistance has the greatest potential in the long-term management of wilt. Although sources of wilt resistance were first noted some 60 years ago (McKerral, 1923), subsequent work on resistance has been rather limited (Ayyar and Iyer, 1936; Mathur *et al.,* 1960; Sharma and Khare, 1969; Singh *et al.,* 1974b). Large-scale, systematic screening of *Cicer* germplasm has been conducted recently at ICRISAT (Haware and Nene, 1980b; Kumar *et al.,* 1980; Nene and Haware, 1980; Haware *et al.,* 1981), using 'wilt-sick' field nurseries as well as pot- and water-culture tests. Some 50 sources of wilt resistance have now been identified from among more than 7000 accessions. Of the wild species examined, only *C. judaicum* was found to possess resistance to wilt (Nene and Haware, 1980; van der Maesen *et al*., 1980). Resistance appears not to be associated with seed-coat colour, for there is wide variation in this character among resistant accessions (Haware and Nene, 1980b; Haware *et al.,* 1981), although wilt resistance does appear to be more common among desi than among kabuli types (Kumar *et al.,* 1980). An investigation of the nature of wilt resistance was made by Ayyar and Iyer (1936) who found that there was substantially greater formation of suberin in the periphery of the cortex in the resistant cv. Type 468 than in a susceptible cultivar. Since this suberin layer appeared to impede fungal penetration, and post-penetration development of the pathogen was much slower in the resistant cultivar, they concluded that resistance was consequent on the combined effect of two separate mechanisms, one morphological and the other physiological. No recent studies appear to have been made on the subject; although phytoalexins have been isolated from chickpea (Keen, 1975a; Ingham, 1976b), their role in disease resistance remains unestablished. Similarly, very little is known of the mechanism of inheritance of wilt resistance. The published evidence (Ayyar and Iyer, 1936; López García, 1974; Pathak *et al.*, 1975) suggests that either dominant or recessive oligogenes are involved. Smithson *et al.* (unpublished) suggest that two recessive genes are necessary for high-level resistance and that either gene alone can delay wilt.

Multilocational testing of resistant material has often shown strong site effects. Although the 'breakdown' of resistance in cv. BG-212 (ICC 11324) in West Bengal has been attributed to the interaction between *F. oxysporum* f.sp. *ciceris* and *Acrophialophora fusispora* (Purkayastha and Chakraborty, 1981), in other cases it appears that pathogenic variation within *F. oxysporum* f.sp. *ciceris* may be responsible (Chauhan, 1962a; Haware and Nene, 1979a). The implications this may have for breeding are discussed in Section 9.3.

9.2.2 *Ascochyta* blight

A blight of chickpea is caused by the fungus *Ascochyta rabiei* whose various synonyms include *Phyllosticta rabiei* and *Phoma rabiei* (Punithalingam and Holliday, 1972; Khune and Kapoor, 1980). Isolates referable to the related *Phoma medicaginis* have been obtained from blighted chickpea in Kashmir (Haware and Nene, 1981). The perfect state of *A. rabiei, Mycosphaerella rabiei,* has occasionally been reported (Kovačevski, 1936; Gorlenko and Bushkova, 1958; Zachos *et al.,* 1963) but further work on the relationship between the two states appears warranted.

The literature on ascochyta blight, which is very extensive, has been reviewed recently by Nene (1981). The disease has been most intensively studied in the Punjab of India, a part of which is now in Pakistan, where blight was first observed in 1911, though it is probable that it had existed there for very much longer (Sattar, 1933b). The pathogen is widespread in the Mediterranean region and western Asia, where it has presumably co-evolved with *Cicer*; it has spread subsequently into eastern Africa (Clegg, 1947), North America (Morrall and McKenzie, 1974), and Australia (Cother, 1977b). *A. rabiei* attacks all aerial parts of the plant. Lesions on foliage and pods tend to be circular whereas those developing on petioles and stems are usually elongated (Figure 34); the latter often coalesce so that the affected part becomes girdled. If the main stem is girdled at the collar, the whole plant dies but, typically, it is the terminal shoot that is so affected, the tip drooping and becoming withered (Luthra and Bedi, 1932; Sattar, 1933b). Pods of affected plants bear prominent dark-brown lesions in which concentric rings of pycnidia are produced in response in diurnal fluctuations in atmospheric temperature (Hafiz, 1951). Early in the season, individual affected plants are scattered throughout the crop, appearing brown and dried-up. As the disease spreads, whole patches of withered plants are characteristic of blight which can completely devastate a crop. Annual losses of 20–50 per cent are common in some regions (Sattar, 1933a; Kovačevski, 1936). In areas where

Figure 34 Blight lesions on chickpea caused by *Ascochyta rabiei* (photograph courtesy Y. L. Nene)

chickpea is effectively the only crop which can be cultivated, losses from blight can have grave sociological effects (Sattar, 1933a).

A rabiei is a necrotrophic parasite capable of saprophytic survival under dry conditions in infected host debris in the soil for more than two years; viability is lost rapidly at high humidity (65–100 per cent RH) and when buried in soil (Luthra *et al.,* 1935; Kaiser, 1973). The pathogen is also carried over in seed, infected while still in the pod (Luthra and Bedi, 1932; Sattar, 1933a; Halfon-Meiri, 1970; Maden *et al.,* 1975) or contaminated with spores during threshing (Sattar, 1933a). The level of seed transmission, which may be increased by a delay in harvesting or by stacking the crop under humid conditions before threshing (Lukashevich, 1958b), may be as high as 70 per cent (Maden *et al.,* 1975). Superficial and deep-seated infections of seed appear to be equally able to transmit the pathogen (Maden *et al.,* 1975), although severely affected seeds lose their viability (Luthra *et al.,* 1935; Kaiser *et al.,* 1973). The pathogen can retain its viability in seed for at least five years (Kaiser, 1981b). Recent evidence (Weltzien and Kaack, 1981) suggests that infected seed is a more important source of primary inoculum than infected debris. *A rabiei* is thought to be restricted to *Cicer* (Sprague, 1930; Sattar, 1933b; Khachatryan, 1962) and there is no evidence that alternate hosts are involved in survival. Seed infections are probably responsible for the long-distance dissemination of *A. rabiei* (Morrall and McKenzie, 1974). The onset of a blight epidemic and its subsequent rate of secondary spread from infection foci depend heavily on prevailing climatic conditions. Spread is most rapid at atmospheric temperatures around 20 °C at high relative humidity, spore dispersal being favoured by wind-driven rain. Effectively no blight develops at temperatures outside the range 6–32.5 °C, or at wetness periods of less than 6 hours' duration (Luthra *et al.,* 1935; Chauhan and Sinha, 1971; Weltzien and Kaack, 1981). That climatic conditions are the principal determinants of blight severity is illustrated by Luthra *et al.* (1935) who found that blight incidence was correlated with rainfall in the Punjab, where the disease was found to be most severe in districts receiving more than 150 mm rainfall during the growing season (February–April); the disease was sporadic in districts receiving 90–150 mm and, where rainfall was less than 90 mm, blight was of no importance. Similarly, low winter temperatures prevent disease development. In northern Syria, infection does not occur until mid-February when air temperatures first exceed 6 °C; blight epidemics develop in mid-March when temperatures are above 10 °C (Weltzien and Kaack, 1981). On these bases, it appears possible to define zones of blight risk.

Various cultural measures are available, and potentially effective in controlling blight. Eradication of sources of primary inoculum can be achieved through destruction of infected debris, both in the field and on threshing floors, and perhaps by roguing and rotation (Sattar, 1933a; Luthra *et al.,* 1935). There may possibly also be potential in certain soil amendment practices (Lukashevich, 1958a) and in the exploitation of crop heterogeneity in blight management. Luthra *et al.* (1935) found that blight severity in chickpeas intercropped with cereals, or with *Brassica campestris,* was less than in a sole crop, presumably because of a barrier to spore dispersal, though a similar effect was not found in cultivar mixtures (Singh *et al*., 1981c) to which the findings of Jeger *et al.* (1981) are relevant. However, it is apparent that a more promising strategy is the use of clean seed, preferably produced in isolation in areas where *A. rabiei* is absent or ecologically incompetent, supplemented by field inspection and laboratory seed testing (Kaiser, 1981b). Chemical treatment of seed with systemic fungicides, such as benomyl, thiabendazole and tridemorph, can eradicate the pathogen in naturally infected seed (Kaiser *et al.,* 1973; Reddy, 1980). Though certain materials are liable to be phytotoxic, there is potential for their use as a precaution against inadvertent introduction of *A. rabiei* into new areas. Various protectant fungicides have given some control of blight when applied as foliar sprays (Kovačevski, 1936; Lukashevich, 1958a; Solel and Kostrinski, 1964; Vir and Grewal, 1974b), but the high cost of spraying, the scarcity of water in dry regions, and the relative inefficacy of currently available materials (Nene, 1981) make the use of foliar sprays unattractive.

That cultivars of chickpea differ in susceptibility to blight was first suggested by R. S. Singh in 1929 (Sattar, 1933a) and clearly demonstrated by Labrousse (1931). Work in the Punjab led to the identification of three sources of blight resistance, including a black-seeded type, a kabuli type, and a yellow-seeded type later named F8 (Luthra *et al.,* 1938); later, other sources were found (Aujla and Bedi, 1967; Grewal and Vir, 1974). Material with varying levels of resistance has been identified outside the Indian subcontinent, in Russia (Bushkova, 1960), Bulgaria (Solel and Kostrinski, 1964), Iran (Kaiser, 1972a), and Morocco (Boorsma, 1980; Pieters, 1981). With the development of the chickpea improvement programmes of ICRISAT and ICARDA, very extensive evaluations of *Cicer* germplasm have been made and numerous sources of resistance, in both desi and kabuli types, are now available (Singh *et al.,* 1981a, b).

Resistance to blight is partial, and the existence of immunity has not been confirmed. Very little is known of the mechanisms which underlie blight reaction in *Cicer,* although it is firmly established that there is a strong correlation between seed-coat colour and resistance. Resistant genotypes are predominantly black-seeded (Kaiser, 1972a; Singh *et al*., 1981b). Among nearly 5000 accessions, it was found that over 66 per cent of the black-seeded, but only 2 per cent of the lines with seed of other colours were resistant (Singh *et al.,* 1981b). This suggests that a pigment is in some way associated with blight resistance. Various flavonoid compounds have been identified in chickpeas (Wong *et al.,* 1965; Hösel and Barz, 1970; Keen, 1975a; Ingham, 1976b) but as yet there is no evidence that they function as phytoalexins against *A. rabiei.* However, Vir and Grewal (1974c) have shown that after infection phenolic compounds accumulate more rapidly in the resistant cv. I-13 than in a susceptible cultivar, suggesting that phenolics check the advance of the pathogen in the resistant host. There is also evidence that a pre-penetetration resistance operates in some cases. Hafiz (1952) found that penetration in two resistant cultivars (F8 and F10) tended to be delayed relative to the infection process in susceptible cultivars. This effect, which was restricted to adult plants, was thought to be due to a denser pubescence and, consequently, to a greater secretion of malic acid from the glandular hairs, in the resistant types. While dilute malic acid enhances spore germination, high concentrations ($pH < 2.3$) are apparently inhibitory (Sattar, 1933a; Hafiz, 1952). Recent work (Reed *et al.,* 1980) has confirmed that the concentration of acid exudate does vary between cultivars, but the role that glandular pubescence plays in blight resistant in the field remains uncertain. Nevertheless, these studies do suggest that there are distinct components of resistance in chickpea to *A. rabiei,* and the same is apparently true of resistance in *Pisum* to *A. pisi* (Darby *et al.,* 1981). The few investigations which have been made of the inheritance of blight resistance (Hafiz and Ashraf, 1953; Vir *et al.,* 1975; Singh *et al.,* 1981c) each claim to show that resistance is under monogenic control but the evidence is not strong. More convincing are suggestions (Boorsma, 1980; Pieters, 1981) that blight resistance is quantitatively inherited.

The cultivar F8 has been used as a source of blight resistance in the development of new chickpea cultivars, including cvs C12-34 and C235, but in both these genotypes resistance is reported to have broken down (Aziz, 1962; Bedi and Athwal, 1962; Grewal, 1981). There is considerable variation between isolates of *A. rabiei,* in cultural and morphological characters as well as in pathogenicity (Luthra *et al.,* 1939; Aujla, 1964; Arif and Jabbar, 1965; Bedi and Aujla, 1969; Vir and Grewal, 1974a). Claims have been made that physiologic races of the pathogen exist, and that the evolution of new races accounts for the reported breakdown of resistance (Aujla, 1964; Grewal, 1981). Although no studies have been made on the genetics of pathogenicity in *A. rabiei,* evidence from the inoculation of sets of cultivars with a range of isolates suggests a quantitative interaction. There is evidence of a positive correlation between pathogenicity and pycnidial production (Vir and Grewal, 1974a), and a negative correlation between pathogenicity and incubation period (Bedi and Aujla, 1969), though in neither case was this statistically demonstrated nor commented upon by the authors. Variation in malic acid sensitivity between isolates appears to be slight (Arif and Jabbar, 1965) and it remains unknown whether such differences could provide a basis for selective pathogenicity. However, the extent to which pathogenicity is cultivar-specific is debatable. Isolates obtained from the blight-resistant cultivars F8 and C235 appear not to be specifically adapted to their particular cultivars, as suggested by Vir and Grewal's data in which there is little evidence of a reversed ranking characteristic of differential interaction. On present evidence it appears that isolates differ in cultivar non-specific pathogenicity, or aggressiveness, rather than the cultivar-specific pathogenicity, or virulence, shown by distinct physiologic races (Anon, 1973; Scott *et al.,* 1980). This accords with expectation from studies on pathogenic variation in other necrotrophic parasites (Scott and Hollins, 1980). Highly aggressive isolates (e.g. isolate 8 of Aujla, 1964) show increased pathogenicity to susceptible as well as to resistant cultivars. Since there is continuous variation in spectra of aggressiveness, a quantitative rather than a qualitative basis for the recognition of races is required. The implications that this type of pathogenic variation may have for resistance breeding are discussed in Section 9.3.

9.2.3 Stunt

Stunt is caused by pea leafroll virus (PLRV), which belongs to the luteovirus group, transmitted by aphids in the persistent manner (Ashby and Huttinga, 1979; Reddy *et al.,* 1979). The disease was first described from Iran (Kaiser and Danesh, 1971a) and has subsequently been reported in India (Nene and Reddy, 1976a; Reddy *et al.,* 1979) and apparently also in Pakistan, Syria, Turkey, the Lebanon, Ethiopia, the Sudan (Nene, 1978), and North Africa (Reddy *et al.,* 1980). Field incidence is usually low (<5 per cent) but infection of a plant before flowering can cause over 80 per cent loss in its seed yield. Symptoms from natural infection can resemble wilt (Section 9.2.1). Affected plants are typically stunted, through a shortening of the internodes, and proliferation of axillary buds occurs. Foliage becomes chlorotic (in kabuli types) or bronzed (in desi types), and individual leaflets are abnormally small, thickened, and rolled. An important diagnostic feature of stunt is phloem browning, particularly in the collar region; the xylem is not discoloured, unless *Fusarium oxysporum* f.sp. *ciceris* is also present (Kaiser, 1972b; Nene and Reddy, 1976a; Nene *et al.,* 1979; Nene, 1980b).

Pea leafroll virus has a relatively wide natural host range among legumes in which it is transmitted by several species of aphid, including *Aphis craccivora* and *Acyrthosiphon pisum* (Kaiser and Danesh, 1971b; Nene and Reddy, 1976a). Wingless females of *A. pisum* may transmit PLRV more efficiently than winged females, and there is evidence that nymphs are better vectors than adults (Thottappilly, 1969). Lucerne (*Medicago sativa*) is an important reservoir of PLRV (Tinsley, 1959; Kaiser and Danesh, 1971b). It is probable that such alternative hosts are the principal source of primary infection of the chickpea crop in which PLRV is not seedborne. The destruction of alternate hosts and, perhaps, the use of insecticides against vectors are each potential means for the control of stunt, but their practicability has not been investigated. The development of stunt-resistant cultivars appears to hold promise (Reddy *et al.*, 1979).

9.3 DISEASE RESISTANCE AND CHICKPEA IMPROVEMENT

Despite the considerable economic loss caused by the annual recurrence of fusarium wilt and ascochyta blight, disease resistance has until recently received low priority in chickpea breeding programmes. Resistance to both diseases has been recognized for well over 50 years (McKerral, 1923; Labrousse, 1931) yet the few attempts which have been made to develop improved cultivars with resistance (Luthra *et al.*, 1938; Aziz, 1962; Bedi and Athwal, 1962; Singh *et al.*, 1974b) have met with only limited success. In the case of wilt, lack of progress is probably attributable in part to confusion over the disease's aetiology (see Section 9.2.1), while the development of satisfactory blight resistance has perhaps been hampered by its environmental instability. In virtually all cases, breeding has depended on selection within landraces without extensive hybridization. With the advent of chickpea improvement programmes at ICRISAT and ICARDA, the attention given to resistance breeding has escalated; the availability of extensive germplasm collections and the development of more satisfactory procedures for their evaluation are beginning to have impact.

Resistance is available within *C. arietinum* against rust (Mehta and Mundkur, 1946; Bahadur and Sinha, 1970a, b), grey mould (Singh *et al.*, 1981b; Verma *et al.*, 1981b), foot rot (Singh and Bedi, 1974), the collar and root rots caused by *Rhizoctonia solani, Macrophomina phaseolina,* and *Sclerotium rolfsii* (Haware and Nene, 1980b), stunt (Reddy *et al.*, 1979) and perhaps stemphylium blight (Prasad *et al.*, 1969), as well as against wilt and ascochyta blight on which the vast majority of the work has focused. Ascochyta blight resistance has also been found among induced mutant progenies (Haq *et al.*, 1981) and among certain accessions of wild chickpea, including *Cicer arietinum* ssp. *reticulatum, C. bijugum, C. pinnatifidum* and *C. judaicum* (Nene and Haware, 1980; van der Maesen *et al.*, 1980; Singh *et al.*, 1981a); *C. judaicum* also possesses resistance to wilt and stem anthracnose (Nene and Haware, 1980). While the blight resistance of *C. arietinum* ssp. *reticulatum* could be transferred to chickpea cultivars, if it appeared desirable to do so, the utilization of *C. bijugum, C. pinnatifidum,* and *C. judaicum* as donors is more remote; these three species represent chickpeas' tertiary gene pool from which gene transfer is impossible without the use of radical techniques (Ladizinsky and Adler, 1976b). Useful resistance has apparently not been found in *C. echinospermum* which represents the secondary gene pool. While it is premature to judge the value of resistance derived from wide-crossing, there is no doubt that great advances may be expected, both from induced mutation and more effective utilization of resistance currently available within germplasm of *C. arietinum.*

Disease-resistance breeding in the co-operative project between ICARDA and ICRISAT has relied heavily on pedigree selection in both field and greenhouse screening. Sickplots have been developed at Hyderabad for screening against fusarium wilt; with time, other soilborne pathogens, such as *Sclerotium rolfsii, Rhizoctonia solani,* and *Macrophomina phaseolina* have built up, thus enabling concurrent evaluation for resistance against four diseases. Material possessing resitance to wilt and root rot under Hyderabad conditions has been further evaluated at other locations, through the International Chickpea Root Rots and Wilt Nursery (ICRRWN), and the most promising have been used in crosses with high-yielding genotypes from which useful segregants have been selected (Haware and Nene, 1980b; Kumar *et al.*, 1980). Ascochyta blight evaluation has been conducted principally at Tel Hadya, near Aleppo in Syria, alternating with an off-season sowing at Terbol, in the Beqa'a Valley of the Lebanon. An inoculation technique has been developed in which the test material is intersown with a blight-susceptible 'spreader' cultivar; infected debris is scattered between rows, and a spore suspension of *A. rabiei* is sprayed on to plants among which humidity is kept high through supplementary irrigation (Singh *et al.,* 1981a). Using this method, blight resistance has been identified in germplasm accessions as well as in segregating material and such material has been entered in the Chickpea International Ascochyta Blight Nursery (CIABN), conducted since 1977 in Syria, Lebanon, Turkey, Algeria, and Pakistan. Six entries (ILC-72,-191, -194, -202, -3279 and PCh 15) have given resistant reactions in all environments (Singh *et al.,* 1981b). However, experience gained from both the CIABN and ICRRWN has revealed that, in certain cases, resistance effective in one environment has apparently afforded inadequate protection in another. The extent to which pathogenic variation is responsible for such site effects, and the nature of the host-parasite interaction, have vital bearing on resistance breeding methods.

Rust, caused by *Uromyces ciceris-arietini,* is locally damaging to chickpeas, in which it is perhaps the second most important foliar disease (Gallegos *et al.*, 1965; Nene, 1980b). Resistance has been found but cultivar-specific variants of the pathogen are known to exist (Bahadur and Sinha, 1970a, b). While there is evidence that variants of *Fusarium oxysporum* f.sp. *ciceris* are also cultivar-specific (Haware and Nene, 1979a), the epidemiological outcome of using the available, apparently oligogenic, sources of wilt resistance is rather less predictable. As was discussed with reference to pigeon pea wilt (Section 8.2.1), the commercial life of a cultivar possessing race-specific resistance to wilt need not necessarily be terminated when a matching race is recognized. Much is likely to depend on the distribution of races, and their rate of spread, influencing the opportunity for the geographical deployment of race-specific resistance. The nature of the host-parasite relationship in ascochyta blight is currently under examination. Cultivars such as F8 and its derivatives, which were found to possess partial resistance to blight in the Punjab by Luthra *et al.* (1938), were reported to have become highly susceptible by the early 1950s (Bedi and Aujla, 1969; Grewal, 1981). Investigation of the extent of pathogenic variation within *A. rabiei* confirmed that certain isolates were capable of causing severe disease in resistant-host genotypes, including cv. F8 (Aujla, 1964). Subsequent studies (Bedi and Aujla, 1969; Vir and Grewal, 1974a) have concluded that physiologic races were distinguishable, suggesting that such variants are liable to complicate blight-resistance breeding (Hawtin *et al.*, 1980; Grewal, 1981). However, certain evidence (Aujla, 1964; Vir and Grewal, 1974a) suggests rather that there is a constant ranking of isolates. For instance, Aujla's isolate 8 was not only more pathogenic to resistant cultivars but was also relatively more pathogenic to susceptible cultivars, suggesting an increase in host non-specific pathogenicity, or aggressiveness, relative to other isolates. Nevertheless, it is premature to conclude that pathogenicity in *A. rabiei* is host non-specific, and more critical work is warranted.

The mechanism by which more pathogenic variants of *A. rabiei* may arise is also unknown. Although sexual recombination is one possible means, it is apparent that the perfect stage (*Mycosphaerella rabiei*) is uncommon in nature. Heterokaryosis and parasexuality are known to occur in the related *Phoma medicaginis* (= *Ascochyta imperfecta*; Boerema *et al.*, 1965) which is known from chickpea (Haware and Nene, 1981), and these mechanisms, if they occur in *A. rabiei*, would permit the segregation and recombination of genetic factors outside the sexual stage (Sanderson and Srb, 1965). No studies have been made of the inheritance of pathogenicity in *A. rabiei,* yet isolates appear to vary almost continuously, suggesting quantitative control (see Section 10.2.6). Despite claims to the contrary (Hafiz and Ashraf, 1953; Vir *et al.,* 1975; Singh *et al.,* 1981c), the effectively continuous variation in the degree of blight resistance in chickpeas also suggests quantitative inheritance (Boorsma, 1980; Pieters, 1981). Whether or not it is ultimately shown that resistance to blight is under polygenic control, its epidemiological significance will remain uncertain until adequately tested in agriculture.

Strategies for the control of diseases through breeding obviously must depend on the nature of the host-parasite interaction. In the case of wilt, more information is required on the genetics of specificity, so as to confirm that race differences are indeed reflected in genetic differences in the host (Kraft and Haglund, 1978), and thus to aid pyramiding of race-specific genes for use in regions where more than one race is prevalent. Conventional breeding methods, including pedigree selection, are probably satisfactory, particularly where there are opportunities for the accelerated advancement of generations through the use of off-season nurseries, such as is used by ICRISAT in Kashmir. Screening for wilt resistance should utilize several isolates, each with contrasting pathogenicity such as is shown by the Kanpur, Gurdaspur, and Hyderabad isolates of *F. oxysporum* f.sp. *ciceris* held at ICRISAT (Haware and Nene, 1979a). A different strategy is required in the case of blight-resistance breeding. Clearly it is of cardinal importance to establish firmly that blight resistance is indeed a quantitative character, as we have argued here. The following is based on that assumption. We shall also assume that isolates of *A. rabiei* do not interact differentially with chickpea cultivars, for there is no evidence to the contrary. The implications of this are nicely emphasized by van der Plank (1968) who stated that 'races that do not interact differentially with host varieties have little bearing on problems of breeding resistant varieties. . . There is a wide misconception about this. Time and time again one reads in the literature that the author has found evidence that races of the pathogen occur and that the resistance of existing host varieties is therefore in danger of being lost. The fear that resistance will be lost is groundless, unless there is evidence of differential interaction between pathogen and host.' Substantial genotype × environment interaction has been found in each of the international trials of chickpea for most plant characters including seed yield, long known to be a quantitative character, indicating a need to select for local adaptation (Byth *et al.,* 1980). There seems little reason to suppose that the same does not equally apply to blight resistance, the more so since it is known that climatic, rather than pathogenic, variation can cause location effects (Weltzien and Kaack, 1981). Under conditions especially favourable to blight development, such as those encountered in the Pakistan Punjab and in winter-sown crops in the Mediterranean region, the level of blight resistance currently available is liable to provide insufficient protection. The adoption of quantitative breeding strategies to intercross resistant material, followed by a search for transgressive segregants, could be expected to raise the level of blight resistance. Success inevitably depends on the ability to recover resistant segregants within crosses, governed both by the use of appropriate disease pressure and the reliability of scoring. Good progress has already been made. Evidence of transgressive segregation has been found in Morocco (Pieters, 1981) as well as in Syria (Singh *et al.,* 1981c); in the latter programme, however, reliance was placed on unnecessarily susceptible material, a step advocated only in cases where race-specific, usually oligogenic, resistance is known to confer transient protection (Robinson, 1976). Prominence has recently been given to exploiting introgression between kabuli and desi types of chickpea in the hope that crossing between divergent groups would produce transgressive segregants for yield and, perhaps, blight resistance (Hawtin and Singh, 1980). There may also be potential value in applying methods of population improvement and recurrent selection, exploiting natural outcrossing (Niknejad and Khosh-Khui,

1972) or genetic male sterility (Sethi, 1979), to disease-resistance breeding in chickpea. It seems likely that advances might be made in increasing the level and environmental stability of blight resistance by intercrossing genotypes in which distinct components of resistance are recognized (Pieters, 1981), tantamount to a 'pyramiding of resistance mechanisms'. In addition to the mechanisms discussed previously (Section 9.2.2), it seems possible that canopy structure may influence blight susceptibility, as shown to be the case in white mould of beans (Schwartz *et al.,* 1978b). Work on restructuring of the plant habit in chickpeas (IARI, 1979; Byth *et al.,* 1980) should consider the possibility of selecting for a blight-escape mechanism.

Exciting possibilities exist for the improvement of the chickpea crop. Although seed yields are typically low and erratic, making the crop unattractive to commerical growers, the harvest index of chickpea is high (40–60 per cent); currently available cultivars are capable of producing over 4 tonnes of grain per hectare, some six times larger than the average (Nene, 1982). Progress made during the last few years in the improvement of chickpeas suggests that substantial advances are within reach. The assembly and use of a diverse germplasm, coupled with opportunities for the rapid turnover of generations and an international testing of promising material together provide the opportunity for massive genetic improvement. Characters such as 'double-podding' (Sheldrake *et al.,* 1978b), the development of taller, more erect cultivars suited to growth at high plant densities, an increased responsiveness to inputs, and the exploitation of the variation in maturity to permit changes from the traditional time of sowing, could each contribute to increased productivity (Byth *et al.,* 1980). Yet breeders need to develop cultivars which are adapted to the environment and farming system for which they are intended, and resistance to disease, as well as to pests and other stresses, is an important component of such adaptation. Certain combinations of characters are required in particular regions. Thus for peninsular India short-duration desi types with resistance to wilt and the root rots are required, while in northern India and Pakistan medium- to long-duration desi and kabuli types possessing resistance to wilt, the root rots, blight, grey mould, and stunt are appropriate. Opportunities for introducing chickpeas as a winter, rather than as a spring, crop in the Mediterranean region, can best be realized through incorporating resistance to blight, and perhaps tolerance of low temperatures (Singh *et al*., 1981d), into kabuli types. Considerable progress has already been made with respect to the development of resistance in improved genotypes, though there remains scope for raising levels of resistance, especially to blight. For the moment, it seems prudent to reinforce the available blight resistance with other means of disease management, most notably the use of clean seed (Reddy, 1980; Kaiser, 1981b).

Chapter 10

Resistance Breeding and Legume Disease Management

The preceding six chapters have been concerned with specific crop species, their pathology and improvement through disease-resistance breeding. The purpose of this final chapter is to examine the principles of disease control in these food legumes, in a predominantly tropical subsistence setting, and to focus on the art of breeding for resistance as a component of an integrated disease management, and its potential in crop improvement through application of recent advances in agricultural science.

10.1 DISEASE CONTROL IN CROP IMPROVEMENT

We have seen that, in five of the six food legumes covered in this book, grain yields currently obtained on farms are substantially lower than may realistically be expected from small changes to existing farming systems, without the use of large inputs. Soyabeans are the exception, for they are not yet widely adopted as a food legume in the tropics. Thus there is very considerable potential for a spectacular increase in production of this group of ancient but largely neglected crops. The extent to which pathogens are responsible for these low yields remains unquantified, but it is evident that traditional cropping systems tend to be evolutionarily stable (Section 1.3.1), albeit at a low level of productivity. It is equally evident that improvements brought about by the introduction of more productive cultivars and altered agronomic practices will, in many cases, be accompanied by increased disease risk; any yield advantage is unlikely to be sustained without a disease management component to the improvement 'package'. For instance, an altered plant architecture permitting closer spacing is in some cases liable to be accompanied by an increase in susceptibility to foliar pathogens, as suggested by work on pigeon peas (Hammerton, 1975; Tayo, 1982), and off-season irrigation in seasonally arid areas might be expected to influence both pathogen and vector survival and population density. The opportunity for substantial yield increases from earlier, winter sowing of chickpea in the Mediterranean region similarly depends on the satisfactory management of ascochyta blight which is exacerbated by winter sowing (Section 9.2.2). Such innovations might be expected to lead to the emergence of 'new' diseases to a greater extent, perhaps, in regions where the crop was originally domesticated, for it is here that the greatest diversity of parasites occurs. As we shall see in subsequent sections of this chapter, there are many opportunities for the chemical, cultural, and biological protection of food legumes, yet few are likely to be practicable in improved systems of subsistence agriculture. Much the most promising strategy is the use of resistant cultivars: they provide potentially permanent protection against disease, at no cost to the farmer.

10.2 BREEDING LEGUMES FOR RESISTANCE TO DISEASE

The potential for controlling disease of legume crops by host-plant resistance was first realized in the USA by Orton (1902) who reported differences in reaction to fusarium wilt among cowpea cultivars. Wilt resistance was found in pigeon pea in India a few years later (Butler, 1908). Anthracnose-resistant cultivars of bean were to follow (Barrus, 1911). That a single cultivar could possess resistance to more than one pathogen had been shown by Webber and Orton (1902) and by 1921 McRostie (1921) had bred bean cultivars with resistance to three diseases. Subsequent work revealed disease resistance also in chickpea (McKerral, 1923), soyabean (Lehman and Woodside, 1929), and groundnut (Higgins, 1935). Despite these early efforts, however, relatively little progress was made in developing satisfactory resistance in agronomically superior genotypes of any of these legumes for tropical areas until the late 1960s and early 1970s, when an increased awareness of the need to improve this group of food crops led to the establishment of various national and international programmes.

In the rest of this section, we will be concerned with the identification, nature, and expression of resistance; the sources available, and their durability; and the inheritance of resistance and strategies for its proper use in breeding. But first we should briefly consider current concepts and terminology of disease resistance.

10.2.1 Resistance terminology and concepts

Current concepts owe much to the stimulus provided by van der Plank (1963; 1968) whose hypotheses have provided the fillip for extensive debate. While his ideas have been nicely discussed and elegantly elaborated upon by Robinson (1976), imperfect understanding of the original tenets, and misuse of terms, have together contributed to the confusion and controversy which now prevail. Various attempts have been made to standardize disease-resistance terminology; the most useful glossaries are those compiled by the Federation of British Plant Pathologists (Anon, 1973), and by Robinson (1969; 1976), whose usage has generally been followed in this book. The term 'resistance' is considered by Robinson (1969) to refer to the ability of the host to hinder the pathogen though, like Russell (1978), resistance has been used here in a wider sense to embrace also avoidance (or escape) and tolerance, and therefore may be regarded as any inherited characteristic of the host which confers some defence against the pathogen. Its antithesis is 'susceptibility'. Amongst the many terms used to qualify resistance, van der Plank's two epidemiological categories 'vertical' and 'horizontal' have frequently been used in the text where they are treated as synonyms of race-specific and race non-specific resistance, respectively. Similarly, it is considered important to distinguish between vertical (or host-specific) and horizontal (or host non-specific) pathogenicity, often referred to respectively as virulence and aggressiveness. It is recognized that disenchantment with the concept of horizontal resistance stems in part from the impossibility of demonstrating that a particular resistance is indeed, race non-specific (Nelson, 1978; Johnson, 1979). The term 'durable' is therefore preferable to 'horizontal' because it embodies fewer assumptions. Durable resistance is recognized in commercial cultivars which have remained resistant over many years' use in agriculture; the term relates only to proven performance. It need not be permanent, nor does the concept depend on a particular expression, mechanism or mode of inheritance (Johnson, 1979; 1981a, b). Its antithesis is transient resistance, the efficacy of which is short-lived because of the emergence of a resistance-breaking variant of the pathogen. Transient resistance, which is usually but not invariably race-specific, is apparently most often expressed as a complete resistance. The term 'partial resistance' is used here to refer to incomplete resistance without reference to its durability, specificity, or mechanism of operation. The detection of partial resistance sometimes depends upon particular screening conditions, being recognizable chiefly in the field ('field resistance') or only in mature plants ('adult plant resistance'). In this book, 'tolerance' is used to refer to symptomless infection of a host, usually by a virus. This contrasts with 'disease tolerance', said to be shown by a cultivar which develops as severe disease symptoms as another but which suffers less damage in terms of agricultural yield (Robinson, 1969). The term 'combined resistance' refers to the possession of resistance to more than one pathogen, and 'comprehensive resistance' (Robinson, 1976) is an adequate level of resistance to all locally important pathogens.

Failure to distinguish between resistance to infection by a pathogen and resistance to the disease it subsequently causes is the source of considerable confusion in terminology, especially where distinct usage is advocated by different disciplines within plant pathology. This is nicely illustrated by the tendency of virologists to use the terms 'resistance' and 'susceptibility' in the restricted sense of the ability of a plant to act as a host of virus, while the terms 'tolerance' and 'sensitivity' indicate the extent to which the host produces symptoms and becomes diseased (Drijfhout, 1978; J. M. Thresh, personal communication 1981). Such usage is at variance with that used here and underlines the urgent need to standardize terminology.

10.2.2 The nature and expression of resistance

In this section we are concerned with the manner in which inherent differences in susceptibility are expressed, and with the underlying mechanisms govering the major components of disease resistance in legumes.

The expression of resistance to certain pathogens may vary considerably with the growth stage, or with the particular plant organ, of the host plant. For instance, there is evidence that seedling resistance may, in certain cowpea cultivars, be accompanied by adult plant susceptibility to cercospora leaf spot (Schneider and Sinclair, 1975; Ekpo and Esuruoso, 1977), whereas seedling susceptibility is known to be associated with adult plant resistance, in beans to *Rhizoctonia solani* (Bateman and Lumsden, 1965) as well as in certain host-pathogen combinations of bean rust (Atkins, 1973; Ballantyne and McIntosh, 1977), and groundnut rust (Cook, 1980a). That different plant parts may react differently to infection has been shown with halo blight in beans (Hill *et al.*, 1972). There is also obviously a great deal of variation in the degree, or level, of resistance expressed in different host-pathogen combinations. In some diseases there is a continuous gradation between resistance and susceptibility, typical of a quantitative interaction, while in others there is a clear-cut distinction between the two classes, indicative of a qualitative interaction between the host and the pathogen. Of the two, qualitative resistance more frequently confers complete protection, yet there is no sound reason to suppose that equally high levels of resistance under quantitative control cannot also be achieved. Furthermore, lower levels are often sufficient to reduce disease severity to below the economic threshold.

Resistance may be expressed in various ways, divisible into three major categories:

(i) decreased infection frequency, conferred by avoidance mechanisms prior to infection;
(ii) inhibition, or retarded development of symptoms, governed by mechanisms operating after infection;
(iii) decreased damage, or disease tolerance.

Several, possibly independent, mechanisms may be involved in each of these main components of resistance to a particular disease (Russell, 1978). Relatively little attention has been devoted to mechanisms of 'escape', some examples of which are given in Table 25. Most studies of resistance have been devoted to mechanisms of 'true', or 'physiologic', resistance which operate against the establishment and development of the pathogen within the host. Both histological and biochemical changes are involved. Much the most commonly recognized mechanism is hypersensitivity which entails a rapid necrosis of the invaded cell, thus restricting the pathogen which eventually dies. The precise nature of this response, and the rate at which it occurs, is known to vary with the host-pathogen relationship and macroscopic symp-

TABLE 25 The Tendency to Escape Infection as a Component of Disease Resistance in Legumes

Component	Mechanism	Disease	References
Resistance to carry-over	Low incidence of seed transmission	Soyabean mosaic	Goodman and Oard, 1980; Bowers and Goodman, 1982
Resistance to deposition, or settling	Failed spore retention	Groundnut rust	Cook 1980b
	Vector non-preference	Bean curly top	Silbernagel and Jafri, 1974
		Lima bean golden mosaic	Vetten and Allen, in press
Resistance to germination, or penetration	Inhibition of germination	Chickpea blight	Hafiz, 1952
		Chickpea rust	Bahadur and Sinha, 1970b
		Cercospora leaf spot of cowpea	Schneider and Sinclair, 1975
		Phytophthora blight of lima bean	Valenta and Sisler, 1962
	Impaired directed growth and failed appressorium formation	Cercospora leaf spot of groundnut	Abdou *et al.*, 1974
		Groundnut rust	Cook, 1980a
	Stomatal exclusion	Cercospora leaf spot of groundnut	Hemingway, 1957; Gibbons and Bailey, 1967
	Failed cuticular penetration	Septoria leaf spot of cowpea	Rawal and Sohi, 1981
	Impeded or inhibited vector probing	?	Johnson, 1953; Chari *et al.*, 1976; Farrell, 1976c

toms either may or may not be apparent; in the latter case, the response is often referred to as 'immunity'. Different types, or degrees, of hypersensitivity have been recognized in cowpea rust (Heath, 1972; 1974), soyabean rust (McLean, 1979), and anthracnose of cowpea (Skipp, 1975; IITA, 1977). The hypersensitive reaction is also commonly associated with resistance to bacterial pathogens (Klement and Goodman, 1967). Recent evidence from ultrastructural studies suggests that bacteria in intercellular spaces bind to plant cells in resistant but not in susceptible reactions (Roebuck *et al.*, 1978; Slusarenko and Wood, 1981). Hypersensitivity is also an important mechanism of resistance to viruses but little is known of its mode of operation.

Various post-infectional antibiotics (phytoalexins) are commonly associated with hypersensitivity. They occur widely in legumes and many have been isolated and identified. Among these, phaseollin has been shown to accumulate in bean tissue following infection with incomptabile races of anthracnose (Rahe, 1973), rust (Bailey and Ingham, 1971) and halo blight (Stholasula *et al.*, 1971), and hydroxyphaseollin is associated with resistance in soyabean to phytophthora rot (Keen and Horsch, 1972) and bacterial blight infection (Keen and Kennedy, 1974). In certain cases, the difference between compatible and incompatible interactions relates more closely to the timing of phytoalexin accumulation than it does to differences in concentration (Rahe, 1973) while, in other host-pathogen combinations, the converse appears to hold (Gnanamanickam and Patil, 1977). Certain substances produced by the pathogen may play roles as determinants of host response. For instance, an avirulent race of *Phytophthora megasperma* f.sp. *glycinea* is thought to produce a substance capable of eliciting phytoalexin formation in resistant, but not in susceptible, cultivars of soyabean (Keen, 1975b), and phaseotoxin, produced by *Pseudomonas syringae* pv. *phaseolicola*, is capable of suppressing both the hypersensitive response and phytoalexin accumulation, and is thus a determinant of host specificity in halo blight of beans (Patil and Gnanamanickam, 1977). The role of toxins in pathogen recognition of the host is further discussed by Callow (1977) and the extent to which phytoalexins are the cause, not a consequence, of hypersensitive resistance is reviewed by Smith and Ingham (1980) and Mansfield (1982).

Mechanisms not involving hypersensitivity can also confer apparent immunity to both fungi and viruses. Beier *et al.* (1977; 1979) have shown that there are several mechanisms of immunity to cowpea (yellow) mosaic virus. The immune response, which is not merely an exaggerated hypersensitivity, relies on an ability either to prevent or delay virus replication. Restricted virus movement is an important factor in resistance to cowpea chlorotic mottle virus (Wyatt and Kuhn, 1979). Mechanical barriers confer resistance to certain fungal pathogens. Deposits of callose at the point of penetration may lead to the encapsulation of the invading pathogen, as shown from histological studies on bean rust (Hardwick *et al.*, 1971), cowpea rust (Heath, 1971), cowpea anthracnose (Skipp, 1975), and cercospora leaf spot of groundnut

(Abdou *et al*., 1974). Deposits of lignin and suberin can also provide physical barriers to infection, including the resistance of certain chickpea cultivars to wilt (Ayyar and Iyer, 1936) and the resistance of adult soyabean plants to phytophthora rot (Paxton and Chamberlain, 1969). However, the adult plant resistance of beans to *Rhizoctonia solani* appears not to relate to the formation of a mechanical barrier but to a biochemical change in which pectin is converted to calcium pectate, able to resist the enzyme, polygalacturonase, which *R. solani* produces (Bateman and Lumsden, 1965).

So far, we have been concerned with mechanisms of resistance which operate against the pathogen, including those that enable an escape from infection as well as those that entail an active host response in an intimate relationship with the pathogen. There is a third component of resistance, embracing those mechanisms which lead to some escape from damage despite disease. This is usually referred to as tolerance (Schafer, 1971; Robinson, 1976) but this term is widely abused. Not only is the concept of disease tolerance misunderstood but very little is known about its mechanism of operation. Nevertheless, it is apparent that certain plant species including some legumes are able to compensate for damage, both to vegetative and reproductive structures (Ojehomon, 1970; Wien and Tayo, 1978). For instance, damage at or below ground level can be compensated for by the production of adventitious roots, as occurs in beans in response to root rot (*Fusarium solani* f.sp. *phaseoli*) (Wheeler, 1969) as well as in cowpeas with symptoms of stem rot (*Pythium aphanidermatum*) (D. J. Allen, unpublished); the extent to which there are cultivar differences in this ability is unknown. In other cases, physiologically surplus foliage constitutes a compensation mechanism. Whereas pigeon pea cultivars having an indeterminate growth habit can tolerate some defoliation caused by rust (*Uredo cajani*), determinate types are liable to serious damage (Hammerton, 1975; Tayo, 1982). Similarly, alternately branched cultivars of groundnut may be able to compensate for defoliation caused by cercospora leaf spot but the sequentially branched types cannot (Elston *et al*., 1976). However, in certain groundnut cultivars there is also a 'resistance to abscission' which is expressed as a delayed defoliation (Nevill, 1981). It is established that premature defoliation caused by foliar fungi can be accompanied by the increased evolution of ethylene (Williamson, 1950). Recent studies indicate that differences in ethylene sensitivity may be correlated with resistance to pests in cowpea (Wien and Roesingh, 1980) but it remains unknown whether 'resistance to abscission' also relates to ethylene insensitivity in legume diseases. Compensation for disease damage presumably has a physiological basis, independent of 'true' resistance; the extent to which compensatory growth can be made is likely to be governed by the timing and amount of damage incurred as well as by the partitioning and distribution of assimilates within the plant. Finally, it should be borne in mind that compensation for damage can be made by the crop as a whole, as well as by the individual plant.

10.2.3 Identification of resistance

In this section we are concerned with the art of identifying resistance, the success of which jointly depends on the diversity of the available germplasm and the development of an appropriate screening technique. Choice of technique in turn depends on the amount and type of material for evaluation and the facilities available, as well as on knowledge of the biology of the pathogen and the epidemiology of the disease. Laboratory and greenhouse screening procedures ensure more accurate environmental control than can easily be achieved in the field, yet results obtained under controlled conditions may not always accurately reflect field performance. Field evaluation is appropriate for large collections of germplasm, especially where land and labour are relatively abundant. Field screening at sites remote from research centres often has to rely on natural infection, and careful choice of site, sowing date, and plant spacing can be important in ensuring adequate uniformity of disease development. Perhaps the most important supplement to natural infection is the liberal and regular intersowing of susceptible 'infector', or 'spreader', plants, often preferably pre-sown ahead of the test material although doing so tends to complicate land management. Natural infection has enabled the identification of resistance to numerous legume diseases, including bean rust (Ballantyne, 1974b), web blight and golden mosaic of cowpea (Figure 35, Plate 3; IITA, 1980; Vetten and Allen, in press). The detection of partial

Figure 35 The detection of partial resistance to web blight (*Thanatephorus cucumeris*) in cowpea depended on dense sowing at a time when peak susceptibility coincided with heavy rainfall. Onne, Port Harcourt, Nigeria

resistance to web blight (*Thanatephorus cucumeris*) depended on dense sowing at a time when peak susceptibility coincided with heavy rainfall, while cowpea golden mosaic resistance screening depended on sowing to synchronize seedling emergence with peak populations of the whitefly vector.

In many cases, however, artificial inoculation is required. Crude inocula can be used satisfactorily in many cases. Infected leaves, scattered within the test plant canopy, have been employed in screening cowpeas against bacterial blight, septoria leaf spot (Allen *et al*., 1981b), cercospora leaf spot and rust (Williams, 1977a), and infected pods and stems are satisfactory inocula for the establishment of cowpea scab (Allen *et al*., 1981b) and ascochyta blight of chickpea (Singh *et al*., 1981a). Diseased leaflets, infested with the mite vector, are used as inoculum for the evaluation of resistance to pigeon pea sterility mosaic, the detached leaflets being stapled to test seedlings or to spreader rows (Nene and Reddy, 1976a). The release of viruliferous aphids has been used in screening against groundnut rosette (Rossel, 1977). More uniform placement of crude inocula can be achieved by their maceration and subsequent spray application as aqueous suspensions, a method used for bacterial pustule of soyabean (Jones and Hartwig, 1959) and cowpea anthracnose (Asafo-Adjei, 1982). The secondary spread of some of these diseases can be enhanced by supplementary sprinkler irrigation. In the case of soilborne pathogens of the root system, the distribution of propagules in the soil is typically too variable without soil amendment and artificial infestation. 'Sick-plots' have been successfully established for screening against the fusarium wilts of chickpea (Plate 4) and pigeon pea by this means (Nene, 1980a), while seed infestation has been used for screening for resistance to fusarium root rot in beans (Baggett and Frazier, 1973).

Where sufficient laboratory support is available, pure cultures of pathogens can be used as inocula, these being more readily quantified and standardized than crude inocula. Spores, or mycelial fragments, from agar cultures of fungi are most commonly applied as aqueous suspensions to plants by spraying (Figure 36), although spores of some pathogens may be suspended in oil, or applied as a dust in talc, as has been used for bean rust (Allen, 1975c). Cell suspensions prepared from pure cultures of bacterial pathogens are also widely used in spray inoculation, with or without the use of high pressure (Webster *et al*., 1980; Allen *et al*., 1981c). Pure isolates of mechanically transmissible viruses, maintained in a propagative host under controlled conditions, are most commonly employed as field inocula in the form of crude sap in a suitable buffer to which an abrasive, such as carborundum, is added. Virus inocula may be applied to plants as pressurized sprays, used satisfactorily with potyviruses (Burke and Silbernagel, 1974b; Ghanekar, 1980) as well as with viruses having spherical particles (O'Hair *et al*., 1977; D. J. Allen, unpublished), but manual inoculation is more widely used. Although the latter is labour-intensive, and may be inappropriate in some regions where automated inoculators are preferable (Ross, 1978), labour is often readily available and comparatively cheap in developing countries in the tropics, so

Figure 36 Motorized sprayers have been used for the direct inoculation of beans with rust, anthracnose and angular leaf spot, Cali, Colombia (photograph courtesy H. F. Schwartz)

Figure 37 Manual inoculation of bean seedlings with bean common mosaic virus, Cali, Colombia (photograph courtesy H. F. Schwartz)

that manual inoculation can be relied upon, even for large-scale evaluation (Figure 37). Clearly, methods of preparation of inocula and their field application should be appropriate to the scale of operation. Large-scale screening may necessitate special culturing techniques (Schroder *et al.*, 1953; Tribe and Ahmed, 1975) and perhaps tractor-mounted, rather than knapsack, spray inoculation.

The decision as to whether inocula should be applied direct to the test material or only to the spreaders, allowing for natural spread to the test plants, is both important and difficult. Much will depend on the genetic uniformity of the test material, and the rate and pattern of secondary spread. In segregating populations, successful selection depends on the ability to minimize escapes (symptomless susceptible plants) probably to a greater degree than among genetically uniform lines. The use of recognizable, susceptible 'indicator' plants distributed throughout the trial can be useful as a guide to the efficiency of natural spread. Asafo-Adjei (1982) found that, while effective selection for anthracnose resistance in cowpea could be achieved through inoculation of the spreaders alone, direct inoculation of the test material was required for effective screening against susceptibility to cowpea (yellow) mosaic virus. However, the appropriate arrangement of spreader plants may be a more important factor in increasing the precision of selection for resistance than reliance on artificial inoculation alone, as was found to be the case with bacterial pustule of soyabean in southern Nigeria in a season favouring disease development (Kueneman and Allen, 1980). In yet further instances, there is evidence that direct inoculation can mask mechanisms of resistance effective against natural, but not artifical, infection, as suggested by work on field resistance to cowpea mottle virus infection (Allen *et al.*, 1982). Ultimately, the choice between direct and indirect (spreader) inoculation depends on a combination of the epidemiology of the disease including seasonal effects on its development, the nature and level of the resistance sought, the intensity of selection, and the type of material to be evaluated.

Numerous resistance-screening techniques have been developed for the evaluation of legumes under controlled conditions, including both laboratory and greenhouse methods. Such systems of evaluation have the advantage that they tend to be rapid and can usually be employed throughout the year, without reference to season. Implicit in such screening is the accurate control of environment, including the elimination of other pathogens and pests which often complicate field evaluation. Limitations to the adoption of greenhouse and laboratory methods include their cost and, frequently, the poor correspondence of results with those obtained under natural field conditions (Williams, 1974b). However, screenhouse facilities suitable for tropical use need not be especially sophisticated or expensive (Rossel and Ferguson, 1979), and many screening methods have proved successful. Pot-culture techniques, involving the artificial infestation of the rooting medium or the immersion of the root system into inoculum prior to potting, are widely used for screening against soilborne fungi and nematodes, including *Phytophthora drechsleri* f.sp. *cajani* (Figure 38; Kannaiyan *et al*., 1981), *Fusarium solani* (Boomstra *et al*., 1977) and *Meloidogyne* spp. (Wyatt and Fassuliotis, 1979). Bacterial pathogens are most commonly screened against at the seedling stage. Stem inoculation has involved scalpel application of a bacterial smear, often to a cotyledonary scar (Arp *et al*., 1971), or injection of a bacterial suspension with a syringe and hypodermic needle (Saettler, 1971b). Leaves have been pinprick inoculated using multiple needles, a 'florists' frog' (Andrus, 1948; Patel and Jindal, 1970) or clipped with scissors (Webster *et al*., 1980) after their immersion in bacterial suspensions. Leaf water-soaking, either by leaf injection infiltration (Figure 39; Klement, 1963), injecting under pressure (Hagborg, 1970), or atomizing under pressure (Patel and Jindal, 1970), has

Figure 38 Pot culture techniques have been applied to the screening of pigeon pea seedlings for resistance of phytophthora blight, Hyderabad, India (photograph courtesy Y. L. Nene)

Figure 39 The inoculation of cowpea seedlings by injection infiltration in screening for bacterial blight (*Xanthomonas campestris* pv. *vignicola*) resistance, Ibadan, Nigeria

also been employed. The use of more than one technique may, in certain circumstances, assist in the differentiation of distinct resistance mechanisms as well as help to establish the relationship between the resistance of different plant parts (Coyne and Schuster, 1974b; Allen *et al*., 1981c; Patel, 1981).

Because virus infection seldom induces symptoms sufficiently diagnostic for accurate field identification, and because mixed infection is common, the evaluation of resistance to virus diseases is often only reliably conducted under screenhouse conditions, though there are exceptions as noted above. Manual inoculation of seedlings has been used to identify resistance to bean common mosaic virus (Omunyin, 1979), southern bean mosaic and cowpea aphidborne mosaic viruses (Ladipo and Allen, 1979a, b) among many others. For viruses not readily sap-transmissible, techniques such as vector inoculation of seed (Storey and Ryland, 1955; Nutman *et al*., 1964), or of seedlings, and grafting (Rossel, 1977) have each been used in resistance screening. In other cases, detached plant parts have been employed. For instance, detached leaves of beans have been inoculated with powdery mildew, rust (Dundas, 1942), and bean common mosaic virus (CIAT, 1981), and detached groundnut leaves have been used in studies of cercospora leaf spot resistance (Nevill, 1981). Detached petioles and detached hypocotyl tests have been used in screening against anthracnose in cowpea (Figures 40; Skipp, 1975) and beans (Dixon and Doodson, 1973), respecitvely. Such techniques enable the precise control of inoculum concentration, placement and subsequent incubation, allowing for quantification. For spores applied as dusts, the same can be

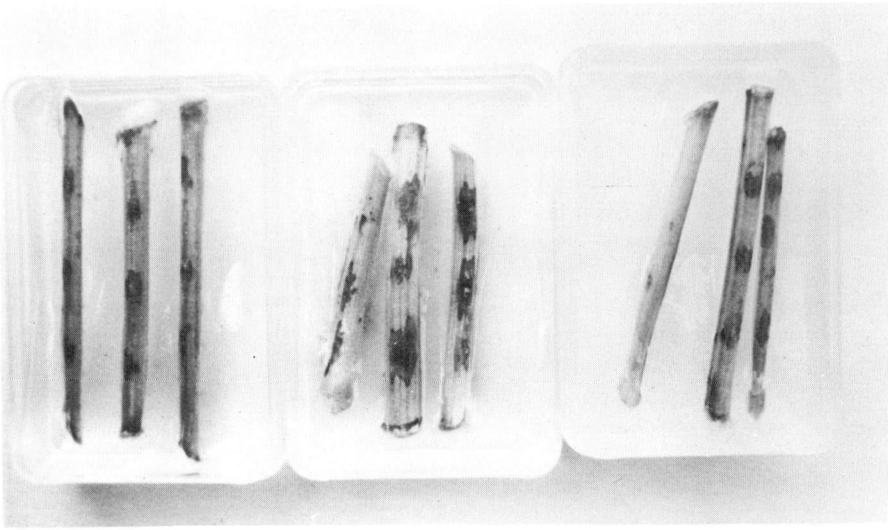

Figure 40 The detached petiole test for identifying anthracnose resistance in cowpea, Ibadan, Nigeria (photograph courtesy R. J. Williams)

achieved by using a spore settling tower (Figure 41; Allen, 1974) for inoculation, and a cyclone separator (Figure 42; Tervet and Cassell, 1951) for collection prior to subsequent measurement of spore yield.

Consideration has been given so far only to techniques of screening against single pathogens whereas breeding programmes often utilize methods for the simultaneous, or sequential, evaluation of germplasm for combined resistance to several diseases. For instance, bean seedlings grown under greenhouse conditions in pots containing soil infested with *Fusarium solani* f.sp. *phaseoli* have been artificially inoculated with bean common mosaic virus and, 2–3 days later, with *Uromyces appendiculatus* (Frazier *et al*., 1961). Field screening methods have been employed in the sequential inoculation of cowpea germplasm with five foliar pathogens (Williams, 1977a), and the artificial establishment of sick-plots has enabled the simultaneous evaluation of chickpea for resistance to wilt and three root rots (Haware and Nene, 1980b). Programmes concerned with breeding for several widely different environments may opt to evaluate early generation materials in simultaneous but separate nurseries in different environments, perhaps utilizing both greenhouse and field screening procedures. For instance, at IITA, early generation material as well as more advanced breeding lines of cowpea have each year been evaluated for resistance to a total of six fungal, two bacterial, and four virus diseases, using greenhouse screening in conjunction with field evaluation at a set of testing locations representative of distinct ecological zones (Allen and Goldsworthy, 1981).

Although simultaneous screening for resistance to several diseases acceler-

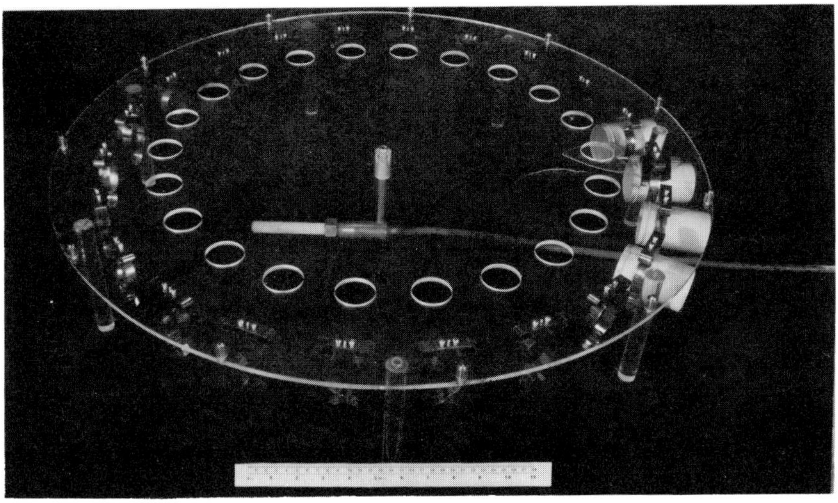

Figure 41 A spore settling tower enables the precise control of inoculum dose, as may be necessary in the quantitative assessment of partial resistance to bean rust

Figure 42 A cyclone separator can be useful in the measurement of spore yield of rusts, including *Uromyces appendiculatus* on beans

ates the identification and use of combined resistance, it also increases the chances of pathogen interactions (Section 3.6). The extent to which such interactions complicate the interpretation of screening results remains unknown but it is clear that such a potential exists (Hedges, 1944). For instance, rust may predispose beans to virus infection (Yarwood, 1951), and soyabean mosaic virus infection predisposes soyabean to *Phomopsis* infection (Hepperly *et al.*, 1979). In other cases, prior infection may apparently protect against a second disease, examples of which include the interactions between sterility mosaic and wilt in pigeon pea (Chadha and Raychaudhuri, 1966), wilt and root knot in soyabean (Goswami and Agarwal, 1978), common mosaic and bacterial blight in bean (Hedges, 1944), and cowpea (yellow) mosaic and anthracnose in cowpea (Asafo-Adjei, 1982). Clearly, there is a need for investigation of the nature of interactions in combined resistance screening, including a study of the effects of direct versus indirect inoculation.

Ultimately, the success of an inoculation technique must be judged by its ability to distinguish degrees of susceptibility and resistance. Systems of rating reactions most often rely on somewhat arbitrary assessment scales which do not always relate closely to crop loss expectation. Choice of scale is therefore subjective, but there is much to be said for a 1–9 scale on which the number of digits (e.g. 1,5,9; 1,3,5,7,9) can be altered to suit circumstances without loss in integrity. The need to standardize systems of rating is especially acute in the management of international disease nurseries and similar trials from which data accumulation is rapid. Such a standardized system has been proposed for rice evaluation (IRRI, 1975).

10.2.4 Sources of resistance

Conventional breeding philosophy dictates that the first requirement in resistance breeding is to find a source of resistance. A new approach (Robinson, 1980) suggests that the breeding process preferably begins with susceptible parents. Here we are concerned with the status of available sources of disease resistance, and the potential for induced mutation and the exploitation of extra-specific variation in the improvement of resistance in the legumes.

It is evident that there is very considerable genetic diversity within the food legumes. Extensive germplasm collection, fostered by the International Bureau of Plant Genetic Resources, has enabled the evaluation of many plant characters including disease resistance (e.g. IITA, 1974). In certain cases, collections from particular regions, including centres of origin (Leppik, 1970), are known to be especially rich in sources of resistance, presumably reflecting the co-evolution of the host and pathogen. The diversity available to legume breeders is probably now substantially larger than that on which even recent genetic improvement is based, and much potential remains for further advances. The available sources of disease resistance are best documented for soyabean (Tisselli *et al.*, 1980) but extensive screening has led to the identification of numerous sources of disease resistance in each of the legume crops with which this book is concerned. In some cases, however, either resistance has not been found, or available sources afford inadequate protection, or are insufficiently durable. In these instances, attempts have sometimes been made to compensate for such shortcomings by unconventional means, including induced mutation and 'wide crossing'.

Artificial mutagenesis, a subject reviewed by Simons (1979), has not been widely used in resistance breeding in the legumes. Attempts to induce mutants with improved resistance have been made with groundnuts against cercospora leaf spot (Cooper and Gregory, 1960), in soyabeans against rust (Tsai *et al.*, 1974), in beans against golden mosaic (Neto *et al.*, 1977), in mung beans against yellow mosaic (Shakoor *et al.*, 1977), in pigeon peas against wilt (Sheriff *et al.*, 1977), and in chickpeas to blight (Haq *et al.*, 1981). While, in some cases, little improvement has been achieved, in others (e.g. Haq *et al.*, 1981) truly resistant mutants have been obtained and are impressive demonstrations of the value of induced mutation in plant breeding (J. B. Smithson, personal communication 1982).

Interspecific hybridization, which has played an important role in the evolution of plants in the wild (Barber, 1970), can also be expected to play a significant part in crop improvement. Interspecific hybridization has been exploited more widely in the improvement of beans than of other food legumes. For instance, *P. coccineus* has been used as a source of resistance to bean yellow mosaic, bacterial brown spot, white mould, and fusarium root rot, while *P. acutifolius* is a donor of common blight resistance (see Section 6.2.7). Resistance, both under qualitative (Dickson and Natti, 1968; Abawi *et al.*, 1978) and quantitative (Coyne *et al.*, 1965a; Wallace and Wilkinson,

1965) genetic control, has been successfully transferred to *P. vulgaris*. There is evident potential for the use of alien germplasm as a source of disease resistance in other legume crops also, but the ease with which such sources may be exploited varies considerably. The crux is clearly the degree of relatedness of the wild source to the cultigen. That taxonomic status is often a poor guide to relationship is well illustrated by the 'intergeneric' hybrids between the pigeon pea (*Cajanus cajan*) and species of *Atylosia* (Kumar *et al.*, 1958) to which the cultigen is closely related and congeneric (Section 8.1). The assigning of taxa to primary, secondary, and tertiary gene pools (Harlan and de Wet, 1971) has much to commend it. Several legume crops are genetically isolated and the prospect for harnessing extra-specific variation remains remote as, for instance, in cowpeas. Similarly, various wild *Glycine* species are potentially valuable sources of rust resistance (Burdon and Marshall, 1981) yet none is interfertile with soyabean (Hymowitz and Newell, 1981). Sterility barriers and differences in ploidy level are obstacles in groundnut breeding to the use of sources of cercospora leaf spot resistance present in *Arachis cardenasii* (Abdou *et al.*, 1974). Nevertheless, a high priority is accorded to interspecific hybridization in groundnut improvement at ICRISAT (Moss, 1980). In the long term, there appear to be exciting opportunities for the application of radical techniques to hybridization (Knott and Dvŏrák, 1976; Razdan and Cocking, 1981), with the prospect that wide crossing will, one day, permit breeding for an absolute, permanent resistance. At present, however, there is no certainty that resistance introduced from alien sources by more conventional means will be more durable, and it has been suggested (Bingham, 1981) that the main objective of new methods of genetic manipulation should be to improve the level of resistance where currently available sources are inadequate.

There is also considerable potential for progress from the genetic diversity available within each of the legume crops. Extensive germplasm collections have been assembled, especially at the international centres, and adequate sources to the majority of the principal diseases have been identified. In most cases, there seems little urgency to harness extra-specific sources. However, in cases where 'good sources' of resistance have been found, they have tended to be used to the complete exclusion of partial resistance. Such overemphasis can be dangerous because 'break down' is a common consequence. The failure to find 'good sources' has often provided the fillip to search among more diverse germplasm, or to develop more sensitive screening techniques to detect partial resistance, as illustrated by work on disease resistance in groundnuts (Section 4.3). The implications that pathogenic variation may have for resistance breeding are discussed in the next section.

10.2.5 Pathogenic variation in perspective

That resistance is impermanent is widely publicized: the 'boom and bust' cycle is notorious, and is at the core of a jaundiced view that disease resistance breeding is unavoidably continuous, because of the genetic ability of pathogens to vary. But variation is a feature of all populations, so that variation in pathogenicity among plant parasites is to be expected. Furthermore, the existence of such variants need not necssarily lead to termination of the commerical life of resistant cultivars; it is the proven transience in agriculture of a particular resistance which ultimately is of concern. Much depends on the degree of specificity shown by the pathogen and its host, a subject well reviewed by Scott *et al.* (1980).

Instances in which resistance has proved transient because of the emergence of a 'matching' variant of the pathogen might be expected to reveal principles of value to breeding, in a manner antithetical to those of breeding for durable resistance (Johnson, 1978; 1981a). Thus, transient resistance is more often a feature of diseases caused by biotrophic than by necrotrophic pathogens; while the former tend to show cultivar-specificity, the latter are more commonly species-specific (Scott *et al.*, 1980). Secondly, transient resistance is more frequently associated with high levels of resistance, including those conferred by the hypersensitive reaction, than with partial resistance. This presumably relates to a greater selection pressure imposed on the pathogen population by complete resistance. And thirdly, transient resistance is more often reported from intensive systems of farming in developed countries than from traditional systems in developing countries, presumably attributable to the greater diversity (Section 1.3.2) of the latter. Resistance-breaking variants were first recognized among legume pathogen populations by Barrus (1918) whose work on bean anthracnose led to one of the earliest demonstrations of variation in plant pathogens. Such variants have subsequently been reported from many other host-pathogen combinations, including bean rust (Harter *et al.*, 1935), downy mildew of soyabean (Dunleavy and Hartwig, 1970), the phytophthora rots of soyabean (Athow *et al.*, 1974), lima bean (Wester and Jorgensen, 1959) and cowpea (Purss, 1972), halo blight of beans (Patel and Walker, 1965), and bean common mosaic (Drijfhout *et al.*, 1978). Such variants are traditionally assigned to physiologic races which, by their definition (Anon, 1973), are taxa characterized by specialization to different cultivars of one host species. Sets of differential cultivars have often been used to distinguish races. Such host-specific pathogenicity, or virulence, is typified by a differential interaction between cultivars and races, often with reversed ranking (Scott *et al.*, 1980). Although quantitative interactions may also be host-specific, none are known among the legume pathogens. Transient resistance is, in most cases, a consequence of a qualitative interaction involving host-specific pathogenicity. The extent to which durable resistance is a quantitative character is uncertain: the basis of durable resistance is not well understood, nor can the absence of specificity be proved. Host non-specific pathogenicity, or aggressiveness, is usually a quantitative character. Quantitative differences in pathogenicity have been demonstrated in the rust fungi (Allen, 1975d; Melching *et al.*, 1979), in *Ascochyta rabiei* (Bedi and Aujla, 1969) and the halo blight bacterium (Szarka and Velich, 1980), and differences in pathogenicity not involving reversed rank order have been found also among legume viruses (Fulton and Allen, 1982); in each of these cases, it is assumed that host non-specific pathogenicity is involved. The durability of resistance presumably relies in part upon the relative specificity of the host-parasite interaction, as well as on the epidemiology of the disease and the way in which the evolution of the pathogen population is managed.

Whereas the concepts of vertical and horizontal resistance are widely recognized, if not so widely accepted, the matching concepts of vertical and horizontal pathogenicity are seemingly much less widely appreciated. The importance of doing so rests in the different implications each has for the plant breeder. Future plant pathologists would do well to pay more attention to the extent of host-specificity of pathogenic variants rather than to remain preoccupied with their mere definition.

10.2.6 Genetics of host-parasite relationships

In the preceding sections, we have discussed variation in the manner that resistance and pathogenicity are expressed, and the practical implications that this has for legume breeding. In this section, we are concerned not with gene function but with the underlying mechanisms of genetic control, of resistance in the host, of pathogenicity in the parasite, and the interactions between these two systems. The genetics of host-parasite relationships is discussed in general terms by Day (1974), and the reader requiring a thorough treatment of the subject should refer to his book. Here, we shall deal briefly with the inheritance of resistance and pathogenicity, as far as possible using examples from legume diseases.

The genetics of disease resistance in the food legumes has been reviewed recently (Meiners, 1981). It is evident that oligogenic ('major gene') inheritance predominates in the literature, but it is much less clear whether this accurately reflects its incidence in nature. It seems probable that resistance under polygenic control, being harder to detect and to demonstrate, is often overlooked, or ignored, by those under pressure to publish. Time will tell whether such bias exists. Oligogenic resistance is reported from all the legume crops against each of the major groups of pathogens. Oligogenic resistance is more often said to be dominant than recessive. Cases in which resistance is known to be a recessive character have been reported from groundnut rosette (Gibbons and Mercer, 1972), cowpea chlorotic mottle (Rogers *et al*., 1973), blackeye cowpea (Walker and Chambliss, 1981), certain genotypes of bean and common mosaic (Ali, 1950), some host-parasite combinations of beans and halo blight (Taylor *et al*., 1978) and of cowpeas and bacterial blight and pustule (Singh and Patel, 1977; Patel, 1978), bacterial pustule of soyabean (Hartwig and Lehman, 1951), and wilt of chickpea (J. B. Smithson, personal communication, 1982). The preponderance of examples from among bacterial and virus diseases rather than from fungal diseases is interesting but possibly spurious. Resistance under oligogenic control is notorious for its transience, but certain single genes are known to govern resistance to a wide spectrum of races, in some cases providing effective protection for many years. Examples include the dominant 'I' gene for resistance to bean common mosaic (Zaumeyer and Meiners, 1975) and the 'Are' gene for anthracnose resistance (Mastenbroek, 1960). There are also instances in which monogenic control is associated with apparently race non-specific resistance, including the single recessive genes which confer resistance to bacterial pustule in the soyabean cv. Clemson Non-Shatter (Hartwig and Lehman, 1951) and to halo blight in the bean accession PI 150414 (Taylor *et al.*, 1978). Thus, oligogenic inheritance of resistance is not in itself a reliable index of transience and crop vulnerability. The few studies made of the genetics of combined resistance (Hubbeling, 1957; Frazier *et al.*, 1961; Raj and Patel, 1978) indicate that no single gene confers resistance to more than one pathogen, yet it seems likely that certain resistance mechanisms do operate against several pathogens.

Polygenic resistance is determined by many genes of individually small effect. It often affords resistance which is apparently race non-specific though its expression tends to be strongly influenced by environment. Examples of polygenic inheritance of resistance include that effective in beans against bacterial blight (Coyne *et al.*, 1965b), the root rots caused by species of *Fusarium, Rhizoctonia, Pythium*, and *Thielaviopsis* (Hassan *et al.*, 1971; Boomstra and Bliss, 1977; Dickson and Boettger, 1977) and golden mosaic (Yoshii *et al.*, 1980), as well as against leaf spot in groundnuts (Nevill, 1980) and, perhaps, blight of chickpea (Boorsma, 1980).

Whereas the breeding of resistant crop cultivars in itself generates knowledge of the genetics of resistance, the production of new races is a natural phenomenon so that knowledge of the genetics of pathogenicity is harder to obtain. Despite the early recognition that pathogens could respond to the use of resistant cultivars (Barrus, 1918), still little is known of the relative importance of the four mechanisms on which the origin of novel pathogenicity depends; these are mutation, heterokaryosis, sexual recombination, and parasexual recombination (Day, 1974). Sexual recombination is a firmly established source of new physiologic races in fungi with perfect states, and heterokaryosis and parasexuality are presumed to serve that function in imperfect fungi. Certainly, heterokaryosis is a basis for the acquisition of virulence in *Thanatephorus cucumeris* (McKenzie *et al.*, 1969). Although parasexuality has been recognized in several fungi, including the legume pathogens *Fusarium oxysporum* (Buxton, 1956) and *Phoma medicaginis* (Sanderson and Srb, 1965), the importance of parasexual recombination in the evolution of pathogenicity remains ill understood (Tinline and MacNeill, 1969).

There is considerable variation in the expression of pathogenicity (Section 10.2.5) which, like resistance, can be under oligogenic or polygenic control. Host-specific pathogenicity, or virulence, is usually governed by a single recessive gene. From work on flax and flax rust, Flor (1942; 1956) proposed that, for each gene that conditions reaction in the host, there is a corresponding gene in the parasite that conditions virulence. Each gene in either member of a host-parasite system may be identified only by its counterpart in the other member of the system. This is known as the gene-for-gene concept. As Day (1974) points out, genetic changes in one member will be balanced by changes in the other, thus tending to restore equilibrium; each is interdependent. Although a gene-for-gene relationship can be assumed to exist in many host-parasite interactions involving specificity, it has been demonstrated in relatively few instances and is unknown in the wild. Examples from the legumes include beans and bean rust (Christ and Groth, 1979), beans and common mosaic (Drijfhout, 1978) and, perhaps, also in clover and *Rhizobium* (Nutman, 1969). Examples wherein single genes confer resistance to a wide spectrum of races, as shown in the soyabean cv. Kanrich to downy mildew (Bernard and Cremeens, 1971), in the bean cv. Cornell 49-242 to anthracnose (Mastenroek, 1960), and in the bean cv. NEP 2 to rust (Ballantyne and McIntosh, 1977), merely indicate that the matching gene for pathogenicity is uncommon, they do not represent exceptions to gene-for-gene theory unless the single gene confers resistance to all races.

The gene-for-gene concept clearly identifies single resistance genes qualitatively as either extremely effective or totally ineffective. However, Riley (1973) warned that it might be imprudent if breeders were to persist in discarding major genes that had been overcome by virulence changes, at least until it had been firmly established that defeated, 'ghost' genes did not contribute to quantitative resistance. Nelson (1978) subsequently proposed that race-specific, oligogenic resistance and race non-specific, polygenic resistance are, in fact, governed not by independent genetic systems but by the same genes, albeit expressed in a different manner in different backgrounds. Although apparently there is evidence that defeated single resistance genes do have a residual effect (Nass *et al.*, 1981), it seems uncertain whether the presence of polygenes in addition to the defeated oligogenes can be entirely ruled out.

Almost nothing is known of the genetics of host non-specific pathogenicity, or aggressiveness. Polygenic control of pathogenicity may be common and many pathogens are known to vary both qualitatively and quantitatively. The way in which genes for aggressiveness may interact with those for virulence is also uncertain. Van der Plank (1968) has claimed that the two characters are negatively correlated, in that 'unnecessary' genes for virulence tend to make a pathogen less able to survive. However, there is no evidence that this 'stabilizing selection' necessarily operates in all host-parasite relationships (Parlevliet, 1981). Genes for aggressiveness in the pathogen are usually presumed to function independently of genes for resistance in the host, but it has been suggested (Parlevliet and Zadoks, 1977) that there is a 'polygene-for-polygene' relationship in quantitative interactions. These concepts have important implications for strategies of disease resistance breeding but their relevance to the host-pathogen interaction of legumes remains unknown.

10.2.7 Resistance breeding strategy

The preponderance of studies in which resistance has been shown to be under oligogenic control has ensured that disease resistance is normally regarded as a highly heritable, qualitative character. Legume breeding programmes, since the first work of Orton (1902), have relied chiefly on conventional methods of improving these predominantly self-pollinated crops. Although bulk and pedigree selection, and backcross procedures, have led to significant improvements in disease resistance, genetic progress tends to be restricted by continued selfing and by the small numbers of parents which can be conveniently handled in crossing, so that the probability of novel recombination is reduced. These limitations may be lessened by the use of population improvement and recurrent selection methods, applied to cross-pollinated crops. Such methods, which should employ mechanisms like genetic male sterility to enhance natural outcrossing, have been applied to the improvement of pigeon peas (Khan, 1973), in which the incidence of natural outcrossing is about 20 per cent, and to disease-resistance breeding in beans (Stoetzer, in Robinson, 1976; Boomstra and Bliss, 1977) and cowpeas (Smithson *et al*., 1980). The principal limitations to conventional strategies have proved to be twofold: that the durability of resistance in some cases is inadequate; and that levels of resistance are sometimes insufficient. Although it is premature to assess whether population improvement and related strategies are likely to overcome these limitations, it is evident that there are many cases in which resistance is now known to be a quantitative character to which these breeding methods are better suited. It would seem prudent to use both approaches concurrently in a resistance breeding programme.

Effective selection for resistance depends to a large extent on the reliability and reproducibility of the screening technique, and in the method of selection and its intensity. Although simple visual assessment methods are necessary, in some host-pathogen combinations there are dangers implicit in the overemphasis of qualitative differences; more stable mechanisms of resistance are often more obscurely expressed. In cases where resistance is commonly transient, greater emphasis should be given to the selection of resistance governed by mechanisms likely to be unmatched by variants of the pathogen, perhaps including the mechanisms of escape from infection (Table 25). There appears to be considerable scope for both stabilizing and raising the level of resistance by intercrossing host genotypes possessing distinct components of resistance to a given disease, a strategy tantamount to a 'pyramiding of mechanisms'. While the pyramiding of separate genes governing a single mechanism has often been advocated, a synthesis of the separate components of resistance has seldom been emphasized (Coyne and Schuster, 1974c; Nevill, 1981; Pieters, 1981) and probably deserves greater emphasis in breeding programmes. The potential for doing so has been repeatedly stated in previous chapters (Sections 4.3, 5.3, and 6.3).

Multilocation testing is usually seen as the cornerstone of resistance breeding strategy at the international institutes (e.g. Dubin and Rajaram, 1981) which often have effectively pan-tropical responsibility for the improvement of a particular crop. The proliferation of international disease nurseries emphasizes the general acceptance of multilocation testing which provides a means of rapid progress, since one season's data from a series of sites can be as useful as those from trials at a single location over several seasons (Day, 1973). But do they measure the same parameters? There are two particular facets of international testing for disease resistance which require closer scrutiny; these are the choice of screening location, and the interpretation of location-specific and location non-specific reactions.

Since the incidence and severity of diseases vary markedly with environment, it is important that testing locations are selected so as to represent contrasting ecological zones, often amounting to distinct pathosystems. Sites in the centre of crop origin, and in recognized 'hot spots' are important. The interpretation of response at these selected locations is more difficult. With diseases in which pathogenic variants are known to be host-specific, durable resistance is an obvious objective. However, the apparent stability of reaction across environments of a particular crop genotype is no guarantee that resistance is in fact durable. Whereas multilocation testing identifies resistance which is effective on a small scale against several pathogen populations in different environments in the short term, durable resistance is recognized only after a long period of testing in large-scale production against a single, albeit changeable, population of the pathogen in one environment (Johnson, 1978; Williams, 1982). The extent to which location non-specific resistance is an index of durability is fundamental to the strategy of multilocation testing, yet the relationship remains uncertain. Similarly, location-specific resistance is an unreliable index of race-specificity (Section 7.3). The quest for location non-specific resistance is part of a philosophy in which wide adaptability is held to be a vital objective of international testing. Yet local adaptation (a greater fitness in a given environment) and wide adaptability (the capacity for genetic change in adaptation) are mutually exclusive attributes (Day, 1973). It may well be preferable for international breeding programmes to adopt a strategy in which early generation segregating material is screened at sites representative of the regions for which the improvement is ultimately intended, with the objective of identifying a comprehensive, durable resistance to all local pathogens, in preference to a strategy in which 'improved' genotypes with wide adaptability are distributed internationally. If the latter approach is 'successful', it may well lead to a dangerous erosion of locally adapted landraces, as suggested by Day (1973), whereas it is these very materials which should be used in the development of more productive cultivars. With such a strategy in mind, attempts have been made in the improvement of cowpeas in West Africa, first to define distinct ecological zones as a basis for the stratification of the breeding programme, then to attempt a regional deployment of certain combined characters including disease resistance (Smithson *et al*., 1980; Allen and Goldsworthy, 1981; Section 7.3).

10.3 INTEGRATED DISEASE MANAGEMENT

Among the components of integrated disease management, we have so far considered only genetic aspects. In this section, we shall briefly discuss the other components, namely chemical, biological, and cultural control measures, in as much as they have relevance to the potential management of legume pathogens in tropical farming systems.

10.3.1 Limitations and potentials of chemical control

The principal limitation to the wider use of chemical control measures in tropical food legumes is assumed to be the prohibitively high cost of the chemicals themselves, as well as of the application equipment. However, few economic appraisals have been made, either of crop loss or of the cost-effectiveness of chemical control of legume diseases (Mukiibi, 1969b; McDonald and Fowler, 1976; Sanders, 1977). Much depends on the value of the crop, as well as on such factors as the persistence and spectrum of activity of the chemical. In general, it seems that foliar spraying to control disease is seldom a viable practice in tropical food legumes, except for seed multiplication (Peregrine, 1971a; McDonald and Fowler, 1976). In dry areas, the scarcity of water adds to the unacceptability of conventional techniques of application. However, there is potential for the application of fungicides as ultra-low-volume sprays (McDonald, 1978), as dusts (Mercer, 1973) and, particularly, as seed dressings. Work on the chemical control of the cercospora leaf spots of groundnut has shown that sulphur dusting can be an economic proposition, especially in regions where the soil is deficient in sulphur which then acts both as a fungicide and fertilizer (Mercer, 1973; Laurence *et al.*, 1976). The practical value of using fungicidal seed dressings has been demonstrated in groundnuts against crown rot (Tarr, 1958), in cowpeas against a seedling mortality (Williams, 1975a), and in chickpea against blight (Reddy, 1980) although, in the latter case, the active chemical is an eradicant which is liable to impair seed germination. Similarly, antibiotics have been applied to seed as eradicants of seedborne bacterial infection (Ralph, 1976), but with limited success.

There is potential for the wider use of insecticides in the control of virus diseases, especially those transmitted in a persistent manner by their vectors. Davies (1975a, b) has shown that foliar application of a systemic aphicide can subsequently lessen the incidence of rosette in groundnuts, and Nene (1973) has suggested that a systemic insecticide in combination with mineral oil would prevent the spread of mung bean yellow mosaic by whiteflies. Rathi (1979) has shown that seed treatment with a granular systemic pesticide can protect against sterility mosaic in pigeon pea. Commonly used systemic insecticides are generally ineffective at preventing the spread of viruses transmitted in a non-persistent manner, largely because of their inability to act fast enough to prevent transmission. However, recent work with pyrethroids has shown that these materials can restrict both the acquisition of and infection by viruses, including a non-persistant potyvirus (Gibson *et al*., 1982), suggesting there is an increasingly wide application for the management of virus diseases by chemical means, should such a practice be economically justified.

A major limitation to the use of pesticides is, of course, their undesirable side-effects, so widely publicized. For our purposes, the more immediately relevant effects of pesticides include the following:

(i) The development of pesticide tolerance, recognized among the cercospora leaf spot fungi in response to benomyl (Clark *et al*., 1974; Littrell, 1974) and in the halo blight bacterium against streptomycin (Russell, 1975);
(ii) Iatrogenic effects (Griffiths, 1981), which are those on non-target organisms leading to the exacerbation of disease; the increased incidence of *Pythium* stem rot in cowpeas treated with benomyl is an example (Williams and Ayanaba, 1975); and
(iii) The detrimental effects on symbiotic organisms, including *Rhizobium* (Fisher and Hayes, 1981) and the vesicular-arbuscular mycorrhizal fungi (Bailey and Safir, 1978).

On balance, the greatest potential for the increased use of pesticides in disease management on tropical food legumes would seem to be as seed dressings. Irrespective of their mode of application, the usage of chemical control measures should seek to supplement, rather than to replace, the other components of integrated disease management.

10.3.2 The potential for biological control

Despite the many examples of hyperparasitism, antagonism, and the like (Section 3.6), almost nothing is known of the contribution of such interactions in the natural suppression of pathogen populations, presumed to occur in traditional legume-farming systems. The potential is suggested by the occasional, often accidental, demonstration that the use of a pesticide exacerbates a disease by eliminating its parasites or natural enemies (Williams and Ayanaba, 1975). Attempts to harness this potential have been made only in a few instances. A general review of the biological control of plant pathogens is available (Baker and Cook, 1974).

Examples of the interactions which could have implications for the biological control of legume pathogens include cases of fungal hyperparasitism of fungal pathogens (Sneh *et al*., 1977; Siddaramaiah *et al*., 1981; Tu and Vaartaja, 1981; Allen, 1982), fungal infection of insect vectors of viruses (Nene, 1972), fungal pathogens of parasitic angiosperms (Zummo, 1977; R. M. Goodman, personal communication 1981), bacterial antagonism of fungal pathogens (Vasudeva *et al*., 1962), virus infection of fungal pathogens (Yarwood and Hecht-Poinar, 1973; Castanho and Butler, 1978), nematode parasitism of fungal pathogens (Bridge *et al*., 1977), insect parasitism of parasitic angiosperms (Singh and Allen, 1979) and insect predation of insect vectors of viruses (Evans, 1954). The few instances in which an attempt has been made to develop a method for the biological control of legume pathogens are divisible into three categories: the coating of seed with preparations of an antagonist, foliar inoculation, and the inoculation or amendment of soil. Singh and Mehrotra (1980b) have shown that the inoculation of chickpea seed with certain isolates of the antagonistic genera *Bacillus* and *Streptomyces* can lead to a significant decrease in the incidence of dry root rot, caused by *Macrophomina phaseolina*. Similarly, seed inoculation with *Bacillus subtilis* has been shown to protect pigeon pea from wilt (Singh *et al*., 1965). Trutmann *et al*. (1982) have evaluated the hyperparasite, *Coniothyrium minitans*, as a biological control agent of the white mould pathogen (*Sclerotinia sclerotiorum*) on aerial parts of bean plants. Although spraying trials established that foliar inoculation with *C. minitans* did not control the disease in the season of application, it appeared that *C. minitans* was effective in decreasing the inoculum potential of *S. sclerotiorum* by its adverse effect on the production of sclerotia. Such an effect might be expected to lessen disease incidence in subsequent seasons. Field inoculation of sclerotia on the soil surface with a pycindial dust preparation of *C. minitans* showed similar potential (Turner and Tribe, 1975). Blackman and Rodriguez-Kabana (1975) were able to demonstrate decreased damage to groundnuts from southern blight (*Sclerotium rolfsii*) by a system of soil infestation with the antagonist, *Trichoderma harzianum*, whereas others have shown that organic amendment of soil can suppress the activity of the pigeon pea wilt pathogen (*Fusarium udum*) by increasing the production of an antibiotic from the antagonist, *Bacillus subtilis* (Vasudeva *et al*., 1963; Singh and Singh, 1981).

Such studies demonstrate that there is a considerable, largely unexploited, potential for the development of biological control measures, perhaps particularly if they are integrated with other disease management practices.

10.3.3 Cultural control practices

The management of plant disease by cultural means embraces a miscellany of agricultural practices, reviewed by Stevens (1960) and Tarr (1972). Here 'cultural control' is used in its broadest sense to include sanitation measures, the use of pathogen-free seed and the modification of agronomic practices to lessen the effects of disease.

The efficacy of sanitation depends both on its agronomic practicability and the precision with which it is carried out. Although sanitation recommendations can be difficult to enforce in subsistence farming, certain traditional practices, such as the burning of crop residues, shifting cultivation, and the cultivation of crop mixtures (Robinson, 1976), can be expected to afford some measure of disease control. Sanitation is likely to be most effective against soilborne pathogens with a slow rate of secondary spread and a narrow host range among annual crops, particularly in regions with a pronounced close season. The burial of infected debris has been recommended in the control of many diseases, including ascochyta blight of chickpea (Sattar, 1933a) and southern blight of groundnut (Boyle, 1956; Clinton, 1957). The removal of diseased plants, or roguing, and the destruction of alternate weed hosts can help to control certain virus diseases (Section 3.1.4) as well as some fungal pathogens (Dhingra and da Silva, 1978).

Since many pathogens are seed-transmitted in legumes, there is obvious potential for their control by the production of pathogen-free seed, especially if seedborne inoculum is the principal means of the pathogen's survival. The importance of clean seed in disease control is well illustrated by the elimination of halo blight through the production of bean seed under arid conditions in the USA (Grogan and Kimble, 1967). Similarly, anthracnose has declined in importance as a bean disease in Britain through measures to ensure clean seed (Dixon and Doodson, 1973), and considerable potential remains for the application of the same principle to other diseases as well (e.g. Kaiser, 1981b). Seed certification schemes and legislation governing the international exchange of seed (Nirula, 1979; Mathys and Baker, 1980), perhaps coupled with improved methods for the detection of bacteria (Wharton, 1967) and viruses (Lister, 1978; Davis *et al*., 1981) in seed, together can be expected further to improve standards of seed health.

The extent to which agronomic practices can be modified so as to lessen the effects of a pathogen obviously depends on the flexibility of the cropping system as a whole; in some cases, the alteration of a particular practice, while controlling one disease, is liable to exacerbate another. Some of the cultural practices which have been used to manage legume diseases include rotation, altered soil management and tillage regime, adjustment to the time of sowing or plant spacing, the use of trap crops, the timing of harvest and post-harvest treatment. We shall examine each of these briefly.

Rotation provides a break in what might otherwise be a continuous crop. Its effectiveness in regulating populations of soilborne pathogens is heavily dependent on the longevity of the pathogen and the width of its host range. Pathogens which produce chlamydospores, oospores, or sclerotia are often able to persist in soil for long periods, making rotation impracticable. However, rotation of soyabeans with resistant crops has been shown to lead to a decline in the population of the cyst nematode (Ross, 1962; Riggs and Hamblen, 1966), and the inclusion of cereal crops in a rotation with lentil can result in a lower incidence of fusarium wilt, caused by *F. oxysporum* f.sp. *lentis* (Kannaiyan and Nene, 1979c). However, a decrease in disease incidence need not necessarily be accompanied by a fall in inoculum level. Nash and Snyder (1962) found that although the number of propagules of *Fusarium solani* f.sp. *phaseoli* was increased in a beans-barley rotation, the incidence of bean root rot was decreased, relative to a bean monoculture. Under certain circumstances, soil management and tillage regimes are known to affect disease incidence. For instance, the application of potash has been shown to check chickpea blight when applied before sowing (Lukashevich, 1958a) and the addition of lime can suppress anthracnose development in soyabean on calcium-deficient soils (Muchovej *et al.*, 1980). Organic amendment of soil is effective against pigeon pea wilt (Vasudeva *et al.*, 1963; Singh and Singh, 1981), and flooding apparently limits populations of nematodes parasitic on mung beans in a rice rotation (Castillo and Litsinger, 1978). Whereas a no-tillage system apparently favours the population growth of root knot nematodes, it tends to limit the population of root lesion nematodes, relative to conventional tillage practices (Caveness, 1974).

Although long association between a crop and its pathogens tends to lead to their synchronization, crops often reach maximum susceptibility, or peak liability, to infection only during defined periods. This offers an opportunity to adjust the sowing date so that the crop escapes severe damage. A good example comes from chickpea in the Mediterranean region where it is sown, not as a winter crop from which the highest yields are obtained, but in the spring when a measure of escape from ascochyta blight is possible (Hawtin *et al.*, 1980; Weltzien and Kaack, 1981). It is sometimes possible to change the time of sowing so as to avoid a period of peak virus vector activity; opportunities for this are evident in certain whitefly-spread diseases (Williams, 1976; Vetten and Allen, in press). Similarly, sowing at a time to avoid periods of heavy rainfall during pod maturation can be important in escaping pod rots, such as that caused by *Choanephora cucurbitarum* in cowpea. Plant spacing can also be modified to lessen the effects of disease. For instance, the avoidance of overcrowding in seedbeds can help to minimize the risk of seedling damping-off, and some benefit can be obtained by wide spacing in lessening the severity of grey mould (Campbell, 1949) and fusarium root rot in beans (Burke, 1965a). Conversely, an increased plant density can help to decrease the incidence of certain virus diseases, including groundnut rosette (A'Brook, 1964; 1968; Davies, 1976; Farrell, 1976a, b) and perhaps soyabean mosaic (Halbert and Irwin, 1981), in each case by influencing the alighting response of the vector. Trap crops are sometimes used to cleanse soil by their ability to induce germination of soilborne propagules pathogenic to the companion crop. Finally, the choice of harvest date and post-harvest treatment can affect the extent of damage. The prompt harvest of soyabean at maturity may help protect against seed deterioration (Ndimande *et al.*, 1981); prompt lifting and careful drying of groundnuts similarly helps to decrease concealed damage (McDonald, 1970b).

10.3.4 The potential of crop mixtures

Despite the preponderance of intercropping practices in subsistence farming in the tropics, there is a dearth of information on the effects of intercropping on disease; the literature relevant to legumes is reviewed in Section 1.3.2. The extent to which a particular crop association affords protection from disease obviously depends on many factors, including the temporal and spatial arrangement of the cropping system, the disease susceptibility of the components, and the epidemiology of the diseases themselves. However, it is not safe to assume that disease severity is necessarily less in mixtures than in monocrops, and much more work is warranted. It seems probable that important lessons are to be learnt from existing systems of traditional farming, as is argued in the next section.

10.3.5 Integrated systems of disease management

Amongst the components of integrated control which we have discussed, disease resistance has been given prominence because it is considered the pivot of improved disease management in the tropical food legumes. In the long term, resistant cultivars are likely to prove of inestimable value because their development costs the farmer nothing nor does their adoption necessarily upset his farming system. However, experience has not always been universally encouraging. Conventional breeding methods have at times seemed to lead inexorably to the 'boom and bust' cycle through the crop uniformity which characterizes sophisticated agriculture in the temperate regions of the world. The ensuing disenchantment has stimulated new thinking, including the pathosystems concept, and has led to the development of more stable strategies of disease management (Robinson, 1976; 1980). Amongst the latter are the development of cultivar mixtures (Wolfe *et al*., 1981) and the pursuit of durable resistance (Johnson, 1978; 1981a). The underlying rationale for the use of cultivar mixtures is that increased genetic diversity affords increased stability, as was discussed in Section 1.3.2.

Subsistence agriculture in the tropics is characterized by complex mixtures which, in some cases, have evolved over thousands of years as balanced, self-supporting tropical agro-ecosystems the components of which are often adapted with surprising precision to the environment (Janzen, 1973; Bunting, 1975). Furthermore, crops in such systems have also achieved an equilibrium with the parasites with which they have evolved, shown by the low levels of disease they normally support: here, too, resistance is durable. Although crop mixtures are often higher yielding than sole crops (Baker, 1980), their productivity is comparatively low. In the face of world food shortage, peasant agriculture is under pressure to change, and to intensify.

Herein lies a paradox; the wheel has gone full circle. Intensive agriculture in developed countries, with its high inputs leading to high productivity, uniformity, and attendant high risk, is looking to old proven cultivars, and to wild plant communities (which in complexity resemble traditional agriculture, in the tropics) for guidance on stability. And tropical subsistence agriculture, being obliged to intensify, is having imposed upon itself alien technology from temperate countries. It would seem prudent to attempt first to analyse the nature of balance in long-established farming systems and their ancient landraces, before it is too late. Information on the relative contributions of host-plant resistance, natural biological control agents, and of simple cultural practices such as the burning of crop debris and the use of intercropping, might then be used to predict the long-term effects of a wide range of methods to identify the most evolutionarily stable, yet productive, strategy with which to replace the existing traditional systems without recourse to heavy input. Successful integration requires that the components of disease management are compatible. This is not invariably the case: fungicides are liable to kill hyperparasites (Wilding and Brobyn, 1980); fungicides and herbicides (Williams and Ayanaba, 1975; Wyse *et al*., 1976), and weeding (Daulton and Curtis, 1964; Mukiibi, 1978) can exacerbate disease. Yet in other instances control practices have the potential to supplement one another. Amongst these, the reinforcement of resistance with strict standards of seed health and with diversified systems of farming would seem to offer good opportunities for the long-term control of most legume pathogens, obviating the need for foliar application of fungicides. However, a wider use of chemicals in tropical agriculture can be expected and, where pest control still necessitates the use of insecticides, as on cowpea in Africa, there may be an opportunity for the combined application of an insecticide and a fungicide.

10.4 FUTURE CHALLENGE

The development of heavy-yielding cultivars capable of responding spectacularly to inputs, principally fertilizer and water, was the basis for the large cereal yield increases which occurred in Asia during the late 1960s. The Green Revolution was a considerable success for the application of relatively standard techniques of plant breeding to the improvement of cereal crops which are the mainstay of world food production. However, there is growing need for yet greater production, particularly in the less developed countries of the tropics. The improved cereal cultivars, coupled with improved farming practices, have made their impact chiefly on large-scale farms on the better soils; as yet, very little impact has been felt by the small-scale subsistence farmer on the poorer soils (Greenland, 1975). Furthermore, the success with cereals has tended to divert attention, and thus land, from the cultivation of the comparatively unimproved grain legumes whose production has remained stagnant. Despite their ancient domestication and their recognized role in the balance of both farming systems and diets, the legumes have received low priority from farmers and, until recently, also from researchers. The cultivars used by traditional farmers, though well-adapted to the local environment and its hazards, are inherently unproductive and yields fickle, so that the grain legumes are seen as high-risk crops. The challenge confronting us now is to devise means whereby the Green Revolution can be brought to the subsistence farmer, without recourse to heavy input, through the development of sustainable cropping systems of which improved cultivars of legumes are likely to be an integral part. Experience gained from the cereals boom seems unlikely to be relevant because, in that case, much emphasis was given to the transfer of technology from temperate agriculture, with its reliance on heavy inputs; alien technology is inappropriate to sustainable systems of cultivation in the tropics, and a radical rethink is required (Janzen, 1973; Greenland, 1975; Tudge, 1977). The places to look for guidance are not the intensive systems of farming in Europe and North America but rather the existing systems of traditional agriculture themselves. We have been unduly slow to realize this, as Bunting (1981) has stressed.

During the last 10–15 years, an unprecendented amount of attention has been devoted to the improvement of grain legumes, especially at the international centres and their predecessors, and it is now no longer true to say, as Borlaug (1973) did a decade ago, that 'neither new high yielding varieties (of grain legumes) nor improved technology have been developed'. That genetically superior cultivars of the major legumes are now available has been repeatedly stated in this book, and the incorporation of disease resistance into a range of plant types possessing heavy-yield potential is an important component of that improvement. But it is true to say that such genetic advance as has been made is yet to be translated into agricultural progress: no spectacular yield increments have occurred outside research stations, though they have been demonstrated in farmers' fields without dependence on heavy inputs. Food legume yields are, on average, in the range of 400–700 kg/ha yet, in all cases, their potential productivity is from three to six times larger than that average. Similarly, improved systems of farming, particularly those involving crop mixtures, are beginning to demonstrate their potential to increase productivity. If the materials and methods now emerging from the international centres are used adequately by national research teams to adapt them to local requirements, then we may expect that greater and more enduring productivity will follow. The extent to which the next phase of the Green Revolution will meet the demand of growing human populations for food remains to be seen.

Appendix. Vernacular and Scientific Names

The following is an alphabetical list of vernacular names and their scientific equivalents of some of the food legumes cultivated in tropical regions.

Adzuki bean	*Vigna angularis* (Willd.) Ohwi and Ohashi
Asparagus bean	*see* Yardlong bean
Bambarra groundnut	*Vigna subterranea* (Thouars) Verdc.
Bean	*Phaseolus vulgaris* L.
Blackeye pea	*Vigna unguiculata* (L.) Walp., cvs with white seed and black hilum
Black gram	*Vigna mungo* (L.) Hepper
Bonavist bean	*see* Hyacinth bean
Butter bean	*see* Lima bean
Catjang	*Vigna unguiculata* (L.) Walp. cv. gr. *biflora* E. Westphal
Chickpea	*Cicer arietinum* L.
Cluster bean	*Cyamopsis tetragonoloba* (L.) Taub.
Common bean	*see* Bean
Cowpea	*Vigna unguiculata* (L.) Walp.
French bean	*see* Bean
Garbanzo	*See* Chickpea
Goa bean	*see* Winged bean
Gram	*see* Chickpea
Green gram	*See* Mung bean
Ground bean	*Macrotyloma geocarpum* (Harms) Maréchal and Baudet
Groundnut	*Arachis hypogaea* L.
Guar	*see* Cluster bean
Haricot bean	*see* Bean
Horse bean	*Mucuna sloanei* Fawcett and Rendle
Horse gram	*Macrotyloma uniflorum* (Lam.) Verdc.

Hyacinth bean	*Lablab purpureus* (L.) Sweet
Jack bean	*Canavalia ensiformis* (L.) DC.
Kersting's groundnut	*see* Ground bean
Kidney bean	*see* Bean
Lentil	*Lens culinaris* Medik.
Lima bean	*Phaseolus lunatus* L.
Mat bean	*see* Moth bean
Monkey nut	*see* Groundnut
Moth bean	*Vigna aconitifolia* (Jacq.) Maréchal
Mung bean	*Vigna radiata* (L.) Wilcz.
Navy bean	*Phaseolus vulgaris* L., cvs with small white seed
Peanut	*see* Groundnut
Pigeon pea	*Cajanus cajan* (L.) Millsp.
Red gram	*see* Pigeon pea
Rice bean	*Vigna umbellata* (Thunb.) Ohwi and Ohashi
Sieva bean	*Phaseolus lunatus* L., small seeded types
Snake bean	*see* Yardlong bean
Snap bean	*Phaseolus vulgaris* L., cvs grown for their green pods
Soyabean	*Glycine max* (L.) Merr.
Sword bean	*Canavalia gladiata* (Jacq.) DC.
Tepary bean	*Phaseolus acutifolius* A. Gray
Urd bean	*see* Black gram
Velvet bean	*Mucuna pruriens* (L.) DC. var. *utilis* (Wall. ex Wight) Baker ex Burck.
Winged bean	*Psophocarpus tetragonolobus* (L.) DC.
Yam bean: African	*Sphenostylis stenocarpa* (Hochst ex Rich.) Harms
Mexican	*Pachyrrhizus erosus* (L.) Urban and *P. tuberosus* (Lam.) *Spreng.*
Yardlong bean	*Vigna unguiculata* (L.) Walp. cv. gr. *sesquipedalis* E. Westphal

References

Abawi, G. S., Crosier, D. C., and Cobb, A. C. (1977). Pod-flecking of snap beans caused by *Alternaria alternata, Pl. Dis. Reptr.*, **61**, 901–905.
Abawi, G. S., Provvidenti, R., Crosier, D. C., and Hunter, J. E. (1978). Inheritance of resistance to white mold disease in *Phaseolus coccineus, J. Hered.*, **69**, 200–202.
Abdou, Y. A-M., Gregory, W. C., and Cooper, W. E. (1974). Sources and nature of resistance to *Cercospora arachidicola* Hori and *Cercosporidium personatum* (Beck and Curtis) Deighton in *Arachis* species, *Peanut Sci.* **1**, 6–11.
Abdullah, H. (1980). A disease of winged bean (*Psophocarpus tetragonolobus*) caused by *Pseudomonas solanacearum* in Malaysia, *Pl. Dis.*, **64**, 798–799.
Abel, G. H. (1977). Brown stem rot of soybean—*Cephalosporium gregatum, Rev. Pl. Path.*, **56**, 1065–1077.
Aboul Ata, A. E., Allen, D. J., Thottappilly, G., and Rossel, H. W. (1982). Variation in the rate of seed transmission of cowpea aphid-borne mosaic virus in cowpea, *Trop. Grain Leg. Bull.*, **25**, 2–7.
Abrams, R. (1967). Studies on natural cross-pollination in pigeon peas (*Cajanus cajan), J. agric. Univ. Puerto Rico,* **51**, 1–3.
Abrams, R., Morales, A., and Julia, F. J. (1978). Status of research on pigeon peas in Puerto Rico, *Trop. Grain Leg. Bull.*, **11/12**, 17–21.
A'Brook, J. (1964). The effect of planting date and spacing on the incidence of groundnut rosette disease and of the vector, *Aphis craccivora* Koch, at Mokwa, Northern Nigeria, *Ann. appl. Biol.*, **54**, 199–208.
A'Brook, J. (1968). The effect of plant spacing on the numbers of aphids trapped over the groundnut crop, *Ann. appl. Biol.*, **61**, 289–294.
Adam, D. B. and Pugsley, A. T. (1935). A yellow bacterium associated with 'halo blight' of beans. *Aust. J. exp. Biol. Med. Sci.*, **13**, 157–164.
Adams, A. N. (1967). The vectors and alternative hosts of groundnut rosette virus in Central Province, Malawi, *Rhod. Zamb. Mal. J. agric. Res.*, **5**, 145–152.
Adams, D. B. and Kuhn, C. W. (1977). Seed transmission of peanut mottle virus in peanuts, *Phytopathology*, **67**, 1126–1129.
Adams, G. C. and Butler, E. E. (1979). Serological relationships among anastomosis groups of *Rhizoctonia solani, Phytopathology*, **69**, 629–633.
Adams, M. W. (1972). Dry beans, in *Genetic Vulnerability of Major Crops,* National Academy of Science, Washington, pp. 224–234.
Adegbola, M. O. K. and Hagedorn, D. J. (1969). Symptomatology and epidemiology of *Pythium* bean blight, *Phytopathology*, **59**, 1113–1118.
Adegbola, M. O. K. and Hagedorn, D. J. (1970). Host resistance and pathogen virulence in *Pythium* blight of bean, *Phytopathology*, **60**, 1477–1479.
Adeniji, M. O., Edwards, D. I., Sinclair, J. B., and Malek, R. B. (1975). Interrelation-

ship of *Heterodera glycines* and *Phytophthora megasperma* var. *sojae* in soybeans, *Phytopathology,* **65**, 722–725.
Adnew, H., Allen, D. J., and Ayanaba, A. A. (1977). *Ann. Rep. for 1976*, International Institute of Tropical Agriculture, Ibadan, Nigeria, p. 33.
Ageeb, O. A. A. and Ayoub, A. T. (1977). Effect of sowing date and soil type on plant survival and grain yield of chickpeas (*Cicer arietinum* L.), *J. agric. Sci., Camb.* **88**, 521–527.
Agrawal, H. and Maat, D. Z. (1964). Serological relationships among polyhedral plant viruses and production of high-titred antisera, *Nature, Lond.,* **202**, 674–674.
Agrawal, S. C. and Kotasthane, S. R. (1971). Resistance in some soybean varieties against *Sclerotium rolfsii* Sacc., *Indian Phytopath.,* **24**, 401–403.
Agudelo, F. V. de (1980). Nematodes, in *Bean Production Problems* (eds H. F. Schwartz and G. E. Gálvez), Centro Internacional de Agricultura Tropical, Cali, Colombia, pp. 315–326.
Ahmad, M. (1978). Whitefly (*Bemisia tabaci*) transmission of a yellow mosaic disease of cowpea (*Vigna unguiculata*), *Pl. Dis. Reptr.,* **62**, 224–226.
Ahmad, S. T., Magoon, M. L., and Mehra, K. L. (1972). Screening of a world collection of cowpea for field resistance to leaf spot and virus diseases, *SABRAO Newsl.,* **4**, 103–106.
Ainsworth, G. C. and Sampson, K. (1950). *The British Smut Fungi* (Ustilaginales), Commonwealth Mycological Institute, Kew.
Aiyer, A. K. Y. N. (1949). Mixed cropping in India, *Indian J. agric. Sci.,* **19**, 439–543.
Akinola, J. O. and Whiteman, P. C. (1972). A numerical classification of *Cajanus cajan* (L.) Millsp. accessions based on morphological and agronomic attributes, *Aust. J. agric. Res.,* **23**, 995–1005.
Alconero, R. and Santiago, A. (1973). *Phaseolus lathyroides* as a reservoir of cowpea mosaic virus in Puerto Rico, *Phytopathology,* **63**, 120–123.
Alcorn, J. L. (1968). Occurrence and host range of *Ascochyta phaseolorum* in Queensland, *Aust. J. biol. Sci.,* **21**, 1143–1151.
Alcorn, J. L., Punithalingam, E., and McCarthy, G. J. P. (1976). Peanut net blotch caused by *Didymosphaeria arachidicola* (Chochrjakov) comb. nov., *Trans. Br. mycol. Soc.,* **66**, 351–355.
Ali, M. A. (1950). Genetics of resistance to the common bean mosaic virus (bean virus 1) in the bean (*Phaseolus vulgaris* L.), *Phytopathology,* **40**, 69–79.
Ali, M. A. Trabulsi, I. Y., and Abd-Elsamea, M. E. (1981). Antagonistic interaction between *Meloidogyne incognita* and *Rhizobium leguminosarum* in cowpea, *Pl. Dis.,* **65**, 432–435.
Allard, R. W. (1954). Sources of root knot nematode resistance in lima beans, *Phytopathology,* **44**, 1–4.
Allard, R. W. (1961). Relationship between genetic diversity and consistence of performance in different environments. Lima beans, *Crop Sci.*, **1**, 127–133.
Allen, D. J. (1970). *Phytophthora nicotianae:* host range studies on the pathogen of zebra disease of *Agave* hybrid no. 11648, *E. Afr. agric. For. J.,* **36**, 119–123.
Allen, D. J. (1974). A technique for measuring sporulation of rust (*Uromyces appendiculatus*) on *Phaseolus vulgaris, A. Rep. Bean Improv. Co-op.,* **17**, 16–18.
Allen, D. J. (1975a). Induced resistance to bean rust and its possible epidemiological significance, *A. Rep. Bean Improv. Co-op.,* **18**, 15–16.
Allen, D. J. (1975b). Additions to the fungi and plant diseases of Malawi, *Soc. Malawi J.,* **28**, 35–44.
Allen, D. J. (1975c). Breeding beans (*Phaseolus vulgaris* L.) for resistance to rust (*Uromyces appendiculatus* (Pers.) Ung.). Ph.D. thesis, Univ. Cambridge, 221 pp.
Allen, D. J. (1975d). Variation in pathogenicity of *Uromyces appendiculatus, A. Rep. Bean Improv. Co-op.,* **18**, 13–14.
Allen, D. J. (1976). Induced resistance to bean rust and its possible epidemiological

significance in mixed cropping, in *Intercropping in Semi-arid Areas* (eds J. H. Monyo, A. D. R. Ker, and M. Campbell), IDRC, Ottawa, p. 46.
Allen, D. J. (1977a). Intercropping and disease, *Ann. Rep. for 1976,* International Institute of Tropical Agriculture, Ibadan, Nigeria, pp. 34–35.
Allen, D. J. (1977b). The international cowpea disease nursery program: towards multiple virus resistance, *Trop. Grain Leg. Bull.*, **8**, 29–32.
Allen, D. J. (1978). Multiple virus resistance in cowpea, *3rd Int. Congr. Pl. Path.,* München, 16–23 Aug. 1978, 58 (abstr.).
Allen, D. J. (1979a). New disease records from grain legumes in tropical Africa, *FAO Pl. Prot. Bull.,* **27**, 134–136.
Allen, D. J. (1979b). Cowpea improvement at IITA, *Proc. Symp. Grain Leg. Improv. E. Africa*, Nairobi, Kenya, 20–24 Aug. 1979.
Allen, D. J. (1980). Identification of resistance to cowpea mottle virus, *Trop. Agric. (Trin.),* **57**, 325–332.
Allen, D. J. (1981). Two bacterial diseases of cowpea in East Africa: a reply, *Trop. Pest Manag.,* **27**, 144.
Allen, D. J. (1982). *Verticillium lecanii* on the bean rust fungus, *Uromyces appendiculatus, Trans. Br. mycol. Soc.,* **79**, 362–364.
Allen, D. J., Anno-Nyako, F. O., Ochieng, R. S., and Ratinam, M. (1981a). Beetle transmission of cowpea mottle and southern bean mosaic viruses in West Africa, *Trop. Agric. (Trin.),* **58**, 171–175.
Allen, D. J., Emechebe, A. M., and Ndimande, B. (1981b). Identification of resistance in cowpea to diseases of the African savannas, *Trop. Agric. (Trin.),* **58**, 267–274.
Allen, D. J. and Goldsworthy, P. R. (1981). Disease resistance and crop improvement in *Vigna unguiculata, International Conference on Tropical Crop Protection,* Lyon, 8–10 July 1981, abstracts p. 40.
Allen, D. J. and Nandra, S. S. (1975). Effects of pH and calcium concentration on the sporulation of *Phytophthora* isolates from *Agave, Pl. Dis. Reptr.,* **59**, 555–558.
Allen, D. J., Nebane, C. L. N., and Raji, J. A. (1981c). Screening for resistance to bacterial blight in cowpea, *Trop. Pest Manag.,* **27**, 218–224.
Allen, D. J., Redden, R., and Jackai, L. (1980). *Trip to Brazil, 17–31 March 1980.* Unpublished report, International Institute of Tropical Agriculture, Ibadan, Nigeria, 61 pp.
Allen, D. J. and Skipp, R. A. (1982). Maize pollen alters the reaction of cowpea to pathogens, *Fld. Crops Res.,* **5**, 265–269.
Allen, D. J., Thottappilly, G., and Rossel, H. W. (1982). Cowpea mottle virus: field resistance and seed transmission in virus-tolerant cowpea, *Ann. appl. Biol.,* **100**, 331–336.
Allen, D. J. and van Damme, P. (1981) On thrips of transmission of cowpea (yellow) mosaic virus, *Trop. Agric. (Trin.),* **58**, 181–184.
Allington, W. B. (1945). Wildfire disease of soybeans, *Phytopathology,* **35**, 857–869.
Allington, W. B. (1946). Bud blight of soybean caused by the tobacco ring-spot virus, *Phytopathology,* **36**, 319–322.
Allington, W. B. and Chamberlain, D. W. (1948). Brown stem rot of soybean, *Phytopathology,* **38**, 793–802.
Allington, W. B. Moorhead, E. L., and Staples, R. (1960). Alfalfa mosaic virus in soybean, *Phytopathology,* **50**, 627 (abstr.).
Almeida, R. T. (1977). *Uromyces appendiculatus* (Pers.) Ung. var. *brasiliensis,* uma nove variedade do fungo da ferrugem do feijoeiro (*Phaseolus vulgaris* L.), *Fitopat. Bras.,* **2**, 243–245.
Altieri, M. A., Francis, C. A., Schoonhoven, A. v., and Doll, J. D. (1978). A review of insect prevalence in maize (*Zea mays* L.) and bean (*Phaseolus vulgaris* L.) polycultural systems, *Fld. Crops Res.,* **1**, 33–49.
Altieri, M. A., Schoonhoven, A. v., and Doll, J. (1977). The ecological role of weeds

in insect pest management systems: a review illustrated by bean (*Phaseolus vulgaris*) cropping systems, *PANS*, **23**, 195–205.

Alvarez, G. L. A. (1960). Phoma canker of pigeon peas in Puerto Rico, *J. Agric. Univ. Puerto Rico,* **44**, 28–30.

Alvarez-Ayala, G. and Schwartz, H. F. (1979). Preliminary investigations of pathogenic variability expressed by *Isariopsis griseola, A. Rep. Bean Improv. Co-op.*, **22**, 86–88.

Amin, K. S., Baldev, B., and Williams, F. J. (1978a). *Phytophthora cajani*, new species causing stem blight on *Cajanus cajan, Mycologia*, **70**, 171–176.

Amin, K. S., Baldev, B., and Williams, F. J. (1978b). Differentiation of *Phytophthora* stem blight from *Fusarium* wilt of pigeon pea by field symptoms, *FAO Pl. Prot. Bull.*, **26**, 123–124.

Amosu, J. O. (1974). The reaction of cowpea (*Vigna unguiculata* (L.) Walp.) to the root knot nematode (*Meloidogyne incognita*) in Western Nigeria, *Niger. Agric. J.*, **11**, 165–169.

Amosu, J. O. and Franckowiak, J. D. (1974). Inheritance of resistance to root knot nematode in cowpea, *Pl. Dis. Reptr.*, **58**, 361–363.

Anahosur, K. H., Bidari, V. B., and Hiremath, R. V. (1974). *Leptosphaerulina* leaf spot of cowpea. A new record to India, *Curr. Sci.,* **43**, 118.

Anahosur, K. H. and Fazalnoor, K. (1972). Studies on foliicolar fungi of soybean crop, *Indian Phytopath.,* **25**, 504–508.

Anahosur, K. H. and Waller, J. M. (1978a). *Phakopsora packyrhizi. CMI Descriptions of Pathogenic Fungi and Bacteria,* No. 589.

Anahosur, K. H. and Waller, J. M. (1978b). *Uredo cajani. CMI Descriptions of Pathogenic Fungi and Bacteria,* No. 590.

Anderson, C. W. (1955). *Vigna* and *Crotalaria* viruses in Florida. II. Notations concerning cowpea mosaic virus (*Marmor vignae), Pl. Dis. Reptr.,* **39**, 349–353.

Andrus, C. F. (1931). Mechanism of sex in *Uromyces appendiculatus* and *U. vignae, J. agric. Res.,* **42**, 559–587.

Andrus, C. F. (1948). A method of testing beans for resistance to bacterial blights, *Phytopathology,* **38**, 757–759.

Andrus, C. F. and Moore L. D. (1935). *Colletotrichum truncatum* (Schw.) n. comb., on garden and lima beans, *Phytopathology,* **25**, 121–125.

Andrus, C. F. and Wade, B. L. (1942). The factorial interpretation of anthracnose resistance in beans, *US Dept. Agric. Tech. Bull.,* **810**, 29 pp.

Angus, A. (1962–66). *Plant Pests and Diseases in Zambia,* Parts 1–7 and supplement. Mt. Makulu Res. Sta., Zambia (mimeo).

Anikster, Y. and Wahl, I. (1979). Co-evolution of the rust fungi on gramineae and liliaceae and their hosts, *A. Rev. Phytopath.,* **17**, 367–403.

Anno-Nyako, F. O. (1980). Studies on the vector relationships of whitefly-transmitted golden mosaic diseases of cowpea and lima bean. M.Sc. thesis, Univ. Science and Technol., Kumasi, Ghana.

Anno-Nyako, F. O., Vetten, H. J., Allen, D. J., and Thottappilly, G. (1983). The relation between cowpea golden mosaic and its vector, *Bemisia tabaci* Genn. *Ann. appl. Biol.,* in press.

Anon (1956). *Rep. Min. Agric. Rhod. Nyasald 1955–56*, pp. 66–70. (*Rev. appl. Mycol.*, **36**, 380.)

Anon (1963). Witches' broom of pigeon pea. *Rep. Dep. Agric. Papua New Guinea for 1960–61*, pp. 74–79. (*Rev. appl. Mycol.*, **43**, 3103.)

Anon (1972). Soybeans and other edible legumes, in *Genetic Vulnerability of Major Crops,* National Academy of Sciences, Washington, pp. 207–252.

Anon (1973). A guide to the use of terms in plant pathology, *Phytopath. Pap.* No. 17, Commonwealth Mycological Institute, Kew.

Anon (1979). *Tropical Legumes: Resources for the Future,* National Academy of Sciences, Washington, 328 pp.

Apple, J. L. (1963). Persistence of *Phytophthora parasitica* var. *nicotianae* in soil, *Pl. Dis. Reptr.*, **47**, 632–634.
Arene, O. B. (1976). Influence of shade and intercropping on the incidence of cassava bacterial blight caused by *Xanthomonas manihotis. Paper at the IDRC-IITA Workshop on Cassava Bacterial Blight,* IITA, Ibadan, 1–4 Nov. 1976.
Arias, E. A. (1980). Chickpea in Mexico, in *Proc. Internat. Wkshop on Chickpea Improv.,* International Crops Research Institute for the Semi-Arid Tropics, Hyderabad, India, pp. 246–248.
Arif, A. G. and Jabbar, A. (1965). A study of physiologic specialisation in *Mycosphaerella rabiei* Koviceveski *Ascochyta rabiei* (Pass.) Lab. the causal organism of gram blight, *W. Pakist. J. agric. Res.,* **3**, 103–121.
Ariyanayagam, R. P. and Spence, H. A. (1978a). A possible gene source for early, day-length neutral pigeon peas, *Cajanus cajan* (L.) Millsp., *Euphytica*, **27**, 505–509.
Ariyanayagam, R. P. and Spence, J. A. (1978b). Pigeon pea improvement activities at the University of the West Indies, Trinidad. *Trop. Grain Leg. Bull.*, **13/14**, 14–18.
Armstrong, G. M. and Armstrong, J. K. (1948). Nonsusceptible hosts as carriers of wilt fusaria, *Phytopathology,* **38**, 808–826.
Armstrong, G. M. and Armstrong, J. K. (1950). Biological races of the *Fusarium* causing wilt of cowpeas and soybeans, *Phytopathology,* **40**, 181–193.
Armstrong, J. K. and Armstrong, G. M. (1951). Physiological races of the *Crotalaria* wilt *Fusarium, Phytopathology,* **41**, 714–721.
Armstrong, J. K. and Armstrong, G. M. (1958). A race of the cotton-wilt *Fusarium* causing wilt of Yelredo soybean and flue-cured tobacco, *Pl. Dis. Reptr.,* **42**, 147–151.
Armstrong, G. M. and Armstrong, J. K. (1964). *Lupinus* species—common hosts for wilt fusaria from alfalfa, bean, *Cassia,* cowpea, lupine, and US cotton, *Phytopathology,* **54**, 1232–1235.
Armstrong, G. M. and Armstrong, J. K. (1965a). A wilt of soybean caused by a new form of *Fusarium oxysporum, Phytopathology,* **55**, 237–239.
Armstrong, G. M. and Armstrong, J. K. (1965b). Wilt of chrysanthemum caused by race 1 of the cowpea *Fusarium, Pl. Dis. Reptr.,* **49**, 673–676.
Armstrong, G. M. and Armstrong, J. K. (1968). Formae speciales and races of *Fusarium oxysporum* causing a tracheomycosis in the syndrome of disease, *Phytopathology,* **58**, 1242–1246.
Armstrong, G. M. and Armstrong, J. K. (1975). Reflections on the wilt fusaria, *A. Rev. Phytopath.,* **13**, 95–103.
Armstrong, G. M. and Armstrong, J. K. (1980). Cowpea wilt *Fusarium oxysporum* f.sp. *tracheiphilum* race 1 from Nigeria, *Pl. Dis.,* **64**, 954–955.
Armstrong, G. M., Armstrong, J. K., and Billington, R. V. (1975). *Fusarium oxysporum* forma specialis *voandzeiae*, a new form species causing wilt of bambarra groundnut, *Mycologia*, **67**, 709–714.
Arneson, P. A. (1970). Chemical control of rust and *Cercospora* leaf spots of peanuts in Honduras and Nicaragua, *Phytopathology,* **60**, 1539 (abstr.).
Arnold, H. C. (1942). Soyabeans, *Rhod. agric. J.*, **39**, 384–390.
Arp. G., Coyne, D. P., and Schuster, M. L. (1971). Disease reaction of bean varieties to *Xanthomonas phaseoli* and *Xanthomonas phaseoli* var. *fuscans* using two inoculation methods, *Pl. Dis. Reptr.,* **55**, 577–579.
Arthur, J. C. (1934). *Manual of the Rusts in the United States and Canada,* Lafayette, Indiana, 438 pp.
Asafo-Adjei, B. (1982). Field screening techniques for resistance to cowpea (yellow) mosaic virus, bacterial pustule and anthracnose in cowpea, M.Sc. thesis, Univ. Guelph.
Ashby, J. W. and Huttinga, H. (1979). Purification and some properties of pea leafroll virus, *Neth. J. Pl. Path.,* **85**, 113–123.

Ashworth, L. J. and Langley, B. C. (1964). The relationship of pod damage to kernel damage by moulds in Spanish peanut, *Pl. Dis. Reptr.*, **48**, 875–878.

Ashworth, L. J., Langley, B. C., and Thames, W. H. (1961). Comparative pathogenicity of *Sclerotium rolfsii* and *Rhizoctonia solani* to Spanish peanut, *Phytopathology*, **51**, 600–605.

Ashworth, L. J., Langley, B. C., Mian, M. A. W., and Wrenn, C. J. (1964). Epidemiology of a seedling disease of Spanish peanut caused by *Aspergillus niger*, *Phytopathology*, **54**, 1161–1166.

Athow, K. L. (1973). Fungal diseases, in *Soybeans: Improvement, Production, and Uses* (ed. B. E. Caldwell), Amer. Soc. Agron., Madison, pp. 459–489.

Athow, K. L. and Bancroft, J. B. (1959). Development and transmission of tobacco ringspot virus in soybean, *Phytopathology*, **49**, 697–701.

Athow, K. L. and Caldwell, R. M. (1954). A comparative study of *Diaporthe* stem canker and pod and stem blight of soybean, *Phytopathology*, **44**, 319–325.

Athow, K. L. and Laviolette, F. A. (1961). The relation of seed-transmitted tobacco ringspot virus to soybean yield, *Phytopathology*, **51**, 341–342.

Athow, K. L., Laviolette, F. A., and Abney, T. S. (1974). Reaction of soybean germplasm strains to four physiologic races of *Phytophthora megasperma* var. *sojae.*, *Pl. Dis. Reptr.*, **58**, 789–792.

Athow, K. L. and Probst, A. H. (1952). The inheritance of resistance to frog-eye leaf spot of soybeans, *Phytopathology*, **42**, 660–662.

Atkins, J. C. (1973). The importance, epidemiology and control of bean rust (*Uromyces appendiculatus*) on white haricot beans (*Phaseolus vulgaris*) in Uganda, M.Sc. thesis, Makerere Univ., Uganda.

Auckland, A. K. (1970). Soyabean improvement in East Africa, in *Crop Improvement in East Africa* (ed. C. L. A. Leakey), Commonwealth Agricultural Bureaux, Farnham Royal, pp. 129–156.

Auckland, A. K. and Singh, K. B. (1977). An international approach to chickpea (*Cicer arietinum* L.) breeding, *SABRAO 3rd Congress*, Canberra 8-10–8-13.

Augustin, E. and Antunes, I. F. (1973). Physiologic specialization of *Uromyces phaseoli typica* in South Brazil from 1968 through 1972, *A. Rep. Bean Improv. Co-op.*, **16**, 68–74.

Augustin, E., Antunes, I. F., and Costa, J. G. C. da (1973). Bean anthracnose race survey in South Brazil, *A. Rep. Bean Improv. Co-op.*, **16**, 42–43.

Augustin, E., Coyne, D. P., and Schuster, M. L. (1972). Inheritance of resistance in *Phaseolus vulgaris* to *Uromyces phaseoli typica* Brazilian rust race B11 and of plant habit, *J. Am. Soc. hort. Sci.*, **97**, 526–529.

Aujla, S. S. (1964). Study on eleven isolates of *Phyllosticta rabiei* (Pass.) Trot., the causal agent of gram blight in the Punjab, *Indian Phytopath.*, **17**, 83–87.

Aujla, S. S. and Bedi, P. S. (1967). Relative reaction of different varieties of gram to blight disease incited by *Phyllosticta rabiei* (Pass.) Trot. in the Punjab, *J. Res., Ludhiana*, **4**, 214–216.

Aulakh, K. S. (1969). *Alternaria* blight of groundnuts, *Pl. Dis. Reptr.*, **53**, 397–398.

AVRDC (1978). *Progress rep. for 1977*, Asian Vegetable Research and Development Center, Shanhua, Taiwan, 90 pp.

AVRDC (1979). *Progress rep. for 1978*, Asian Vegetable Research and Development Center, Shanhua, Taiwan, 173 pp.

Ayala, A. (1962). Occurrence of the nematode *Meloidogyne javanica* on pigeon pea roots in Puerto Rico, *J. Agric. Univ. Puerto Rico*, **46**, 154–156.

Ayala, A. and Ramírez, C. T. (1964). Host-range, distribution, and bibliography of the reniform nematode, *Rotylenchulus reniformis*, with special reference to Puerto Rico, *J. Agric. Univ. Puerto Rico*, **48**, 140–161.

Aykroyd, W. R. and Doughty, J. (1964). *Legumes in Human Nutrition*, FAO, Rome, 138 pp.

Ayonoadu, U. W. U. (1974). Races of bean anthracnose in Malawi, *Turrialba,* **24**, 311–314.
Ayyar, V. R. and Iyer, R. B. (1936). A preliminary note on the mode of inheritance of reaction to wilt in *Cicer arietinum, Proc. Indian Acad. Sci. B.,* **3**, 438–443.
Aziz, M. A. (1962). C 727—a new blight-resistant gram variety for barani areas, *W. Pak. J. agric. Res.,* **1**, 165–166.
Backman, P. A. and Rodriguez-Kabana, R. (1975). A system for the growth and delivery of biological control agents to the soil, *Phytopathology,* **65**, 819–821.
Baggett, J. R. and Frazier, W. A. (1957). The inheritance of resistance to bean yellow mosaic virus in *Phaseolus vulgaris, Proc. Am. Soc. hort. Sci.*, **70**, 325–333.
Baggett, J. R. and Frazier, W. A. (1973). Seasonal pattern of fusarium root rot *Phaseolus coccineus* L., *Pl. Dis. Reptr.,* **43**, 137–143.
Baggett, J. R. and Frazier, W. A. (1973). Seasonal pattern of Fusarium root rot development in resistant vs. susceptible cultivars of *Phaseolus vulgaris* L., *Hortsci.,* **8**, 214–216.
Bahadur, P. and Sinha, S. (1970a). Physiologic specialisation in *Uromyces ciceris-arietini, Indian Phytopath.,* **23**, 626–628.
Bahadur, P. and Sinha, S. (1970b). Studies on spore germination of *Uromyces ciceris-arietini*, influence of leaf exudates of gram, *Indian Phytopath.*, **23**, 644–649.
Bailey, A. G. (1966). *A Checklist of Plant Diseases in Nigeria,* Fed. Dept. agric. Res., Ibadan, Nigeria, Mem. No. 96, 37 pp.
Bailey, J. A. (1973). Production of antifungal compounds in cowpea (*Vigna sinensis*) and pea (*Pisum sativum*) after virus infection, *J. Gen. Microbiol.,* **75**, 119–123.
Bailey, J. A. (1974). The relationship between symptom expression and phytoalexin concentration in hypocotyls of *Phaseolus vulgaris* infected with *Colletotrichum lindemuthianum, Physiol. Pl. Path.*, **4**, 477–488.
Bailey, J. A. and Deverall, B. J. (1971). Formation and activity of phaseollin in the interaction between bean hypocotyls (*Phaseolus vulgaris*) and physiological races of *Colletotrichum lindemuthianum, Physiol. Pl. Path.,* **1**, 435–449.
Bailey, J. A. and Ingham, J. L. (1971). Phaseollin accumulation in bean (*Phaseolus vulgaris*) in response to infection by tobacco necrosis virus and the rust *Uromyces appendiculatus, Physiol. Pl. Path.,* **1**, 451–456.
Bailey, J. E. and Safir, G. R. (1978). Effect of benomyl on soybean endomycorrhizae, *Phytopathology,* **68**, 1810–1812.
Baker, E. F. I. (1978). Mixed cropping in Northern Nigeria 1. Cereals and groundnuts, *Expl. Agric.,* **14**, 293–298.
Baker, E. F. I. (1980). Mixed cropping in Northern Nigeria IV. Extended trials with cereals and groundnuts, *Expl. Agric.,* **16**, 365–369.
Baker, K. F. and Cook, R. J. (1974). *Biological Control of Plant Pathogens,* W. H. Freeman, San Francisco, 433 pp.
Baker, R. E. D. and Dale, W. T. (1951). Fungi of Trinidad and Tobago. *Mycol. Pap.* No. 33, Commonwealth Mycological Institute, Kew, 123 pp.
Balasubramanian, R. (1979). A new type of alternariosis in *Arachis hypogaea* L., *Curr. Sci.,* **48**, 76–77.
Baldev, B. and Amin, K. S. (1974). Studies on the existence of races in *Fusarium udum* causing wilt of *Cajanus cajan, SABRAO J.,* **6**, 201–205.
Ballantyne, B. (1970). Field reactions of bean varieties to summer death in 1970, *Pl. Dis. Reptr.,* **54**, 903–905.
Ballantyne, B. (1971a). Diseases of dry beans, *Agric. Gaz. N.S.W.,* **82**, 270–279.
Ballantyne, B. (1971b). Summer death of beans, *Agric. Gaz. N.S.W.,* **82**, 295–297.
Ballantyne, B. (1974a). Resistance to rust in beans, *A. Rep. Bean Improv. Co-op.,* **17**, 19–20.
Ballantyne, B. (1974b). Resistance to rust (*Uromyces appendiculatus*) in beans (*Phaseolus vulgaris*), *Proc. Linn. Soc. N.S.W.,* **98**, 107–121.

Ballantyne, B. (1974c). Development of a set of international differential varieties and a standard nomenclature of races, *Proc. bean rust workshop*, CIAT, Cali, Colombia, 1974.

Ballantyne, B. and McIntosh, R. A. (1976). Resistance to rust, caused by *Uromyces appendiculatus* in bean, *A. Rep. Bean Improv. Co-op.*, **19**, 15–16.

Ballantyne, B. and McIntosh, R. A. (1977). The genetic basis of rust resistance in bean, 3rd Int. Congr. SABRAO, Canberra, *Plant breeding papers 1.*, **4**(a), 13–16.

Bancroft, J. B. (1971). Cowpea chlorotic mottle virus, *CMI/AAB Descriptions of Plant Viruses No. 49.*

Baniecki, J. F. and Bloss, H. E. (1969). The basidial stage of *Phymatotrichum omnivorum, Mycologia,* **61**, 1054–1059.

Bannerot, H. and Richter, R. (1968). Étude de la descendance de deux croisements de haricot pour leur résistance a différentes races physiologiques d'anthracnose, *Annls Amél. Pl.,* **18**, 171–179.

Barber, H. N. (1970). Hybridization and evolution of plants, *Taxon,* **19**, 154–160.

Barker, K. P., Lehman, P. S., and Huisingh, D. (1971). Influence of nitrogen and *Rhizobium japonicum* on the activity of *Heterodera glycines, Nematologica,* **17**, 377–385.

Barnes, J. M. (1970) Aflatoxin as a health hazard, *J. appl. Bact.*, **33**, 285–298.

Barrett, J. A. (1980). Pathogen evolution in multilines and variety mixtures, *Pflkrankh.,* **87**, 383–396.

Barrett, J. A. and Wolfe, M. S. (1978). Multilines and super-races—a reply, *Phytopathology,* **68**, 1535–1537.

Barros, O., Cardeñosa, R., and Skiles, R. L. (1957). The severity and control of angular leaf spot of beans in Colombia, *Phytopathology,* **47**, 3 (abstr.).

Barros, O., Cardona, C., Cardeñosa, R., and Skiles, R. L. (1958a). Angular leaf spot of bean in Colombia, *Pl. Dis. Reptr.,* **42**, 420–424.

Barros, O., Cardona, C., Cardeñosa, R., and Skiles, R. L. (1958b). The control of angular leaf spot of bean in Colombia. *FAO Pl. Prot. Bull.,* **6**, 97–101.

Barrus, M. F. (1911). Variation of varieties of beans in their susceptibility to anthracnose, *Phytopathology,* **1**, 190–195.

Barrus, M. F. (1918). Varietal susceptibility of beans to strains of *Colletotrichum lindemuthianum* (Sacc. and Magn.) B. and C, *Phytopathology,* **8**, 589–614.

Bateman, D. F. and Lumsden, R. D. (1965). Relation of calcium content and nature of the pectic substances in bean hypocotyls of different ages to susceptibility to an isolate of *Rhizoctonia solani, Phytopathology,* **55**, 734–738.

Bates, G. R. (1955). News from Southern Rhodesia, *Commonw. Phytopath. News,* **1**, 43–44.

Bates, G. R. (1957). *Rep. Min. Agric. Rhod. Nyasald 1955–6,* pp. 79–86. (*Rev. appl. Mycol.*, **37**, 134–135.)

Bates, G. R. (1961). *Rep. Min. Agric. Rhod. Nyasald 1959–1960*, pp. 50–57. (*Rev. appl. Mycol.*, **41**, 126.)

Bates, J. J. and Tweedy, B. G. (1971). The effect of oxycarboxin on the bean rust disease caused by *Uromyces phaseoli, Phytopathology,* **61**, 884 (abstr.).

Batista, A. C., Bezerra, J. C., de Ponte, J. J., and Vasconcelos, I. (1966). *Entyloma vignae* n.sp. Excepcional e curioso fungo de 'carvâo' assinalado sobre leguminose no Brasil, *Atas Inst. Micol.,* **3**, 146–149.

Batista, A. C. and Vital, A. F. (1952). Monografia das espécies de *Phyllosticta* em Pernambuco, *Bol. Sec. Agric. Pernambuco,* **19**, 1–80. (*Rev. appl. Mycol.,* **33**, 117–118.)

Bawden, F. C. and Van der Want, J. P. H. (1949). Bean stipple-streak caused by a tobacco necrosis virus, *Tijdschr. Pl Ziekt.,* **55**, 142–150.

Bedi, K. S. and Athwal, D. S. (1962). C235 is the answer to blight, *Indian Farming,* **12**(9), 20–22.

Bedi, P. S. and Aujla, S. S. (1969). Variability in *Phyllosticta rabiei* (Pass.) Trot; the incitant of blight disease of gram in the Punjab, *J. Res., Ludhiana*, **6**, 103–106.

Beebe, S. E., Bliss, F. A., and Schwartz, H. F. (1981). Root rot resistance in common bean germplasm of Latin American origin, *Pl. Dis.*, **65**, 485–489.

Behncken, G. M. and Maleevsky, L. (1977). Detection of cowpea aphid-borne mosaic virus in Queensland, *Aust. J. exp. Agric. Anim. Husb.*, **17**, 674–678.

Beier, H., Bruening, G., Russell, M. L., and Tucker, C. L. (1979). Replication of cowpea mosaic virus in protoplasts isolated from immune lines of cowpea, *Virology*, **95**, 165–175.

Beier, H., Siler, D. J., Russell, M. L., and Bruening, G. (1977). Survey of susceptibility to cowpea mosaic virus among protoplasts and intact plants from *Vigna sinensis* lines, *Phytopathology*, **67**, 917–921.

Bell, D. K. (1967). Effects of soil temperature and plant age on the severity of *Cylindrocladium* rot of peanut, *Pl. Dis. Reptr.*, **51**, 986–988.

Bell, D. K. and Sobers, E. K. (1966). A peg, pod and root necrosis of peanuts caused by a species of *Calonectria, Phytopathology*, **56**, 1361–1364.

Benedict, W. C. (1964). Studies on the effect of *Pseudomonas glycinea* on *Septoria glycines* development on foliage of the Harosoy soybean grown under controlled environmental conditions, *Can. J. Bot.*, **42**, 1135–1142.

Beniwal, S. P. S. and Srivastava, S. K. (1968). Downy mildew of lentil (*Lens esculenta*), *Pl. Dis. Reptr.*, **52**, 817.

Benlloch, M. (1949). Observaciones fitopatológicas en el año 1948. *Bol. Pat. veg. Ent. agric., Madr.*, **16**, 203–242. (*Rev. appl. Mycol.*, **29**, 201.)

Bennett. C. W. (1952). Origin and distribution of new or little-known virus diseases, *Pl. Dis. Reptr. Suppl.*, **221**, 43–46.

Bennett, C. W. (1967). Epidemiology of leafhopper-transmitted viruses, *A. Rev. Phytopath.*, **5**, 87–108.

Berchoux, C. de (1958). Etude sur la résistance de l'arachide à la rosette en Haute-Volta, *Oléagineux*, **13**, 237–239.

Berchoux, C. de (1960). La rosette de l'arachide en Haute-Volta, *Oléagineux*, **15**, 229–233.

Bergeson, G. B., Athow, K. L., Laviolette, F. A., and Thomasine, M. (1964). Transmission, movement, and vector relationships of tobacco ringspot virus in soybean, *Phytopathology*, **54**, 723–728.

Bernard, R. L. and Cremeens, C. R. (1971). A gene for general resistance to downy mildew of soybeans, *J. Hered.*, **62**, 359–362.

Beute, M. K. (1980). *Cylindrocladium* black rot (CBR) disease of peanut (*Arachis hypogaea*), in *Proc. Intern. Wkshop groundnuts* (ed. R. W. Gibbons), ICRISAT, Hyderabad pp. 171–176.

Beute, M. K. and Lockwood, J. L. (1968). Mechanism of increased root rot in virus-infected peas, *Phytopathology*, **58**, 1643–1651.

Bhelwa, P. W. (1962). Seed decay, seedling blight and root rot of *Cicer arietinum* caused by *Phytophthora cryptogaea, Diss. Abstr.*, **23**, 389–390. (*Rev. appl. Mycol.*, **42**, 287.)

Bingefors, S. (1967). International dispersal of nematodes, *Neth. J. Pl. Path.*, **73**, (Suppl. 1), 44–60.

Bingham, J. (1981). The achievements of conventional plant breeding, *Phil. Trans. R. Soc. Lond. B*, **292**, 441–455.

Birat, R. B. S. (1969). Effects of monoculture on the population of *Meloidogyne javanica, Nematologica*, **15**, 154–155.

Birch, R. G., Alvarez, A. M., and Patil, S. S. (1981). A bacterial leaf spot caused in yam bean by *Pseudomonas syringae* pv. *phaseolicola, Phytopathology*, **71**, 1289–1293.

Bird, J., Sánchez, J., Rodríguez, R. L., and Juliá, F. J. (1975). Rugaceous (whitefly-transmitted) viruses in Puerto Rico, in *Tropical Diseases of Grain Legumes* (eds J. Bird and K. Maramorosch), Academic Press, New York, pp. 3–25.

Bisht, N. S. and Banerjee, A. K. (1965). Occurrence of two new virus diseases in Uttar Pradesh, *Labdev J. Sci. Technol.,* **3**, 271–272. (*Rev. appl. Mycol.,* **47**, 81.)

Bissonnette, H. L. and Johnson, H. G. (1972). Pinto bean rust control with fungicides, *Univ. Minn. agric. ext. serv., Pl. Path.,* **20**.

Bitancourt, A. A. and Jenkins, A. E. (1940). Novas especies de 'Elsinoe' and 'Sphaceloma' sobre hospedes de importancia economica, *Arq. Inst. biol., S. Paulo,* **11**, 45–58. (*Rev. appl. Mycol.,* **20**, 427.)

Blamey, F. P. C., Chapman, J., and Young, B. W. (1977). Epiphytology of *Phoma* web blotch and *Cercospora* leafspot in Spanish groundnuts, *Phytophylactica,* **9**, 63–64.

Bliss, F. A. (1977). Recurrent selection in beans, *A. Rep. Bean Improv. Co-op.,* **20**, 50–52.

Bliss, F. A. and Robertson, D. G. (1971). Genetics of host reaction in cowpea to cowpea yellow mosaic virus and cowpea mottle virus, *Crop Sci.,* **11**, 258–262.

Boccas, B. R. (1981). Interspecific crosses between closely related heterothallic *Phytophthora* species, *Phytopathology,* **71**, 60–65.

Bock, K. R. (1971). Notes on East African plant virus diseases. 1. Cowpea mosaic virus, *E. Afr. agric. For. J.*, **37**, 60–62.

Bock, K. R. (1973a). East African strains of cowpea aphid-borne mosaic virus, *Ann. appl. Biol.*, **74**, 75–83.

Bock, K. R. (1973b). Peanut mottle virus in East Africa, *Ann. appl. Biol.*, **74**, 171–179.

Bock, K. R. and Conti, M. (1974). Cowpea aphid-borne mosaic virus. *CMI/AAB Descriptions of Plant Viruses No. 134.*

Bock, K. R., Guthrie, E. J., Figueiredo, G., and Woods, R. D. (1981). Purification and some properties of Tephrosia symptomless virus, *Ann. appl. Biol.*, **97**, 277–283.

Bock, K. R., Guthrie, E. J., and Meredith, G. (1978). Viruses occurring in East Africa that are related to peanut mottle virus, *Ann. appl. Biol.*, **89**, 423–428.

Bock, K. R., Guthrie, E. J., Meredith, G. C., and Njuguna, J. G. M. (1976). Viruses of other leguminous crop plants. Record of research, *Ann. Rep. E. Afr. agric. For. Res. Org. for 1974*, pp. 125–126.

Bock, K. R., Guthrie, E. J., Meredith, G., and Njuguna, J. G. M. (1977). Groundnut viruses. Record of research, *Ann. Rep. E. Afr. agric. For. Res. Org. for 1975*, pp. 117–118.

Bock, K. R., Guthrie, E. J., and Pearson, M. N. (1975). Notes on East African plant virus diseases. 9. Cucumber mosaic virus, *E. Afr. agric. For. J.*, **41**, 81–84.

Bock, K. R. and Kuhn, C. W. (1975). Peanut mottle virus. *CMI/AAB Descriptions of Plant Viruses No. 141.*

Bock, K. R. and Ngugi, E. (1972). Groundnut viruses. Record of research, *Ann. Rep. E. Afr. agric. For. Res. Org. for 1971*, pp. 169–170.

Bock, K. R., Ngugi, E., Ambetsa, T., and Mwathi, G. K. (1970). Groundnut viruses. Record of research, *Ann. Rep. E. Afr. agric. For. Res. Org. for 1969*, p. 85.

Bock, K. R., Ngugi, E., Ambetsa, T., and Mwathi, G. K. (1971). Cowpea viruses. Record of research, *Ann. Rep. E. Afr. agric. For. Res. Org. for 1970*, pp. 72–73.

Boerema, G. H. (1972). *Ascochyta phaseolorum* synonymous with *Phoma exigua, Neth. J. Pl. Path.*, **78**, 113–115.

Boerema, G. H., Dorenbosch, M. M. J., and Leffring, L. (1965). A comparative study of the black stem fungi on lucerne and red clover and the footrot fungus on pea, *Neth. J. Pl. Path.*, **71**, 79–89.

Boerma, H. R., Kuhn, C. W., and Harris, H. B. (1975). Inheritance of resistance to cowpea chlorotic mottle virus (soybean strain) in soybeans, *Crop Sci.*, **15**, 849–850.

Boettinger, C. F. and Bowers, J. L. (1975). Hot water treatment a possible control for seed-borne bacterial blight in southern peas, *Hortsci.*, **10**, 148.

Boewe, G. H. (1948). Bacterial canker of cowpea in Illinois, *Pl. Dis. Reptr.*, **32**, 275.

Bolkan, H. A. (1980). Roots rots, in *Bean Production Problems* (eds H. F. Schwartz and G. E. Gálvez), Centro Internacional de Agricultura Tropical, Cali, Colombia, pp. 65–100.

Bonde, M. R., Melching, J. S., and Bromfield, K. R. (1976). Histology of the suscept-pathogen relationship between *Glycine max* and *Phakopsora pachyrhizi*, the cause of soybean rust, *Phytopathology*, **66**, 1290–1294.

Booker, R. H. (1963). The effect of sowing date and spacing on rosette disease of groundnut in Northern Nigeria, with observations on the vector *Aphis craccivora, Ann. appl. Biol.*, **52**, 125–131.

Boomstra, A. G. and Bliss, F. A. (1977). Inheritance of resistance to *Fusarium solani* f.sp. *phaseoli* in beans (*Phaseolus vulgaris* L.) and breeding strategy to transfer resistance, *J. Am. Soc. hort. Sci.*, **102**, 186–188.

Boomstra, A. G., Bliss, F. A., and Beebe, S. E. (1977). New sources of *Fusarium* root rot resistance in *Phaseolus vulgaris* L., *J. Am. Soc. hort. Sci.*, **102**, 182–185.

Boorsma, P. A. (1980). Variability in chickpea for blight resistance, *FAO Pl. Prot. Bull.*, **28**, 110–113.

Booth, C. (1971). *The genus Fusarium.* Commonwealth Mycological Institute, Kew, 237 pp.

Booth, C. (1975). The present status of *Fusarium* taxonomy, *A. Rev. Phytopath.*, **13**, 83–93.

Booth, C. (1978). *Fusarium udum. CMI Descriptions of Pathogenic Fungi and Bacteria No. 575.*

Booth, C. and Pirozynski, K. A. (1967). *Leptosphaerulina trifolii. CMI Descriptions of Pathogenic Fungi and Bacteria No. 146.*

Boquet, D., Williams, C., and Birchfield, W. (1976). Inheritance of resistance to the Wartelle race of root-knot nematode in soybeans, *Crop Sci.*, **16**, 783–785.

Borlaug, N. E. (1973). Building a protein revolution on grain legumes, in *Nutritional Improvement of Food Legumes by Breeding* (ed. M. Milner), John Wiley, New York, pp. 7–11.

Bos, L. (1964). Tentative list of viruses reported from naturally infected leguminous plants, *Neth. J. Pl. Path.,* **70**, 161–174.

Bos, L. (1970). Bean yellow mosaic virus. *CMI/AAB Descriptions of Plant Viruses No. 40.*

Bos, L. (1971). Bean common mosaic virus. *CMI/AAB Descriptions of Plant Viruses No. 73.*

Bos, L. (1972). Soybean mosaic virus. *CMI/AAB Descriptions of Plant Viruses No. 93.*

Bos, L. (1976). Applied plant virus research: an analysis, and a scheme of organisation, *Neth. J. Pl. Path.,* **82**, 24–38.

Bos, L. (1977). Seed-borne viruses, in *Plant Health and Quarantine in International Transfer of Genetic Resources* (eds W. B. Hewitt and L. Chiarappa), CRC Press, Cleveland, Ohio pp. 39–69.

Bos. L., Hagedorn, D. J., and Quantz, L. (1960). Suggested procedures for international identification of legume viruses, *Tijdschrift over Plantenzickten,* **66**, 328–343.

Bos, L. and Maat, D. Z. (1974). A strain of cucumber mosaic virus, seed transmitted in beans, *Neth. J. Pl. Path.,* **80**, 113–123.

Bose, R. D. (1938). The rotation of tobacco for the prevention of wilt disease in pigeon-peas (*Cajanus cajan* (Linn.) Millsp.), *Agric. Livest. India*, **8**, 653–668.

Boswell, T. E. (1968). Some effects of *Pratylenchus brachyurus* upon Spanish peanuts, *Nematologica,* **14**, 3–4 (abstr.).

Bowers, G. R. and Goodman, R. M. (1979). Soybean mosaic virus: infection of soybean seed parts and seed transmission, *Phytopathology,* **69**, 569–572.

Bowers, G. R. and Goodman, R. M. (1982). New sources of resistance to seed transmission of soybean mosaic virus in soybeans, *Crop Sci.,* **22**, 155–156.

Bowyer, J. W. (1974). Tomato big bud, legume little leaf and lucerne witches' broom: three diseases associated with different mycoplasma-like organisms in Australia, *Aust. J. agric. Res.,* **25**, 449–457.

Bowyer, J. W. and Atherton, J. G. (1970). Observations on the relationship between mycoplasma-like bodies and host cells of legume little-leaf diseased plants, *Aust. J. biol. Sci.,* **23**, 115–125.

Bowyer, J. W., Dale, J. L., and Behncken, G. M. (1980). Glycine mosaic virus: a comovirus from Australian native *Glycine* species, *Ann. appl. Biol.,* **95**, 385–390.

Boyle, L. W. (1953). Heat canker: a primary phase of collar rot on peanuts, *Phytopathology,* **43**, 571–576.

Boyle, L. W (1956). Fundamental concepts in the development of control measures for southern blight and root rot of peanuts, *Pl. Dis. Reptr.,* **40**, 661–665.

Bozarth, R. F. and Shoyinka, S. A. (1979). Cowpea mottle virus. *CMI/AAB Descriptions of Plant Viruses No. 212.*

Braak, J. P. and Kooistra, E. (1975). A successful cross between *Phaseolus vulgaris* L. and *Phaseolus ritensis* Jones with the aid of embryo culture, *Euphytica,* **24**, 669–679.

Braithwaite, C. W. D. (1981). Diseases of pigeon pea in the Caribbean area, in *Proc. Internat. Wkshop on Pigeon Peas,* **1**, 129–135, International Crops Research Institute for the Semi-Arid Tropics, Hyderabad, India.

Brantley, B. B. and Kuhn, C. W. (1970). Inheritance of resistance to southern bean mosaic virus in southern pea, *Vigna sinensis, J. Am. Soc. hort. Sci.,* **95**, 155–158.

Brantley, B. B., Kuhn, C. W., and Sowell, G. (1965). Effect of cucumber mosaic virus on southern pea (*Vigna sinensis*), *Proc. Am. Soc. hort. Sci.*, **87**, 355–358.

Brasier, C. M. (1971). Induction of sexual reproduction in single A_2 isolates of *Phytophthora* by *Trichoderma viride, Nature New Biology,* **231**, 283.

Brasier, C. M. (1975). Stimulation of sex organ formation in *Phytophthora* by antagonistic species of *Trichoderma.* II. Ecological implications, *New Phytol.,* **74**, 195–198.

Bravo, A., Wallace, D. H. and Wilkinson, R. E. (1969). Inheritance of resistance to *Fusarium* root rot of beans, *Phytopathology,* **59**, 1930–1933.

Bredemann, G. and Koolman, H. (1927). Zur immunitätszüchtung bei *Phaseolus vulgaris* gegenüber *Colletotrichum lindemuthianum* und seinen biotypen, *Zschr. Pfl. Züchtg.,* **12**, 209–217.

Bridge, J. (1973). *Hoplolaimus seinhorsti,* an endoparasitic nematode of cowpea in Nigeria, *Pl. Dis. Reptr.,* **57**, 798–799.

Bridge, J. (1981). Nematodes, in *Pest Control in Tropical Grain Legumes,* Centre for Overseas Pest Research, ODA, London, pp. 111–125.

Bridge, J., Bos, W. S., Page, L. J., and McDonald, D. (1977). The biology and possible importance of *Aphelenchoides arachidis,* a seed-borne endoparasitic nematode of groundnuts from northern Nigeria, *Nematologica*, **23**, 253–259.

Brierley, P. and Smith, F. F. (1962). Three cowpea mosaic viruses from gladiolus, *Pl. Dis. Reptr.,* **46**, 335–337.

Brittingham, W. H. (1946). A key to the horticultural groups of varieties of the southern pea, *Vigna sinensis, Proc. Am. Soc. hort. Sci.,* **48**, 478–480.

Brock, R. D. (1951). Resistance to angular leaf spot among varieties of beans, *Aust. Inst. agric. Sci. J.,* **17**, 25–30.

Bromfield, K. R. (1974). Current distribution of rust of groundnut and known sources of resistance, *FAO Pl. Prot. Bull.,* **22**, 29–31.

Bromfield, K. R. (1976). World soybean rust situation, in *World Soybean Res.,* pp. 491–500.

Bromfield, K. R. (1977). Soybean rust and the pathogen, some needed research, in *Rust of Soybean: the problem and research needs* (eds R. E. Ford and J. B. Sinclair), INTSOY, Illinois, pp. 34–39.

Bromfield, K. R. (1978). Review of research on soybean rust, in *Proc. Workshop Soybean Rust in W. Hemisphere* (ed. N. G. Vakili), USDA, Mayaguez, pp. 16–23.
Bromfield, K. R. and Bailey, W. K. (1972). Inheritance of resistance to *Puccinia arachidis* in peanut, *Phytopathology*, **62**, 748 (abstr.).
Bromfield, K. R. and Cevario, J. (1970). Greenhouse screening of peanut (*Arachis hypogaea*) for resistance to peanut rust (*Puccinia arachidis*), *Pl. Dis. Reptr.*, **54**, 381–383.
Bromfield, K. R. and Hartwig, E. E. (1980). Resistance to soybean rust and mode of inheritance, *Crop Sci.*, **20**, 254–255.
Bromfield, K. R., Melching, J. S., and Kingsolver, C. H. (1980). Virulence and aggressiveness of *Phakopsora pachyrhizi* isolates causing soybean rust, *Phytopathology*, **70**, 17–21.
Brown, J. F. and Sharp, E. L. (1970). The relative survival ability of pathogenic types of *Puccinia striiformis* in mixtures, *Phytopathology*, **60**, 529–533.
Browning, J. A. (1974). Relevance of knowledge about natural ecosystems to development of pest management programs for agro-ecosystems, *Proc. Am. Phytopath. Soc.*, **1**, 191–199.
Browning, J. A. and Bustamante, E. (1973). Evidence for environmental races of *Puccinia graminis avenae*, *Proc. 2nd Int. Congr. Pl. Path.* Minneapolis, abstr. 0717.
Browning, J. A. and Frey, K. J. (1969). Multiline cultivars as a means of disease control, *A. Rev. Phytopath.*, **7**, 355–382.
Bruner, S. C. and Jenkins, A. E. (1933). Identity and host relations of the *Elsinoe* of lima beans, *J. agric. Res.*, **47**, 783–789.
Brunt, A. A. (1975). Tropical leguminous crops, *Rep. Glasshouse Crops Res. Inst. for 1974*, 120.
Brunt, A. A. and Kenten, R. H. (1973). Cowpea mild mottle, a newly recognized virus infecting cowpeas (*Vigna unguiculata*) in Ghana, *Ann. appl. Biol.*, **74**, 67–74.
Brunt, A. A. and Kenten, R. H. (1974). Cowpea mild mottle virus, *CMI/AAB Descriptions of Plant Viruses No. 140.*
Brunt, A. A. and Phillips, S. (1978). Cowpea mild mottle virus. *Rep. Glasshouse Crops Res. Inst. for 1977*, 133–134.
Brunt, A. A. and Phillips, S. (1981a). 'Fuzzy vein', a disease of tomato (*Lycopersicon esculentum*) in Western Nigeria induced by cowpea mild mottle virus, *Trop. Agric. (Trin.)*, **58**, 177–180.
Brunt, A. A. and Phillips, S. (1981b). Plant virus diseases in Fiji. *Rep. Glasshouse Crops Res. Inst. for 1979*, 157–158.
Buddenhagen, I. W. (1965). The relation of plant-pathogenic bacteria to the soil, in *Ecology of Soil-borne Plant Pathogens, Prelude to Biological Control* (eds K. F. Baker and W. C. Snyder), John Murray, London, pp. 269–282.
Buddenhagen, I. W. (1977). Resistance and vulnerability of tropical crops in relation to their evolution and breeding, *Ann. N. Y. Acad. Sci.*, **287**, 309–326.
Buddenhagen, I. W. (1978). Varietal improvement in relation to stability and environmental balance in tropical peasant agriculture, *3rd Int. Congr. Pl. Path.*, München, 16–23 Aug. 1978, 298 (abstr.).
Bunting, A. H. (1950). Technical aspects of the East African groundnut organisation, *Ann. appl. Biol.*, **37**, 699–704.
Bunting, A. H. (1961). Some problems of agricultural climatology in tropical Africa, *Geography*, **46**, 283–294.
Bunting, A. H. (1975). Time. phenology and the yields of crops, *Weather*, **30**, 312–325.
Bunting, A. H. (1981). People, food, agriculture and biology: a letter to an Academy, *Biologist*, **28**, 161–165.
Bunting, A. H. and Elston, J. (1980). Ecophysiology of growth and adaptation in the

groundnut: an essay on structure, partition and adaptation, in *Advances in Legume Science* (eds R. J. Summerfield and A. H. Bunting), HMSO, London, pp. 495–500.
Burden, R. S. and Bailey, J. A. (1975). Structure of the phytoalexin from soybean, *Phytochemistry,* **14**, 1389–1390.
Burdon, J. J. (1978). Mechanisms of disease control in heterogeneous plant populations—an ecologist's view, in *Plant Disease Epidemiology* (eds P. R. Scott and A. Bainbridge), Blackwell Sci. Publ., Oxford, pp. 193–200.
Burdon, J. J. and Marshall, D. R. (1981). Evaluation of Australian native species of *Glycine* for resistance to soybean rust, *Pl. Dis.* **65**, 44–45.
Burdon, J. J. and Shattock, R. C. (1980). Disease in plant communities, *Appl. Biol.,* **5**, 145–219.
Burke, D. W. (1957). Incidence of bacterial pathogens in dry beans in irrigated districts of Nebraska, Wyoming and Colorado in 1954 and 1955, *Pl. Dis. Reptr.*, **41**, 488–490.
Burke, D. W. (1964). Time of planting in relation to disease incidence and yields of beans in central Washington, *Pl. Dis. Reptr.,* **48**, 789–793.
Burke, D. W. (1965a). Plant spacing and *Fusarium* root rot of beans, *Phytopathology*, **55**, 757–759.
Burke, D. W. (1965b). The near immobility of *Fusarium solani* f. *phaseoli* in natural soils, *Phytopathology,* **55**, 1188–1190.
Burke, D. W. (1965c). *Fusarium* root rot of beans and behavior of the pathogen in different soils, *Phytopathology,* **55**, 1122–1126.
Burke, D. W. (1968). Root growth obstructions and *Fusarium* root rot of beans, *Phytopathology,* **58**, 1575–1576.
Burke, D. W. and Barker, A. W. (1966). Importance of lateral roots in *Fusarium* root rot of beans, *Phytopathology,* **56**, 292–294.
Burke, D. W. and Silbernagel, M. J. (1974a). Hypersensitive resistance to bean common mosaic virus in dry field beans, *A. Rep. Bean Improv. Co-op.,* **17**, 22.
Burke, D. W. and Silbernagel, M. J. (1974b). Yield response among bean selections inoculated with two prevalent strains of bean common mosaic virus, *A. Rep. Bean Improv. Co-op.,* **17**, 23.
Burke, D. W. and Silbernagel, M. J. (1980). New sources of multiple resistance to root rot pathogens, *A. Rep. Bean Improv. Co-op.*, **23**, 69.
Burkholder, W. H. (1918). The production of an anthracnose-resistance white marrow bean, *Phytopathology,* **8**, 353–359.
Burkholder, W. H. (1923). The gamma strain of *Colletotrichum lindemuthianum* (Sacc. and Magn.) B. and C., *Phytopathology*, **13**, 316–323.
Burkholder, W. H. (1944). *Xanthomonas vignicola* sp. nov. pathogenic on cowpeas and beans, *Phytopathology,* **34**, 430–432.
Burkholder, W. H. and Bullard, E. T. (1946). Varietal susceptibility of beans to *Xanthomonas phaseoli* var. *fuscans, Pl. Dis. Reptr.,* **30**, 446–448.
Burkholder, W. H. and Starr, M. P. (1948). The generic and specific characters of phytopathogenic species of *Pseudomonas* and *Xanthomonas, Phytopathology,* **38**, 494–502.
Bushkova, L. N. (1960). Susceptibility of chickpea varieties to ascochytosis, *Zashch. Rast., Mosk.,* **5**, 55. (*Rev. appl. Mycol.* **39**, 717.)
Bustillo, B. A. (1972). Influence of *Rotylenchulus reniformis* infection on replication of cowpea chlorotic mottle virus in 'Davis' soybean, *J. Nemat.*, **4**, 220–221 (abstr.).
Butcher, C. L., Dean, L. L., and Guthrie, J. W. (1969). Effectiveness of halo blight control in Idaho bean seed crops, *Pl. Dis. Reptr.,* **53**, 894–896.
Butler, E. E. (1957). *Rhizoctonia solani* as a parasite of fungi, *Mycologia,* **49**, 354–373.
Butler, E. J. (1906). The wilt disease of pigeon pea and pepper, *Agric. J., India,* **1**, 25–36.

Butler, E. J. (1908). Selection of pigeon pea for wilt disease, *Agric. J. India,* **3**, 182–183.
Butler, E. J. (1910). The wilt disease of pigeon pea and the parasitism of *Neocosmospora vasinfecta* Smith, *Mem. Dept. Agric. India*, Bot. Ser. 2, 1–64.
Butler, E. J. and Jones, S. G. (1949). *Plant Pathology,* Macmillan, London, 979 pp.
Butler, G. M. (1958). *Uromyces* in a population of *Euphorbia cyparissias, Trans. Br. mycol. Soc.,* **41**, 401–404.
Buxton, E. W. (1955). *Fusarium* diseases of peas, *Trans. Br. mycol. Soc.,* **38**, 309–316.
Buxton, E. W. (1956). Heterokaryosis and parasexual recombination in pathogenic strains of *Fusarium oxysporum, J. Gen. Microbiol.,* **15**, 133–139.
Byth, D. E., Green, J. M., and Hawtin, G. C. (1980). ICRISAT/ICARDA chickpea breeding strategy, in *Proc. Internat. Wkshop on Chickpea Improv.,* International Crops Research Institute for the Semi-Arid Tropics, Hyderabad, India, pp. 11–27.
Byth, D. E., Wallis, E. S., and Saxena, K. B. (1981). Adaptation and breeding strategies for pigeon pea, in *Proc. Internat. Wkshop on Pigeonpeas, 1,* 460–465. International Crops Research Institute for the Semi-Arid Tropics, Hyderabad, India.
Cafati, C. R. and Saettler, A. W. (1980). Transmission of *Xanthomonas phaseoli* in seed of resistant and susceptible *Phaseolus* genotypes, *Phytopathology*, **70**, 638–640.
Callow, J. A. (1977). Recognition, resistance and the role of plant lectins in host-parasite interactions, *Adv. bot. Res.,* **4**, 1–49.
Campbell, L. (1949). Gray mold of beans in Western Washington, *Pl. Dis. Reptr.,* **33**, 91–93.
Campbell, W. W. and Brett, C. H. (1966). Varietal resistance of beans to the Mexican bean beetle, *J. econ. Ent.,* **59**, 899–902.
Capoor, S. P. (1952). Observation on the sterility disease of pigeon pea, *Indian J. agric. Sci.,* **22**, 271–274.
Cardenas, F., Adams, M. W., and Andersen, A.L. (1964). The genetic system for reaction of field beans (*Phaseolus vulgaris* L.) to infection by three physiologic races of *Colletotrichum lindemuthianum, Euphytica,* **13**, 178–186.
Cardona-Alvarez, C. and Skiles, R. L. (1958). Floury leaf spot (Mancha Harinosa) of bean in Colombia, *Pl. Dis. Reptr.,* **42**, 778–780.
Cardona-Alvarez, C. and Walker, J. C. (1956). Angular leaf spot of bean, *Phytopathology,* **46**, 610–615.
Carter, J. B. H. (1973). The influence of the testa, damage and seed dressing on the emergence of groundnut (*Arachis hypogaea* L.), *Ann. appl. Biol.,* **74**, 315–323.
Caspary, W. and Nienhaus, F. (1976). Inhibierender Einfluss von *Ditylenchus dipsaci* auf die Tabaknekrosevirus-Infektion in *Phaseolus vulgaris, Zeitschrift für Pflanzenkrankheiten und Pflanzenschutz,* **83**, 669–673. (*Rev. Pl. Path.,* **56**, 3314.)
Cass Smith, W. P., Harvey, H. L., and Goss, O. M. (1951). 'Westralia': a new rust-resistant pole bean, *W. Aust. Dep. Agric. J.,* **28**, 317–324.
Castano, B. and Butler, E. E. (1978). Rhizoctonia decline: a degenerative disease of *Rhizoctonia solani, Phytopathology,* **68**, 1505–1510.
Castillo, M. B. and Litsinger, J. A. (1978). Plant parasitic nematodes of mung bean in Philippines, in *Proc. 1st Internat. Mungbean Symp.,* Asian Vegetable Research and Development Centre, Taiwan, pp. 195–200.
Castillo, M. B., Morrison, L. S., Russell, C. C., and Banks, D. J. (1973a). Resistance to *Meloidogyne hapla* in peanut, *J. Nemat.,* **5**, 281–285.
Castillo, M. B., Russell, C. C., and Morrison, L. S. (1973b). Development of *Meloidogyne hapla* in peanut, *Phytopathology*, **63**, 583–585.
Caveness, F. E. (1963). Cowpeas and plant pathogenic nematodes in Western Nigeria, in *Proc. 1st Niger. Grain Legume Conf.* (ed. W. M. Steele), Institute for Agric. Res., Samaru, Nigeria, pp. 63–74.

Caveness, F. E. (1974). Plant-parasitic nematode population differences under no-tillage and tillage soil regimes in Western Nigeria, *J. Nemat.*, **6**, 138.

Caveness, F. E. (1975). Screening cowpea germplasm for resistance to root-knot nematodes at IITA, *Nematropica*, **5**, 21.

Caveness, F. E., Gilmer, R. M., and Williams, R. J. (1975). Transmission of cowpea mosaic virus by *Xiphenema basiri* in Western Nigeria, in *Nematode Vectors of Plant Viruses* (eds. F. Lamberti, C. E. Taylor, and J. W. Seinhorst), Plenum Press, London and New York.

Caviness, C. E. and Walters, H. J. (1971). Effect of *Phytophthora* rot on yield and chemical composition of soybean seed, *Crop Sci.,* **11**, 83–84.

Caviness, C. E., Walters, H. J., and Johnson, D. L. (1970). A partially male sterile strain of soybean, *Crop Sci.,* **10**, 107–108.

CGIAR (1976). *Consultative Group on International Agricultural Research,* CGIAR, New York.

Chadha, K. C. and Raychaudhuri, S. P. (1966). Interaction between sterility virus and *Fusarium udum* Butl. in pigeon pea, *Indian J. agric. Sci.,* **36**, 133–139.

Chalal, C. S. and Chohan, J. S. (1971). *Puccinia* rust on groundnut, *FAO Pl. Prot. Bull.,* **19**, 90.

Chamberlain, D. W. (1951). Sclerotinia stem rot of soybeans, *Pl. Dis. Reptr.,* **35**, 490–491.

Chamberlain, D. W. (1956). Methods of inoculation for wildfire of soybean and the effect of bacterial pustule on wildfire development, *Phytopathology,* **46**, 96–98.

Chamberlain, D. W. and Bernard, R. L. (1968). Resistance to brown stem rot in soybeans, *Crop Sci.,* **8**, 728–729.

Chamberlain, D. W. and Gray, L. E. (1974). Germination, seed treatment, and micro-organisms in soybean seed produced in Illinois, *Pl. Dis. Reptr.,* **58**, 50–54.

Chamberlain, D. W. and McAlister, D. F. (1954). Factors affecting the development of brown stem rot of soybean, *Phytopathology,* **44**, 4–6.

Chambers, A. Y. and Andes, J. O. (1962). Stem anthracnose (*Colletotrichum truncatum*) and its control on lima beans, *Bull. Tenn. agric. Exp. Sta*., 338 (*Rev. appl. Mycol.,* **42**, 227.)

Chambers, A. Y. and Epps, J. M. (1966). Nature of resistance in soybean varieties and breeding lines resistant to *Heterodera glycines, Pl. Dis. Reptr.*, **51**, 771–774.

Chan, K. L., Cheng, Y. W., and Chen, S. H. (1971). The breeding of 'Tainung 4' soybean, *J. Taiwan agric. Res.,* **20**, 56–60.

Chant, S. R. (1959). Viruses of cowpea, *Vigna unguiculata* L. (Walp), in Nigeria, *Ann. appl. Biol.,* **47**, 565–572.

Chant, S. R. (1960). The effect of infection with tobacco-mosaic and cowpea yellow-mosaic viruses on the growth rate and yield of cowpea in Nigeria, *Emp. J. exp. Agric.,* **28**, 114–120.

Chant, S. R. (1962). Further studies on the host range and properties of Trinidad cowpea mosaic virus, *Ann. appl. Biol.,* **50**, 159–162.

Chari, M. S., Patel, G. J., Patel, P. N., and Raj, S. (1976). Evaluation of cowpea lines for resistance to aphid, *Aphis craccivora* Koch, *Gujarat agric. Univ. Res. J.,* **1**, 130–132.

Charrier, A. and Bannerot, H. (1970). Contribution à l'étude des races physiologiques de l'anthracnose du haricot, *Ann. Phytopath.,* **2**, 489–506.

Chauhan, R. K. S. and Sinha, S. (1971). Effect of varying temperature, humidity and light during incubation in relation to disease development in blight of gram (*Cicer arietinum*) caused by *Ascochyta rabiei, Proc. Nat. Acad. India,* B, **37**, 473–482.

Chauhan, S. K. (1962a). Physiologic variations in *Fusarium orthoceras* App. and Wr. var. *ciceri* Padwick causing wilt of gram (*Cicer arietinum* L.), *Proc. Nat. Acad. Sci. India* B, **32**, 78–84.

Chauhan, S. K. (1962b). A note on soil reaction in relation to *Fusarium* wilt of gram (*Cicer arietinum* L.), *Proc. Nat. Acad. Sci. India,* B, **32**, 385–386.

Chauhan, S. K. (1963a). Incidence of *Fusarium* wilt of gram (*Cicer arietinum* L.) in relation to soil moisture, *Agra Univ. J. Res.*, **12**, 271–274.

Chauhan, S. K. (1963b). Influence of different soil temperatures on the incidence of *Fusarium* wilt of gram (*Cicer arientinum* L.), *Proc. Nat. Acad. Sci., India*, B, **33**, 552–554.

Chaves, G. (1980). Anthracnose, in *Bean Production Problems* (eds H. F. Schwartz and G. E. Gálvez), Centro Internacional de Agricultura Tropical, Cali, Colombia, pp. 37–54.

Chenulu, W., Sachchidananda, J., and Mehta, S. C. (1968). Studies on a mosaic disease of cowpea from India, *Phytopath. Z.,* **63**, 381–387.

Cheo, C. C. and Jenkins, A. E. (1945). *Elsinoe* and *Sphaceloma* diseases in Yunnan, China, particularly hyacinth bean scab and scab of castor bean, *Phytopathology,* **35**, 339–352.

Cho, E. K., Chung, B. J., and Lee, S. H. (1977). Studies on identification and classification of soybean virus diseases in Korea. II. Etiology of a necrotic disease of *Glycine max., Pl. Dis. Reptr.,* **61**, 313–317.

Cho, E. K. and Goodman, R. M. (1979). Strains of soybean mosaic virus: classification based on virulence in resistant soybean cultivars, *Phytopathology,* **69**, 467–470.

Chohan, J. S. (1965). Collar-rot of groundnut (*Arachis hypogaea* L.) caused by *Aspergillus niger* van Tiegham and *Aspergillus pulverulentus* (McAlpine) Thom, in the Punjab, *J. Res., Ludhiana,* **2**, 25–32.

Chohan, J. S. and Gupta, V. K. (1968). Aflaroot, a new disease of groundnut, caused by *Aspergillus flavus* Link., *Indian J. agric. Sci.,* **38**, 568–570.

Chohan, J. S., Sandhu, R. S., and Dalal, J. L. (1969). Screening of world collection of groundnut (*Arachis hypogaea* L.) germplasm for resistance to collar-rot disease caused by *Aspergillus niger* Van Tiegham in Punjab, *Indian J. agric. Sci.*, **40**, 546.

Chorin, M. (1961). Powdery mildew on leaves of groundnuts, *Bull. Res. Coun. Israel, Sect. D,* **10**, 148–149. (*Rev. appl. Mycol.,* **41**, 193.)

Chorin, M. and Frank, Z. R. (1961). *Diplodia* dry rot of groundnuts, *Israel J. agric. Res.,* **11**, 209.

Christ, B.J. and Groth, J. V. (1979). Gene-for-gene interaction in bean-bean rust system, *Phytopathology,* **69**, 1024 (abstr.).

Christen, R. and Echandi, E. (1967). The most common physiologic races of bean rust *Uromyces phaseoli* var. *phaseoli* in Costa Rica and evaluation of the resistance to rust of some bean plantations, *Turrialba,* **17**, 7–10.

Christensen, C. M. and Dorworth, C. E. (1966). Influence of moisture content, temperature, and time on invasion of soybeans by storage fungi, *Phytopathology,* **56**, 412–418.

Christou, T. (1962). Penetration and host-parasite relationships of *Rhizoctonia solani* in the bean plant, *Phytopathology,* **52**, 381–389.

Chu Chou, M. and Preece, T. F. (1968). The effect of pollen grains on infection caused by *Botrytis cinerea, Ann. appl. Biol.,* **62**, 11–22.

Chupp, C. (1954). *A Monograph of the Fungus genus Cercospora*, Ithaca, New York, 667 pp.

CIAT (1980a). *Ann. Rep. Bean Program for 1979,* Centro Internacional de Agricultura Tropical, Cali, Colombia, 111 pp.

CIAT (1980b). *Proc. Regional Workshop on Potential for Field Beans in Eastern Africa,* Centro Internacional de Agricultura Tropical, Cali, Colombia, 226 pp.

CIAT (1981). *CIAT Rep. 1981: Highlights of activities in 1980*, Centro Internacional de Agricultura Tropical, Cali, Colombia, 112 pp.

Claflin, L. E., Stuteville, D. L., and Armbrust, D. V. (1973). Windblown soil in the epidemiology of bacterial leaf spot of alfalfa and common blight of beans, *Phytopathology,* **63**, 1417–1419.

Clark, E., Backman, P. A., and Rodriguez-Kabana, R. (1974). *Cercospora* and *Cercosporidium* tolerance to benomyl and related fungicides in Alabama peanut fields, *Phytopathology,* **64**, 1476–1477.

Clark, M. F. and Adams, A. N. (1977). Characteristics of the microplate method of enzyme-linked immunosorbent assay for the detection of plant viruses, *J. Gen. Virol.,* **34**, 475–483.

Clark, W. J., Harris, F. A., Maxwell, F. G., and Hartwig, E. E. (1972). Resistance of certain soybean cultivars to bean leaf beetle, striped blister beetle and bollworm, *J. econ. Ent.,* **65**, 1669–1672.

Clayton, E. E. (1936). Water soaking of leaves in relation to development of the wildfire disease of tobacco, *J. agric. Res.,* **52**, 239–269.

Clegg, C. G. (1947). Notes on chickpea and its cultivation in the Lake Province, Tanganyika, *E. Afr. agric. J.,* **13**, 27–28.

Clinton, P. K. S. (1957). A note on a wilt of groundnuts due to *Sclerotium rolfsii* Sacc., in Tanganyika, *E. Afr. agric. J.,* **22**, 137–141.

Clinton, P. K. S. (1960). Seed-bed pathogens of groundnut in Sudan and attempt to control with an artificial testa, *Emp. J. exp. Agric.,* **28**, 211–222.

CMI (1970, 1971, 1979). A bibliography of lists of plant diseases and fungi, *Rev. Pl. Path.,* **49**, 103–108; **50**, 1–7; **58**, 305–308.

Cobley, L. S. and Steele, W. M. (1976). *An Introduction to the Botany of Tropical Crops,* 2nd edn, Longman, London, 371 pp.

Cockbain, A. J. and Jones, P. (1980). Bean yellow vein-banding and its helper viruses, *Virology Group Meeting, Fed. Br. Pl. Path.,* London, 19 Dec. 1980.

Cocking, E. C. (1981). Opportunities from the use of protoplasts, *Phil. Trans. R. Soc. Lond.,* B **292**, 557–568.

Coker, R. D. (1980). Aflatoxin: past, present and future, *Trop. Sci.,* **21**, 143–162.

Cole, D. L. and Grimmer, C. J. (1977). The effects of foliar disease on groundnut yield, *Rhod. agric. J.,* **74**, 19–20 (abstr.).

Cole, H. (1966). Angular leaf spot associated with severe defoliation of red kidney beans (*Phaseolus vulgaris*) in Pennsylvania, *Pl. Dis. Reptr.,* **50**, 494.

Conover, R. A. (1948). Studies of two viruses causing mosaic diseases of soybean, *Phytopathology,* **38**, 724–735.

Conti, M. (1981). Reference to CAMV omitted in article on cowpea virus, *Pl. Dis.,* **65**, 459.

Cook, M. (1972). Screening of peanut for resistance to peanut rust in the greenhouse and field, *Pl. Dis. Reptr.,* **56**, 382–386.

Cook, M. (1980a). Host-parasite relations in uredial infections of peanut by *Puccinia arachidis, Phytopathology,* **70**, 822–826.

Cook, M. (1980b). Peanut leaf wettability and susceptibility to infection by *Puccinia arachidis, Phytopathology,* **70**, 826–830.

Cook, M. (1981). Susceptibility of peanut leaves to *Cercosporidium personatum, Phytopathology,* **71**, 787–791.

Cook, R. J. and Snyder, W. C. (1965). Influence of host exudates on growth and survival of germlings of *Fusarium solani* f. *phaseoli* in soil, *Phytopathology*, **55**, 1021–1025.

Cooper, W. E. (1952). Control of peanut root knot by soil fumigation and by crop rotation, *Phytopathology.* **42**, 282–283 (abstr.).

Cooper, W. E. and Gregory, W. C. (1960). Radiation-induced leafspot resistant mutants in the peanut, *Agron. J.*, **52**, 1–4.

Corbin, E. J. (1975). *Present Status of Chickpea Research in Australia*, International Workshop on grain legumes, ICRISAT, Hyderabad, India, pp. 87–93.

Costa, A. S. (1955). Studies on abutilon mosaic in Brazil, *Phytopath. Z.,* **24**, 97–112.

Costa, A. S. (1965). Three whitefly-transmitted virus diseases of beans in Saô Paulo, Brazil, *FAO Pl. Prot. Bull.*, **13**, 121–130.
Costa, A. S. (1975). Increase in the populational density of *Bemisia tabaci,* a threat of widespread virus infection of legume crops in Brazil, in *Tropical Diseases of Legumes* (eds J. Bird and K. Maramorosch), Academic Press, New York, pp. 27–49.
Costa, A. S. (1976). Whitefly-transmitted plant diseases, *A. Rev. Phytopath,* **14**, 429–449.
Costa, A. S. and Carvalho, A. M. B. (1961). Studies on Brazilian tobacco streak, *Phytopath. Z.*, **42**, 113–138.
Costa, C. L. and Cupertino, F. P. (1976). Avaliaçâo das perdas na produçâo do feijoeiro causados pelo virus do mosaico dourado, *Fitopat. Bras.*, **1**, 18–25.
Cother, E. J. (1977a). Isolation of important pathogenic fungi from seeds of *Cicer arietinum, Seed Sci. Technol.*, **5**, 593–597.
Cother, E. J. (1977b). Identification and control of root-rot fungi in *Cicer arietinum* (chickpea), *Pl. Dis Reptr.*, **61**, 736–740.
Cother, E. J. and Griffin, D. M. (1973). Formation of chlamydospores by *Phytophthora drechsleri, Trans. Br. mycol. Soc.*, **61**, 379–402.
Cox, R. S. (1949). Further studies on stem anthracnose of lima bean in North Carolina, *Phytopathology,* **39**, 5 (abstr.).
Cox, R. S. and Hyre, R. A. (1951). Overwintering of *Phytophthora phaseoli* on infected lima bean pods in the soil, *Pl. Dis. Reptr.*, **35**, 354.
Coyne, D. P. (1964). Species hybridization in *Phaseolus, J. Hered.*, **55**, 5–6.
Coyne, D. P. and Schuster, M. L. (1969). Moderate tolerance of bean varieties to brown spot bacterium (*Pseudomonas syringae), Pl. Dis. Reptr.*, **53**, 677–680.
Coyne, D. P. and Schuster, M. L. (1973). *Phaseolus* germplasm tolerant to common blight bacterium (*Xanthomonas phaseoli), Pl. Dis. Reptr.*, **57**, 111–114.
Coyne, D. P. and Schuster, M. L. (1974a). Inheritance and linkage relations of reaction to *Xanthomonas phaseoli* (E. F. Smith) Dowson (common blight), stage of plant development and plant habit in *Phaseolus vulgaris* L., *Euphytica*, **23**, 195–204.
Coyne, D. P. and Schuster, M. L. (1974b). Differential reaction of pods and foliage of beans (*Phaseolus vulgaris*) to *Xanthomonas phaseoli, Pl. Dis. Reptr.*, **58**, 278–282.
Coyne, D. P. and Schuster, M. L. (1974c). A morphological and physiological component genetic approach to breeding for tolerance to the bacterial pathogens, *Pseudomonas phaseolicola* and *Xanthomonas phaseoli, A. Rep. Bean Improv. Co-op.*, **17**, 23–24.
Coyne, D. P. and Schuster, M. L. (1975). Genetic and breeding strategy for resistance to rust (*Uromyces phaseoli* (Reben) Wint.) in beans (*Phaseolus vulgaris* L.), *Euphytica,* **24**, 795–803.
Coyne, D. P., Schuster, M. L., and Erwin, C. (1979). Reaction of *Phaseolus vulgaris* germplasm to a new virulent strain of *Pseudomonas phaseolicola, A. Rep. Bean Improv. Co-op.*, **22**, 20–21.
Coyne, D. P., Schuster, M. L., and Harris, L. (1965b). Inheritance, heritability, and response to selection for common blight (*Xanthomonas phaseoli*) tolerance in *Phaseolus vulgaris* field bean crosses, *Proc. Am. Soc. hort. Sci.*, **86**, 373–379.
Coyne, D. P., Schuster, M. L., and Shaughnessy, L. (1966). Inheritance of reaction to halo blight and common blight bacteria in a *Phaseolus vulgaris* variety cross, *Pl. Dis. Reptr.*, **50**, 29–32.
Coyne, D. P., Schuster, M. L., and Young, J. O. (1965a). A genetic study of bacterial wilt (*Corynebacterium flaccumfaciens* var. *aurantiacum*) tolerance in *Phaseolus vulgaris* crosses and the development of tolerance to two bacterial diseases in beans, *Proc. Am. Soc. hort. Sci.*, **87**, 279–285.
Crane, J. L. and Crittenden, H. W. (1966). The relation of soybean seed infection by

Cercospora kikuchii to the duration of certain plant development stages, *Pl. Dis. Reptr.,* **50**, 464–468.

Crispin, A. and Dongo, S. (1962). New physiologic races of bean rust, *Uromyces phaseoli typica,* from Mexico, *Pl. Dis. Reptr.,* **46**, 411–413.

Crittenden, H. W., Hastings, K. M., and Moore, D. M. (1966). Soybeans losses caused by tobacco ringspot virus, *Pl. Dis. Reptr.,* **50**, 910–913.

Cromwell, R. O. (1917). Fusarium-blight, or wilt disease, of the soybean, *J. agric. Res.,* **8**, 421–439.

Cross, J. E., Kennedy, B. W., Lambert, J. W., and Cooper, R. L. (1966). Pathogenic races of the bacterial blight pathogen of soybeans, *Pseudomonas glycinea, Pl. Dis. Reptr.,* **50**, 557–560.

Crossan, D. F. (1958). The relationships of seven species of *Ascochyta* occurring in North Carolina, *Phytopathology,* **48**, 248–255.

Culp, T. W., Bailey, W. K., and Hammons, R. O. (1968). Natural hybridization of peanuts, *Arachis hypogaca* L., in Virginia, *Crop Sci.*, **8**, 109–111.

Culp, T. W. and Troutman, J. L. (1967). Reductions in yield and quality of peanuts, *Arachis hypogaea,* by stunt virus, *Pl. Dis. Reptr.,* **51**, 856–860.

Culp, T. W. and Troutman, J. L. (1968). Varietal reaction of peanuts, *Arachis hypogaea,* to stunt virus disease, *Pl. Dis. Reptr.,* **52**, 914–918.

Cummins, D. G. and Smith, D. H. (1973). Effect of *Cercospora* leafspot of peanuts on forage yield and quality and on seed yield, *Agron. J.*, **65**, 919–921.

Curtis, D. L. (1968). The relation between the date of heading of Nigerian sorghums and the duration of the growing season, *J. appl. Ecol.,* **5**, 215–226.

Daft, G. C. and Leben, C. (1972). Bacterial blight of soybeans, epidemiology of blight outbreaks, *Phytopathology,* **62**, 57–62.

Dahiya, J. S., Cooksey, C. J., Garrett, P. J., and Strange, R. N. (1982). Identification of pinostrobin chalcone and two novel stilbene carboxylic acids as phytoalexins in leaves of pigeon pea (*Cajanus cajan* L.), *Contrib. to Brit. Soc. Pl. Path.,* Imperial Coll., London, 29 April 1982.

Dale, W. T. (1949). Observations on a virus disease of cowpea in Trinidad, *Ann. appl. Biol.,* **36**, 327.

Dale, W. T. (1953). Transmission of plant viruses by biting insects, with particular reference to cowpea mosaic virus, *Ann. appl. Biol.,* **40**, 384–392.

Dalela, G. G. (1962). A technique for estimating crop losses in the rust (*Uromyces ciceris-arietinis* (Grogn.) Jacz.) of gram (*Cicer arietinum* L.), *Agra Univ. J. Res.,* **11**, 117–121.

Daniel, I. and Berchoux, C. de (1965). Sur la résistance au virus dans la rosette de l'arachide, *Oléagineaux,* **20**, 373–376.

da Ponte, J. J. (1966). Uma nova enfermidade do feijao de corda *Vigna sinensis, Bol. Soc. Cearense Agron.,* **7**, 35–38. (*Rev. appl. Mycol.,* **47**, 2337.)

da Ponte, J. J. (1976). Doenças do feijâo macássar, *Vigna sinensis* (L.) Savi Unpublished report.

da Ponte, J. J. and da Silva, A. T. (1977). Sources of resistance to *Fusarium* rot in *Vigna sinensis.* I. Behaviour of four cowpea cultivars, *Fitopat. Bras.,* **2**, 155–158. (*Rev. Pl. Path.,* **57**, 1527.)

da Ponte, J. J. and Santos, A. A. (1973). A murcha bacteriana do feijâo macássar, *Vigna sinensis* Endl., *Bol. Soc. Cearense Agron.*, **14**, 23–27.

da Ponte, J. J., Vasconcelos, I., Paiva, J. B., de Castro, F. E., and Sobral, C. A. M. (1976). Incidencia do carvâo (*Entyloma vignae*) em cultivares de feijâo macássar (*Vigna sinensis*), procedentes da Nigéria, *Summa Phytopathologica,* **2**, 50–52. (*Rev. Pl. Path.,* **55**, 4955.)

Darby, P., Matthews, P., and Lewis, B. G. (1981). The mechanisms of resistance to *Ascochyta pisi* in *Pisum*, *A. Rep. John Innes Institute for 1980*, 30–31.

Das, G. N. and Sen Gupta, P. K. (1961). A *Stemphylium* leaf spot disease of gram, *Pl. Dis. Reptr.,* **45**, 979.

Dastur, J. E. (1935). Gram-wilts in the Central Provinces, *Agric. Livest. India,* **5**, 615–627.
Daub, M. E. and Hagedorn, D. J. (1979). Resistance of *Phaseolus* line WBR 133 to *Pseudomonas syringae, Phytopathology,* **69**, 946–951.
Daulton, R. A. C. and Curtis, R. F. (1964). The effects of *Tagetes* spp. on *M. javanica* in southern Rhodesia, *Nematologica,* **9**, 357–362.
Davies, J. C. (1972). Studies of the ecology of *Aphis craccivora* Koch. (Hem., Aphidae), the vector of rosette disease of groundnuts, in Uganda, *Bull. ent. Res.,* **62**, 169–181.
Davies, J. C. (1975a). Insecticides for the control of the spread of groundnut rosette disease in Uganda, *PANS,* **21**, 1–8.
Davies, J. C. (1975b). Menazon control of rosette disease of groundnuts, *Trop. Agric. (Trin.),* **52**, 359–367.
Davies, J. C. (1976). The incidence of rosette disease in groundnuts in relation to plant density and its effect on yield, *Ann. appl. Biol.,* **82**, 489–501.
Davies, J. C. and Kasule, F. K. (1964). The control of groundnut rosette disease in Uganda, *Trop. Agric. (Trin.),* **41**, 303–309.
Davis, J. H. C., Amézquista, M. C., and Muñoz, J. G. (1981). Border effects and optimum plot sizes for climbing beans (*Phaseolus vulgaris*) and maize in association and monoculture, *Expl. Agric.,* **17**, 127–135.
Davis, R. F., Weber, Z., Pospieszny, H., Silbernagel, M., and Hampton, R. O. (1981). Seedborne cucumber mosaic virus in selected *Phaseolus vulgaris* germ plasm and breeding lines in Idaho, Washington, and Oregon, *Pl. Dis.,* **65**, 492–494.
Davison, A. D. and Vaughan, E. K. (1963). A simplified method for identification of races of *Uromyces phaseoli* var. *phaseoli, Phytopathology,* **53**, 456–459.
Davison, A. D. and Vaughan, E. K. (1964). Effect of urediospore concentration on determination of races of *Uromyces phaseoli* var. *phaseoli, Phytopathology,* **54**, 336–338.
Day, P. R. (1973). Genetic variability of crops, *A. Rev. Phytopath.,* **11**, 293–312.
Day, P. R. (1974). *Genetics of Host-parasite Interaction,* W. H. Freeman, San Francisco, 238 pp.
De, D. N. (1974). Pigeon pea, in *Evolutionary Studies in World Crops: Diversity and Change in the Indian Subcontinent* (ed. J. B. Hutchinson), Cambridge Univ. Press, pp. 79–87.
Deakin, J. R. and Dukes, P. D. (1975). Breeding snap beans for resistance to diseases caused by *Rhizoctonia solani* Kuehn, *Hortsci.,* **10**, 269–271.
Debrot, E. A. and Ordosgoitti, A. (1975). Estudios sobre un mosaico amarillo de la soya en Venezuela, *Agron. Trop. (Maracay),* **25**, 435–449.
Debrot, E. C. and de Rojas, C. E. B. (1967). El virus del mosaico del frijol, *Vigna sinensis* Endl. (cowpea mosaic virus) en Venezuela, *Agron. Trop.,* **17**, 3–15.
Deighton, F. C. (1945). *Rep. Dep. Agric. S. Leone, 1944,* p. 9. (*Rev. appl. Mycol.,* **25**, 253.)
Deighton, F. C. (1967a). Floury leaf spot of french bean caused by *Ramularia phaseoli* (Drummond) comb. nov., *Trans. Br. mycol. Soc.,* **50**, 123–127.
Deighton, F. C. (1967b). Studies on *Cercospora* and allied genera. ii. *Passalora, Cercosporidium* and some species of *Fusicladium* on *Euphorbia, Mycol. Pap. 112,* Commonwealth Mycological Institute, Kew, 80 pp.
Deighton, F. C. (1967c). New names in *Mycosphaerella (M. arachidis* and *M. pruni-persici)* and validation of *M. rosicola, Trans. Br. mycol. Soc.,* **50**, 328–329.
Deighton, F. C. (1971). Brown leaf mould of *Canavalia* caused by *Stenella canavaliae* (H. and P. Syd.) comb. nov., *Trans. Br. mycol. Soc.,* **56**, 411–418.
Deighton, F. C. (1973). Studies on *Cercospora* and allied genera. iv. *Cercosporella* Sacc., *Pseudocercosporella* gen. nov. and *Pseudocercosporidium* gen. nov., *Mycol. Pap. 133,* Commonwealth Mycological Institute, Kew, 62 pp.
Deighton, F. C. (1974). Studies on *Cercospora* and allied genera. v. *Mycovellosiella*

Rangel, and a new species of *Ramulariopsis, Mycol. Pap. No. 137,* Commonwealth Mycological Institute, Kew, 75 pp.
Deighton, F. C. (1976). Studies on *Cercospora* and allied genera. vi. *Pseudocercospora* Speg., *Pantospora* Cif. and *Cercoseptoria* Petr. *Mycol. Pap. No. 140*, Commonwealth Mycological Institute, Kew 168 pp.
Deighton, F. C. (1977). *Cercospora solimani* Speg., a synonym of *Phaeoisariopsis griseola* (Sacc.) Ferr., *Trans. Br. mycol. Soc.*, **68**, 282–284.
Deighton, F. C. (1979). Studies on *Cercospora* and allied genera, vii. New species and redispositions. *Mycol. Pap. No. 144,* Commonwealth Mycological Institute, Kew, 56 pp.
de Jager, C. P. (1979). Cowpea severe mosaic virus. *CMI/AAB Descriptions of Plant Viruses No. 209.*
Delassus, M. (1978). Soybean diseases in Western Cameroon, in *Diseases of Tropical Food Crops* (eds H. Maraite and J. A. Meyer), UCL, Louvain-la-Neuve, Belgium, pp. 191–198.
Demeure, Y. (1975). Résistance à la sécheresse, en zone sahélienne, du nématode phytoparasite *Scutellonema cavenessi* Sher, 1963, *Cahiers ORSTOM, Ser. Biol.*, **10**(3), 283–292. (*Helminth. Abstr.* B **45**, 1512.)
Demski, J. W. and Harris, H. B. (1974). Seed transmission of virus in soybean, *Crop Sci.,* **14**, 888–890.
Demski, J. W., Kahn M. A., Wells, H. D., and Miller, J. D. (1981). Peanut mottle virus in forage legumes, *Pl. Dis.*, **65**, 359–362.
Demski, J. W. and Kuhn, C. W. (1975). Resistant and susceptible reactions of soybeans to peanut mottle virus, *Phytopathology,* **65**, 95–99.
Demski, J. W. and Phillips, D. V. (1974). Reactions of soybean cultivars to powdery mildew, *Pl. Dis. Reptr.,* **58**, 723–726.
Demski, J. W., Smith, D. H., and Kuhn, C. W. (1975). Incidence and distribution of peanut mottle virus in peanut in the United States, *Peanut Sci.*, **2**, 91–93.
Dennis, R W. G. (1968). *British Ascomycetes,* Lehre, Verlag Von J. Cramer, 455 pp.
Desai, M.V. and Prasad, N. (1955). A new *Colletotrichum* from India, *Indian Phytopath.*, **8**, 52–57.
Deshpande, R. B., Jeswani, L. M., and Joshi, A. B. (1963). Breeding wilt resistant varieties of pigeon pea, *Indian J. Genet. Pl. Breed.*, **23**, 58–63.
Deslandes, J. A. (1979). Ferrugen da soja e de outras leguminosas causada por *Phakopsora pachyrhizi* no Est. Minas Gerais, *Fitopat. Bras.*, **4**, 337–339. (*Rev. Pl. Path.*, **59**, 3467.)
Deverall, B. J., Keogh, R. C., and McLeod, S. (1977). Responses of soybean to infection by, and to germination fluids from, urediniospores of *Phakopsora pachyrhizi, Trans. Br. Mycol. Soc.,* **69**, 411–415.
Devine, T. E. and Breithaupt, B. H. (1980). Significance of incompatibility reactions of *Rhizobium japonicum* strains with soybean host genotypes, *Crop Sci.,* **20**, 269–271.
Dey, P. K. (1919). Studies in the physiology of parasitism. V. Infection by *Colletotrichum lindemuthianum, Ann. Bot.,* **33**, 305–312.
Dhéry, M. and Gillier, P. (1971). Un nouveau pas dans la lutte contre la rosette de l'arachide. Résultats obtenus en Haute Volta avec les nouveaux hybrides, *Oléagineux,* **26**, 243–251.
Dhingra, O. D. and da Silva, J. F. (1978). Effect of weed control on the internally seed-borne fungi in soybean seeds, *Pl. Dis. Reptr.,* **62**, 513–516.
Dhingra, O. D., Nicholson, J. F., and Sinclair, J. B. (1973). Influence of temperature on recovery of *Aspergillus flavus* from soybean seed, *Pl. Dis. Reptr.,* **57**, 185–187.
Dias, I. R. and Costa, J. G. C. da (1968). Identificaçâo das raças fisiológicas da ferrugen (*Uromyces phaseoli typica* Arth.) do feijoeiro (*Phaseolus vulgaris* L.) em duas regiôes fisiográficas do Rio Grande do Sul, Brasil, *Pesq. agropec. bras., Sér. Agron.,* **3**, 165–170.

Diaz, A. (1974). Wild hosts of cowpea mosaic virus, *Phytopathology,* **64**, 767 (abstr.).
Dickinson, C. H. (1979). External synergisms among organisms inducing disease, in *Plant Disease, an advanced treatise. iv. How pathogens induce disease* (eds J. G. Horsfall and E. B. Cowling), Academic Press, New York, pp. 97–111.
Dickson, J. G. (1956). Peanut diseases, in *Diseases of Field Crops,* McGraw-Hill, New York, 517 pp., pp. 396–400.
Dickson, M. H. and Boettger, M. A. (1977). Breeding for multiple root rot resistance in snap beans, *J. Am. Soc. hort. Sci.,* **102**, 373–377.
Dickson, M. H. and Natti, J. J. (1966). Tolerance to halo blight race 2 in beans, *A. Rep. Bean Improv. Co-op.,* **9**, 19–20.
Dickson, M. H. and Natti, J. J. (1968). Inheritance of resistance of *Phaseolus vulgaris* to bean yellow mosaic virus, *Phytopathology,* **58**, 1450.
Diercks, R. and Klewitz, R. (1962). Zur samenübertragbarkeit einer an Ackerbohnen vorkommenden Her Kunft des Stengelächens *Ditylenchus dipsaci* (Kühn), *Nematologica,* **7**, 155–163.
Dillon-Weston, W. A. R. (1931). The effect of ultra violet radiation on the uredinio-spores of some physiologic forms of *Puccinia graminis tritici, Scient. Agric.,* **12**, 81–87.
Dixon, G. R. and Doodson, J. K. (1973). A technique used for testing dwarf French bean cultivars for resistance to anthracnose (*Colletotrichum lindemuthianum*), *J. natn. Inst. agric. Bot.,* **13**, 87–89.
Dixon, G. R. and Doodson, J. K. (1975). Techniques used for testing dwarf French bean cultivars for resistance to grey mould (*Botrytis cinerea*). *J. natn. Inst. agric. Bot.,* **13**, 338–341.
Dodman, R. L., Barker, K. R., and Walker, J. C. (1968). Modes of penetration by different isolates of *Rhizoctonia solani, Phytopathology,* **58**, 31–33.
Doidge, E. M. (1926). A preliminary study of the South African rust fungi, *Bothalia,* **2**, 1–228.
Doku, E. V. and Karikari, S. K. (1971) Bambarra groundnut, *Econ. Bot.,* **25**, 255–262.
Dovlo, F. E., Williams, C. E., and Zoaka, L. (1976). *Cowpeas: home preparation and use in West Africa,* International Development Research Centre, Ottawa, Canada, 96 pp.
Doupnik, B. (1969). Aflatoxins produced on peanut varieties previously reported to inhibit production, *Phytopathology,* **59**, 1554.
Drijfhout, E. (1978). *Genetic interaction between* Phaseolus vulgaris *and bean common mosaic virus with implications for strain identification and breeding for resistance,* Centre for Agricultural Publishing and Documentation, Wageningen, The Netherlands, 98 pp.
Drijfhout, E. and Bos, L. (1977).The identification of two new strains of bean common mosaic virus, *Neth. J. Pl. Path.,* **83**, 13–25.
Drijfhout, E., Silbernagel, M. J., and Burke, D. W. (1978). Differentiation of strains of bean common mosaic virus, *Neth. J. Pl. Path.,* **84**, 13–26.
Drinkall, M. J. (1978). False rust disease of the winged bean, *PANS,* **24**, 160–166.
Dropkin, V. H. (1959). Varietal response of soybeans to *Meloidogyne*—a bioassay for separating races of root-knot nematodes, *Phytopathology,* **49**, 18–23.
Dubern, J. and Dollet, M. (1980). Groundnut eyespot virus, a new member of the potyvirus group, *Ann. appl. Biol.,* **96**, 193–200.
Dubin, H. J. and Rajaram, S. (1981). The strategy of the International Maize and Wheat Improvement Center (CIMMYT) for breeding disease resistant wheat: an international approach, in *Strategies for the Control of Cereal Disease* (eds J. F. Jenkyn and R. T. Plumb), Blackwell, Oxford, pp. 27–35.
Duffus, J. E. (1971). Role of weeds in the incidence of virus diseases, *A. Rev. Phytopath.,* **9**, 319–340.
Dukes, P. D., Fery, R. L., and Hamilton, M. G. (1979). Comparison of plant resis-

tance and a nematicide for control of southern root-knot nematodes in southern peas, *Vigna unguiculata, Phytopathology,* **69**, 526–527 (abstr.).
Duncan, L. W., Caveness, F. E., and Perez, A. T. (1979). The susceptibility of winged bean (*Psophocarpus tetragonolobus)* to the root-knot nematodes, *Meloidogyne incognita,* race 2, and *M. javanica, Trop. Grain Leg. Bull.,* **15**, 30–34.
Dundas, B. (1939). Inheritance of resistance to powdery mildew in runner beans (*Phaseolus coccineus*), tepary beans (*P. acutifolius*), yard long beans (*Vigna sesquipedalis*) and cowpeas (*Vigna sinensis*), *Phytopathology,* **29**, 824 (abstr.).
Dundas, B. (1942). Breeding beans for resistance to powdery mildew and rust, *Phytopathology,* **32**, 828 (abstr.).
Dundas, B. and Scott, G. W. (1939). Physiologic strains of bean rust, *Phytopathology,* **29**, 820–821.
Duniway, J. M. (1975). Limiting influence of low water potential on the formation of sporangia by *Phytophthora drechsleri* in soil, *Phytopathology,* **65**, 1089–1093.
Dunlap, A. A. (1943). Two bacterial diseases in Texas, *Pl. Dis Reptr.,* **27**, 274.
Dunleavy, J. M. (1957). The grasshopper as a vector of tobacco ringspot virus in soybean, *Phytopathology,* **47**, 681–682.
Dunleavy, J. M. (1963). A vascular disease of soybeans caused by *Corynebacterium* sp. *Pl. Dis. Reptr.,* **47**, 612–613.
Dunleavy, J. M. (1964). A new bacterial disease of soybeans, *Phytopathology,* **54**, 891–892.
Dunleavy, J. M. (1971). Races of *Peronospora manshurica* in the United States, *Am. J. Bot.,* **58**, 209–211.
Dunleavy, J. M. (1973). Viral diseases, in *Soybeans: Improvement, Production and Uses* (ed. B. E. Caldwell), Amer. Soc. Agron., Madison, pp. 505–526.
Dunleavy, J. M. (1978). Soybean seed yield losses caused by powdery mildew, *Crop Sci.,* **18**, 337–339.
Dunleavy, J. M. and Hartwig, E. E. (1970). Sources of immunity from and resistance to nine races of the soybean downy mildew fungus, *Pl. Dis. Reptr.,* **54**, 901–902.
Durbin, R. D. (1959). Factors affecting the vertical distribution of *Rhizoctonia solani* with reference to CO_2 concentration, *Am. J. Bot.,* **46**, 22–25.
Dwivedi, R. P. (1977). Studies on *Thanatephorus* species attacking cowpea (*Vigna catjang* Walp.) plants. Ph.D thesis, Kanpur Univ., India.
Dwivedi, R. P. and Saksena, H. K. (1975). Web blight disease of arhar (*Cajanus cajan* (L.) Millsp.) caused by *Thanatephorus cucumeris, Ind. J. Farm Sci*., **3**, 113.
Dye, D. W., Bradbury, J. F., Goto, M., Hayward, A. C., Lelliott, R. A., and Schroth, M. N. (1980). International standards for naming pathovars of phytopathogenic bacteria and a list of pathovar names and pathotype strains, *Rev. Pl. Path.,* **59**, 153–168.
Ebbels, D. L. and Allen, D. J. (1979). A supplementary and annotated list of plant diseases, pathogens and associated fungi in Tanzania. *Phytopath. Pap. no. 22*, Commonwealth Mycological Institute, Kew.
Ebbels, D. L. and Billington, R. V. (1972). *Fusarium* wilt of *Voandzeia subterranea* in Tanzania, *Trans. Br. mycol. Soc.,* **58**, 336–338.
Echandi, E. (1965). Basidiospore infection by *Pellicularia filamentosa* (= *Corticium microsclerotia*), the incitant of web blight of common bean, *Phytopathology,* **55**, 698–699.
Echandi, E. (1970). Wilt of chickpeas or garbanzo beans (*Cicer arietinum*) incited by *Fusarium oxysporum, Phytopathology,* **60**, 1539 (abstr.).
Echandi, E. (1976). Principales enfermedades de hongo del frijol (*Phaseolus vulgaris*) en los trópicos Americanos en diferentes zonas ecologicas, *Fitopat. Bras.,* **1**, 171–177.
Echandi, E. and Hebert, T. T. (1971). Stunt of beans incited by peanut stunt virus, *Phytopathology,* **61**, 328–330.

Edje, O. T., Mughogho, L. K., and Rao, Y. P. (1976a). Effects of defoliation on bean yield, *A. Rep. Bean Improv. Co-op.*, **19**, 29–31.

Edje, O. T., Mughogho, L. K., and Rao, Y. K. (1976b). Effects of mixed cropping of maize and beans on seed yield, *A. Rep. Bean Improv. Co-op.*, **19**, 31–34.

Ekpo, E. J. A. (1979). Effect of *Xanthomonas vignicola* on the yield of cowpea (*Vigna unguiculata*), *Af. J. agric. Sci.*, **5**, 67–69.

Ekpo, E. J. A. (in press). Comparative virulence of *Xanthomonas vignicola* (bacterial blight) and *X.* sp. (bacterial pustule) in cowpea (*Vigna unguiculata*), *Niger. J. agric. Sci.*

Ekpo, E. J. A. and Esuruoso, O. F. (1977). Factors affecting spore germination in cowpea isolates of *Cercospora cruenta* Sacc., *Phytopath. Z.*, **89**, 249–255.

Ekpo, E. J. A. and Saettler, A. W. (1974). Distribution pattern of bean common mosaic virus in developing bean seed, *Phytopathology*, **64**, 269–270.

Ekpo, E. J. A. and Saettler, A. W. (1976). Pathogenic variation in *Xanthomonas phaseoli* and *X. phaseoli* var. *fuscans*, *Pl. Dis. Reptr.*, **60**, 80–83.

Ellis, M. A., Machado, C. C., Prasartsee, C., and Sinclair, J. B. (1974). Occurrence of *Diaporthe phaseolorum* var. *sojae (Phomopsis* sp.) in various soybean seedlots, *Pl. Dis. Reptr.*, **58**, 173–176.

Ellis, M. A. and Paschal, H. (1979). Effect of fungicide seed treatment on internally seedborne fungi, germination and field emergence of pigeon pea (*Cajanus cajan), Seed Sci. and Technol.*, **7**, 75–81.

Ellis, M. A. and Sinclair, J. B. (1976). Effect of benomyl field sprays on internally-borne fungi, germination, and emergence of later-harvested soybeans seeds, *Phytopathology*, **66**, 680–682.

Ellis, M. A. and Smith, R. S. (1978). Effect of incubation temperature on recovery of internally seedborne fungi and germination of pigeon pea seeds, *Trop. Grain Leg. Bull.*, **13/14**, 22–25.

Ellis, M. A., Tenne, F. D., and Sinclair, J. B. (1977). Effect of antibiotics and high temperature storage on decay of soybean seeds by *Bacillus subtilis, Seed Sci. and Technol.*, **5**, 753–761.

Ellis, M. B. (1957). Some species of *Corynespora. Mycol. Pap. No. 65,* Commonwealth Mycological Institute, Kew, 15 pp.

Ellis, M. B. (1971). *Dematiaceous Hyphomycetes,* Commonwealth Mycological Institute, Kew, 608 pp.

Ellis, M. B. (1976). *More Dematiaceous Hyphomycetes,* Commonwealth Mycological Institute, Kew, 507 pp.

Ellis, M. B. and Holliday, P. (1971). *Corynespora cassiicola. CMI Descriptions of Pathogenic Fungi and Bacteria No. 303.*

Ellis, M. B. and Waller, J. M. (1974). *Sclerotinia fuckeliana. CMI Descriptions of Pathogenic Fungi and Bacteria No. 431.*

Elliston, J. E., Kuć, J., and Williams, E. B. (1971). Induced resistance to bean anthracnose at a distance from the site of the inducing interaction, *Phytopathology*, **61**, 1110–1112.

Elston, H., Harkness, C., and McDonald, D. (1976). The effects of *Cercospora* leaf disease on the growth of groundnuts (*Arachis hypogaea*) in Nigeria, *Ann. appl. Biol.*, **83**, 39–51.

Emechebe, A. M. (1980). Scab disease of cowpea (*Vigna unguiculata*) caused by a species of the fungus *Sphaceloma, Ann. appl. Biol.*, **96**, 11–16.

Emechebe, A. M. (in press). Brown blotch of cowpea in northern Nigeria, *Samaru J. agric. Res.*, **1**.

Emechebe, A. M. and McDonald, D. (1979). Seed-borne pathogenic fungi and bacteria of cowpea in northern Nigeria, *PANS*, **25**, 401–404.

Endo, B. Y. (1959). Responses of root-lesion nematodes, *Pratylenchus brachyurus* and *P. zeae,* to various plants and soil types, *Phytopathology,* **49**, 417–421.

Enyi, B. A. C. (1975). Effects of defoliation on growth and yield in groundnut (*Arachis hypogaea*), cowpeas (*Vigna unguiculata*), soyabean (*Glycine max*) and green gram (*Vigna aureus*), *Ann. appl. Biol.*, **79**, 55–56.

Epps, J. M. and Duclos, L. A. (1970). Races of the soybean cyst nematode in Missouri and Tennessee, *Pl. Dis. Reptr.*, **54**, 319–320.

Epton, H. A. S. and Deverall, B. J. (1965). Physiological races of *Pseudomonas phaseolicola* causing halo blight of bean, *Pl. Path.*, **14**, 53–54.

Erwin, D. C. (1958a). Verticillium wilt of *Cicer arietinum* in S. California, *Pl. Dis. Reptr.*, **42**, 1111.

Erwin, D. C. (1958b). *Fusarium lateritium* f. *ciceri*, incitant of fusarium wilt of *Cicer arietinum* L., *Phytopathology*, **48**, 498–501.

Erwin, D. C. and McCormick, W. H. (1971). Germinating oospores produced by *Phytophthora megasperma* var. *sojae*, *Mycologia*, **63**, 972–977.

Erwin, D. C. and Snyder, W. C. (1958). Yellowing of garbanzo beans, *Calif. Agric.*, **12**(11), 6.

Esuruoso, O. F. (1975). Seed-borne fungi of cowpea (*Vigna unguiculata*) in Western Nigeria, *Niger. J. Pl. Prot.*, **1**, 87–90.

Evans, A. C. (1954). Groundnut rosette disease in Tanganyika. 1. Field studies, *Ann. appl. Biol.*, **41**, 189–206.

Evans, A. M. (1975). Species hybridization in the genus *Vigna*. *Proc. IITA Collaborators' Meeting on Grain Legume Improvement*, International Institute of Tropical Agriculture, Ibadan, Nigeria, pp. 31–34.

Evans, A. M. (1976). Beans, in *Evolution of Crop Plants* (ed. N. W. Simmonds), Longmans, London, pp. 168–172.

Evans, A. M. (1979). The evolution of grain legumes—implications in breeding, *Proc. Symp. Grain Leg. Improv. E. Af.*, Nairobi, Kenya, 20–24 Aug. 1979.

Evans, A. M. (1980a). Structure, variation, evolution, and classification in *Phaseolus*, in *Advances in Legume Science* (eds R. J. Summerfield and A. H. Bunting), HMSO, London, pp. 337–347.

Evans, A. M. (1980b). Germplasm collections of *Phaseolus* beans, *A. Rep. Bean Improv. Co-op.*, **23**, 7–9.

Evans, A. M. and Davis, J. H. C. (1978). Breeding *Phaseolus* beans as grain legumes for Britain, *Appl. Biol.*, **3**, 1–42.

Fagbenle, H. H. and Ford, R. E. (1970). Tobacco streak virus isolated from soybeans, *Glycine max.*, *Phytopathology*, **60**, 814–820.

Fahim, M., Abdu, Y. A. M., and Karara, M. (1969). *Cercospora* leaf spot of cowpea in UAR, *J. Phytopath. UAR*, **1**, 47–50. (*Rev. Pl. Path.*, **49**, 2245.)

Fajardo, T. G. (1930). Studies on the mosaic disease of the bean (*Phaseolus vulgaris* L.), *Phytopathology*, **20**, 469–494.

Fakorede, M. A. B. (1976). Temperature requirements of *Colletotrichum lindemuthianum* isolated from cowpeas (*Vigna unguiculata*), *Niger. J. Pl. Prot.*, **2**, 88–97.

Fakorede, M. A. B., Barker, L. N., and Franckowiak, J. D. (in press). Inheritance of resistance to stem anthracnose of cowpea (*Vigna unguiculata* (L.) Walp.), *Niger. Agric. J.*

FAO (1976). *Production Yearbook for 1975*, vol. 29, Food and Agriculture Organization of the United Nations, Rome, 555 pp.

FAO (1981). *Production Yearbook for 1980*, Food and Agriculture Organization ot the United Nations, Rome, 296 pp.

Faris, D. G. (1965). The origin and evolution of the cultivated forms of *Vigna sinensis*, *Can. J. Genet. Cytol.*, **7**, 433–452.

Farrell, J. A. K. (1976a). Effects of groundnut sowing date and plant spacing on rosette virus disease in Malawi, *Bull. ent. Res.*, **66**, 159–171.

Farrell, J. A. K. (1976b). Effects of groundnut crop density on the population dynam-

ics of *Aphis craccivora* Koch (Hemiptera, Aphididae) in Malawi, *Bull. ent. Res.*, **66**, 317–329.
Farrell, J. A. K. (1976c). Effects of intersowing with beans on the spread of groundnut rosette virus by *Aphis craccivora* Koch (Hemiptera, Aphididae) in Malawi, *Bull. ent. Res.*, **66**, 331–333.
Farrell, J. A. K., Bailey, B. E., and Mills W. R. (1967). The effects of time of planting, spacing and fungicide on *Cercospora* leaf spots of groundnuts in Malawi, *Rhod. Zamb. Mal. J. agric. Res.*, **5**, 241–247.
Fassuliotis, G., Deakin, J. R., and Hoffman, J. C. (1970). Root-knot nematode resistance in snap beans; breeding and nature of resistance, *J. Am. Soc. hort. Sci.*, **95**, 640–645.
Fawcett, H. S. (1931). The importance of investigation on the effects of known mixtures of microorganisms, *Phytopathology*, **21**, 545–550.
Feakin, S. D. (1973). (ed.), *Pest Control in Groundnuts,* PANS Manual No. 2, Centre of Overseas Pest Control, London, 197 pp.
Fennell, J. L. (1948). New cowpeas resistant to mildew, *J. Hered.*, **39**, 274–279.
Ferraz, S. (1980). Angular leaf spot, in *Bean Production Problems* (eds H. F. Schwartz and G. E. Gálvez), Centro Internacional de Agricultura Tropical, Cali, Colombia, pp. 55–64.
Fery, R. L. (1980). Genetics of *Vigna,* in *Horticultural Reviews* (ed. J. Janick), **2**, 311–394, Westport, Conn.
Fery, R. L. and Cuthbert, F. P. (1972). Assocation of plant density, cowpea curculio damage and *Choanephora* pod rot in southern peas, *J. Am. Soc. hort. Sci.*, **97**, 800–802.
Fery, R. L. and Dukes, P. D. (1977). An assessment of 2 genes for *Cercospora cruenta* leaf-spot resistance in the southern pea, *Vigna unguiculata, Hortsci.*, **12**, 454–456.
Fery, R. L. and Dukes, P. D. (1980). Inheritance of root-knot resistance in the cowpea (*Vigna unguiculata* (L.) Walp.), *J. Am. Soc. hort. Sci.*, **105**, 671–674.
Fery, R. L., Dukes, P. D., and Cuthbert, F. P. (1976). The inheritance of *Cercospora* leaf spot resistance in the southern pea (*Vigna unguiculata* (L.) Walp.), *J. Am. Soc. hort. Sci.*, **101**, 148–149.
Fery, R. L., Dukes, P. D., Cuthbert, F. P. (1977). Yield-loss of southern pea (*Vigna unguiculata*) caused by *Cercospora* leafspot, *Pl. Dis. Reptr.*, **61**, 741–743.
Firman, I. D. (1972). A list of fungi and plant parasitic bacteria, viruses and nematodes in Fiji. *Phytopath. Pap. No. 15,* Commonwealth Mycological Institute, Kew, 36 pp.
Fischer, H. U. and Lockhart, B. E. (1976a). A strain of cowpea aphid-borne mosaic virus isolated from cowpeas in Morocco, *Phytopath. Z.*, **85**, 43–48.
Fischer, H. U. and Lockhart, B. E. (1976b). A strain of cucumber mosaic virus isolated from cowpeas in Morocco, *Phytopath. Z.*, **85**, 132–138.
Fischer, H. U. and Lockhart, B. E. (1978). Host range and properties of peanut stunt virus from Morocco, *Phytopathology,* **68**, 289–293.
Fisher, D. J. and Hayes, A. L. (1981). Effects of some fungicides used against cereal pathogens on the growth of *Rhizobium trifolii* and its capacity to fix nitrogen in white clover, *Ann. appl. Biol.*, **98**, 101–107.
Fisher, H. H. (1952). New physiologic races of bean rust (*Uromyces phaseoli typica*), *Pl. Dis. Reptr.*, **36**, 103–105.
Fisher, N. M. (1974). A comparison of the relative seed yields of eight bean cultivars in pure stand and in mixtures with maize, *A. Rep. Bean Improv. Co-op.*, **17**, 38–40.
Fisher, N. M. (1977). Studies in mixed cropping. 1. Seasonal differences in relative productivity of crop mixtures and pure stands in the Kenya highlands, *Expl. Agric.*, **13**, 177–184.
Flentje, N. T. (1957). Studies on *Pellicularia filamentosa* (Pat.) Rogers, III. Host

penetration and resistance, and strain specialization, *Trans. Br. mycol. Soc.*, **40**, 322–336.

Flentje, N. T., Dodman, R. L., and Kerr, A. (1963). The mechanism of host penetration by *Thanatephorus cucumeris, Aust. J. Biol. Sci.*, **16**, 784–799.

Flentje, N. T. and Saksena, H. K. (1957). Studies on *Pellicularia filamentosa* (Pat.) Rogers. II. Occurrence and distribution of pathogenic strains, *Trans. Br. mycol. Soc.*, **40**, 95–108.

Flor, H. H. (1942). Inheritance of pathogenicity in *Melampsora lini, Phytopathology*, **32**, 653–669.

Flor, H. H. (1956). The complementary genic systems in flax and flax rust, *Adv. Genet.*, **8**, 29–54.

Flowers, R. A. and Littrell, R. H. (1973). Influence of *Rhizoctonia solani* on population densities of *Pythium aphanidermatum* in soil, *2nd Int. Congr. Pl. Path.*, Minneapolis, 1058 (abstr.).

Ford, E. J., Gold, A. H., and Snyder, W. C. (1970). Induction of chlamydospore formation in *Fusarium solani* by soil bacteria, *Phytopathology*, **60**, 479–484.

Foster, R. E. and Walker, J. C. (1947). Predisposition of tomato to *Fusarium* wilt, *J. agric. Res.*, **74**, 165–185.

Foudin, A. S. and Macko, V. (1974). Identification of the self-inhibitor and some germination characteristics of peanut rust uredospores, *Phytopathology*, **64**, 990–993.

Fouilloux, G. (1976). L'anthracnose du haricot (*Colletotrichum lindemuthianum* Sacc. and Magn.): nouvelles sources de résistance et nouvelles races physiologiques, *Annls Amél. Pl.*, **26**, 443–453.

Fouilloux, G. (1979). New races of bean anthracnose and consequences on our breeding programs, in *Diseases of Tropical Food Crops* (eds H. Maraite and J. A. Meyer), Univ. Catholique de Louvain, Louvain-la-Neuve, Belgium, pp. 221–235.

Fouilloux, G. and Bannerot, H. (1977). RH_{13} a four-diseases resistant line, *A. Rep. Bean Improv. Co-op.*, **20**, 59.

Fowler, A. M. (1970a). *Cercospora* leaf spots of groundnuts, *Samaru agric. Newsl.*, **12**, 22–25.

Fowler, A. M. (1970b). The epidemiology of *Cercospora* leaf-spot diseases of groundnuts, *Samaru agric. Newsl.*, **12**, 66–69.

Fowler, A. M. (1971). Losses caused by leaf-spot diseases of groundnuts, *Samaru agric. Newsl.*, **13**, 11–13.

Francis, C. A. (1978). Multiple cropping potentials of beans and maize, *Hort. sci*., **13**, 12–17.

Francki, R. I. B. and Randles, J. W. (1970). Lettuce necrotic yellows virus. *CMI/AAB Descriptions of Plant Virus Diseases no. 26.*

Frank, Z. R. (1972). Notes on soil management in relation to *Pythium* rot of peanut pods, *Pl. Dis Reptr.*, **56**, 600–601.

Frank, Z. R. and Krikun, J. (1969). Evaluation of peanut (*Arachis hypogaea*) varieties for verticillium wilt resistance, *Pl. Dis. Reptr.*, **53**, 744–746.

Frazier, W. A., Hendrix, J. W., and Kikuta, K. (1948). Breeding rust-resistant pole green beans for Hawaii, *Proc. Am. Soc. hort. Sci.*, **51**, 468–470.

Frazier, W. A., Vaughan, E. K., McWhorter, F. P., Baggett, J. R., Hikida, H. R., and Kleczinski, D. L. (1961). Testing for combined resistance to certain diseases in snap beans, *Proc. Am. Soc. hort. Sci.*, **78**, 398–309.

French, E. R. and Kennedy, B. W. (1963). The role of *Fusarium* in the root rot complex of soybean in Minnesota, *Pl. Dis. Reptr.*, **47**, 672–676.

Frezzi, M. (1969). *Ascochyta arachidis* Woron., hongo parásito del mani (*Arachis hypogaea* L.) y su forma sexual *Mycosphaerella argentinensis* n.sp. en Argentina, *Revta Invest. agrop., Ser. 5*, **6**, 147–153. (*Rev. Pl. Path*., **49**, 1228.)

Frezzi, M. J. (1950). Las especies de 'Phytophthora' en la Argentina, *Rev. Invest. agric., B. Aires,* **4**, 47–133. (*Rev. appl. Mycol.,* **30**, 433.)

Fromme, F. D. (1924). The rust of cowpeas, *Phytopathology,* **14**, 67–79.

Fromme, F. D. and Wingard, S. A. (1921). Varietal susceptibility of beans to rust, *J. agric. Res.,* **21**, 385–404.

Fulton, J. P. and Allen, D. J. (1982). Identification of resistance to cowpea severe mosaic virus, *Trop. Agric. (Trin.),* **59**, 66–68.

Fulton, J. P., Gamez, R., and Scott, H. A. (1975). Cowpea chlorotic mottle and bean yellow stipple viruses, *Phytopathology,* **65**, 741–742.

Fulton, J. P. and Scott, H. A. (1974). Virus-vectoring efficiences of two species of leaf-feeding beetles, *Proc. Amer. Phytopath. Soc.,* **1**, 159 (abstr.).

Fulton, J. P. and Scott, H. A. (1977). Bean rugose mosaic and related viruses and their transmission by beetles, *Fitopat. Bras.,* **2**, 9–16.

Fulton, J. P. and Scott, H. A. (1979). A serogrouping concept for legume comoviruses, *Phytopathology,* **69**, 305–306.

Fulton, R. W. (1971). Tobacco streak virus. *CMI/AAB Descriptions of Plant Viruses no. 44.*

Gallegos, H. M. L., Reynoso, E. L., and Ortega, J. (1965). Observations on chickpea rust in Central Mexico, *Phytopathology*, **55**, 125 (abstr.).

Gálvez, G. E. (1980). Aphid-transmitted viruses, in *Bean Production Problems* (eds H. F. Schwartz and G. E. Gálvez), Centro Internacional de Agricultura Tropical, Cali, Colombia, pp. 211–238.

Gálvez, G. E. and Cárdenas, M. R. (1980). Whitefly-transmitted viruses, in *Bean Production Problems* (eds H. F. Schwartz and G. E. Gálvez), Centro Internacional de Agricultura Tropical, Cali, Colombia, pp. 261–289.

Gálvez, G. E., Guzmán, P., and Castaño, M. (1980). Web blight, in *Bean Production* Problems (eds H. F. Schwartz and G. E. Gálvez), Centro Internacional de Agricultura Tropical, Cali, Colombia, pp. 101–110.

Gámez, R. (1971). Los virus del fríjol en Centroamérica. I. Transmisión por moscas blancas (*Bemisia tabaci* Genn.) y plantas hospendantes del virus del mosaico dorado, *Turrialba,* **21**, 22–27.

Gámez, R. (1980). Beetle-transmitted viruses, in *Bean Production Problems* (eds H. F. Schwartz and G. E. Gálvez), Centro Internacional de Agricultura Tropical, Cali, Colombia, pp. 239–259.

Garber, R. J. and Odland, T. E. (1926). Natural crossing in soybean, *J. Am. Soc. Agron.,* **18**, 967–970.

Garcia, R. and Mitchell, D. J. (1975). Interactions of *Pythium myriotylum* with *Fusarium solani, Rhizoctonia solani,* and *Meloidogyne arenaria* in pre-emergence damping-off of peanut, *Pl. Dis. Reptr.,* **59**, 665–669.

Gardner, M. W. (1924). A native weed host for bacterial blight of bean, *Phytopathology,* **14**, 340.

Gardner, M. W. (1925). *Cladosporium* spot of cowpea, *Phytopathology*, **15**, 453–462.

Gardner, M. W. and Kendrick, J. B. (1923). Bacterial spot of cowpea, *Science,* **57**, 275–276.

Gardner, M. W. and Kendrick, J. B. (1925). Bacteria spot of cowpea and lima bean, *J. agric. Res.,* **31**, 841–863.

Garren, K. H. (1964). Inoculum potential and differences among peanuts in susceptibility to *Sclerotium rolfsii, Phytopathology,* **54**, 279–281.

Garren, K. H. (1970). *Rhizoctonia solani* versus *Pythium myriotylum* as pathogens of peanut pod breakdown, *Pl. Dis. Reptr.,* **54**, 840–843.

Garren, K. H. and Higgins, B. B. (1947). Fungi associated with runner peanut seeds and their relation to concealed damage, *Phytopathology,* **37**, 512–522.

Garren, K. H. and Jackson, C. R. (1973). Peanut diseases, in *Peanuts: Culture and Uses,* Am. Peanut Res. Educ. Assoc. Virginia, pp. 429–494.

Garren, K. H., Porter, D. M., and Allison, A. H. (1971). *Cylindrocladium* black rot of peanuts in Virginia, *Pl. Dis. Reptr.,* **55**, 419–421.

Garren, K. H. and Wilson, C. (1951). Peanut diseases, in *The Peanut, the Unpredictable Legume,* National Fertiliser Assoc., Washington DC, pp. 262–324.

Garrett, S. D. (1970). *Pathogenic Root-infecting Fungi,* Cambridge Univ. Press, 294 pp.

Gatehouse, A. M. R., Gatehouse, J. A., Dobie, P., Kilminster, A. M., and Boulter, D. (1979). Biochemical basis of insect resistance in *Vigna unguiculata, J. Sci. Fd and Agric.,* **30**, 948–958.

Gathuru, E. M. (1979). A pigeon pea virus disease in Kenya, *Proc. Symp. Grain Leg. Improv. E. Af.,* Nairobi, Kenya, 20–24 Aug. 1979.

Gathuru, E. M. and Buruchara, R. A. (1979). A cowpea yellow mottle disease of wide occurrence in Kenya, *Proc. Symp. Grain Leg. Improv. E. Af.,* Nairobi, Kenya, 20–24 Aug. 1979.

Gay, J. D. (1969a). Effects of temperature and moisture on snap bean damping-off caused by three isolates of *Pythium myriotylum, Pl. Dis. Reptr.,* **53**, 707–709.

Gay, J. D. (1969b). Effect of seed maturation on the infectivity of cowpea chlorotic mottle virus, *Phytopathology*, **59**, 802–804.

Gay, J. D. (1976), 'Worthmore' southern pea, *Hort. Sci*., **11**, 621–622.

Gerdemann, J. W. (1954). Pathogenicity of *Leptodiscus terrestris* on red clover and other Leguminosae, *Phytopathology,* **44**, 451–455.

Germani, G. and Dhery, M. (1973). Observations et expérimentations concernant le rôle des nématodes dans deux affections de l'arachide en Haute Volta: la chlorose et le clump, *Oléagineux*, **28**, 235–242.

Ghanekar, A. M. (1980). Groundnut virus research at ICRISAT, in *Proc. Intern. Wkshop Groundnuts*, ICRISAT, Hyderabad, pp. 211–216.

Ghanekar, A. M., Reddy, D. V. R., Iizuka, N., Amin, P. W., and Gibbons, R. W. (1979). Bud necrosis of groundnut (*Arachis hypogaea*) in India caused by tomato spotted wilt virus, *Ann. appl. Biol.,* **93**, 173–179.

Ghanekar, A. M. and Schwenk, F. W. (1974). Seed transmittion and distribution of tobacco streak virus in six cultivars of soybean, *Phytopathology,* **64**, 112–114.

Gibbons, R. W. (1966). *Mycosphaerella* leafspots of groundnuts, *FAO Pl. Prot. Bull.,* **14**, 25–30.

Gibbons, R. W. (1980a). Adaptation and utilisation of groundnuts in different environments and farming systems, in *Advances in Legume Science* (eds R. J. Summerfield and A. H. Bunting), HMSO, London, pp. 483–493.

Gibbons, R. W. (1980b). The ICRISAT groundnut program, in *Proc. Intern. Wkshop Groundnuts*, ICRISAT, Hyderabad, pp. 12–16.

Gibbons, R. W. and Bailey, B. E. (1967). Resistance to *Cercospora arachidicola* in some species of *Arachis, Rhod. Zamb. Mal. J. agric. Res.,* **5**, 57–59.

Gibbons, R. W., Bunting, A. H., and Smartt, J. (1972). The classification of varieties of groundnut (*Arachis hypogaea* L.), *Euphytica,* **21**, 78–85.

Gibbons, R. W., Laurence, R. C. N., Norse, D., and Cox, R. A. J. (1975). Groundnut drying trials in Malawi, *Trop. Sci.,* **17**, 15–24.

Gibbons, R. W. and Mercer, P. C. (1972). Peanut disease control in Malawi, Central Africa, *J. Am. peanut Res. Educ. Assoc.,* **4**, 58–66.

Gibbs, A. J., Harrison, B. D., Watson, D. W., and Wildy, P. (1966). What's in a virus name? *Nature (Lond.),* **209**, 450–454.

Gibbs, A. J. and Harrison, B. D. (1970). Cucumber mosaic virus. *CMI/AAB Descriptions of Plant Viruses no. 1.*

Gibbs, A. and Watson, L. (1980). Viruses and legumes, in *Advances in Legume Science* (eds R. J. Summerfield and A. H. Bunting), HMSO, London, pp. 235–241.

Gibson, I. A. S. (1953). Crown rot, a seedling disease of groundnuts caused by *Aspergillus niger, Trans. Br. mycol. Soc.,* **36**, 198–209.

Gibson, I. A. S. and Clinton, P. K. S. (1953). Pre-emergence seed-bed losses in groundnuts at Urambo, Tanganyika Territory, *Emp. J. exp. Agric.,* **21**, 226–235.

Gibson, R. W., Rice, A. D., and Sawicki, R. M. (1982). Effects of the pyrethroid

deltamethrin on the acquisition and inoculation of viruses by *Myzus persicae, Ann. appl. Biol.*, **100**, 49–54.
Gill, C. C. (1965). Increased multiplication of viruses in rusted bean and sunflower tissue, *Phytopathology*, **55**, 141–147.
Gill, J. S. and Swarup, G. (1977). Pathogenic effect of *Tylenchorhynchus vulgaris* on gram, *Indian J. Nemat.*, **7**, 155–156.
Gillier, P. (1978). Nouvelles limites des cultures d'arachides résistantes à la sécheresse et à la rosette, *Oléagineux,* **33**, 25–28.
Gilmer, R. M., Whitney, W. K., and Williams, R. J. (1974). Epidemiology and control of cowpea mosaic in Western Nigeria, *Proc. 1st IITA Grain Leg. Improv. Workshop*, International Institute of Tropical Agriculture, Ibadan, Nigeria, p. 269 (abstr.).
Givord, L. (1981). Southern bean mosaic virus isolated from cowpea (*Vigna unguiculata)* in the Ivory Coast, *Pl. Dis.*, **65**, 755–756.
Gnanamanickam, S. S. and Patil, S. S. (1977). Accumulation of anti-bacterial isoflavonoids in hypersensitively responding bean leaf tissues inoculated with *Pseudomonas phaseolicola, Physiol. Pl. Path.,* **10**, 159–168.
Golden, A. M., Epps, J. M. Riggs, R. D., Duclos, L. A., Fox, J. A., and Bernard, R. L. (1970). Terminology and identity of infraspecific forms of the soybean cyst nematode (*Heterodera glycines*), *Pl. Dis. Reptr.,* **54**, 544–546.
Goldsworthy, P. R. and Heathcote, R. (1964). Fertilizer trials with groundnuts in northern Nigeria, *Emp. J. exp. Agric.,* **31**, 351–366.
Good, J. M. (1973). Nematodes, in *Soybeans: Improvement, Production, and Uses* (ed. B. E. Caldwell), Amer. Soc. Agron., Madison, pp. 527–543.
Goode, M. J. (1961). A new race of bean rust in Arkansas, *Pl. Dis Reptr.,* **45**, 690–691.
Gooding, G. V. and Lucas, G. B. (1959). Factors influencing sporangial formation and zoospore activity in *Phytophthora parasitica* var. *nicotianae, Phytopathology,* **49**, 277–281.
Gooding, H. J. (1962). The agronomic aspects of pigeon peas, *Fld. Crop Abstr.*, **15**, 1–5.
Goodman, R. M. (1981). Geminiviruses, *J. Gen. Virol.,* **54**, 9–21.
Goodman, R. M. and Bird, J. (1978). Bean golden mosaic virus. *CMI/AAB Descriptions of Plant Viruses No. 192.*
Goodman, R. M., Bowers, G. R., and Paschal, E. H. (1979). Identification of soybean germplasm lines and cultivars with low incidence of soybean mosaic virus transmission through seeds, *Crop Sci.,* **19**, 264–267.
Goodman, R. M. and Nene, Y. L. (1976). Virus diseases of soybeans, *Expanding the use of soybeans: Proc. Conf. for Asia and Oceania,* Chiang Mai, Thailand. INTSOY series no. 10.
Goodman, R. M. and Oard, J. H. (1980). Seed transmission and yield losses in tropical soybeans infected by soybean mosaic virus, *Pl. Dis.,* **64**, 913–914.
Goodman, R. M., Sinclair , J. M., and Hepperly, P.R. (1981). Plant pathology and the International Soybean Program (INTSOY), *Pl. Dis*., **65**, 214–222.
Gorlenko, M. V. and Bushkova, L. N. (1958). Perfect state of the causal agent of ascochytosis of chickpea, *Pl. Prot., Moscow,* **3**, 60. (*Rev. appl. Mycol.,* **37**, 695.)
Goss, R. W. (1940). The relation of temperature to common and halo blight of beans, *Phytopathology,* **30**, 258–264.
Goswami, B. K. and Agarwal, D. K. (1978). Interrelationships between species of *Fusarium* and root-knot nematode. *Meloidogyne incognita*, in soybean, *Nematol. Medit.,* **6**, 125–128. (*Helminth. Abstr.,* B **48**, 570.)
Goswami, R. N. (1975). Rust, a new menace to groundnut in North-East India, *Indian Phytopath.,* **27**, 238.
Goth, R. W. and Zaumeyer, W. J. (1963). Occurrence of *Phyllosticta* leaf spot in beans in 1963, *Pl. Dis. Reptr.,* **47**, 1079.

Goth, R. W. and Zaumeyer, W. J. (1965). Reactions of bean varieties to four races of anthracnose, *Pl. Dis. Reptr.,* **49**, 815–818.

Graham, J. H. (1953). Cultural and epiphytotic relationships of three bacterial pathogens of soybeans, *Phytopathology,* **43**, 193–194.

Graham, P. H. (1978). Problems and potential of field beans (*Phaseolus vulgaris* L.) in Latin America, *Fld. Crops Res.,* **1**, 295–317.

Granada, G. A. (1979a). Machismo disease of soybeans: 1. Symptomatology and transmission, *Pl. Dis. Reptr.,* **63**, 47–50.

Granada, G. A. (1979b). Machismo disease of soybeans: 11. Suppressive effects of tetracycline on symptom development, *Pl. Dis. Reptr.,* **63**, 309–312.

Granada, G. A. (1980). Mycoplasma disease in Colombia, in *Bean Production Problems* (eds H. F. Schwartz and G. E. Gálvez), Centro Internacional de Agricultura Tropical, Cali, Colombia, pp. 204–210.

Grau, C. R. and Bissonnette, H. L. (1974). *Whetzelinia* stem rot of soybean in Minnesota, *Pl. Dis. Reptr.,* **58**, 693–695.

Green, G. J. (1973). Environmental races of *Puccinia graminis tritici* in North America and their epidemiological potential, *Proc. 2nd Int. Congr. Pl. Path.,* Minneapolis, abstr. 0720.

Green, J. M. (1978). Pigeon pea research at ICRISAT, *Trop. Grain Leg. Bull.,* **11/12**, 22–23.

Green, J. M., Sharma, D., Reddy, L. J., Saxena, K. B., Gupta, S. C., Jain, K. C., Reddy, B. V. S., and Rao, M. R. (1981). Methodology and progress in the ICRISAT pigeon pea breeding program, in *Proc. Internat. Wkshop on Pigeon Peas,* **1**, 437–449. International Crops Research Institute for the Semi-Arid Tropics, Hyderabad.

Greenland, D. J. (1975). Bringing the green revolution to the shifting cultivator, *Science,* **190**, 841–844.

Greenway, P. J. (1945). Origins of some East African food plants, part III, *E. Afr. agric. J.,* **10**, 177–180.

Gregory, W. C. and Gregory, M. P. (1976). Groundnut, *Arachis hypogaea* (Leguminosae -Papilionatae), in *Evolution of Crop Plants* (ed. N. W. Simmonds), Longman, London, pp. 151–154.

Gregory, W. C., Krapovickas, A., and Gregory, M. P. (1980). Structure, variation, evolution, and classification in *Arachis,* in *Advances in Legume Science* (eds R. J. Summerfield and A. H. Bunting), HMSO, London, pp. 469–481.

Grewal, J. S. (1981). Evidence of physiologic races in *Ascochyta rabiei, Proc. Wkshop on Ascochyta Blight and Winter Sowing of Chickpea,* 4–7 May 1981, International Center for Agricultural Research in the Dry Areas, Aleppo.

Grewal, J. S., Pal. M., and Kulshrestha, D. D. (1974). A new record of wilt of gram caused by *Fusarium solani, Curr. Sci.,* **43**, 767.

Grewal, J. S. and Vir, S. (1974). Varietal resistance of gram to *Ascochyta* blight, *Indian Phytopath.,* **27**, 643–645.

Griffiths, E. (1981). Iatrogenic plant diseases, *A. Rev. Phytopath.,* **19**, 69–82.

Grigg, D. B. (1974). *The Agricultural Systems of the World: an Evolutionary Approach,* Cambridge Univ. Press, Cambridge.

Grogan, R. G. and Kimble, K. A. (1967). The role of seed contamination in the transmission of *Pseudomonas phaseolicola* in *Phaseolus vulgaris, Phytopathology,* **57**, 28–31.

Grogan, R. G. and Walker, J. C. (1948). The relation of common mosaic to black root of bean, *J. agric. Res.,* **77**, 315–331.

Groth, J. V. (1976). Mutilines and 'super-races': a simple model, *Phytopathology,* **66**, 937–939.

Groth, J. V. (1978). Rebuttal to 'multilines and super-races - a reply', *Phytopathology,* **68**, 1538–1539.

Groth, J. V. and Mogen, B. D. (1978). Completing the life cycle of *Uromyces phaseoli* var. *typica* on bean plants, *Phytopathology,* **68**, 1674–1677.
Groth, J. V. and Urs, N. V. R. R. (1982). Differences among bean cultivars in receptivity to *Uromyces phaseoli* var. *typica, Phytopathology,* **72**, 374–378.
Guerra, E. and Dongo, S. (1973). Determinación de razas fisiológicas del hongo *Uromyces phaseoli* var. *typica* Arth. en el Perú, *Invest. Agropec.*, **3**, 92–94.
Guillemin, R. (1952). Études agronomiques sur l'arachide en A. E. F., Oubangui-Chari, Tchad, *Oléagineux,* **7**, 699–704.
Gupta, S. C. and Sinha, S. (1951). Further additions to the synchytria of India, *Indian Phytopath.*, **4**, 7–10.
Guthrie, J. W. (1968). The serological relationship of races of *Pseudomonas phaseolicola, Phytopathology*, **58**, 716–717.
Guthrie, J. W. (1970). Factors influencing halo blight transmission from externally contaminated *Phaseolus vulgaris* seed, *Phytopathology,* **60**, 371–372.
Guthrie, J. W., Huber, D. M., and Fenwick, H. S. (1965). Serological detection of halo blight, *Pl. Dis. reptr.*, **49**, 297–299.
Habish, H. A. (1972). Aflatoxin in haricot bean and other pulses, *Expl. Agric.*, **8**, 135–137.
Hadley, B. A., Beute, M. K., and Leonard, K. J. (1979). Variability of *Cylindrocladium crotalariae* response to resistant host plant selection pressure in peanut, *Phytopathology,* **69**, 1112–1114.
Hafiz, A. (1951). Cultural studies on *Ascochyta rabiei* with special reference to zonation, *Trans. Br. mycol. Soc.*, **34**, 259–269.
Hafiz, A. (1952). Basis of resistance in gram to *Mycosphaerella* blight, *Phytopathology,* **42**, 422–424.
Hafiz, A. and Ashraf, M. (1953). Studies on the inheritance of resistance to *Mycosphaerella* blight in gram, *Phytopathology*, **43**, 580–581.
Hagborg, W. A. F. (1970). A device for injecting solutions and suspensions into thin leaves of plants, *Can. J. Bot.*, **48**, 1135–1136.
Hagedorn, D. J., Crill, J. P., and Davis, J. R. (1969). Diseases of corn and soybeans in Rio Grande do Sul, Brazil in 1968, *Pl. Dis. Reptr.*, **53**, 165–168.
Hagedorn, D. J. and Rand, R. E. (1976). Resistance of *Phaseolus coccineus* to bean bacterial brown spot, *A. Rep. Bean Improv. Co-op.*, **19**, 41–43.
Hagedorn, D. J. and Rand, R. E. (1977). The first bacterial brown spot resistant bush bean, *A. Rep. Bean Improv. Co-op*., **20**, 67–68.
Hagedorn, D. J., Rand, R. E., and Ercolani, G. L. (1972). Survival of *Pseudomonas syringae* on hairy vetch in relation to epidemiology of bacterial brown spot of bean, *Phytopathology,* **62**, 762 (abstr.).
Hague, N. G. M. (1980). Nematodes of legume crops, in *Advances in Legume Science* (eds R. J. Summerfield and A. H. Bunting), HMSO, London, pp. 199–205.
Halbert, S. E. and Irwin, M. E. (1981). Effect of soybean canopy closure on landing rates of aphids with implications for restricting spread of soybean mosaic virus, *Ann. appl. Biol.,* **98**, 15–19.
Halbert, S. E., Irwin, M. E., and Goodman, R. M. (1981). Alate aphid (Homoptera: Aphididae) species and their relative importance as field vectors of soybean mosaic virus, *Ann. appl. Biol.*, **97**, 1–9.
Halfon-Meiri, A. (1970). Infection of chickpea seeds by *Ascochyta rabiei* in Israel, *Pl. Dis. Reptr.*, **54**, 442–445.
Hamblin, J. (1980). Breeding legumes for improved systems of crop protection, in *Advances in Legume Science* (eds R. J. Summerfield and A. H. Bunting), HMSO, London, pp. 243–250.
Hamblin, J. and Kent, S. P. (1973). A possible role of phytohaemagglutinin in *Phaseolus vulgaris* L., *Nature New Biology,* **245**, 28–30.
Hamilton, R. I., Edwardson, J. R., Francki, R. I. B., Hsu, H. T., Hull, R., Koenig, R.,

and Milne, R. G. (1981). Guidelines for the identification and characterization of plant viruses, *J. gen. Virol.,* **54**, 223–241.
Hammerton, J. L. (1975). Effects of defoliation on pigeon peas (*Cajanus cajan), Expl. Agric.,* **11**, 177–182.
Hammons, R. O. (1976). Peanuts: genetic vulnerability and breeding strategy, *Crop Sci.,* **16**, 527–530.
Hammons, R. O. (1977). Groundnut rust in the United States and the Caribbean, *PANS,* **23**, 300–304.
Hampton, R. O (1975).The nature of bean yield reduction by bean yellow and bean common mosaic virus, *Phytopathology,* **65**, 1342–1346.
Hampton, R., Beczner, L., Hagedorn, D., Bos, L., Inouye, T., Barnett, O., Musil, M., and Meiners, J. (1978). Host reactions of mechanically transmissible legume viruses of the northern temperate zone, *Phytopathology*, **68**, 989–997.
Hansford, C. G. (1937). Annotated host list of Uganda parasitic fungi and plant diseases – Part III, *E. Afr. agric. J.*, **3**, 79–84.
Haq, M. A., Shakoor, A., Sadik, M., and Mahmud-ul-Hassan (1981). Induction of *Ascochyta* blight resistant mutants in chickpea, *Mutation Breed. Newsl.,* **17**, 5–6.
Haque, S. Q. and Persad, G. C. (1975). Some observations of the seed-transmission of beetle-transmitted cowpea mosaic virus, in *Tropical Diseases of Legumes* (eds J. Bird and K. Maramorosch), Academic Press, New York, pp. 119–121.
Haragopal, T. and Nayudu, M. V. (1971). A new sap transmissible groundnut chlorotic spot virus, *Phytopath. Z.,* **71**, 33–41.
Hardwick, N. V., Greenwood, A. D., and Wood, R. K. S. (1971). The fine structure of the haustorium of *Uromyces appendiculatus* in *Phaseolus vulgaris, Can. J. Bot.,* **49**, 383–390.
Hare, W. W. (1953). A new race of *Fusarium* causing wilt of cowpea, *Phytopathology*, **43**, 291 (abstr.).
Hare, W. W. (1957). Mississippi Crowder, a new disease-resistant cowpea, *Phytopathology,* **47**, 565–566.
Hare, W. W. (1959). Resistance to root-knot nematode in cowpea, *Phytopathology,* **49**, 318 (abstr.).
Hare, W. W. (1967). A combination of disease resistance in a new cowpea, Mississippi Silver, *Phytopathology,* **57**, 460 (abstr.).
Harlan, J. R. (1976). Disease as a factor in plant evolution, *A Rev. Phytopath.*, **14**, 31–51.
Harlan, J. R. and de Wet, J. M. J. (1971). Towards a rational classification of cultivated plants, *Taxon*, **20**, 509–517.
Harris, H. B. and Kuhn, C. W. (1971). Influence of cowpea chlorotic mottle virus (soybean strain) on agronomic performance of soybeans, *Crop Sci.,* **11**, 71–72.
Harrison, A. L. (1972). Observation on the development and spread of peanut rust in South Texas in 1971, *Pl. Dis. Reptr.*, **56**, 873–874.
Harrison, A. L. (1973). Control of peanut leaf rust alone or in combination with *Cercospora* leaf spot, *Phytopathology*, **63**, 668–673.
Harrison, A. N. and Gudauskas, R. T. (1968). Effects of some viruses on growth and seed production of two cowpea cultivars, *Pl. Dis. Reptr.*, **52**, 509–512.
Harrison, B. D., Finch, J. T., Gibbs, A. J., Hollings, M., Shepherd, R. J., Valenta, V., and Wetter, C. (1971). Sixteen groups of plant viruses, *Virology*, **45**, 356–363.
Harrison, B. D., Barker, H., Bock, K. R., Guthrie, E. J., Meredith, G., and Atkinson, M. (1977). Plant viruses with circular single-stranded DNA, *Nature, Lond.*, **270**, 760–762.
Harter, L. L. (1939). Physiologic races of the fungus causing bean rust, *Phytopathology*, **29**, 9 (abstr.).
Harter, L. L., Andrus, C. F., and Zaumeyer, W. J. (1935). Studies on bean rust caused by *Uromyces phaseoli typica, J. agric. Res.*, **50**, 737–759.

Harter, L. L. and Zaumeyer, W. J. (1931). *Pythium butleri* – the cause of a bean wilt, *Phytopathology*, **21**, 991–994.
Harter, L. L. and Zaumeyer, W. J. (1941). Differentiation of physiologic races of *Uromyces phaseoli typica* on bean, *J. agric. Res.*, **62**, 717–731.
Hartwig, E. E. (1959). Effect of target spot on yield of soybeans, *Pl. Dis. Reptr.*, **43**, 504–505.
Hartwig, E. E. (1975). Breeding soybeans resistant to diseases, in *Internat. Workshop on Grain Legumes*, International Crops Research Institute for the Semi-Arid Tropics, Hyderabad, pp. 305–310.
Hartwig, E. E. (1978). Breeding soybeans for rust resistance, in *Proc. Workshop Soybean Rust in W. Hemisphere* (ed. N. G. Vakili), USDA, pp. 26–28.
Hartwig, E. E. and Johnson, H. W. (1953). Effect of bacterial pustule disease on yield and chemical composition of soybeans, *Agron. J.*, **45**, 22–23.
Hartwig, E. E. and Lehman, S. G. (1951). Inheritance of resistance to bacterial pustule disease in soybeans, *Agron. J.*, **43**, 226–229.
Hasan, A. and Khan, A. M. (1979). Varietal screening for anthracnose disease of arhar, *Acta Bot. Indica*, **7**, 126–128. (*Rev. Pl. Path.*, **59**, 4859.)
Hassan, A. A., Wilkinson, R. E., and Wallace, D. H. (1971). Relationship between genes controlling resistance to *Fusarium* and *Thielaviopsis* root rots in beans, *J. Am. Soc. hort. Sci.*, **96**, 631–632.
Hassan, H. N. and Beute, M. K. (1977). Evaluation of resistance to *Cercospora* leaf spot in peanut potentially useful in a breeding program, *Peanut Sci.*, **4**, 78–83.
Haware, M. P., Kumar, J., and Reddy, M. V. (1980). Disease resistance in kabuli-desi chickpea introgression, in *Proc. Internat. Wkshop in Chickpea Improv.*, International Crops Research Institute for the Semi-Arid Tropics, Hyderabad, pp. 67–74.
Haware, M. P. and Nene, Y. L. (1976). Some uncommon but potentially serious diseases of chickpea, *Trop. Grain Leg. Bull.*, **5**, 26–30.
Haware, M. P. and Nene, Y. L. (1979a). Physiologic races of the chickpea wilt pathogen, *Internat. Chickpea Newsl.*, **1**, 7–8.
Haware, M. P. and Nene, Y. L. (1979b). Symptomless carriers of the chickpea wilt fungus, *Internat. Chickpea Newsl.*, **1**, 8.
Haware, M. P. and Nene, Y. L. (1980a). Influence of wilt at different growth stages on yield loss in chickpea, *Trop. Grain Leg. Bull.*, **19**, 38–40.
Haware, M. P. and Nene, Y. L. (1980b). Sources of resistance to wilt and root rots of chickpea, *Internat. Chickpea Newsl.*, **3**, 11–12.
Haware, M. P. and Nene, Y. L. (1981). *Phoma* blight – a new disease of chickpea, *Pl. Dis.*, **65**, 282.
Haware, M. P., Nene, Y. L., and Rajeshwari, R. (1978). Eradication of *Fusarium oxysporum* f. sp. *ciceri* transmitted in chickpea seed, *Phytopathology*, **68**, 1364–1367.
Haware, M. P., Nene, Y. L., and Rao, N. (1981). Additional sources of resistance to wilt and root rots of chickpea, *Internat. Chickpea Newsl.*, **4**, 18.
Haware, M. P. and Pavgi, M. S. (1976a). Cytology of chlamydospore development in *Protomycopsis* species, *Cytologia*, **41**, 459–465.
Haware, M. P. and Pavgi, M. S. (1976b). Chlamydospore germination in *Protomycopsis* species, *Mycopathologia*, **59**, 105–111.
Hawthorne, P. L. (1943). The breeding and improvement of edible cowpeas, *Proc. Am. Soc. hort. Sci.*, **42**, 562–564.
Hawtin, G. C. (1974). The exploitation of genetic variation in *Glycine max* (L.) Merrill. Ph.D. thesis, Univ. Cambridge, 250 pp.
Hawtin, G. C. and Singh, K. B. (1980). Kabuli-desi introgression: problems and prospects, in *Proc. Internat. Wkshop in Chickpea Improv.*, International Crops Research Institute for the Semi-Arid Tropics, Hyderabad, pp. 51–60.

Hawtin, G. C., Singh, K. B., and Saxena, M. C. (1980). Some recent developments in the understanding and improvement of *Cicer* and *Lens,* in *Advances in Legume Science* (eds R. J. Summerfield and A. H. Bunting), HMSO, London, pp. 613–623.

Hayes, R. T. (1932). Groundnut rosette disease in the Gambia, *Trop. Agric. (Trin.),* **9**, 211–217.

Hayward, A. C. and Waterston, J. M. (1965a). *Corynebacterium flaccumfaciens. CMI Descriptions of Pathogenic Fungi and Bacteria No. 43.*

Hayward, A. C. and Waterson, J. M. (1965b). *Xanthomonas phaseoli. CMI Descriptions of Pathogenic Fungi and Bacteria No. 48.*

Hayward, A. C. and Waterston, J. M. (1965c). *Xanthomonas phaseoli* var. *fuscans. CMI Descriptions of Pathogenic Fungi and Bacteria No. 49.*

Heath, M. C. (1971). Haustorial sheath formation on cowpea leaves immune to rust infection, *Phytopathology,* **61**, 383–388.

Heath, M. C. (1972). Ultrastructure of host and non-host reactions to cowpea rust, *Phytopathology,* **62**, 27–38.

Heath, M. C. (1974). Light and electron microscope studies of the alterations of host and non-host plants with cowpea rust - *Uromyces phaseoli* var. *vignae, Physiol. Pl. Path.,* **4**, 403–414.

Heath, M. C. and Heath, I.B. (1971). Ultrastructure of an immune and a susceptible reaction of cowpea leaves to rust infection, *Physiol. Pl. Path.,* **1**, 277–287.

Hedges, F. (1944). Association of *Xanthomonas phaseoli* and the common bean mosaic virus, *Marmor phaseoli.* 1. Effect on pathogenicity of the seedborne infective agents, *Phytopathology,* **34**, 662–693.

Helms, K., Grylls, N. E., and Purss, G. S. (1961). Peanut plants in Queensland infected with tomato spotted wilt virus, *Aust. J. agric. Res.,* **12**, 239–246.

Hemingway, J. S. (1954). *Cercospora* leafspots of groundnuts in Tanganyika, *E. Afr. agric. J.,* **19**, 263–271.

Hemingway, J. S. (1955). The prevalence of two species of *Cercospora* on groundnuts, *Trans. Br. mycol. Soc.,* **38**, 243–246.

Hemingway, J. S. (1957). The resistance of groundnuts to *Cercospora* leafspots, *Emp. J. exp. Agric.,* **25**, 60–68.

Hemmi, T. (1955). Notes on new or noteworthy plant diseases in Japan - 1. (*Rev. appl. Mycol.,* **36**, 9–10.)

Hendrickx, F. L. (1939). Observations phytopathologiques à la station de Mulungu en 1938, *Publ. Inst. nat. Étud. agron. Congo Belge,* pp. 117–128. (*Rev. appl. Mycol.,* **19**, 329–330.)

Hepperly, P. R., Bowers, G. R., Sinclair, J. B., and Goodman, R. M. (1979). Predisposition to seed infection by *Phomopsis sojae* in soybean plants infected by soybean mosaic virus, *Phytopathology,* **69**, 846–848.

Hepperly, P. R., Kirkpatrick, B. L., and Sinclair, J. B. (1980). *Abutilon theophrasti:* wild host for three fungal parasites of soybean, *Phytopathology,* **70**, 307–310.

Hepperly, P. R. and Sinclair, J. B. (1981). Relationships among *Cercospora kikuchii,* other seed mycoflora, and germination of soybeans in Puerto Rico and Illinois, *Pl. Dis.,* **65**, 130–132.

Herrera, W. T., Zandstra, H. G., and Liboon, S. P. (1978). The management of mungbean in rice-based cropping systems, *Proc. 1st Internat. Mungbean Symp.*, Asian Vegetable Research and Development Centre, Taiwan, pp. 115–119.

Higgins, B. B. (1935). Breeding peanuts for disease resistance, *Phytopathology,* **25**, 971–972 (abstr.).

Higgins, B. B. (1951). Origin and early history of the peanut, in *The Peanut, the Unpredictable Legume,* National Fertilizer Assoc., Washington DC, pp. 18–27.

Higgins, B. B. (1963). Les maladies de l'arachide aux Etats-Unis, *Oléagineux,* **11**, 213–230.

Hikida, H. R. (1961). Race 33 of *Uromyces phaseoli* var. *typica* Arth., a distinct physiologic race of bean rust from Oregon, *Pl. Dis. Reptr.,* **45**, 388.

Hildebrand, A. A. (1959). A root and stalk rot of soybean caused by *Phytophthora megasperma, Can. J. Bot.*, **37**, 927–957.
Hill, G. C. H. (1981). Chickpea research in Zambia, 1977–1980, *Internat. Chickpea Newsl.*, **4**, 9–10.
Hill, J. H., Lucas, B. S., Benner, H. I., Tachibana, H., Hammond, R. B., and Pedigo, L. P. (1980). Factors associated with the epidemiology of soybean mosaic virus in Iowa, *Phytopathology,* **70**, 536–540.
Hill, K., Coyne, D. P., and Schuster, M. L. (1972). Leaf, pod, and systemic chlorosis reactions in *Phaseolus vulgaris* to halo blight controlled by different genes, *J. Am. Soc. hort. Sci.*, **97**, 494–498.
Hiratsuka, N. (1936). *Phakopsora* in Japan, *Bot. Mag., Tokyo,* **49**, **50**. (*Rev. appl. Mycol.*, **15**, 401.)
Hirumi, H., Maramorosch, K., and Hichez, E. (1973). Rhabdovirus and mycoplasma-like organism: natural dual infection of *Cajanus cajan, Phytopathology,* **63**, 202 (abstr.).
Ho, H. H. (1969). Notes on the behaviour of *Phytophthora megasperma* var. *sojae* in soil, *Mycologia,* **61**, 835–838.
Ho, H. H. and Hickman, C. J. (1967a). Asexual reproduction and behaviour of zoospores of *Phytophthora megasperma* var. *sojae, Can. J. Bot.*, **45**, 1963–1981.
Ho. H. H. and Hickman, C. J. (1967b). Factors governing zoospore responses of *Phytophthora megasperma* var. *sojae* to plants roots, *Can. J. Bot.*, **45**, 1983–1994.
Hobbs, H. A. (1981). Transmission of bean curly dwarf mosaic virus and bean mild mosaic virus by beetles on Costa Rica, *Pl. Dis.*, **65**, 491–492.
Hobbs, H. A. and Fulton, J. P. (1979). Beetle transmission of cowpea chlorotic mottle virus, *Phytopathology,* **69**, 255–256.
Hoch, H. C., Hagedorn, D. J., Pinnow, D. L., and Mitchell, J. E. (1975). Role of *Pythium* spp. as incitants of bean root and hypocotyl rot in Wisconsin, *Pl. Dis. Reptr.*, **59**, 443–447.
Hocking, D. (1967). A new virulent form of *Phaeoisariopsis griseola* causing circular leaf spot of french bean, *Pl. Dis. Reptr.*, **51**, 276–278.
Hoffmaster, D. E. (1944). Bacterial canker of cowpea, *Phytopathology,* **34**, 439–441.
Hoitink, H. A. J., Pelletier, R. L., and Coulson, J. G. (1966). Toxemia of halo blight of beans, *Phytopathology,* **56**, 1062–1065.
Holliday, P. (1970). *Fusarium oxysporum* f.sp. *tracheiphilum. CMI Descriptions of Pathogenic Fungi and Bacteria No. 220.*
Holliday, P. (1980). *Fungus Diseases of Tropical Crops,* Cambridge Univ. Press, 607 pp.
Holliday, P. and Punithalingham, E. (1970). *Macrophomina phaseolina. CMI Descriptions of Pathogenic Fungi and Bacteria No. 275.*
Hollings, M., Pawley, R. R., and Bock, K. R. (1981). French bean, *Rep. Glasshouse Crops Res. Inst. for 1979,* 159.
Holmes, F. O. (1951). Indications of a New World origin of tobacco mosaic virus, *Phytopathology,* **41**, 341–349.
Honma, S. (1956). A bean interspecific hybrid, *J. Hered.*, **47**, 217–220.
Honma, S. and Heeckt, O. (1958). Bean interspecific hybrid involving *Phaseolus coccineus* and *P. lunatus, Proc. Am. Soc. hort. Sci.*, **72**, 360–364.
Horn, N. Y., Newsom, L. D., and Jensen, R. L. (1973). Economic injury thresholds of bean pod mottle and tobacco ringspot virus infection of soybeans, *Pl. Dis. Reptr.*, **57**, 811–813.
Horsfall, J. G. and Cowling, E. B. (eds) (1980). *Plant disease, an Advanced Treatise.* V. How plants defend themselves, Academic Press, New York, 534 pp.
Hösel, W. and Barz, W. (1970). Flavonoide aus *Cicer arietinum* L., *Phytochemistry,* **9**, 2053–2056.
Houston, B. R. (1945). Culture types and pathogenicity of isolates of *Corticium solani, Phytopathology,* **35**, 371–393.

Howard, A., Howard, G. L. C. and Khan, A. R. (1915). Some varieties of Indian gram (*Cicer arietinum* L.), *Mem. Dept. Agric. India (Bot)*, **7**, 213–235.
Howland, A. K. and Macartney, J. C. (1966). East African bean rust studies, *E. Af. agric. For. J.*, **32**, 208–210.
Hsi, E. C. H. (1966). Observations on an outbreak of *Aspergillus* crown rot of Valencia peanuts in New Mexico, *Pl. Dis. Reptr.*, **50**, 175–177.
Hubbeling, N. (1957). New aspects of breeding for disease resistance in beans (*Phaseolus vulgaris* L), *Euphytica*, **6**, 111–141.
Hubbeling, N. (1969). Genes for resistance to bean common mosaic and strains of the virus, *A. Rep. Bean Improv. Co-op.*, **12**, 26–27.
Hubbeling, N. (1973). Field resistance in *Phaseolus* beans to *Pseudomonas phaseolicola*, *Meded. Fak. Landb., Gent*, **38**, 1351–1363.
Hubbeling, N. (1977). The new jota race of *Colletotrichum lindemuthianum*, *A. Rep. Bean Improv. Co-op.*, **20**, 58.
Hull, R. (1964). Spread of groundnut rosette virus by *Aphis craccivora* Koch, *Nature, Lond.*, **202**, 213–214.
Hull, R. and Adams, A. N. (1968). Groundnut rosette and its assistor virus, *Ann. appl. Biol.*, **62**, 139–145.
Hunt, P. (1968). Cuticular penetration by germinating uredospores, *Trans. Br. mycol. Soc.*, **51**, 103–112.
Hunter, J. E., Dickson, M. H., and Cigna, J. A. (1981). Limited-term inoculation: a method to screen bean plants for partial resistance to white mold, *Pl. Dis.*, **65**, 414–417.
Hutchinson, J. B. (1970). The evolutionary diversity of the pulses, *Proc. Nutr. Soc.*, **29**, 49–55.
Hutton, D. G. and Hammerton, J. L. (1975). Investigating the role of *Rotylenchulus reniformis* in a decline of pigeon pea. *Nematropica*, **5**, 24. (*Helminth. Abstr.*, B **45**, 634.)
Hutton, D. G., Wilkinson, R. E., and Mai, W. F. (1973). Effect of two plant parasitic nematodes on *Fusarium* dry rot of beans, *Phytopathology*, **63**, 749–751.
Hutton, E. M. and Grylls, N. E. (1956). Legume 'little leaf', a virus disease of subtropical pasture species, *Aust. J. agric. Res.*, **7**, 85–97.
Hwang, S. C. and Ko, W. H. (1978). Biology of chlamydospores, sporangia, and zoospores of *Phytophthora cinnamomi* in soil, *Phytopathology*, **68**, 726–731.
Hymowitz, T. (1970). On the domestication of the soybean, *Econ. Bot.*, **24**, 408–421.
Hymowitz, T. (1976). Soybeans, in *Evolution of Crop Plants* (ed. N. W. Simmonds), Longman, London, pp. 159–162.
Hymowitz, T. and Newell, C. A. (1980). Taxonomy, speciation, domestication, dissemination, germplasm resources and variation in the genus *Glycine*, in *Advances in Legume Science* (eds R. J. Summerfield and A. H. Bunting), HMSO, London, pp. 251–264.
Hymowitz, T. and Newell, C. A. (1981). Taxonomy of the genus *Glycine*, domestication and uses of soybeans, *Econ. Bot.*, **35**, 272–288.
IARI (1979). *A. Rep. for 1977*, Indian Agricultural Research Institute, New Delhi, 154 pp.
ICRISAT (1976). *A. Rep. for 1975–76*, International Crops Research Institute for the Semi-Arid Tropics, Hyderabad, 233 pp.
ICRISAT (1977). *A. Rep. Pulse Pathology*, Part A: pigeon pea pathology, International Crops Research Institute for the Semi-Arid Tropics, Hyderabad, 221 pp.
ICRISAT (1978). *A. Rep. for 1977–78*, International Crops Research Institute for the Semi-Arid Tropics, Hyderabad, 295 pp.
ICRISAT (1980). *A. Rep. for 1978–79*, International Crops Research Institute for the Semi-Arid Tropics, Hyderabad, 288 pp.
Ie, T. S. (1970). Tomato spotted wilt virus. *CMI/AAB Descriptions of Plant Viruses No. 39.*

Igbozurike, M. U. (1971). Ecological balance in tropical agriculture, *Geogr. Rev.*, **61**, 521–529.
IITA (1973). *Rep. Grain Legume Improv. Program for 1972,* International Institute of Tropical Agriculture, Ibadan, Nigeria, 83 pp.
IITA (1974). *Cowpea Germplasm Catalog No. 1,* International Institute of Tropical Agriculture, Ibadan, Nigeria, 209 pp.
IITA (1976). *A Rep. for 1975*, International Institute of Tropical Agriculture, Ibadan, Nigeria, 219 pp.
IITA (1977). *A. Rep. for 1976*, International Institute of Tropical Agriculture, Ibadan, Nigeria, 126 pp.
IITA (1978). *A. Rep. for 1977*, International Institute of Tropical Agriculture, Ibadan, Nigeria, 98 pp.
IITA (1979). *A. Rep. for 1978*, International Institute of Tropical Agriculture, Ibadan, Nigeria, 129 pp.
IITA (1980). *A. Rep. for 1979*, International Institute of Tropical Agriculture, Ibadan, Nigeria, 152 pp.
IITA (1981). *A. Rep. for 1980*, International Institute of Tropical Agriculture, Ibadan, Nigeria, 185 pp.
Iizuka, N. (1973). Seed transmission of viruses in soybean, *Bull. Tohoku Natl. Agric. Exp. Stn.*, **46**, 131–141. (*Rev. Pl. Path.*, **53**, 4660.)
Ilyas, M. B., Dhingra, O. D., Ellis, M. A., and Sinclair, J. B. (1975). Location of mycelium of *Diaporthe phaseolorum* var. *sojae* and *Cercospora kikuchii* in infected soybean seeds, *Pl. Dis. Reptr.*, **59**, 17–19.
Ingham, J. L. (1976a). Induced isoflavonoids from fungus-infected stems of pigeon pea (*Cajanus cajan*), *Z. Naturf.*, C **31**, 504–508.
Ingham, J. L. (1976b). Induced and constitutive isoflavonoids from stems of chickpeas (*Cicer arietinum* L.) inoculated with spores of *Helminthosporium carbonum* Ellstrup, *Phytopath. Z.*, **87**, 353–367.
Ingham, J. L., Keen, N. T., and Hymowitz, T. (1977). A new isoflavone phytoalexin from fungus-inoculated stems of *Glycine wightii, Phytochemistry,* **16**, 1943.
Innes, N. L., Hardwick, R. C., Taylor, J. D., and Walkey, D. G. A. (1977). Research on *Phaseolus vulgaris* at the National Vegetable Research Station, Wellesbourne, England, *A. Rep. Bean Improv. Co-op.*, **30**, 106–107.
Innes, N. L. and Walkey, D. G. A. (1980). The genetics of resistance to two strains of bean common mosaic virus in three cultivars of *Phaseolus vulgaris* L., *J. agric. Sci., Camb.*, **95**, 619–630.
IRRI (1975). *Standard Evuluation System for Rice,* International Rice Research Institute, Los Baños, Philippines.
Irwin, M. E. and Goodman, R. M. (1981). Ecology and control of soybean mosaic virus, in *Plant Diseases and Vectors: Ecology and Epidemiology* (eds K. Maramorosch and K. F. Harris), Academic Press, New York, pp. 181–220.
Iwaki, M. and Auzay, H. (1978). Virus diseases of mungbean in Indonesia, *Proc. 1st International Mungbean Symp.* Asian Vegetable Research and Development Centre, Taiwan, pp. 169–172.
Iwaki, M., Roechan, M., Hibino, H., Tochihara, H., and Tantera, D. M. (1980). A persistent aphidborne virus of soybean, Indonesian soybean dwarf virus, *Pl. Dis.*, **64**, 1027–1030.
Iwaki, M., Thongmeearkom, P., Honda, Y., Prommin, M., and Hibi, T. (1981). Three whitefly-borne virus diseases of legumes in Thailand, *Workshop Pathogens Transmitted by Whiteflies,* Oxford, 31 July 1981 (abstr.).
Jackson, C. R. (1962). *Aspergillus* crown rot of peanut in Georgia, *Pl. Dis. Reptr.*, **46**, 888–892.
Jackson, C. R. (1965). Reduction of *Sclerotium bataticola* infection of peanut kernels by *Aspergillus flavus, Phytopathology,* **55**, 934.

Jackson, C. R. and Bell, D. K. (1968). *Leptosphaerulina crassiasca* (Sechet) comb. nov., the cause of leaf scorch and pepper spot of peanut, *Oléagineux,* **23**, 387–388.

Jackson, H. R. and Wallen, V. R. (1975). Microdensitometer measurements of sequential aerial photographs of field beans infected with bacterial blight, *Phytopathology,* **65**, 961–968.

Jacoway, J. A. and Owen, J. H. (1951). Stem blight of peanut, *Phytopathology*, **41**, 19 (abstr.).

Jain, H. K. and Mehra, K. L. (1980). Evolution, adaptation, relationships, and uses of the species of *Vigna* cultivated in India, in *Advances in Legume Science* (eds R. J. Summerfield and A. H. Bunting), HMSO, London, pp. 459–468.

Jamal, A. (1976). Studies on the relationship between *Meloidogyne incognita* and behaviour of *Cicer arietinum* roots, *Curr. Sci.,* **45**, 230–231.

James, W. C. and Teng, P. S. (1979). The quantification of production constraints associated with plant diseases, *App. Biol.,* **4**, 201–267.

Janarthanan, R., Navaneethan, G., Subramanian, K. S., and Samuel, G. S. (1972). Some observations on the transmission of sterility mosaic disease of pigeon pea, *Curr. Sci.,* **41**, 646–647.

Janarthanan, R., Samuel, G. S., Subramanian, K. S., Naveneethan, G., and Kandasamy, T. K. (1973). A report on the survey of sterility mosaic disease incidence on redgram in Tamilnadu, *Madras agric. J.,* **60**, 41–44.

Jansen, W. P. and Staples, R. (1970a). Effect of cowpeas and soybeans as source or test plants of cowpea mosaic virus on vector efficiency and retention of infectivity of the bean leaf beetle and the spotted cucumber beetle, *Pl. Dis. Reptr.,* **54**, 1053–1055.

Jansen, W. P. and Staples, R. (1970b). Transmission of cowpea mosaic virus by the Mexican bean beetle, *J. econ. Ent.,* **63**, 1719–1720.

Jansen, W. P. and Staples, R. (1971). Specificity of transmission of cowpea mosaic virus by species within the subfamily Galerucinae, family Chrysomelidae, *J. econ. Ent.,* **64**, 365–367.

Janzen, D.H. (1973). Tropical agroecosystems, *Science,* **182**, 1212–1219.

Janzen, D. H., Juster, H. B. and Liener, I. E. (1976). Insecticidal action of the phytohaemagglutinin in black beans on a bruchid beetle, *Science* **192**, 795–796.

Janzen, D. H., Juster, H. B., and Bell, E. A. (1977). Toxicity of secondary compounds to the seed-eating larvae of the bruchid beetle *Callosobruchus maculatus, Phytochemistry,* **16**, 223–227.

Jaspers, E. M. J. and Bos, L. (1980). Alfalfa mosaic virus. *CMI/AAB Descriptions of Plant Viruses No. 229.*

Javaid, I. and Ashraf, M. (1978). Some observations on soybean diseases in Zambia and occurrence of *Pyrenochaeta glycines* on certain varieties, *Pl. Dis. Reptr.,* **62**, 46–47.

Jeger, M. J., Griffiths, E., and Jones, D. G. (1981). Effects of cereal cultivar mixtures on disease epidemics caused by splash-dispersed pathogens, in *Strategies for the Control of Cereal Disease* (eds J. F. Jenkyn and R. T. Plumb), Blackwell, Oxford, pp. 81–88.

Jehle, R. A. and Wood, J. I. (1926). Diseases of field and vegetable crops in the United States in 1925, *Pl. Dis Reptr. Suppl.,* **45**, 152. (*Rev. appl. Mycol.,* **5**, 597–599.)

Jenkins, A. E. (1931a). Scab of *Canavalia* caused by *Elsinoe canavaliae., J. agric. Res.,* **42**, 1–12.

Jenkins, A. E. (1931b). Lima bean scab caused by *Elsinoe, J. agric. Res.,* **42**, 13–23.

Jenkins, A. E. (1933a). Application of the terms 'anthracnose' and 'scab' to plant diseases caused by *Sphaceloma* and *Gloeosporium, Phytopathology,* **23**, 389–395.

Jenkins, A. E. (1933b). Further studies of lima-bean scab, *Phytopathology,* **23**, 662–666.

Jenkins, A. E. (1951). *Sphaceloma* scab, a new disease of soybeans discovered by plant pathologists in Japan, *Pl. Dis. Reptr.*, **35**, 110–11.
Jenkins, A. E. and Bitancourt, A. A. (1941). Revised descriptions of the genera *Elsinoë* and *Sphaceloma, Mycologia,* **33**, 338–340.
Jenkins, S. F., Hammons, R. O., and Dukes, P. D. (1966). Disease reaction and symptom expression of seventeen peanut cultivars to bacterial wilt, *Pl. Dis. Reptr.,* **50**, 520–523.
Jenkins, W. A. (1938). Two fungi causing leaf spot of peanut, *J. agric. Res.,* **56**, 317–332.
Jenkins, W. A. (1941). A histological study of snap bean diseases affected with black root, *J. agric. Res.,* **62**, 683–690.
Jensen, J. H. and Livingston, J. E. (1944). Variation in symptoms produced by isolates of *Phytomonas medicaginis* var. *phaseolicola, Phytopathology,* **34**, 471–480.
Jensen, R. E. and Boyle, L. W. (1966). A technique for forcasting leafspot on peanuts, *Pl. Dis. Reptr.,* **50**, 810–814.
Jhooty, J. S. and Munshi, G. D. (1980). Identity of powdery mildew of cowpea in Punjab, *Indian Phytopath.,* **33**, 106–107.
Jindal, J. K., Patel, P. N., and Khan, A. M. (1981). Variability in xanthomonads of grain legumes 11. Pathogenic variability in *Xanthomonas phaseoli* mungbean strain, *X. vignicola* and *X. phaseoli* var. *sojense, Phytopath. Z.*, **100**, 1–9.
Johnson, A. W., Rosebery, G., and Parker, C. (1976). A novel approach to *Striga* and *Orobanche* control using synthetic germination stimulants, *Weed Res.,* **16**, 223.
Johnson, B. (1953). The injurious effects of the hooked epidermal hairs of French beans (*Phaseolus vulgaris* L.) on *Aphis craccivora* Koch, *Bull. ent. Res.,* **44**, 779–788.
Johnson, E. M. and Valleau, W. D. (1949). Synonymy in some common species of *Cercospora, Phytopathology,* **39**, 763–770.
Johnson, H. W. and Jones, J. P. (1962). Purple stain of guar, *Phytopathology*, **52**, 269–272.
Johnson, J. C. (1969). 'Halo-less' halo blight of French bean in Queensland, *Qd J. Agric. Anim. Sci.,* **26**, 293–302.
Johnson, R. (1978). Practical breeding for durable resistance to rust diseases in self-pollinating cereals, *Euphytica,* **27**, 529–540.
Johnson, R. (1979). The concept of durable resistance, *Phytopathology,* **69**, 198–199.
Johnson, R. (1981a). Durable resistance: a definition of, genetic control, and attainment in plant breeding, *Phytopathology*, **71**, 567–568.
Johnson, R. (1981b). Durable disease resistance, in *Strategies for the Control of Cereal Disease* (eds J. F. Jenkyn and R. T. Plumb), Blackwell Scientific Publ., Oxford, pp. 55–63.
Johnson, R. and Allen, D. J. (1975). Induced resistance to rust diseases and its possible role in the resistance of multiline varieties, *Ann. appl. Biol.,* **80**, 359–363.
Johnson, R. and Taylor, A. J. (1976). Effects of resistance induced by non-virulent races of *Puccinia striiformis, Cereal Rusts Conf.,* Interlaken, Switzerland, September 1976.
Jones, E. D. (1960). Aecial stage of bean rust found in New York State, *Pl. Dis. Reptr.,* **44**, 809.
Jones, J. P. (1961). A weed host of *Xanthomonas phaseoli* var. *sojense, Phytopathology,* **51**, 206.
Jones, J. P. (1968). Survival of *Cercospora kikuchii* on soybean stems in the field, *Pl. Dis. Reptr.,* **52**, 931–934.
Jones, J. P. and Hartwig, E.E. (1959). A simplified method for field inoculation of soybeans with bacteria, *Pl. Dis. Reptr.,* **43**, 946.
Jones, R. A. C. (1981). The ecology of viruses infecting wild and cultivated potatoes in the Andean region of South America, in *Pests, Pathogens and Vegetation* (ed. J. M. Thresh), Pitman, London, pp. 89–107.

Joshi, A. B. (1957). Genetics of resistance to diseases and pests, *Indian J. Genet. Pl. Breed.*, **17**, 305–317.

Joshi, M. M. and Singh, R. S. (1969). A *Botrytis* gray mold of gram, *Indian Phytopath.*, **22**, 125–128.

Kahn, R. P. and Latterell, F. M. (1955). Symptoms of bud-blight of soybeans caused by the tobacco and tomato-ringspot viruses, *Phytopathology*, **45**, 500–502.

Kaiser, S. A. K. M. and Sen Gupta, P. K. (1976). Serological and electrophoretic studies of three *formae speciales* of *Fusarium oxysporum, Trans. Br. mycol. Sco.*, **67**, 33–37.

Kaiser, S. A. K. M. and Sen Gupta, P. K. (1977). Inhibition of wilt symptoms by *Fusarium oxysporum* f.sp. *udum* in pigeon pea (*Cajanus cajan*) induced by other *formae speciales* of *Fusarium oxysporum, Phytopath. Medit.*, **16**, 1–4.

Kaiser, W. J. (1972a). Occurrence of three fungal diseases of chickpea in Iran, *FAO Pl. Prot. Bull.*, **20**, 73–78.

Kaiser, W. J. (1972b). Diseases of food legumes caused by pea leafroll virus in Iran, *FAO Pl. Prot. Bull.*, **20**, 127–132.

Kaiser, W. J. (1973). Factors affecting growth, sporulation, pathogenicity, and survival of *Ascochyta rabiei, Mycologia,* **65**, 444–447.

Kaiser, W. J. (1976). Important diseases and pests of bean (*Phaseolus vulgaris*), lima bean (*Phaseolus lunatus*) and pigeon pea (*Cajanus cajan*) in Africa, *Af. J. Pl. Prot.*, **1**, 97–102.

Kaiser, W. J. (1979). Natural infection of cowpea and mung bean by alfalfa mosaic virus in Iran, *Pl. Dis. Reptr.*, **63**, 414–418.

Kaiser, W. J. (1981a). Diseases of chickpea, lentil, pigeon pea and tepary bean in Continental United States and Puerto Rico, *Econ. Bot.*, **35**, 300–320.

Kaiser, W. J. (1981b). Control of ascochyta blight through clean seed, in *Proc. Wkshop on Ascochyta Blight and Winter Sowing of Chickpea,* 4–7 May 1981, International Center for Agricultural Research in the Dry Areas, Aleppo, Syria.

Kaiser, W. J. and Danesh, D. (1971a). Etiology of virus-induced wilt of *Cicer arietinum, Phytopathology,* **61**, 453–457.

Kaiser, W. J. and Danesh, D. (1971b). Biology of four viruses affecting *Cicer arietinum* in Iran, *Phytopathology,* **61**, 372–375.

Kaiser, W. J., Danesh, D., Okhovat, M., and Mossahebi, H. (1968). Diseases of pulse crops (edible legumes) in Iran, *Pl. Dis. Reptr.,* **52**, 687–691.

Kaiser, W. J. and Hannan, R. M. (1981). Pre-emergence damping-off of chickpea in the Palouse region of eastern Washington caused by *Pythium ultimum, Phytopathology,* **71**, 230 (abstr.).

Kaiser, W. J. and Melendez, P. L. (1978). *A Phytophthora parasitica* stem canker disease of pigeon pea in Puerto Rico, *Pl. Dis. Reptr.,* **62**, 240–242.

Kaiser, W. J. and Mossahebi, G. H. (1975). Studies with cowpea aphid-borne mosaic virus and its effect on cowpea in Iran, *FAO Pl. Prot. Bull.,* **23**, 33–39.

Kaiser, W. J., Okhovat, M., and Mossahebi, G. H. (1973). Effect of seed-treatment fungicides on control of *Ascochyta rabiei* in chickpea seed infected with the pathogen, *Pl. Dis. Reptr.,* **57**, 742–746.

Kaiser, W. J. and Ramos, A. H. (1979). Two bacterial diseases of cowpea in East Africa, *Pl. Dis. Reptr.,* **63**, 304–308.

Kaiser, W. J. and Ramos, A. H. (1980). Occurrence of *Pseudomonas syringae* on bean and soybean in Kenya, *Pl. Dis.,* **64**, 593–595.

Kaiser, W. J. and Vakili, N. G. (1978). Insect transmission of pathogenic *Xanthomonas phaseoli, X. phaseoli* f.sp. *vignicola* to bean and cowpea in Puerto Rico, *Phytopathology,* **68**, 1057–1063.

Kamat, M. N. and Patel, M. K. (1948). Some new hosts of *Oidiopsis taurica* (Lév.) Salmon in Bombay, *Indian Phytopath.,* **1**, 153–158.

Kandaswamy, T. K. and Natarajan, C. (1974). A note on phyllody disease on Bengal gram (*Cicer arietinum* L.), *Madras agric. J.,* **61**, 1019–1020.

Kandaswamy, T. K. and Ramakrishnan, K. (1960). An epiphytotic of pigeon pea sterility mosaic at Coimbatore, *Madras agric. J.*, **47**, 440–441.
Kannaiyan, J. (1981). Diseases of chickpea in Malawi, *Internat. Chickpea Newsl.*, **4**, 16.
Kannaiyan, J. and Nene, Y. L. (1977) *Alternaria* leaf spot of pigeon pea, *Trop. Grain Leg. Bull.*, **9**, 34.
Kannaiyan, J. and Nene, Y. L. (1979a). Occurrence of powdery mildew on *Atylosia* species, *Trop. Grain Leg. Bull.*, **15**, 22–23.
Kannaiyan, J. and Nene, Y. L. (1979b). Association of different *Fusarium* species with wilt disease of pigeon pea, *Trop. Grain Leg. Bull.*, **15**, 26–27.
Kannaiyan, J. and Nene, Y. L. (1979c). Effect of crop rotation and planting time on the incidence of wilt in lentil, *Indian J. Pl. Prot.*, **7**, 114–118.
Kannaiyan, J. and Nene, Y. L. (1981). Influence of wilt at different growth stages on yield loss in pigeon pea, *Trop. Pest Manag.*, **27**, 141.
Kannaiyan, J., Nene, Y. L., Raju, T. N., and Sheila, V. K. (1981). Screening for resistance to *Phytophthora* blight of pigeon pea, *Pl. Dis.*, **65**, 61–62.
Kannaiyan, J., Ribeiro, O. K., Erwin, D. C., and Nene, Y. L. (1980). *Phytophthora* blight of pigeon pea in India, *Mycologia*, **72**, 169–181.
Kanwar, J. S. (1963). Investigations on sulphur in soils I. Sulphur deficiency in groundnut soils of Samrala (Ludhiana), *Indian J. agric., Sci.*, **33**, 196–198.
Kaplan, L., Lynch, T. F., and Smith, C. E. (1973). Early cultivated beans (*Phaseolus vulgaris*) from an intermontane Peruvian valley, *Science*, **179**, 76–77.
Karan, D. and Rao, P. N. (1968). Some new hosts to *Leptosphaerulina* McAlp. from India III, *Mycopath. Mycol. appl.*, **35**, 193–196.
Kasasian, L. (1971). *Orobanche* spp., *PANS*, **17**, 35–41.
Kassanis, B. (1970). Tobacco necrosis virus. *CMI/AAB Descriptions of Plant Viruses No. 14.*
Kassanis, B. and Varma, A. (1975). Sunnhemp mosaic virus. *CMI/AAB Descriptions of Plant Viruses No. 153.*
Katan, J. (1971). Symptomless carriers of the tomato *Fusarium* wilt pathogen, *Phytopathology*, **61**, 1213–1217.
Katznelson, H. (1950). The detection of internally-borne bacterial pathogenes of beans by a rapid phage-plaque count technique, *Science*, **112**, 645–647.
Katznelson, H., Sutton, M. D., and Bayley, S. T. (1954). The use of bacteriophage of *Xanthomonas phaseoli* in detecting infection in beans, with observations on its growth and morphology, *Can. J. Microbiol.*, **1**, 22–29.
Kaufmann, M. J. and Gerdemann, J. W. (1958). Root and stem rot of soybean caused by *Phytophthora sojae* n.sp., *Phytopathology*, **48**, 201–208.
Kayumbo, H. Y., Finlay, R. C., and Doto, S. A. (1976). Effect of spraying on yield of cowpeas grown in monoculture and under maize, sorghum or millet, in *Intercropping in Semi-Arid Areas* (eds J. H. Monyo, A. D. R. Ker, and M. Campbell), IDRC, Ottawa, pp. 44–45.
Keeling, B. L. (1972). Studies on nature of soybean resistance to seed rot caused by *Pythium*. *Phytopathology*, **62**, 768 (abstr.).
Keen, N. T. (1975a). The isolation of phytoalexins from germinating seeds of *Cicer arietinum, Vigna sinensis, Arachis hypogaea,* and other plants, *Phytopathology*, **65**, 91–92.
Keen, N. T. (1975b). Specific elictors of plant phytoalexin production: determinants of race specificity in pathogens? *Science*, **187**, 74–75.
Keen, N. T. and Horsch, R. (1972). Hydroxyphaseollin production by various soybean tissues: a warning against use of 'unnatural' host-parasite systems, *Phytopathology*, **62**, 439–442.
Keen, N. T. and Kennedy, B. W. (1974). Hydroxyphaseollin and related isoflavonoids in the hypersensitive response of soybeans against *Pseudomonas glycinea, Physiol. Pl. Path.*, **4**, 173–185.
Keenan, J. G., Moore, H. D., Oshima, N., and Jenkins, L. E. (1974). Effect of bean

root rot on dry and Pinto bean production in Southwestern Colorado, *Pl. Dis. Reptr.,* **58**, 890–892.

Kelman, A. and Person, L. H. (1961). Strains of *Pseudomonas solanacearum* differing in pathogenicity to tobacco and peanut, *Phytopathology,* **51**, 158–161.

Kendrick, J. B. (1931). Seed transmission of cowpea *Fusarium* wilt, *Phytopathology,* **21**, 979–983.

Kendrick, J. B. (1934). Seed transmission of *Fusarium* yellows of beans. *Phytopathology,* **24**, 1139 (abstr.).

Kendrick, J. B. (1936). A cowpea resistant to *Fusarium* wilt and nematode root knot, *Phytopathology,* **26**, 98 (abstr.).

Kendrick, J. B. and Gardner, M. W. (1924). Soybean mosaic: seed transmission and effect on yield, *J. agric. Res.,* **27**, 91–98.

Kendrick, J. B. and Snyder, W. C. (1942), *Fusarium* yellows of beans, *Phytopathology,* **32**, 1010–1014.

Kenknight, G. (1942). Peanut diseases in certain south Texas counties in 1942, *Pl. Dis. Reptr.,* **26**, 499–501.

Kennedy, B. W. (1964). Moisture content, mold invasion, and seed viability of stored soybeans, *Phytopathology,* **54**, 771–774.

Kennedy, B. W. and Cross, J. E. (1968). Inoculation procedures for comparing reaction of soybeans to bacterial blight, *Pl. Dis. Reptr.,* **50**, 560–565.

Kennedy, B. W. and Ercolani, G. L. (1978). Soybean primary leaves as a site for epiphytic multiplication of *Pseudomonas glycinea, Phytopathology,* **68**, 1196–1201.

Kennedy, B. W. and Tachibana, H. (1973). Bacterial diseases, in *Soybeans: Improvement, Production, and Uses* (ed. B. E. Caldwell), Amer. Soc. Agron., Madison, pp. 491–504.

Kennedy, G. G. (1976). Host plant resistance and the spread of plant viruses, *Env. Ent.,* **5**, 827–832.

Kennedy, J. S. (1950). Aphid migration and the spread of plant viruses, *Nature, Lond.,* **165**, 1024.

Kennedy, J. S., Booth, C. O., and Kershaw, W. J. S. (1961). Host finding by aphids in the field. III. Visual attraction, *Ann. appl. Biol.,* **49**, 1–21.

Keogh, R. C., Deverall, B. J., and McLeod, S. (1980). Comparison of histological and physiological responses to *Phakopsora pachyrhizi* in resistant and susceptible soybean, *Trans. Br. mycol. Soc.,* **74**, 329–333.

Kernkamp, M. F., de Zeeuw, P. J., Chen, S. M., Ortega, B. C., Tsiang, C. T., and Khan, A. M. (1952). Investigation on physiologic specialization and parasitism of *Rhizoctonia solani, Minn. Agric. Exp. Sta. Tech. Bull.,* **200**, 36 pp.

Kerr, A. and Flenje, N. T. (1957). Host infection in *Pellicularia filamentosa* controlled by chemical stimuli, *Nature, Lond.,* **179**, 204.

Kerr, E. D., Steadman, J. R., and Nelson, L. A. (1978). Estimation of white mold disease reduction of yield and yield components of dry edible beans, *Crop Sci.,* **18**, 275–279.

Keswani, C. L. and Ondieki, J. J. (1976). Rust on groundnut in Kenya and Tanzania, *FAO Pl. Prot. Bull.,* **24**, 135.

Ketring, D. L., Benedict, C. R., and Yeager, M. (1976). Growing season and location effects on water uptake and drying rates of peanut seeds from genotypes resistant and susceptible to invasion by *Aspergillus flavus* Lk. ex Fr., *Agron. J.,* **68**, 661–665.

Khachatryan, M. S. (1962). Some problems of the biology and dynamics of development of ascochytosis of chick pea in the Armenian SSR, *Izv. Akad. Nank Armyan. SSR, biol. Sci.,* **15**, 23–30. (*Rev. appl. Mycol.,* **42**, 426.)

Khan, I. U. (1980). Chickpea pathology in Pakistan, in *Proc. Internat. Wkshop on Chickpea Improv.,* International Crops Research Institute for the Semi-Arid Tropics, Hyderabad, India, p. 257.

Khan, M. W. and Singh, R. K. (1974) Anthracnose of arhar incited by *Colletotrichum truncatum. Indian Phytopath.,* **27**, 622–624.

Khan, T. N. (1973). A new approach to the breeding of pigeon pea (*Cajanus cajan* Millsp.): formation of composites, *Euphytica,* **22**, 373–377.
Khan, T. N. and Ashley, J. M. (1975). Factors affecting plant stand in pigeon pea, *Expl. Agric.,* **11**, 315–322.
Khan, T. N. and Rachie, K. O. (1972). Preliminary evaluation and utilization of pigeon pea germplasm in Uganda, *E. Afr. agric. For. J.,* **38**, 78–82.
Kheswalla, K. F. (1941). Foot rot of gram (*Cicer arietinum* L.) caused by *Operculella padwickii* nov. gen., nov. spec., *Indian J. agric. Sci.,* **11**, 316–318.
Khune, N. N., Charde, J. D., and Rant, J. G. (1979). Chickpea, a new host record of *Leveillula taurica, Indian Phytopath.,* **32**, 655.
Khune, N. N. and Kapoor, J. N. (1980). *Ascochyta rabiei* synonymous with *Phoma rabiei, Indian Phytopath.,* **33**, 119–120.
Kiihl, R. A. S. and Hartwig, E. E. (1979). Inheritance of reaction to soybean mosaic virus in soybeans, *Crop Sci.,* **19**, 372–375.
Kilpatrick, R. A. and Johnson, H. W. (1956). Purple stain of legume seeds caused by *Cercospora* species, *Phytopathology,* **46**, 201–204.
Kincaid, R. R. and Gammon, N. (1954). Incidence of tobacco blackshank related to soil pH, *Pl. Dis. Reptr.,* **38**, 852–853.
Kirby, R. S. and Archer, W. A. (1927). Diseases of cereal and forage crops in the United States in 1926, *Pl. Dis. Reptr., Suppl.,* **53**, 110–208.
Kishun, R., Sohi, H. S., and Rawal, R. D. (1980). Screening of cowpea germplasm against bacterial blight, *Indian Phytopath.,* **33**, 506–507.
Kistler, H. C. and van Etten, H. D. (1981). Phaseollin metabolism and tolerance in *Fusarium solani* f.sp. *phaseoli, Physiol. Pl. Path.,* **19**, 257–271.
Kitajima, E. W. (1980). General review of mycoplasma-like diseases, in *Bean Production Problems: Disease, Insect, Soil and Climatic Constraints of* Phaseolus vulgaris (eds H. F. Schwartz and G. E. Gálvez), Centro Internacional de Agricultura Tropical, Cali, Colombia, pp. 199–203.
Kitazawa, K., Tsuchiya, S., Kodama, F., Kamjaipai, W., Ogoshi, A., and Yanagita, K. (1978). *Phytophthora* stem rot of adzuki bean (*Phaseolus angularis*) caused by *Phytophthora vignae* Purss, *Ann. Phytopath. Soc. Japan,* **44**, 528–531. (*Rev. Pl. Path.,* **58**, 4568.)
Klas, F. E. (1979). Cowpea severe mosaic virus, a constraint on the improvement of cowpea cultivation in Suriname, *Paramaribo Agric. Exp. Sta. Bull.,* **104**.
Klein, H. H. (1959). Etiology of the *Phytophthora* disease of soybeans, *Phytopathology,* **49**, 380–383.
Klement, Z. (1963). Rapid detection of the pathogenicity of phytopathogenic pseudomonads, *Nature, Lond.,* **199**, 299.
Klement, Z. and Goodman, R. N. (1967). A hypersensitive reaction to infection by bacterial plant pathogens, *A. Rev. Phytopath.,* **5**, 17–44.
Klesser, P. J. (1966). Groundnut ring spot virus - a new sap-transmissible virus of *Arachis hypogaea* and *A. monticola, S. Afr. J. agric. Sci.,* **9**, 711–717.
Kmetz, K. T., Schmitthenner, A. F., and Ellett, C. W. (1978). Soybean seed decay: prevalence of infection and symptom expression caused by *Phomopsis* sp., *Diaporthe phaseolorum* var. *sojae* and *D. phaseolorum* var. *caulivora, Phytopathology,* **68**, 836–841.
Knott, D. R. and Dvořák, J. (1976). Alien germplasm as a source of resistance to disease, *A. Rev. Phytopath.,* **14**, 211–235.
Ko, W. H., Chang, H. S., and Su, H. J. (1978). Isolates of *Phytophthora cinnamomi* from Taiwan as evidence for an Asian origin of the species, *Trans. Br. mycol. Soc.,* **71**, 496–499.
Kochman, J. K. (1979). The effect of temperature on development of soybean rust (*Phakopsora pachyrhizi*), *Aust. J. agric. Res.,* **30**, 273–277.
Kolawole, K. B. (1976). A short progress report on transfer of *Cercospora* resistant traits to the cultivated *Arachis hypogaea* L., *Samaru agric. Newsl.,* **18**, 40–43.

Kopek, J. A. and Scott, H. A. (1979). Southern bean mosaic virus in Mexican bean beetle and bean leaf beetle regurgitants, *Phytopathology,* **69**, 1034 (abstr.).

Koshimizu, Y. and Iizuka, N. (1963). Studies on soybean virus diseases in Japan, *Bull. Tôhoku agric. Exp. Sta.,* **27**, 1–103.

Koshy, P. K. and Swarup, G. (1971a). Investigations on the life history of the pigeon-pea cyst nematode, *Heterodera cajani, Indian J. Nemat.,* **1**, 44–51. (*Helminth. Abstr.,* B **40**, 684.)

Koshy, P. K. and Swarup, G. (1971b). On the number of generations of *Heterodera cajani,* the pigeon-pea cyst nematode in a year, *Indian J. Nemat.,* **1**, 88–90. (*Helminth. Abstr.,* B **40**, 685.)

Koshy, P. K. and Swarup, G. (1971c). Factors affecting emergence of larvae from cysts of *Heterodera cajani* Koshy, 1967, *Indian J. Nemat.,* **1**, 209–219.

Koshy, P. K. and Swarup, G. (1971d). Susceptibility of aerial parts of pigeon-pea seedlings to *Heterodera cajani* larvae, *Indian J. Nemat.,* **1**, 245–246. (*Helminth. Abstr.,* B **41**, 505.)

Koshy, P. K. and Swarup, G. (1972). Susceptibility of plants to pigeon-pea cyst nematode, *Heterodera cajani, Indian J. Nemat.,* **2**, 1–6. (*Helminth. Abstr.,* B **42**, 501.)

Koshy, P. K. Swarup, G., and Sethi, C. L. (1970). Further notes on the pigeon-pea cyst nematode, *Heterodera cajani, Nematologica,* **16**, 477–482. (*Helminth. Abstr.,* B **41**, 504.)

Kotasthane, S. R., Gupta, O., and Singh, B. R. (1980). A new record of parasitic *Cuscuta* on pigeon pea (*Cajanus cajan* (L.) Millsp.), *Trop. Grain Leg. Bull.,* **19**, 16–18.

Kovačevski, I. C. (1936). The blight of chick pea, *Mycosphaerella rabiei* n.sp., *Min. Agric. nat. Domains, Sofia. (Rev. appl. Mycol.,* **15**, 700.)

Kowal, J. M. and Kassam, A. H. (1978). *Agricultural Ecology of Savanna; a Study of West Africa,* Clarendon Press, Oxford, 403 pp.

Kowal, J. M. and Knabe, D. (1972). *An Agroclimatological Atlas of the Northern States of Nigeria*, Ahmadu Bello Univ. Press, Samaru, 112 pp.

Kraft, J. M. (1969). Chickpea, a new host of *Fusarium solani* f.sp. *pisi, Pl. Dis. Reptr.*, **53**, 110–111.

Kraft, J. M. and Haglund, W. A. (1978). A reappraisal of the race classification of *Fusarium oxysporum* f.sp. *pisi, Phytopathology,* **68**, 273–275.

Krauss, F. G. (1932). The pigeon pea (*Cajanus indicus*): its improvement, culture, and utilization in Hawaii, *Bull. Hawaii agric. Exp. Sta. No. 6.4.*

Kraybill, H. F., and Shimkin, M. B. (1964). Carcinogenesis related to foods contaminated by processing and fungal metabolites, *Adv. Cancer Res.*, **8**, 191–248.

Kreitlow, K. W., Boyd, H. C., Chamberlain, D. W., and Dunleavy, J. M. (1957). A bibliography of viruses infecting the soybean (*Glycine max* (L.) Merr.), *Pl. Dis. Reptr.,* **41**, 579–588.

Kruger, J., Hoffman, G. M., and Hubbeling, N. (1977). The Kappa race of *Colletotrichum lindemuthianum* and sources of resistance to anthracnose in *Phaseolus* beans, *Euphytica,* **26**, 23–25.

Kuan, Ta-Li and Erwin, D. C. (1980). Formae speciales differentiation of *Phytophthora megasperma* isolates from soybean and alfalfa, *Phytopathology,* **70**, 333–338.

Kueneman, E. A. and Allen, D. J. (1980). A technique for screening populations of soybean for resistance to bacterial pustule, *Trop. Grain Leg. Bull.,* **19**, 34–37.

Kuhn, C. W. (1964). Separation of cowpea virus mixtures, *Phytopathology,* **54**, 739–740.

Kuhn, C. W. (1965). Symptomatology, host range, and effect on yield of a seed-transmitted peanut virus, *Phytopathology,* **55**, 880–884.

Kuhn, C. W. (1968). Identification and specific infectivity of a soybean strain of cowpea chlorotic mottle virus, *Phytopathology,* **58**, 1441–1442.

Kuhn, C. W. and Adams, D. B. (1976). Synthesis and movement of southern bean mosaic virus in cowpea hosts with virus-induced necrotic local lesions, *Phytopathology,* **66**, 1298–1301.

Kuhn, C. W. and Brantley, B. B. (1963). Cowpea resistance to the cowpea strain of southern bean mosaic virus, *Pl. Dis. Reptr.*, **47**, 1094–1096.

Kuhn, C. W., Brantley, B. B., and Sowell, G. (1965). Immunity to bean yellow mosaic virus in cowpea, *Pl. Dis. Reptr.,* **49**, 879–881.

Kuhn, C. W., Brantley, B. B., and Sowell, G. (1966). Southern pea virus: identification, symptomatology, and sources of resistance, *Georgia Agric. Exp. Sta. Bull. NS. 157.*

Kuhn, C. W. and Dawson, W. O. (1973). Multiplication and pathogenesis of cowpea chlorotic mottle virus and southern bean mosaic virus in single and double infections in cowpea. *Phytopathology*, **63**, 1380–1385.

Kuhn, C. W., Demski, J. W., and Harris, H. B. (1972). Peanut mottle virus in soybean, *Pl. Dis. Reptr.,* **56**, 146–147.

Kuhn, C. W., Hammons, R. O., and Sowell, G. (1964). A ringspot disease of peanuts, *Pl. Dis. Reptr.,* **48**, 729–732.

Kuhn, C. W., Paguio, O.R., and Adams, D. B. (1978). Tolerance in peanuts to peanut mottle virus, *Pl. Dis. Reptr.,* **62**, 365–368.

Kuhn, C. W., Sowell, G., Chalkley, J. H., and Stubbs, H. F. (1968). Screening for immunity to peanut mottle virus, *Pl. Dis. Reptr.,* **52**, 467–468.

Kuhn, C. W. and Wyatt, S. D. (1979). A variant of cowpea chlorotic mottle virus obtained by passage through beans, *Phytopathology,* **69**, 621–624.

Kuhn, P. J. and Smith, D. A. (1979). Isolation from *Fusarium solani* f.sp. *phaseoli* of an enzymic system responsible for keivitone and phaseollidin detoxification, *Physiol. Pl. Path.,* **14**, 179–190.

Kuijt, J. (1977). Haustoria of phanerogamic parasites, *A. Rev. Phytopath.*, **15**, 91–118.

Kulkarni, H. Y. (1973). Notes on East African plant virus diseases. 4. Bean common mosaic virus, *E. Af. agric. For. J.,* **39**, 72–76.

Kulkarni, Y. S., Patel, M. K., and Abhyankar, S. G. (1950). A new bacterial leaf-spot and stem canker of pigeon pea, *Curr. Sci.,* **19**, 384.

Kumar, J., Haware, M. P., and Nene, Y. L. (1980). *Fusarium* wilt resistant lines developed at ICRISAT, *Internat. chickpea Newsl.,* **3**, 5.

Kumar, L. S. S., Thombre, M. V., and D'Cruz, R. (1958). Cytological studies of an intergeneric hybrid of *Cajanus cajan* (Linn.) Millsp. and *Atylosia lineata* W. and A., *Proc. Indian Acad. Sci.,* B **47**, 252–262.

Kurata, H. and Kuribayashi, K. (1954). Soybean scab caused by *Sphaceloma glycines* sp. nov., *Ann. Phytopath. Soc. Jap.,* **18**, 119–121.

Kurstak, E. (ed.) (1981). *Handbook of Plant Virus Infections Comparative Diagnosis,* Elsevier, Amsterdam, 943 pp.

Kushalappa, A. C., Bartz, J. A., and Norden, A. J. (1979). Susceptibility of pods of different peanut genotypes to *Aspergillus flavus* group fungi, *Phytopathology,* **69**, 159–162.

Kuz'mina, G. (1966). Antibiotics against wilt, *Zasch. Rast., Mosk.,* **11**, 31. (*Rev. appl. Mycol.,* **45**, 3438.)

Kvíčala, B. A., Smrž, J., and Blanco, N. (1973). A beetle-transmitted virus disease of cowpea in Cuba, *FAO Pl. Prot. Bull.,* **21**, 27–30.

Kwon, S. H. and Oh, J. H. (1979). Inheritance of resistance to necrotic strain of SMV in soybean, *Soybean Genetics Newsl.,* **6**, 66–68.

Labrousse, F. (1931). L'anthracnose du pois-chiche, *Rev. Path. Vég. Ent. Agric.,* **28**, 226–231. (*Rev. appl. Mycol.,* **11**, 150–151.)

Lackey, C. F. (1946). Reaction of dodders to stems of other dodders and to their own stems, *Phytopathology,* **36**, 386–388.

Lackey, C. F. (1966). Recovery of a virus resembling tobacco streak from dodder growing on a desert shrub in southern California, *Pl. Dis. Reptr.*, **50**, 479–482.

Lackey, J. A. (1977). *Neonotonia*, a new generic name to include *Glycine wightii* (Arnott) Verdcourt (Leguminosae, Papilionoideae), *Phytologia*, **37**, 209–212.

Lackey, J. A. (1981). Phaseoleae, in *Advances in Legume Systematics* (eds R. M. Polhill and P. H. Raven), HMSO, London, pp. 301–327.

Ladipo, J. L. (1976). A veinbanding strain of cowpea aphid-borne mosaic virus in Nigeria, *Niger. J. Sci.*, **10**, 77–88.

Ladipo, J.L. (1977). Seed transmission of cowpea aphid-borne mosaic virus in some cowpea cultivars, *Niger. J. Pl. Prot.*, **3**, 3–10.

Ladipo, J. L. and Allen, D. J. (1979a). Identification of resistance to southern bean mosaic virus in cowpea, *Trop. Agric. (Trin.)*, **56**, 33–40.

Ladipo, J. L. and Allen, D. J. (1979b). Identification of resistance to cowpea aphid-borne mosaic virus, *Trop. Agric. (Trin.)*, **56**, 353–359.

Ladizinsky, G. and Adler, A. (1976a). The origin of chickpea, *Cicer arietinum* L., *Euphytica*, **25**, 211–217.

Ladizinsky, G. and Adler, A. (1976b). Genetic relationships among the annual species of *Cicer* L., *Theor. Appl. Genet.*, **48**, 197–203.

Lampard, J. F. (1974). Dimethylhomopterocarpin: an antifungal compound in *Canavalia ensiformis* and *Vigna unguiculata* following infection, *Phytochemistry*, **13**, 291–292.

Lamptey, P. N. L. and Hamilton, R. I. (1974). A new cowpea strain of southern bean mosaic virus from Ghana, *Phytopathology*, **64**, 1100–1104.

Lana, A. F. (1980). Properties of a virus occurring in *Arachis hypogaea* in Nigeria, *Phytopath. Z.*, **97**, 169–178.

La Prade, J. C. and Bartz, J. A. (1972). Mechanical resistance of selected genotypes of dried peanuts to colonization by strains of aflatoxin-producing *Aspergillus* sp., *Phytopathology* **62**, 771 (abstr.).

Larsh, H. W. (1943). Bacterial crown canker of cowpea in Oklahoma, *Pl. Dis. Reptr.*, **27**, 555.

Latham, D. H. (1934). Life history of a *Cercospora* leaf spot fungus of cowpea, *Mycologia*, **26**, 516–527.

Laundon, G. F. (1967). Terminology in the rust fungi, *Trans. Br. mycol. Soc.*, **50**, 189–194.

Laundon, G. F. and Rainbow, A. F. (1971). *Uromyces mucunae. CMI Descriptions of Pathogenic Fungi and Bacteria No. 290.*

Laundon, G. F. and Waterston, J. M. (1965a). *Puccinia arachidis. CMI Descriptions of Pathogenic Fungi and Bacteria No. 53.*

Laundon, G. F. and Waterston, J. M. (1965b). *Uromyces appendiculatus. CMI Descriptions of Pathogenic Fungi and Bacteria No. 57.*

Laundon, G. F. and Waterston, J. M. (1965c). *Uromyces viciae-fabae. CMI Descriptions of Pathogenic Fungi and Bacteria No. 60.*

Laurence, R. C. N., Gibbons, R. W., and Young, C. T. (1976). Changes in yield, protein, soil and maturity of groundnut cultivars with the application of sulphur fertilizers and fungicides, *J. agric. Sci., Camb.*, **86**, 245–250.

Laviolette, F. A. and Athow, K. L. (1977). Three new physiologic races of *Phytophthora megasperma* var. *sojae, Phytopathology*, **67**, 267–268.

Laviolette, F. A., Athow, K. L., Probst, A. H., and Wilcox, J. R. (1970). Effect of bacterial pustule on yield of soybeans, *Crop Sci.*, **10**, 150–151.

Leach, J. G., Lilly, V. G., Wilson, H. A., and Purvis, M. R. (1957). Bacterial polysaccharides: the nature and function of the exudate produced by *Xanthomonas phaseoli, Phytopathology*, **47**, 113–120.

Leach, R. and Wright, J. (1930). Collar and stem canker of (*Cajanus indicus*) pigeon

pea, caused by a species of *Physalospora, Mem. Imper. Coll. Trop. Agric. Trinidad* (Mycol. Ser.), **1**, 1–12. (*Rev. appl. Mycol.,* **9**, 622.)
Leakey, C. L. A. (1964). *Dactuliophora*, a new genus of mycelia sterilia from tropical Africa, *Trans. Br. mycol. Soc.*, **47**, 341–350.
Leakey, C. L. A. (1970a). The improvement of beans (*Phaseolus vulgaris*) in East Africa, in *Crop Improvement in East Africa* (ed. C. L. A. Leakey), Commonwealth Agricultural Bureaux, Farnham Royal, pp. 99–128.
Leakey, C. L. A. (1970b). Need one grow pure lines in developing countries? *A. Rep. Bean Improv. Co-op.,* **13**, 62–63.
Leakey, C. L. A. (1973). A note on *Xanthomonas* blight of beans (*Phaseolus vulgaris* (L.) Savi) and prospects for its control by breeding for tolerance, *Euphytica,* **22**, 132–140.
Leakey, C. L. A. (1979). Heirloom information - bean anthracnose, *A. Rep. Bean Improv. Co-op.,* **22**, 22–24.
Leakey, C. L. A., Atkins, J., and Magara, J. (1972). Bean rust studies in Uganda, *A. Rep. Bean Improv. Co-op.,* **15**, 60–63.
Leakey, C. L. A. and Rubaihayo, P. R. (1970). Soya beans as a potentially valuable crop for agricultural diversification in Central Uganda, *E. Afr. agric. For. J.,* **36**, 39–44.
Leakey, C. L. A. and Simbwa-Bunnya, M. (1972). Races of *Colletotrichum lindemuthianum* and implications for bean breeding in Uganda, *Ann. appl. Biol.*, **70**, 25–34.
Leben, C., Rusch, V. and Schmitthenner, A. F. (1968). The colonization of soybean buds by *Pseudomonas glycinea* and other bacteria, *Phytopathology,* **58**, 1677–1681.
Le Clerg, E. L. (1939). Methods of determination of physiologic races of *Rhizoctonia solani* on the basis of parasitism on several crop plants, *Phytopathology,* **29**, 609–616.
Lee, Y-S. and Ross, J. P. (1972). Top necrosis and cellular changes in soybean doubly infected by soybean mosaic and bean pod mottle viruses, *Phytopathology,* **62**, 839–845.
Lefebvre, C. L. and Sherwin, H. S. (1945). Observations on the bacterial canker of cowpeas, *Phytopathology,* **35**, 487 (abstr.).
Lefebvre, C. L. and Sherwin, H. S. (1950). Inheritance of resistance to bacterial canker in cowpea, *Vigna sinensis, Phytopathology,* **40**, 17–18 (abstr.).
Lefebvre, C. L. and Stevenson, J. A. (1945). The fungus causing zonate leafspot of cowpea, *Mycologia,* **37**, 37–45.
Lefebvre, C. L. and Weimer, J. L. (1939). *Choanephora cucurbitarum* attacking cowpeas, *Phytopathology,* **29**, 898–901.
Lehman, S. G. (1928). Frog-eye leaf spot of soybean caused by *Cercospora diazu* Miura, *J. agric. Res.,* **36**, 811–833.
Lehman, S. G. and Wolf, F. A. (1926a). *Pythium* root rot of soybean, *J. agric. Res.*, **33**, 375–380.
Lehman, S. G. and Wolf, F. A. (1926b) Soybean anthracnose, *J. agric. Res.,* **33**, 381–390.
Lehman, S. G. and Woodside, J. G. (1929). Varietal resistance of soybean to the bacterial pustule disease, *J. agric. Res.,* **39**, 795–805.
Lemos, J. W. V. and da Ponte, J. J. (1978). Cultivares de feijâo-de-corda, *Vigna sinensis* (L.) Savi, resistentes à meloidoginose, *Bol. Soc. Cearense Agron.*, **19**, 1–19. (*Helminth. Abstr.* B **49**, 121.)
Leng, E. R. (1968). Soybeans - potential for extension to areas of protein shortage, *Econ. Bot.,* **22**, 37–41.
Lenné J. M. (1981). Sphaceloma scab on *Zornia latifolia, Pl. Dis.,* **65**, 162–164.
Leonard, K. J. (1969). Factors affecting rates of stem rust increase in mixed plantings of susceptible and resistant oat varieties, *Phytopathology,* **59**, 1845–1850.

Leppik, E. E. (1970). Gene centers of plants as sources of disease resistance, *A. Rev. Phytopath.*, **8**, 323–344.

Licha-Baquero, M. (1980). The witches' broom disease of pigeon pea (*Cajanus cajan* (L.) Millsp.) in Puerto Rico, *J. Agric. Univ. Puerto Rico,* **64**, 424–441.

Lim, S. M. (1979). Evaluation of soybean for resistance to *Septoria* brown spot, *Pl. Dis. Reptr.,* **63**, 242–245.

Lima, J. A. A. and Nelson, M. R. (1977). Etiology and epidemiology of mosaic of cowpea in Ceara, Brazil, *Pl. Dis. Reptr.,* **61**, 864–867.

Lima, J. A. A., Purcifull, D. E., and Hiebert, E. (1979). Purification, partial characterization, and serology of blackeye cowpea mosaic virus, *Phytopathology,* **69**, 1252–1258.

Lin, M. T. (1979). Purification and serology of legume and corn viruses in Brazil, *Fitopat. Bras.,* **4**, 203–213.

Lin, M. T., Anjos, J. R. N., and Rios, G. P. (1981a). Serological grouping of cowpea severe mosaic virus isolates from central Brazil, *Phytopathology,* **71**, 435–438.

Lin, M. T., Anjos, J. R. N., and Rios, G. P. (1982). Cowpea severe mosaic virus in five legumes in central Brazil, *Pl. Dis.,* **66**, 67–70.

Lin, M. T., Kitajima, E. W., and Rios, G. P. (1981b). Serological identification of several cowpea viruses in Central Brazil, *Fitopat. Bras.,* **6**, 73–85.

Lin, S-Y. (1966). Studies on the physiologic races of soybean rust fungus, *Phakopsora pachyrhizi* Syd., *J. Taiwan agric. Res.,* **15**, 24–28.

Lin, Y-S. and Cook, R. J. (1977). Root rot of lentils, caused by *Fusarium roseum* 'Avenaceum', *Pl. Dis. Reptr.,* **61**, 752–755.

Ling, L. (1940). Seedling stem blight of soybean caused by *Glomerella glycines, Phytopathology,* **30**, 345–347.

Lister, R. M. (1978). Application of the enzyme-linked immunosorbent assay for detecting viruses in soybean seed and plants, *Phytopathology,* **68**, 1393–1400.

Lister, R. M. and Thresh, J. M. (1955). A mosaic disease of leguminous plants caused by a strain of tobacco mosaic virus, *Nature, Lond.,* **175**, 1047.

Litsinger, J. A., Quirino, C. B., Lumaban, M. D., and Bandong, J. P. (1978). The grain Legume pest complex of rice-based cropping systems at three locations in the Philippines, in *Pests of Grain Legumes: Ecology and Control.* (eds S. R. Singh, H. F. Van Emden, and T. Ajibola Taylor), pp. 309–320. Academic Press, London.

Littlefield, L. J. (1969). Flax rust resistance induced by prior inoculation with an avirulent race of *Melampsora lini, Phytopathology*, **59**, 1323–1328.

Littrell, R. H. (1974). Tolerance in *Cercospora arachidicola* to benomyl and related fungicides, *Phytopathology*, **64**, 1377–1378.

Liu, P. S. W. (1977). A supplement to a host list of plant diseases in Sabah, Malaysia, *Phytopath. Pap. No. 21*, Commonwealth Mycological Institute, Kew, 50 pp.

Lo, T. C. and Han, Y. H. (1969). Studies on witches' broom of soybean, *Bot. Bull. Acad. sin. Taipei, N.S.*, **10**, 10–22. (*Rev. appl. Mycol.*, **48**, 3748 (abstr.).)

Locci, R. and Bisiach, M. (1970). The *Phaseolus vulgaris-Uromyces appendiculatus* complex. I. Examination of the uredospore infection process by scanning electron microscopy, *Riv. Patol. veg., Pavia Ser. IV,* **6**, 21–36. (*Rev. Pl. Path.*, **50**, 386.)

Lockhart, B. E. L. and Fischer, H. V. (1974). Chronic infection by seedborne common mosaic virus in Morocco, *Pl. Dis. Reptr.*, **58**, 307–308.

Lockhart, B. E. L. and Fischer, H. V. (1976). Identification of soybean mosaic virus in Morocco, *Pl. Dis. Reptr.*, **60**, 260–261.

Lockwood, J. L., Yoder, D. L., and Smith, N. A. (1970). *Thielaviopsis basicola* root rot of soybeans in Michigan, *Pl. Dis. Reptr.*, **54**, 849–850.

Loden, H. D. and Hildebrand, E. M. (1950). Peanuts, especially their diseases, *Econ. Bot.*, **4**, 354–379.

Long, D. W. (1963). Inhibition of *Fusarium* wilt symptoms in cowpea by species of *Cephalosporium, Phytopathology*, **53**, 881 (abstr.).

López García, H. (1974). Herencia del carácter 'resistencia a rabia' (*Fusarium* sp.) en garbanzo (*Cicer arietinum*), tajo condiciones de campo, *Agric. Técn. México*, **3**, 286–289. (*Rev. Pl. Path.*, **54**, 1077.)
López Rosa, J. H. (1969). *Phoma* sp., the causal agent of pigeon pea canker, *Phytopathology*, **59**, 1348 (abstr.).
Lu, Y. C. (1970). Mutation breeding for rust resistance in soybeans, *Proc. Symp. on Impoving Plant Protein by Nuclear Techniques*, I.A.E.H. and FAO, Vienna, 8–12 June 1970, pp. 185–187.
Lukashevich, A. I. (1958a). Control measures against ascochytosis of chick pea, *J. agric. Sci., Moscow*, **5**, 131–135. (*Rev. appl. Mycol.*, **37**, 627.)
Lukashevich, A. I. (1958b). Peculiarities of the parasitism of the causal agent of ascochytosis of chick pea, and their role in the accumulation of infection, *Rep. Acad. Sci. Ukr.*, 1958(**7**), 788–792.
Lush, W. M. (1979). Foral morphology of wild and cultivated cowpeas, *Econ. Bot.*, **33**, 442–447.
Lush, W. M. and Evans, L. T. (1981). The domestication and improvement of cowpeas (*Vigna unguiculata* (L.) Walp.), *Euphytica*, **30**, 579–587.
Luthra, J. C. and Bedi, K. S. (1932). Some preliminary studies on gram-blight with reference to its cause and mode of perennation, *Ind. J. agric. Sci.*, **2**, 499–515.
Luthra, J. C., Sattar, A., and Bedi, K. S. (1935). Life history of gram blight (*Ascochyta rabiei* (Pass.) Lab. = *Phyllosticta rabiei* (Pass.) Trot.) on gram (*Cicer arietinum* L.) and its control in the Punjab, *Agric. Livest., India*, **5**, 489–498.
Luthra, J. C., Sattar, A., and Bedi, K. S. (1938). The control of the blight disease of gram by resistant types, *Curr. Sci.*, **7**, 45–47.
Luthra, J. C., Sattar, A., and Bedi, K. S. (1939). Variation in *Ascochyta rabiei* (Pass.) Lab., the causal fungus of the blight of gram (*Cicer arientinum* L.), *Ind. J. agric. Sci.*, **9**, 791–805.
Luttrell, E. S. and Boyle, L. W. (1960). Leaf spot of peanut in Georgia caused by *Leptosphaerulina arachidicola, Pl. Dis. Reptr.*, **44**, 609–611.
Luttrell, E. S. and Weimer, J. L. (1952). *Macrophomina* stem canker and ashy stem blight of cowpea, *Pl. Dis. Reptr.*, **36**, 194–195.
Lyon, F. M. and Wood, R. K. S. (1975). Production of phaseollin, coumestrol, and related compounds in bean leaves inoculated with *Pseudomonas* species, *Physiol. Pl. Path.*, **6**, 117–124.
Macartney, J. C. (1966). The selection of haricot bean varieties suitable for canning, *E. Afr. agric. For. J.*, **32**, 214–219.
Machado, P. F. R. and Pinchinat, A. M. (1975). Herencia de la reacción del fríjol común a la infección por el virus del mosaico rugoso, *Turrialba*, **25**, 418–419.
Mackie, W. W. (1934). Breeding for resistance in Blackeye cowpeas to *Fusarium* rot, charcoal rot, and nematode root knot, *Phytopathology*, **24**, 1135 (abstr.).
Mackie, W. W. (1939). Breeding for resistance in Blackeye cowpeas to cowpea wilt, charcoal rot, and root-knot nematode, *Phytopathology,* **29**, 826 (abstr.).
Macko, V., Staples, R. C., Gershon, H., and Renwick, J. A. A. (1970). Self-inhibitor of bean rust uredospores: methyl 3, 4-dimethoxycinnamate, *Science (Washington),* **170**, 539–540.
Maden, S., Singh, D., Mathur, S. B., and Neergaard, P. (1975). Detection and location of seed-borne inoculum of *Ascochyta rabiei* and its transmission in chickpea (*Cicer arietinum*), *Seed Sci. and Technol.*, **3**, 667–681.
Maduewesi, J. N. C. (1975). Host range and thermal inactivation of a cowpea isolate of *Sclerotium rolfsii, Niger. J. Pl. Prot.*, **1**, 23–28.
Maheshwari, R., Allen, P. J., and Hildebrandt, A. C. (1967). Physical and chemical factors controlling the development of infection structures from urediospore germ tubes of rust fungi, *Phytopathology,* **57**, 855–862.
Mali, V. R. and Kulthe, K. S. (1980). A seedborne potyvirus causing mosaic of cowpea in India, *Pl. Dis.,* **64**, 925–928.

Malik, R. P. (1945). Collar rot of pigeon pea caused by *Pythium aphanidermatum*, *Indian J. agric. Sci.,* **15**, 92–93.
Mallaiah, K. V. (1976). A note on the seasonal changes in the incubation time of groundnut rust, *Curr. Sci.,* **45**, 26.
Mallaiah, K. V. and Rao, A. S. (1976). Some observations on pepper spot disease of groundnut, *Curr. Sci.,* **45**, 737–738.
Mallaiah, K.V. and Rao, A. S. (1982). Aerial dissemination of urediniospores of groundnut rust, *Trans. Br. mycol. Soc.,* **78**, 21–28.
Mann, H. H. (1947). Pulse grain crops in the Middle East, *Emp. J. exp. Agric.,* **15**, 249–259.
Manning, H. L. (1956). The statistical assessment of rainfall and its application in Uganda agriculture, *Proc. Roy. Soc. (B),* **144**, 460–480.
Mansfield, J. W. (1982). The role of phytoalexins in disease resistance, in *Phytoalexins* (eds J. A. Bailey and J. W. Mansfield), Blackie, Glasgow, pp. 253–288.
Maramorosch, K. (1974). Mycoplasmas and rickettsiae in relation to plant diseases, *A. Rev. Microbiol.,* **28**, 301–324.
Maramorosch, K., Hirumi, H., Kumura, M., Bird, J., and Vakili, N.G. (1974). Pigeon pea witches' broom disease, *Phytopathology,* **64**, 582–583 (abstr.).
Maramorosch, K., Kimura, M., and Nene, Y. L. (1976). Mycoplasma-like organisms associated with pigeon pea rosette disease in India, *FAO Pl. Prot. Bull.,* **24**, 33–35.
Marasas, W. F. O., Pauer, G. D., and Boerema, G. H. (1974). A serious leaf blotch disease of groundnuts (*Arachis hypogaea* L.) in southern Africa caused by *Phoma arachidicola* sp. nov., *Phytophylactica,* **6**, 195–201.
Marchetti, M. A., Melching, J. S., and Bromfield, K. R. (1976). The effects of temperature and dew period on germination and infection by uredospores of *Phakopsora pachyrhizi, Phytopathology,* **66**, 461–463.
Marchetti, M. A., Uecker, F. A., and Bromfield, K. R. (1975). Uredial development of *Phakopsora pachyrhizi* in soybeans, *Phytopathology,* **65**, 822–823.
Marcus, C. P. (1952). A new physiologic race of rust (*Uromyces phaseoli typica* Arthur) causing losses to beans in Maryland, *Phytopathology,* **42**, 342 (abstr.).
Maréchal, R., Mascherpa, J-M., and Stainier, F. (1978). Etude taxonomique d'un groupe complexe d'espèces des genres *Phaseolus* et *Vigna* (Papilionaceae) sur la base de données morphologiques et polliniques, traitées par l'analyse informatique, *Boissiera,* **28**, 1–273.
Markham, P. G. (1980). The 'yellows' plant diseases and their etiology, *Contrib. Plant Disease Etiology: 100 years of Koch's postulates*, Fed. Br. Pl. Path., London, 17–19 Dec.
Marshall, D. R. and Pryor, A. J. (1978). Multiline varieties and disease control. I. The 'dirty crop' approach with each component carrying a unique single resistance gene, *Theor. Appl. Genet.,* **51**, 177–184.
Marshall, D. R. and Pryor, A. J. (1979). Multiline varieties and disease control. II. The 'dirty crop' approach with components carrying two or more genes for resistance, *Euphytica,* **28**, 145–159.
Martin, P. M. D. and Gilman, G. A. (1976). A consideration of the mycotoxin hypothesis with special reference to the mycoflora of maize, sorghum, wheat and groundnuts, *Rep. Trop. Prod. Inst.,* **G. 105**, vii and 112.
Martinez, A., Coiduras, A., Moreno, M. T., and Cubero, J. (1979). Quantitative inheritance and barriers to crossability in *Cicer arientinum* L., *Theor. Appl. Genet.,* **55**, 5–16.
Mastenbroek, C. (1960). A breeding programme for resistance to anthracnose in dry shell haricot beans, based on a new gene, *Euphytica,* **9**, 177–184.
Mathur, R. S., Jain, J. S., and Atheya, S. C. (1960). Resistance of gram varieties to *Fusarium* wilt in Uttar Pradesh, 1949–1958, *Curr. Sci.,* **29**, 403.
Mathur, S. B. (1963). Effect of soil temperature on root-rot of guar (*Cyamopsis*

psoralioides DC.) and wilt of gram (*Cicer arietinum* L.) caused by *Sclerotium rolfsii* Sacc., *Agra Univ. J. Res.*, **12**, 261–269.

Mathys, G. and Baker, E. A. (1980). An appraisal of the effectiveness of quarantines, *A. Rev. Phytopath.*, **18**, 85–101.

Mayee, C. D., Godbole, G. M., and Patil, F. S. (1977). Appraisal of groundnut rust in India: problems and approach, *PANS*, **23**, 162–165.

Mazzani, B., Allievi, J., and Bravo, P. (1972). Relacion entre la incidencia de manchas foliares por *Cercospora* spp. y algunas caracteristicas varietales del mani, *Agron. Trop.*, **22**, 119–132.

Mazzani, B. and Hinojosa, S. (1961). Differencias varietales de susceptibilidad a la roya del mani en Venezuela, *Agron. Trop.*, **11**, 41–45.

McCarthy, E. V. and Swinburne, T. R. (1980). Interaction of *Pseudomonas* with *Botrytis cinerea* on *Phaseolus vulgaris*, *Contrib. Plant Disease Etiology: 100 Years of Koch's postulates,* Fed. Br. Pl. Path., London, 17–19 Dec.

McDonald, D. (1969). Groundnut pod diseases, *Rev. appl. Mycol.*, **48**, 465–474.

McDonald, D. (1970a). Fungal infection of groundnut fruit before harvest, *Trans. Br. mycol. Soc.*, **54**, 453–460.

McDonald, D. (1970b). Fungal infection of groundnut fruit after maturity and during drying, *Trans. Br. mycol. Soc.*, **54**, 461–472.

McDonald, D. (1973). Trials with a copper-sulphur dust for control of *Cercospora* leaf spots of groundnuts in Nigeria, *Samaru Misc. Pap.* No. 45, Ahmadu Bello Univ., Nigeria.

McDonald, D. (1978). *The Groundnut Disease Situation in Northern Nigeria*, Inst. Agric. Res., Ahmadu Bello Univ., Samaru, Zaria, 47 pp.

McDonald, D. and Emechebe, A. M. (1978). Occurrence and preliminary survey of peanut rust in Nigeria, *Pl. Dis. Reptr.*, **62**, 5–8.

McDonald, D. and Fowler, A. M. (1976). Control of *Cercospora* leaf-spot disease of groundnuts (*Arachis hypogaea*) in Nigeria, *Niger. J. Pl. Prot.*, **2**, 43–59.

McDonald, D. and Harkness, C. (1965). Growth of *Aspergillus flavus* and production of aflatoxin in groundnuts. Part VIII, *Trop. Sci.*, **7**, 122–137.

McDonald, D. and Raheja, A. K. (1980). Pests, diseases, resistance, and crop protection in groundnuts, in *Advances in Legume Science* (eds R. J. Summerfield and A. H. Bunting), HMSO, London, pp. 501–514.

McDonald, J. G. and Hamilton, R. I. (1972). Distribution of southern bean mosaic virus in the seed of *Phaseolus vulgaris*, *Phytopathology*, **62**, 387.

McGuire, J. M. and Cooper, W. E. (1965). Interaction of heat injury and *Diplodia gossypina* and other etiological aspects of collar rot of peanut, *Phytopathology*, **55**, 231–236.

McKenzie, A. R., Flentje, N. T., Stretton, H. M., and Mayo, M. J. (1969). Heterokaryon formation and genetic recombination within one isolate of *Thanatephorus cucumeris, Aust. J. biol. Sci.*, **22**, 895–904.

McKerral, A. (1923). A note on *Fusarium* wilt of gram in Burma and measures taken to combat it, *Agric. J. India,* **18**, 608–613.

McLaughlin, M. R., Mignucci, J. S., and Milbrath, G. M. (1977). *Microsphaera diffusa,* the perfect stage of the soybean powdery mildew pathogen, *Phytopathology,* **67**, 726–729.

McLaughlin, M. R., Thongmeearkom, P., Milbrath, G. M., and Goodman, R. M. (1977). Isolation and some properties of a yellow subgroup member of cowpea mosaic virus from Illinois, *Phytopathology,* **67**, 844–847.

McLean, R. J. (1979). Histological studies of resistance to soybean rust, *Phakopsora pachyrhizi* Syd., *Aust. J. agric. Res.*, **30**, 77–84.

McLean, R. and Byth, D. E. (1976). Resistance of soybean to rust in Australia, *APPS Newsletter,* **5**, 34–36. (*Rev. Pl. Path.*, **56**, 4267.)

McMillan, R. T. (1972). A new race of bean rust in pole beans in Florida, *Pl. Dis. Reptr.*, **56**, 759.
McNeill, K. E. (1976). Studies on the nature of resistance in protopea (cowpea) to *Verticillium* wilt. Ph.D thesis, Mississippi State Univ. (*Diss. Abstr. Int. B. Sci. Eng.*, **36**, 1005.)
McRae, W. (1932). Report of the Imperial Mycologist, *Scient. Repts. Imper. Inst. Agric. Res.*, Pusa, 1930–31, pp. 73–86. (*Rev. appl. Mycol.*, **11**, 424–426.)
McRae, W. and Shaw, F. J. F. (1933). *Influence of Manures on the Wilt Disease of* Cajanus indicus *Spreng. and the Isolation of Types Resistant to the Disease,* Sci. Monog. No. 7, the Imperial Council of Agricultural Research, Delhi, 68 pp.
McRostie, G. P. (1919). Inheritance of anthracnose resistance as indicated by a cross between a resistant and a susceptible bean, *Phytopathology,* **9**, 141–148.
McRostie, G. P. (1921). Inheritance of disease resistance in the common bean, *J. Am. Soc. Agron.,* **13**, 15–32.
McVey, D. V. (1965). Inoculation and development of rust on peanuts grown in the greenhouse, *Pl. Dis. Reptr.,* **49**, 191–192.
Medina, A. C. and Grogan,R. G. (1961). Seed transmission of bean mosaic viruses, *Phytopathology,* **51**, 452–456.
Mehrotra, R. S. (1972). Behaviour of zoospores of *Phytophthora megasperma* var. *sojae* and *P. drechsleri* in soil, *Can. J. Bot.,* **50**, 2125–2130.
Mehta, P. R. and Mundkur, B. B. (1946). Some observations on the rust of gram (*Cicer arietinum* L.), *Indian J. agric. Sci.,* **16**, 186–192.
Meiners, J. P. (1979a). Sources of resistance to U.S. bean rust - update, *A. Rep. Bean Improv. Co-op.,* **22**, 62–63.
Meiners, J. P. (1979b). Results - uniform snap bean rust nursery, 1978, *A. Rep. Bean Improv. Co-op.,* **22**, 85–86.
Meiners, J. P. (1981). Genetics of disease resistance in edible legumes, *A. Rev. Phytopath.,* **19**, 189–209.
Meiners, J. P. and Elden, T. C. (1980). Resistance to insects and diseases in *Phaseolus,* in *Advances in Legume Science* (eds R. J. Summerfield and A. H. Bunting), HMSO, London, pp. 359–364.
Meiners, J. P. and Gillaspie, A. G. (1980). Screening for resistance in the field to peanut stunt virus in snap beans, *A. Rep. Bean Improv. Co-op.,* **23**, 26–29.
Meiners, J. P., Gillaspie, A. G., Lawson, R. H., and Smith, F. F. (1978). Identification and partial characterization of a strain of bean common mosaic virus from *Rhynchosia minima, Phytopathology,* **68**, 283–287.
Meiners, J. P., Hilty, J. W., and Mullins, C. A. (1975). Field tests in Tennessee for rust resistance in snap beans, *A. Rep. Bean Improv. Co-op.,* **18**, 43–44.
Meiners, J. P., Waterworth, H. E., Lawson, R. H., and Smith, F. F. (1977). Curly dwarf mosaic disease of beans from El Salvador, *Phytopathology,* **67**, 163–168.
Melching, J. S., Bromfield, K. R., and Kingsolver, C. H. (1979). Infection, colonization, and uredospore production on Wayne soybean by four cultures of *Phakopsora pachyrhizi,* the cause of soybean rust, *Phytopathology,* **69**, 1262–1265.
Meléndez, P. L. (1978). The history of *Phakopsora* rust in Puerto Rico, in *Proc. Workshop Soybean Rust in W. Hemisphere* (ed. N. G. Vakili), USDA, Mayaguez, pp. 1–3.
Mendgen, K. (1973). Feinbau der Infektionsstrukturen von *Uromyces phaseoli, Phytopath. Z.,* **78**, 109–120.
Mercer, P. C. (1973). The control of *Cercospora* leafspot on groundnuts in Malawi and its effect on the parameters of yield, *PANS*, **19**, 201–207.
Mercer, P. C. (1977). Pests and diseases of groundnuts in Malawi. I. Virus and foliar diseases, *Oléagineux,* **32**, 483–488.
Mercer, P. C. (1978). Pests and diseases of groundnuts in Malawi. II. Disorders of pods, kernels and seedlings, *Oléagineux*, **33**, 119–122.

Mercer, P. C., Wood, R. K. S., and Greenwood, A. D. (1974). Resistance to anthracnose of French bean, *Physiol. Pl. Path.,* **4**, 291–306.

Mercer, P. C., Wood, R. K. S., and Greenwood, A. D. (1975). Ultrastructure of the parasitism of *Phaseolus vulgaris* by *Colletotrichum lindemuthianum, Physiol. Pl. Path.,* **5**, 203–214.

Messieha, M. (1969). Transmission of tobacco ringspot virus by thrips, *Phytopathology,* **59**, 943–945.

Mew, I-P. C., Wang, T. C., and Mew, T. W. (1975). Inoculum production and evaluation of mung bean varieties for resistance to *Cercospora canescens, Pl. Dis. Reptr.,* **59**, 397–401.

Mew, T. W. and Kennedy, B. W. (1971). Growth of *Pseudomonas glycinea* on the surface of soybeans leaves, *Phytopathology,* **61**, 715–716.

Meyer, W. A., Sinclair, J. B., and Khare, M. N. (1974). Factors affecting charcoal rot of soybean seedlings, *Phytopathology,* **64**, 845–849.

Milbrath, G. M. and Soong, M-M (1976). A local lesion assay for soybean mosaic virus using *Phaseolus vulgaris* L. cv. Top Crop, *Phytopath. Z.,* **87**, 255–259.

Milbrath, G. M. and Tolin, S. A. (1977). Identification, host range and serology of peanut stunt virus isolated from soybean, *Pl. Dis. Reptr.,* **61**, 637–640.

Milbrath, J. A. (1944). Studies on the control of bean rust, *Phytopathology,* **34**, 936 (abstr.).

Miller, D. E. and Burke, D. W. (1977). Effect of temporary excessive wetting on soil aeration and *Fusarium* root rot of beans, *Pl. Dis. Reptr.,* **61**, 175–179.

Miller, L. I. (1949). Cultural and parasitic races of *Cercospora arachidicola* and *C. personata, Phytopathology,* **39**, 15 (abstr.).

Minchin, F. R., Summerfield, R. J., Hadley, P., and Roberts, E. H. (1980). Growth, longevity and nodulation of roots in relation to seed yield in chickpeas (*Cicer arietinum*), *Expl. Agric.,* **16**, 241–261.

Mink, G. I. (1972). Peanut stunt virus. *CMI/AAB Descriptions of Plant Viruses No. 92.*

Mink, G. I., Saksena, K. M., and Silbernagel, M. J. (1966). Purification of the bean red node strain of tobacco streak virus, *Phytopathology,* **56**, 645–649.

Minton, N.A. (1963). Effects of two populations of *Meloidogyne arenaria* on peanut roots, *Phytopathology,* **53**, 79–81.

Minton, N. A. and Hammons, R. O. (1975). Evaluation of peanut for resistance to the peanut root-knot nematode, *Meloidogyne arenaria, Pl. Dis. Reptr.,* **59**, 944–945.

Minton, N. A. and Parker, M. B. (1974). Yields of six soybean cultivars following fumigation of soil infested with root-knot nematodes, *Phytopathology*, **64**, 219–221.

Minton, N. A., Parker, M. B., and Flowers, R. A. (1975). Response of soybean cultivars to *Meloidogyne incognita* and to the combined effects of *M. arenaria* and *Sclerotium rolfsii, Pl. Dis. Reptr*., **59**, 920–923.

Miranda, C. S. (1967). Origen de *Phaseolus vulgaris* L. (Fríjol Comun), *Agrociencia*, **1**, 99–109.

Miranda, C. S. and Evans, A. M. (1973). Exploring the genetic isolating mechanisms between *Phaseolus vulgaris* L. and *P. coccineus* Lam., *A. Rep. Bean Improv. Co-op*., **16**, 39–41.

Mircetich, S. M. and Zentmyer, G. A. (1966). Production of oospores and chlamydospores of *Phytophthora cinnamomi* in roots and soil, *Phytopathology*, **56**, 1076–1078.

Mishra, J. N. (1955). Wilt of gram (*Cicer arietinum* L.) in Bihar, *Curr. Sci*., **24**, 210.

Misra, A. K. and Misra, A. P. (1975). Groundnut rust in Bihar – varietal reaction, *Indian Phytopath*., **28**, 557–558.

Misra, D. P. and Gurha, S. N. (1980). Losses due to yellow mosaic in winter crop of pigeon pea in India, *Trop. Grain Leg. Bull*., **19**, 18–19.

Misra, D. P. and Mehra, R. C. S. (1969). *Choanephora cucurbitarum* on *Cajanus cajan* in India, *Indian Phytopath.*, **22**, 515–517.

Mitchell, J. W., Zaumeyer, W. J., and Preston, W. H. (1954). Absorption and translocation of streptomycin by bean plants and its effect on the halo and common blight organisms, *Phytopathology*, **44**, 25–30.

Mitra, M. (1934). Wilt disease of *Crotalaria juncea* Linn. (Sannhemp), *Indian J. agric. Sci.*, **4**, 701–714.

Mixon, A. C. and Rogers, K. M. (1973). Peanut accessions resistant to seed infection by *Aspergillus flavus*, *Agron. J.*., **65**, 560–562.

Mixon, A. C. and Rogers, K. M. (1975). Factors affecting *Aspergillus flavus* Lk. ex Fr. colonization of resistant and susceptible genotypes of *Arachis hypogaea* L., *Peanut Sci.*, **2**, 18–22.

Mode, C. J. (1958). A mathematical model for the co-evolution of obligate parasites and their hosts, *Evolution*, **12**, 158–165.

Mohanty, U. N. (1946). The wilt disease of pigeon pea (*Cajanus cajan* (L.) Millsp.) with special reference to the distribution of the causal organism in the host tissue, *Indian J. agric. Sci.*, **16**, 379–390.

Mok, D. W. S., Mok, M. C., and Rabakoarihanta, A. (1978). Interspecific hybridization of *Phaseolus vulgaris* with *P. lunatus* and *P. acutifolius, Theor. Appl. Genet.*, **52**, 209–215.

Monteith, J. L. (1972). Solar radiation and productivity in tropical ecosystems, *J. appl. Ecol.*, **9**, 747–766.

Moore, B. J. (1973). Quail pea mosaic virus: a new member of the comovirus group, *Pl. Dis. Reptr.*, **57**, 311–315.

Moore, B. J. and Scott, H. A. (1981). Quail pea mosaic virus. *CMI/AAB Descriptions of Plant Viruses No. 238.*

Moore, W. F. (1974). Studies on techniques of hybridization and the inheritance of resistance to verticillium wilt in protopea (cowpea). Ph.D thesis, Mississippi State Univ. (Diss. Abstr. Int. B Sci. Eng., **35**, 2013–2014.)

Mordue, J. E. M. (1971a). *Colletotrichum lindemuthianum. CMI Descriptions of Pathogenic Fungi and Bacteria No. 316.*

Mordue, J. E. M. (1971b). *Colletotrichum capsici. CMI Descriptions of Pathogenic Fungi and Bacteria No. 317.*

Mordue, J. E. M. (1971c). *Glomerella cingulata. CMI Descriptions of Pathogenic Fungi and Bacteria No. 315.*

Mordue, J. E. M. (1974a). *Thanatephorus cucumeris. CMI Descriptions of Pathogenic Fungi and Bacteria No. 406.*

Mordue, J. E. M. (1974b). *Corticium rolfsii. CMI Descriptions of Pathogenic Fungi and Bacteria No. 410.*

Mordue, J. E. M. and Holliday, P. (1976). *Sclerotinia sclerotiorum. CMI Descriptions of Pathogenic Fungi and Bacteria No. 513.*

Moreno, M-T. and Cubero, J. I. (1978). Variation in *Cicer arietinum* L., *Euphytica*, **27**, 465–485.

Moreno, R. (1975). Diseminación de *Ascochyta phaseolorum* en variedades de frijol de costa bajo diferentes sistemas de cultivo, *Turrialba*, **25**, 361–364.

Moreno, R. (1977). Efecto de diferentes sistemas de cultivo sobra la severidad de la mancha angular del frijol (*Phaseolus vulgaris* L.) causado por *Isariopsis griseola* Sacc., *Agronomia Costarricense*, **1**, 39–42.

Moreno, R. A. (1979). Crop protection implications of cassava intercropping, in *Intercropping with Cassava* (eds E. Weber, B. Nestel, and M. Campbell), IDRC-142 e, Ottawa, pp. 113–127.

Morgan, F. and Dunleavy, J. M. (1966). Brown stem rot of soybeans in Mexico, *Pl. Dis. Reptr.*, **50**, 598–599.

Morrall, R. A. A. and McKenzie, D. L. (1974). A note on the inadvertent introduction to North America of *Ascochyta rabiei*, a destructive pathogen of chickpea, *Pl. Dis. Reptr.*, **58**, 342–345.

Mortimer, A. M., Shaw, D. S., and Sansome, E. R. (1977). Genetical studies of secondary homothallism in *Phytophthora dreschsleri*, *Arch. Microbiol.*, **111**, 255–259.

Morwood, R. B. (1945). Peanut diseases, *Qld. agric. J.* **61**, 266–271.

Moss, J. P. (1980). Wild species in the improvement of groundnuts, in *Advances in Legume Science* (eds. R. J. Summerfield and A. H. Bunting), HMSO, London, pp. 525–533.

Moss, R. P. (ed.) (1968). *The Soil Resources of Tropical Africa*, Cambridge Univ. Press, 226 pp.

Mound, L. A. (1962). Studies on the olfaction and colour sensitivity of *Bemisia tabaci* (Genn.), *Entom. Exp. Appl.*, **5**, 99–104.

Mound, L. A. (1981). Whitefly biology, *Workshop Pathogens Transmitted by Whiteflies*, Oxford, 31 July 1981 (abstr.).

Mountain, W. B. (1965). Pathogensis by soil nematodes, in *Ecology of Soilborne Plant Pathogens, Prelude to Biological Control* (eds K. F. Baker and W. C. Snyder), John Murray, London, pp. 285–301.

Msuku, W. A. B. and Edje, O. T. (1982). Effect of mixed cropping of maize and bean on bean diseases, *A. Rep. Bean Improv. Co-op.*, **25**, 16–18.

Muchovej, J. J., Muchovej, R. M. C., Dhingra, O. D., and Maffia, L. A. (1980). Suppression of anthracnose of soybeans by calcium, *Pl. Dis.*, **64**, 1088–1089.

Mughogho, L. K. (1970). *Phaseolus vulgaris* beans improvement project, *Res. Bull. Bunda Coll. Agric.*, **1**, 17–18.

Mughogho, L. K., Edje, O. T., and Ayonoadu, U. W. U. (1972). Bean improvement programme in Malawi, *A. Rep. Bean Improv. Co-op.*, **15**, 69–71.

Mukelar, A. (1978). Mungbean scab in Indonesia, *Proc. 1st Internat. Mungbean Symp.*, Asian Vegetable Research and Development Centre, Taiwan, pp. 161–164.

Mukerji, K. G. (1968). *Leveillula taurica. CMI Descriptions of Pathogenic Fungi and Bacteria No. 182.*

Mukherjee, D., De, T. K., and Parui, N. R. (1971). A note on the screening of arhar against wilt disease, *Indian Phytopath.*, **24**, 598–601.

Mukherjee, D., Lambert, J. W., Cooper, R. L., and Kennedy, B. W. (1966). Inheritance of resistance to bacterial blight (*Pseudomonas glycinea* Coeper) in soybeans (*Glycine max* L.), *Crop Sci.*, **6**, 324–326.

Mukiibi, J. (1969a). Morphogenesis of sclerotia of *Dactuliophora tarrii, Trans. Br. mycol. Soc.*, **52**, 496–498.

Mukiibi, J. (1969b). Control of leaf diseases of cowpeas caused by fungal pathogens, *Contrib. 12th Meet. Spec. Comm. Agric. Bot.*, Kampala, Sept. 1969.

Mukiibi, J. (1976a). Preliminary observations on the germination of sclerotia of *Dactuliophora tarrii* pathogen of cowpeas, *E. Afr. agric. For. J.*, **42**, 44–46.

Mukiibi, J. (1976b). Possible relationship between intercropping and plant disease problems in Uganda, in *Intercropping in Semi-arid Areas* (eds J. H. Monyo, A. D. Ker, and M. Campbell), IDRC, Ottawa, p. 45.

Mukiibi, J. (1978). The effect of some crop protection measures on diseases, yield and quality of groundnuts, *Proc. 10th Meet. Assoc. Adv. Agric. Sci. Afr.*, Univ. Ibadan, Nigeria, April 1978.

Mukiibi, J. K. (1979). Some effects of mancozeb, DDT and weeding on disease and pest incidence and on yield of cowpeas *Vigna unguiculata* (L.) Walp., *Proc. Symp. Grain Leg. Improv. E. Africa*, Nairobi, Kenya, 20–24 Aug. 1979.

Mukunya, D. M. (1974). Bean diseases in Kenya, *A. Rep. Bean Improv. Co-op.*, **17**, 57–59.

Mulder, J. L. and Holliday, P. (1974a). *Mycosphaerella arachidis. CMI Descriptions of Pathogenic Fungi and Bacteria No. 411.*

Mulder, J. L. and Holliday, P. (1974b). *Mycosphaerella berkeleyi. CMI Descriptions of Pathogenic Fungi and Bacteria No. 412.*

Mulder, J. L. and Holliday, P. (1975a). *Cercospora canescens. CMI Descriptions of Pathogenic Fungi and Bacteria No. 462.*

Mulder, J. L. and Holliday, P. (1975b). *Cercospora cruenta. CMI Descriptions of Pathogenic Fungi and Bacteria No. 463.*

Mulder, J. L. and Holliday, P. (1975c). *Cercospora kikuchii. CMI Descriptions of Pathogenic Fungi and Bacteria No. 466.*

Muller, A. S. (1953). A foliar disease of legumes in Central America, *FAO Pl. Prot. Bull.*, **1**, 83–84.

Mundkur, B. B. (1935). Influence of temperature and maturity on the incidence of sannhemp and pigeon pea wilt at Pusa, *Indian J. agric Sci.*, **5**, 609–618.

Murakishi, H. H. (1951). Purple seed stain of soybean, *Phytopathology*, **41**, 305–318.

Muras, V. A. (1963). On bacterial wilt of soybeans in the Ukraine, *J. Microbiol. (Kiev)*, **25**, 42–49. (*Rev. appl. Mycol.*, **43**, 2457.)

Murayama, D. (1966). On the witches' broom diseases of sweet potato and leguminous plants in the Ryukyu Islands, *Mem. Fac. Agric. Hokkaido Univ.*, **6**, 81–103. (*Rev. appl. Mycol.*, **46**, 724.)

Murthy, G. S. and Bagyaraj, D. J. (1980). Flavanol and alkaloid content of pigeon pea cultivars resistant and susceptible to *Fusarium udum, Indian Phytopath.*, **33**, 633–634.

Murty, B. R. (1975). Biology of adaptation in chickpea, in *Internat. Workshop on Grain Legumes*, International Crops Research Institute for the Semi-Arid Tropics, Hyderabad, India, pp. 239–251.

Musselman, L. J. (1980). The biology of *Striga, Orobanche*, and other root-parasitic weeds, *A. Rev. Phytopath.*, **18**, 463–489.

Mutitu, E. W. and Mukunya, D. M. (1979). Epidemiology and control of bean scab caused by *Elsinoe phaseoli* Jenkins in Kenya, *Proc. Symp. Grain Leg. Improv. E. Africa*, Nairobi, 20–24 August, 1979.

Mutsaers, H. J. W., Mbouémboué, P., and Boyomo, M. (1981). Traditional food crop growing in the Yaoundé area (Cameroon). 1. Synopsis of the system, *Agro-Ecosystems*, **6**, 273–287.

Naito, N. (1951). On the influence of temperature to germination of urediniospores of *Uromyces appendiculatus* (Pers.) Fr. parasitic on bean, *Tech. Bull. Kagawaken agric. Coll.*, **1**, 14–17. (*Rev. appl. Mycol.*, **32**, 165.)

Narayan, R. K. J. and Macefield, A. J. (1976). Adaptive responses and genetic divergence in a world germplasm collection of chickpea (*Cicer arietinum* L.), *Theor. Appl. Genet.*, **47**, 179–187.

Narayanan, A. and Sheldrake, A. R. (1979). Pigeon pea (*Cajanus cajan*) as a winter crop in peninsular India, *Expl. Agric.*, **15**, 91–95.

Narayanasamy, P. and Durairaj, P. (1971). A new blight disease of soybean, *Madras agric. J.*, **58**, 711–712.

Narayanasamy, P. and Ramakrishnan, K. (1965). Studies of the sterility mosaic disease of pigeon pea. I. Transmission of the disease, *Proc. Indian Acad. Sci.*, B. **62**, 73–86.

Narayanasamy, P., Seshadri, A. R., and Ramakrishnan, K. (1963). A preliminary note on suspected nematode transmission of red gram sterility mosaic virus, *Madras agric. J.*, **50**, 109–110.

Nariani, T. K. and Dhingra, K. L. (1963). A mosaic disease of groundnut (*Arachis hypogaea*), *Indian J. agric. Sci.*, **33**, 25–27.

Nash, S. M. and Alexander, J. V. (1965). Comparative survival of *Fusarium solani* f. *cucurbitae* and *F. solani* f. *phaseoli* in soil, *Phytopathology*, **55**, 963–966.

Nash, S. M., Christou, T., and Snyder, W. C. (1961). Existence of *Fusarium solani* f. *phaseoli* as chlamydospores in soil, *Phytopathology*, **51**, 308–312.

Nash, S. M. and Snyder, W. C. (1962). Quantitative estimations by plate counts of propagules of the bean root rot *Fusarium* in field soils, *Phytopathology*, **52**, 567–572.
Nash, S. M. and Snyder, W. C. (1964). Dissemination of the root rot *Fusarium* with bean seed, *Phytopathology*, **54**, 880.
Nass, H. A., Pedersen, W. L., MacKenzie, D. R., and Nelson, R. R. (1981). The residual effects of some 'defeated' powdery mildew resistance genes in isolines of Chancellor winter wheat, *Phytopathology*, **71**, 1315–1318.
Nattrass, R. M. (1961). *Host Lists of Kenya Fungi and Bacteria*. Mycol. Pap. No. 81, Commonwealth Mycological Institute, Kew, 46 pp.
Nayudu, M. V. (1963). *Leptosphaerulina arachidicola* on groundnut, *Indian Phytopath*., **16**, 384–386.
Ndimande, B. N., Wien, H. C., and Kueneman, E. A. (1981). Soybean seed deterioration in the tropics. I. The role of physiological factors and fungal pathogens, *Fld. Crops Res*., **4**, 113–121.
Nelson, R. R. (1978). Genetics of horizontal resistance to plant diseases, *A. Rev. Phytopath*., **16**, 359–378.
Nema, K. G., Jain, A. C., and Asthana, R. P. (1955). Further studies on *Aspergillus* blight of groundnut seedlings: its occurrence and control, *Indian Phytopath*., **8**, 13–21.
Nene, Y. L. (1972). *A Survey of Viral Diseases of Pulse Crops in Uttar Pradesh*, G. B. Pant Univ. Agric. and Tech., Pantnagar, 191 pp.
Nene, Y. L. (1973). Control of *Bemisia tabaci* Genn., a vector of several plant viruses, *Indian J. agric. Sci.*, **43**, 433–436.
Nene, Y. L. (1978). *Pulse pathology progress rep. 3*, International Crops Research Institute for the Semi-Arid Tropics, Hyderabad, India.
Nene, Y. L. (ed.) (1980a). *Proc. Consultants Group Discussion on Resistance to soil-borne Diseases of Legumes*, International Crops Research Institute for the Semi-Arid Tropics, Hyderabad, India, 167 pp.
Nene, Y. L. (1980b). Diseases of chickpea, in *Proc. Internat. Wkshop on Chickpea Improv*., International Crops Research Institute for the Semi-Arid Tropics, Hyderabad, India, pp. 171–178.
Nene, Y. L. (1981). A review of *Ascochyta* blight of chickpea (*Cicer arietinum* L.), in *Proc. Wkshop on Ascochyta Blight and Winter Sowing of Chickpea*, 4–7 May 1981, International Center for Agricultural Research in the Dry Areas, Aleppo, Syria.
Nene, Y. L. (1982). The outlook for chickpea and pigeonpea, *Span*, **25**, 14–16.
Nene, Y. L. and Haware, M. P. (1980). Screening chickpea for resistance to wilt, *Pl. Dis*., **64**, 379–380.
Nene, Y. L., Haware, M. P., and Reddy, M. V. (1979). *Diagnosis of some wilt-like disorders of chickpea* (Cicer arietinum *L.*), Info. Bull. No. 3, International Crops Research Institute for the Semi-Arid Tropics, Hyderabad, India.
Nene, Y. L. and Kannaiyan, J. (1982). Screening pigeonpea for resistance to *Fusarium* wilt, *Pl. Dis*., **66**, 306–307.
Nene, Y. L., Kannaiyan, J., and Reddy, M. V. (1981). Resistance to major pigeonpea diseases, in *Proc. Internat. Wkshop in Pigeonpeas*, **1**, International Crops Research Institute for the Semi-Arid Tropics, Hyderabad, India, pp. 121–128.
Nene, Y. L. and Rathi, Y. P. S. (1972). Diseases of pigeon pea, in *A Survey of Viral Diseases of Pulses Crops in Uttar Pradesh*, G. B. Pant Univ. Agric. and Tech., Pantnagar, India, pp. 154–162.
Nene, Y. L. and Reddy, M. V. (1976a). Preliminary information on chickpea stunt, *Trop. Grain Leg. Bull.*, **5**, 31–32.
Nene, Y. L. and Reddy, M. V. (1976b). Screening for resistance to sterility mosaic of pigeonpea, *Pl. Dis. Reptr*., **60**, 1034–1036.

Nene, Y. L. and Reddy, M. V. (1976c). A new technique to screen pigeon pea for resistance to sterility mosaic, *Trop. Grain Leg. Bull.*, **5**, 23.

Neto, A. T., Ando, A., and Costa, A. S. (1977). Bean breeding program at CENA. III. New results in attempts to induce mutants resistant or tolerant to golden mosaic virus in dry beans (*Phaseolus vulgaris* L.), *A Rep. Bean Improv. Co-op.*, **20**, 86.

Netto, A. J., Athow, K. L., and Vieira, C. (1967). Physiologic races of bean rust, *Uromyces phaseoli typica*, in the State of Minas Gerais, Brazil, *A Rep. Bean Improv. Co-op.*, **10**, 22–23.

Neveling, C. H. (1956). Report of the Secretary for Agriculture for the year ended 31 August 1955, *Farming in South Africa*, **31**, 359. (*Rev. appl. Mycol.*, **35**, 587

Nevill, D. J. (1978). Breeding groundnuts (*Arachis hypogaea* L.) for resistance to foliar pathogens. Ph.D thesis, Univ. Cambridge, 198 pp.

Nevill, D. J. (1980). Studies of resistance to foliar pathogens, in *Proc. Internat. Wkshop Groundnuts*, ICRISAT, Hyderabad, India, pp. 199–202.

Nevill, D. J. (1981). Components of resistance to *Cercospora arachidicola* and *Cercosporidium personatum* in groundnuts, *Ann. appl. Biol.*, **99**, 77–86.

Nevill, D. J. and Evans, A. M. (1980). The effect of host development on the field assessment of disease resistance to *Cercospora* leaf spots in groundnuts (*Arachis hypogaea* L.), *J. agric. Sci., Camb.*, **94**, 229–237.

Newhook, F. J., Waterhouse, G. M., and Stamps, D. J. (1978). Tabular key to the species of *Phytophthora* de Bary, *Mycol. Pap.*, **143**, Commonwealth Mycological Institute, Kew.

Newton, W. and Peiris, J. W. L. (1953). Virus diseases of plants in Ceylon, *FAO Pl. Prot. Bull.*, **2**, 17–21.

Ngundo, B. W. (1977). Screening of bean cultivars for resistance to *Meloidogyne* spp. in Kenya, *Pl. Dis. Reptr.*, **61**, 991–993.

Ngundo, B. W. and Taylor, D. P. (1974). Effects of *Meloidogyne* spp. on bean yields in Kenya, *Pl. Dis. Reptr.*, **58**, 1020–1023.

Nicholson, J. F., Dhingra, O. D., and Sinclair, J. B. (1972). Internal seed-borne nature of *Sclerotinia sclerotiorum* and *Phomopsis* sp. and their effects on seed quality, *Phytopathology*, **62**, 1261–1263.

Nicholson, J. F. and Sinclair, J. B. (1971). Amsoy soybean seed germination inhibited by *Pseudomonas glycinea, Phytopathology*, **61**, 1390–1393.

Nigram, S. N., Dwivedi, S. L., and Gibbons, R. W. (1980). Groundnut breeding at ICRISAT, in *Proc. Intern. Wkshop Groundnuts* (ed. R. W. Gibbons), ICRISAT, Hyderabad, India, pp. 62–68.

Niknejad, M. and Khosh-Khui, M. (1972). Natural cross-pollination of gram (*Cicer arietinum* L.), *Indian J. agric. Sci.*, **42**, 273–274.

Nirula, K. K. (1979). Quarantine regulations of seed imports at the International Crops Research Institute for the Semi-Arid Tropics, *FAO Pl. Prot. Bull.*, **27**, 119–122.

Nirwan, R. S. and Upadhyaya, J. (1972). *Phytophthora* blight of cowpea new to India, *Indian Phytopath.*, **25**, 162–163.

Noble, M. and Richardson, M. J. (1968). An annotated list of seed-borne diseases. *Phytopath. Pap. No. 8*, Commonwealth Mycological Institute, Kew, 191 pp.

Norden, A. J. (1980). Crop improvement and genetic resources in groundnuts, in *Advances in Legume Science* (eds. R. J. Summerfield and A. H. Bunting), HMSO, London, pp. 515–523.

Norse, D. (1974). Plant diseases in Barbados. *Phytopath. Pap. No. 18*, Commonwealth Mycological Institute, Kew, 38 pp.

Nour, M. A. (1957). Control of powdery mildew diseases in the Sudan, with special reference to broad bean, *Emp. J. exp. Agric.*, **25**, 119–131.

Nour, M. A. and Nour, J. J. (1962). A mosaic of *Dolichos lablab* and diseases of other crops caused by alfalfa mosaic virus in the Sudan, *Phytopathology*, **52**, 427–432.

Nowell, W. (1924). Diseases of leguminous plants, in *Diseases of Crop-plants in the Lesser Antilles*, The West India Committee, London, pp. 338–343.
Nowsher, A. Z. M., Khan, A., and Islam, S. (1975). *Colletotrichum lindemuthianum* on *Dolichos lablab* from Bangladesh, *Bandladesh J. Bot.*, **4**, 121–123. (*Rev. Pl. Path.*, **56**, 3352.)
Nutman, F. J., Roberts, F. M., and Williamson, J. G. (1964). Studies on varietal resistance in the groundnut (*Arachis hypogaea* L.) to rosette disease, *Rhod. J. agric. Res.*, **2**, 63–77.
Nutman, P. S. (1969). Genetics of symbiosis and nitrogen fixation in legumes, *Proc. Roy. Soc. B.*, **172**, 417–437.
O'Brien, R. G. (1974). Peanut rust on the Atherton Tableland, *Qd agric. J.*, **100**, 108–110.
Ochieng, R. S. (1978). Studies on the bionomics of cowpea leaf beetle, *Ootheca mutabilis*, in *Pests of Grain Legumes: Ecology and Control* (eds S. R. Singh, H. F. van Emden, and T. Ajibola Taylor), Academic Press, London, pp. 187–191.
Ogbuji, R. O. (1978). Reactions of cowpea to *Meloidogyne incognita* in Eastern Nigeria, *PANS*, **24**, 32–34.
Ogle, H. J. and Brown, J. F. (1970). Relative ability of two strains of *Puccinia graminis tritici* to survive when mixed, *Ann. appl. Biol.*, **66**, 273–279.
Ogle, H. J. and Johnson, J. C. (1974). Physiologic specialization and control of bean rust (*Uromyces appendiculatus*) in Queensland, *Qd. J. agric. anim. Sci.*, **31**, 71–82.
O'Hair, S. K., Miller, J. C., and Toler, R. W. (1981). Reaction of cowpea introductions to infection with the cowpea strain of southern bean mosaic virus, *Pl. Dis.*, **65**, 251–252.
O'Hair, S. K., Toler, R. W., and Miller, J. C. (1977). Artists' airbrush inoculations as a tool for screening *Vigna unguiculata* (L.) Walp. accessions for reaction to the cowpea strain of southern bean mosaic virus (SBMV-CS), *Hortsci.*, **12**, 402 (abstr.).
Ojehomon, O. O. (1970). Effect of continuous removal of open flowers on the seed yield of two varieties of cowpea, *Vigna unguiculata* (L.) Walp, *J. agric. Sci., Camb.*, **84**, 375–381.
Okigbo, B. N. (1978). Grain legumes in the agriculture of the tropics, in *Pests of Grain Legumes: Ecology and Control* (eds S. R. Singh, H. F. van Emden, and T. Ajibola Taylor), Academic Press, London, pp. 1–11.
Okonkwo, S. N. C. and Nwoke, F. I. O. (1975). Observations on haustorial development in *Striga gesnerioides* (Willd.) Vatke (Scrophulariaceae), *Ann. Bot. n.s.*, **39**, 979–981.
Okusanya, B. A. M. and Watson, M. A. (1966). Host range and some properties of groundnut rosette virus, *Ann. appl. Biol.*, **58**, 377–387.
Oldfield, G. N. (1970). Mite transmission of plant viruses, *A. Rev. Entom.*, **15**, 343–380.
Oliari, L., Vieira, C., and Wilkinson, R. E. (1973). Physiologic races of *Colletotrichum lindemuthianum* in the state of Minas Gerais, Brazil, *Pl. Dis. Reptr.*, **57**, 870–872.
Olive, E. W. (1911). Origin of heteroecism in the rusts, *Phytopathology*, **1**, 139–149.
Olive, L. S. and Walker, E. A. (1944). A severe leafspot of soybean caused by *Phyllosticta sojaecola, Pl. Dis. Reptr.*, **28**, 1122–1123.
Omer, M. E. H. and Wood, R. K. S. (1969). Growth of *Pseudomonas phaseolicola* in susceptible and in resistant bean plants, *Ann. appl. Biol.*, **63**, 103–116.
Omunyin, M. E. (1979). Screening for resistance to bean common mosaic virus in food beans, *Proc. Symp. Grain Leg. Improv. E. Africa*, Nairobi, Kenya, 20–24 Aug. 1979.
Ondieki, J. J. (1973). Host lists of Kenya fungi and bacteria (a ten-year supplement 1961–1970), *E. Afr. agric. For. J.*, **28**, spec. suppl., 31 pp.
Onesirosan, P. T. (1975). Seedborne and weedborne inoculum in web blight of cowpea, *Pl. Dis. Reptr.*, **59**, 338–339.

Onesirosan, P. T. (1977). Comparison of *Rhizoctonia solani* isolates from web blight and basal canker of cowpea and from soil, *Pl. Soil*, **46**, 135–143.

Onesirosan, P. T. and Barker, L. N. (1971). Stem anthracnose of cowpeas in Nigeria, *Pl. Dis. Reptr*., **55**, 820–822.

Onesirosan, P. T. and Sagay, S. O. (1975). Survival of two pathogens of cowpea over the dry season, *Pl. Dis. Reptr*., **59**, 1018–1020.

Onim, J. F. M. (1980). Foliar fungicides for the control of *Mycovellosiella cajani* causing the leaf spot of pigeonpea, *Indian J. agric. Sci*., **50**, 615–619.

Onim, J. F. M. and Rubaihayo, P. R. (1976). Screening pigeon pea for resistance to *Mycovellosiella cajani*, *SABRAO J*., **8**, 121–125.

Onions, A. H. S. (1966). *Aspergillus flavus. CMI Descriptions of Pathogenic Fungi and Bacteria No. 91*.

Onuorah, P. E. (1973). *Pythium* seed decay and stem rot of cowpea (*Vigna sinensis* (Linn.) Savi) in southern Nigeria, *Pl. Soil*, **39**, 187–191.

Oostenbrink, M. (1969). In *Nematodes of Tropical Crops* (Ed. J. E. Peachey), Commonwealth Agricultural Bureaux, Farnham Royal, pp. 81–83.

Orellana, R. G. (1966). A new occurrence of tobacco ringspot of guar in the United States, *Pl. Dis. Reptr.* **50**, 7–10.

Orellana, R. G. (1981). Resistance to bud blight in introductions from the germplasm of wild soybean, *Pl. Dis*., **65**, 594–595.

Orozco-Sarria, S. H. and Cardona-Alvarez, C. (1959). Evidence of seed transmission of angular leaf spot of bean, *Phytopatholgoy*, **49**, 159.

Orton, W. A. (1902). The wilt disease of the cowpea and its control, *US Dep. Agric. Bur. Pl. Ind. Bull*., **17**, 9–20.

Oshima, N. and Dickens, L. E. (1971). Effects of copper sprays on secondary spread of common bacterial blight of beans, *Pl. Dis Reptr*., **55**, 609–610.

Oso, B. A. (1972). Conidial germination in *Cercospora arachidicola* Hori, *Trans. Br. mycol. Soc*., **59**, 169–172.

Oyekan, P. O. (1975). Occurrence of cowpea wilt caused by *Fusarium oxysporum* f.sp. *tracheiphilum* in Nigeria, *Pl. Dis. Reptr*., **59**, 488–490.

Oyekan, P. O. (1977). Reaction of some cowpea varieties to *Fusarium oxysporum* f.sp. *tracheiphilum* in Nigeria, *Trop. Grain. Leg. Bull*., **8**, 47–49.

Oyekan, P. O. (1978). Soybean wilt caused by *Neocosmospora vasinfecta, 3rd Int. Congr. Pl. Path.*, München, 16–23 Aug. 1978 (abstr.), p. 124.

Oyekan, P. O. (1979). Chemical control of web blight and leaf spot of cowpea in Nigeria, *Pl. Dis. Reptr.*, **63**, 574–577.

Oyekan, P. O., Onesirosan, P. T., and Williams, R. J. (1976). Screening for resistance in cowpea to web blight, *Trop. Grain Leg. Bull*., **3**, 6–8.

Padwick, G. W. (1940a). The genus *Fusarium*. III. A critical study of the fungus causing wilt of gram (*Cicer arietinum* L.) and of the related species of the subsection Orthocera, with special relation to the variability of key characteristics, *Indian J. agric. Sci*., **10**, 241–284.

Padwick, G. W. (1940b). The genus *Fusarium*. V. *Fusarium udum* Butler, *F. vasinfectum* Atk. and *F. lateritium* Nees var. *uncinatum* Wr., *Indian J. agric. Sci*., **10**, 863–878.

Padwick, G. W. and Bhagawagar, P. R. (1943). Wilt of gram in relation to date of sowing, *Indian J. agric. Sci*., **13**, 289–290.

Paguio, O. R. and Kuhn, C. W. (1973). Strains of peanut mottle virus, *Phytopathology*, **63**, 976–980.

Pal, M., Grewal, J. S., and Sarbhoy, A. K. (1970). A new stem rot of arhar caused by *Phytophthora, Indian Phytopath*., **23**, 583–587.

Papavizas, G. C., Adams, P. B., and Lewis, J. A. (1968). Survival of root-infecting fungi in soil. V. Saprophytic multiplication of *Fusarium solani* f.sp. *phaseoli* in soil, *Phytopathology*, **58**, 414–420.

Park, D. (1965). Survival of microorganisms in soil, in *Ecology of Soil-borne Plant Pathogens* (eds K. F. Baker and W. C. Snyder), John Murray, London, pp. 82–97.
Parker, C. (1965). The *Striga* problem – a review, *PANS* (C), **11**, 99–111.
Parker, M. C. (1936). Inheritance of resistance to the common mosaic virus in the bean, *J. agric. Res.*, **52**, 895–915.
Parker, M. C. (1966). Use of excised bean cotyledons for detecting the presence of bacterial blight organisms, *A Rep. Bean Improv. Co-op.*, **9**, 28–29.
Parlevliet, J. E. (1981). Stabilizing selection in crop pathosystems: an empty concept or a reality? *Euphytica*, **30**, 259–269.
Parlevliet, J. E. and Zadoks, J. C. (1977). the integrated concept of disease resistance, a new view including horizontal and vertical resistance in plants, *Euphytica*, **26**, 5–21.
Parmeter, J. R. (1970). Rhizoctonia solani, *Biology and Pathology*, Univ. California Press, Berkeley, 255 pp.
Parmeter, J. R., Sherwood, R. T., and Platt, W. D. (1969). Anastomosis grouping among isolates of *Thanatephorus cucumeris, Phytopathology*, **59**, 1270–1278.
Partridge, J. E. and Keen, N. T. (1976). Association of the phytoalexin kievitone with single-gene resistance of cowpeas to *Phytophthora vignae, Phytopathology*, **66**, 426–429.
Paschal, E. H. and Ellis, M. A. (1978). Variation in seed quality characteristics of tropically grown soybeans, *Crop Sci.*, **18**, 837–840.
Patel, G. J., Shah, H. M., and Patel, D. J. (1977). Screening of cowpea cultivars against root-knot nematodes, *Ind. J. Nemat.*, **7**, 169–170. (*Helminth. Abstr.*, B **49**, 477.)
Patel, M. K., Dhande, G. W., and Kulkarni, Y. S. (1953). Bacterial leaf-spot of *Cyamopsis tetragonoloba* (L.) Taub., *Curr. Sci.*, **22**, 183.
Patel, M. K., Kulkarni, Y. S., and Dhande, G. W. (1949). A new *Synchytrium* on *Phaseolus mungo, Curr. Sci.*, **18**, 171.
Patel, P. N. (1978). Bacterial pustule disease in Tanzania: pathogenic variability and host resistance, *3rd Int. Congr. Pl. Path.*, München, 16–23 Aug. 1978, 72 (abstr.).
Patel, P. N. (1981). Pathogen variability and host resistance in bacterial pustule disease of cowpea in Africa, *Trop. Agric. (Trin.)*, **58**, 275–280.
Patel, P. N. (1982). Genetics of cowpea reactions to two strains of cowpea mosaic virus from Tanzania. *Phytopathology*, **72**, 460–466.
Patel, P. N. and Jindal, J. K. (1970). Studies on resistance to bacterial diseases in India. II. Resistance in cowpea to bacterial blight (canker) disease, *Indian Phytopath. Soc. Bull.*, **6**, 28–34.
Patel, P. N. and Jindal, J. K. (1972a). Occurrence of bacterial blight disease on mothbean (*Phaseolus aconitifolius*) in India, *Indian Phytopath.*, **25**, 464–466.
Patel, P. N. and Jindal, J. K. (1972b). Bacterial leaf spot and halo blight diseases of mung bean (*Phaseolus aureus*) and other legumes in India, *Indian Phytopath.*, **25**, 517–525.
Patel, P. N. and Jindal, J. K. (1972c). Occurrence of fuscous and common blight diseases of bean (*Phaseolus vulgaris*) in India, *Indian Phytopath.*, **25**, 387–391.
Patel, P. N., Jindal, J. K., and Rao, M. V. B. (1972). Studies on resistance in crops to bacterial diseases in India. III. Pustule resistance in soybean, *Indian Phytopath.*, **25**, 168–171.
Patel, P. N. and Walker, J. C. (1963). Relation of air temperature and age and nutrition of the host to the development of halo and common bacterial blights of bean, *Phytopathology*, **53**, 407–411.
Patel, P. N. and Walker, J. C. (1965). Resistance in *Phaseolus* to halo blight, *Phytopathology*, **55**, 889–894.
Patel, P. N. and Walker, J. C. (1966). Inheritance of tolerance to halo blight in bean, *Phytopathology*, **56**, 681–682.

Pathak, M. M., Singh, K. P., and Lal, S. B. (1975). Inheritance of resistance to wilt (*Fusarium oxysporum* f. *ciceri*) in gram, *Indian J. Farm Sci.*, **3**, 10–11. (*Rev. Pl. Path.*, **56**, 936.)

Patil, B. C. and Sable, J. E. (1973). A note on the screening of tur against wilt disease, *Punjabrao Krishi Vidyapeeth Res. J.*, **2**, 73–76. (*Rev. Pl. Path.*, **53**, 3260.)

Patil, S. S. and Gnanamanickam, S. S. (1976). Suppression of bacterially-induced hypersensitive reaction and phytoalexin accumulation in bean by phaseotoxin, *Nature*, Lond. **259**, 486–487.

Patil, S. S., Hayward, A. C., and Emmons, R. (1974). An ultraviolet-induced non-toxigenic mutant of *Pseudomonas phaseolicola* of altered pathogenicity, *Phytopathology*, **64**, 590–595.

Pavgi, M. S. and Singh, R. A. (1965). Some parasitic fungi on pigeon peas from India, *Mycopath. Mycol. Appl.*, **27**, 97–106. (*Rev. appl. Mycol.*, **45**, 1588.)

Pavgi, M. S. and Thirumalachar, M. J. (1953). Angular black spot disease of mung beans, *Nature, Lond.*, **172**, 314–315.

Paxton, J. D. and Chamberlain, D. W. (1969). Phytoalexin production and disease resistance in soybeans as affected by age, *Phytopathology*, **59**, 775–777.

Paxton, J. D. and Rogers, D. P. (1974). Powdery mildew of soybean, *Mycologia*, **66**, 894–896.

Payak, M. M. (1962). Natural occurrence of gram rust in uredial stage on *Trigonella polycerata* L., in Simla hills, *Curr. Sci.*, **31**, 433–434.

Peachey, J. E. (ed.) (1969). *Nematodes of Tropical Crops*, Commonwealth Agricultural Bureaux, Farnham Royal, 355 pp.

Peacock, F. C. (1956). The reniform nematode in the Gold Coast, *Nematologica*, **1**, 307–310.

Pegg, K. G. and Alcorn, J. L. (1967). *Ascochyta* disease of French beans, *Qd agric. J.*, **93**, 321–323.

Peregrine, W. T. H. (1971a). A preliminary note on chemical control of bean anthracnose in Malawi, *PANS*, **17**, 47–50.

Peregrine, W. T. H. (1971b). Groundnut rust (*Puccinia arachidis*) in Brunei, *PANS*, **17**, 318–319.

Peregrine, W. T. H. and Siddiqi, M. A. (1972). A revised and annotated list of plant diseases in Malawi. *Phytopath. Pap. No. 16*, Commonwealth Mycological Institute, Kew, 51 pp.

Pereira, A. A. and Chaves, G. M. (1977). Differential varieties and a ternary system of nomenclature to designate races of *Uromyces phaseoli typica* Arth., *A. Rep. Bean Improv. Co-op.*, **20**, 85.

Perrin, R. M. (1977). Pest management in multiple cropping sytems, *Agroecosystems*, **3**, 93–118.

Perring, F. H. and Walters, S. M. (1962). *Atlas of the British Flora*, Nelson.

Perry, D. A. (1967). Premature death of groundnut plants in northern Nigeria, *Expl. Agric.*, **3**, 211–214.

Pettit, R. E., Taber, R. A., and Harrison, A. L. (1968). *Leptosphaerulina-Cercospora* complex on peanuts in Texas, *Phytopathology*, **58**, 1063 (abstr.).

Pettit, R. E., Taber, R. A., and Harrison, A. L. (1973). *Ascochyta* web-blotch of peanuts, *Phytopathology*, **63**, 447 (abstr.).

Pfender, W. F., Hine, R. B., and Stanghellini, M. E. (1977). Production of sporangia and release of zoospores by *Phytophthora megasperma* in soil, *Phytopathology*, **67**, 657–663.

Phatak, H. C. (1974). Seed-borne plant viruses – identification and diagnosis in seed health testing, *Seed Sci. and Technol.*, **2**, 22–39.

Phatak, H. C., Diaz-Ruiz, J. R., and Hull, R. (1976). Cowpea ringspot virus: a seed-transmitted cucumovirus, *Phytopath. Z.*, **87**, 132–142.

Phillips, D. V. (1972). A soybean disease caused by *Neocosmospora vasinfecta, Phytopathology*, **62**, 612–615.

Phillips, D. V. and Boerma, H. R. (1981). *Cercospora sojina* race 5: a threat to soybeans in the southeastern United States, *Phytopathology*, **71**, 334–336.
Pieczarka, D. J. and Abawi, G. S. (1978). Effect of interation between *Fusarium, Pythium*, and *Rhizoctonia* on severity of bean root rot, *Phytopathology*, **68**, 403–408.
Pierce, W. H. (1935). The inheritance of resistance to common bean mosaic in field and garden beans, *Phytopathology*, **25**, 875–883.
Pierce, H. W. and Hungerford, C. W. (1929). A note on the longevity of the bean mosaic virus, *Phytopathology*, **19**, 605–606.
Pierce, R. E. (1971). Phytoalexin induction in beans resistant or susceptible to *Fusarium* and *Thielaviopsis, Phytopathology*, **61**, 322–327.
Pierce, R. E. (1975). Observations on the golden mosaic of bean (*Phaseolus vulgaris* L.) in Jamaica, in *Tropical Diseases of Grain Legumes* (eds J. Bird and K. Maramorosch), Academic Press, New York, pp. 55–59.
Pieters, R. (1981). Horizontal resistance to *Ascochyta* blight. *Proc. Wkshop on* Ascochyta *Blight and Winter Sowing of Chickpea*, 4–7 May 1981, International Center for Agricultural Research in Dry Areas, Aleppo, Syria.
Pinstrup-Andersen, P., Londono, N. de, and Infante, M. (1976). A suggested procedure for estimating yield and production losses in crops, *PANS*, **22**, 359–365.
Pio-Ribeiro, G. and Kuhn, C. W. (1980). Cowpea stunt: heterogeneous and differential reactions of cowpea cultivars, *Phytopathology*, **70**, 244–249.
Pio-Ribeiro, G., Wyatt, S. D., and Kuhn, C. W. (1978). Cowpea stunt: a disease caused by a synergistic interaction of two viruses, *Phytopathology*, **68**, 1260–1265.
Pitkethley, R. N. (1973). Peanut rust in Australia, *Pl. Dis. Reptr*., **57**, 803.
Polach, F. J. and Abawi, G. S. (1975). The occurrence and biology of *Botryotinia fuckeliana* on beans in New York, *Phytopathology*, **65**, 657–660.
Pollard, D. G. (1955). Feeding habits of the cotton whitefly, *Bemisia tabaci* Genn. (Homoptera: Aleyrodidae), *Ann. appl. Biol.*, **43**, 664–671.
Porter, C. L. (1924). Concerning the characters of certain fungi as exhibited by their growth in the presence of other fungi, *Am. J. Bot*., **11**, 168–187.
Porter, D. M. (1970). Peanut wilt caused by *Pythium myriotylum, Phytopathology*, **60**, 393–394.
Porter, D. M. (1980). *Sclerotinia* blight of groundnut – a disease of major importance in the USA, in *Proc. Intern. Wkshop Groundnuts* (ed. R. W. Gibbons), ICRISAT, Hyderabad, pp. 177–185.
Porter, D. M. and Beute, M. K. (1974). *Sclerotinia* blight of peanuts, *Phytopathology*, **64**, 263–264.
Porter, D. M., Beute, M. K., and Wynne, J. C. (1975b). Resistance of peanut germplasm to *Sclerotinia sclerotiorum*, *Peanut Sci*., **2**, 78–80.
Porter, D. M., Garren, K. H., and van Schaik, P. H. (1975a). Pod breakdown resistance in peanuts, *Peanut Sci*., **2**, 15–18.
Porter, D. M. and Hammons, R. O. (1975). Differences in plant and pod reaction of peanut lines to infection by *Diplodia gossypina*, *Peanut Sci*., **2**, 23–25.
Powell, N. T. (1971). Interactions between namatodes and fungi in disease complexes, *A. Rev. Phytopath*., **9**, 253–274.
Powell, N. T. (1979). Internal synergisms among organisms inducing disease, in *Plant Disease, an Advanced Treatise* (eds J. G. Horsfall and E. B. Cowling), vol. IV, Academic Press, New York, pp. 113–133.
Prabhu, A. S. and Albuquerque, F. C. (1980). Observations on the chlamydospore germination of *Entyloma vignae*, the causal agent of cowpea smut. Personal Communication, 1980, data in press.
Prakash, J. and Joshi, R. D. (1980). Some aspects of seed transmission of cowpea banding mosaic virus in cowpea, *Seed Sci. and Technol*., **8**, 393–399.
Prasad, H., Haider, M. G., and Prasad, K. D. (1969). Blight disease of gram, *Indian Phytopath*., **22**, 405–406.

Prasad, K. and Weigle, J. L. (1976). Association of seed coat factors with resistance to *Rhizoctonia solani* in *Phaseolus vulgaris, Phytopathology*, **66**, 342–345.

Prasad, N., Agarwal, J. P., and Agnihotri, J. P. (1962). The genus *Protomycopsis* in India, *Indian Phytopath*., **15**, 24–27.

Prasad, N. and Padwick, G. W. (1939). The genus *Fusarium*. II. A species of *Fusarium* as a cause of wilt of gram (*Cicer arietinum* L.), *Indian J. agric. Sci*., **9**, 371–380.

Preston, N. W. (1975a). Antifungal compounds in tropical legumes, *Proc. IITA Collab. Meet. Grain Legume Improv*., International Institute of Tropical Agriculture, Ibadan, Nigeria, pp. 77–81.

Preston, N. W. (1975b). 2'-O-Methylphaseollidinisoflavan from infected tissue of *Vigna unguiculata*, *Phytochemistry*, **14**, 1131–1132.

Preston, N. W. (1977). Cajanone: an antifungal isoflavanone from *Cajanus cajan*, *Phytochemistry*, **16**, 143–144.

Preston, N. W., Chamberlain, K., and Skipp, R. A. (1975). A 2-arylbenzofuran phytoalexin for cowpea (*Vigna unguiculata*), *Phytochemistry*, **14**, 1843–1844.

Price, T. V. (1976). Diseases of the winged bean in Papua New Guinea, *Austr. Pl. Path. Soc. Newsletter*, **5**, 209 (abstr.).

Price, T. V. (1977). Powdery mildew, a new disease of the winged bean (*Psophocarpus tetragonolobus*) in Papua New Guinea, *Pl. Dis. Reptr*., **61**, 384–385.

Price, T. V. and Munroe, P. E. (1978a). Fungi associated with collar rot of winged bean in Papua New Guinea, *PANS*, **24**, 53–56.

Price, T. V. and Munroe, P. E. (1978b). *Pseudocercospora psophocarpi* on winged bean in Papua New Guinea, *Trans. Br. mycol. Soc*., **70**, 47–55.

Provvidenti, R. (1975). Resistance in *Glycine max* to isolates of bean yellow mosaic virus in New York state, *Pl. Dis. Reptr*., **59**, 917–919.

Provvidenti, R. and Braverman, S. W. (1976). Seed transmission of bean common mosaic virus in phasemy bean, *Phytopathology*, **66**, 1274–1275.

Provvidenti, R. and Cobb, E. D. (1975). Seed transmission of bean common mosaic virus in tepary bean, *Pl. Dis. Reptr*., **59**, 966–969.

Pulver, E. L., Brockman, F., Nangju, D., and Wien, H. C. (1978). IITA's programme on N_2 fixation, in *Isotopes in biological dinitrogen fixation*, International Atomic Energy Agency, Vienna, pp. 269–284.

Punithalingam, E. (1968). *Uromyces ciceris-arietini. CMI Descriptions of pathogenic Fungi and Bacteria No. 178*.

Punithalingam, E. (1974). Studies on Sphaeropsidales in culture. II. *Mycol. Pap. 136*, Commonwealth Mycological Institute, Kew.

Punithalingam, E. (1976). *Botryodiplodia theobromae. CMI Descriptions of Pathogenic Fungi and Bacteria No. 519*.

Punithalingam, E. and Holliday, P. (1972a). *Diaporthe phaseolorum. CMI Descriptions of Pathogenic Fungi and Bacteria No. 336*.

Punithalingam, E. and Holliday, P. (1972b). *Ascochyta rabiei. CMI Descriptions of Pathogenic Fungi and Bacteria No. 337*.

Punithalingam, E. and Holliday, P. (1972c). *Septoria glycines. CMI Descriptions of Pathogenic Fungi and Bacteria No. 339*.

Purivirojkul, W. and Poehlman, J. M. (1977). Injury in mungbean from natural infection with cucumber mosaic virus, *Crop Sci*., **17**, 654–656.

Purkayastha, R. P. and Chakraborty, B. N. (1981). *Acrophialophora* wilt of gram in India, *Pl. Dis*., **65**, 362.

Purseglove, J. W. (1968). *Tropical Crops: Dicotyledons*, Longmans, London, 719 pp.

Purss, G. S. (1953). Stem rot of cowpeas, *Qld agric. J*.., **77**, 139–142.

Purss, G. S. (1957). Stem rot: a disease of cowpeas caused by an undescribed species of *Phytophthora, Qld J. agric. Sci*., **14**, 125–154.

Purss, G. S. (1958). Studies on varietal resistance to stem rot (*Phytophthora vignae* Purss) in the cowpea, *Qld J. agric. Sci*., **15**, 1–14.

Purss, G. S. (1960). Further studies on the control of pre-emergency rot and crown rot of peanuts, *Qld J. agric. Sci.*, **17**, 1–14.
Purss, G. S. (1961). Wilt of peanut (*Arachis hypogaea* L.) in Queensland, with particular reference to *Verticillium* wilt, *Qld J. agric. Sci.*, **18**, 453–462.
Purss, G. S. (1962). Peanut diseases in Queensland, *Qld agric. J.*., **88**, 540–553.
Purss, G. S. (1963). Caloona – stem rot resistant cowpea, *Qld agric. J.*, **89**, 756–759.
Purss, G. S. (1972). Pathogenic specialization in *Phytophthora vignae, Aust. J. agric. Res.*, **23**, 453–456.
Putter, C. A. J. (1978). Pathosystems management under subsistence farming conditions, *3rd Int. Congr. Pl. Path.*, München, p. 318 (abstr.).
Quebral, F. C. and Pua, D. R. (1976). Screening of soybeans against *Sclerotium rolfsii, Trop. Grain Leg. Bull.*, **6**, 22–23.
Quinones, S. S., Dunleavy, J. M., and Fisher, J. W. (1971). Performance of three soybean varieties inoculated with soybean mosaic virus and bean pod mottle virus, *Crop Sci.*, **11**, 662–664.
Rachie, K. O., Rawal, R., Franckowiak, J. D., and Akinpelu, M. A. (1975). Two outcrossing mechanisms in cowpeas, *Vigna unguiculata* (L.) Walp., *Euphytica*, **24**, 159–163.
Rachie, K. O. and Roberts, L. M. (1974). Grain legumes of the lowland tropics, *Adv. Agron.*, **26**, 1–132.
Rachie, K. O. and Silvestre, P. (1977). Grain legumes, in *Food Crops of the Lowland Tropics* (eds C. L. A. Leakey and J. B. Wills), Oxford Univ. Press, pp. 41–74.
Rachie, K. O. and Wurster, R. T. (1971). The potential of pigeon pea (*Cajanus cajan* Millsp.) as a horticultural crop in East Africa, *Acta Hort.*, **21**, 172–178.
Raemaekers, R. and Preston, G. (1977). Groundnut rust occurrence and foliar disease control in Zambia, *PANS*, **23**, 166–170.
Rahe, J. E. (1973). Occurrence and levels of the phytoalexin phaseollin in relation to delimitation at sites of infection of *Phaseolus vulgaris* by *Colletotrichum lindemuthianum*, *Can. J. Bot.*, **51**, 2423–2430.
Rahe, J. E., Kúc, J., Chuang, C-M., and Williams, E. B. (1969). Induced resistance in *Phaseolus vulgaris* to bean anthracnose, *Phytopathology*, **59**, 1641–1645.
Raheja, A. K. and Leleji, O. I. (1974). An aphid-borne virus disease of irrigated cowpea in Northern Nigeria, *Pl. Dis. Reptr.*, **58**, 1080–1084.
Raheja, P. C. and Das, G. P. (1957). Development studies in crop plants. II. Effect of cultural treatments on the incidence of gram wilt, *Indian J. agric. Sci.*, **27**, 237–250.
Raj, S. and Patel, P. N. (1977). Studies on resistance in crops to bacterial diseases in India. IX. Sources of multiple disease resistance in cowpea, *Indian Phytopath.*, **30**, 207–212.
Raj. S. and Patel, P. N. (1978). Studies on resistance in crops to bacterial diseases in India. X. Inheritance of multiple disease resistance in cowpea, *Indian Phytopath.*, **31**, 294–299.
Ralph, W. (1976). Pelleting seed with bacteriocides: the effect of streptomycin on seed-borne halo-blight of French bean, *Seed Sci. and Technol.*, **4**, 325–332.
Ramakrishnan, T. S. (1947). Studies in the genus *Colletotrichum*. III, *Proc. Indian Acad. Sci. B*, **25**, 15–27.
Ramakrishnan, T. S. and Sundaram, N. V. (1954). Notes on some fungi from South India. IV, *Indian Phytopath.*, **7**, 140–151.
Ramanujam, S. (1974). Grain legume improvement in India. *Proc. 1st IITA Grain Legume Improv. Wkshop*, Oct./Nov. 1973, International Institute of Tropical Agriculture, Ibadan, Nigeria, pp. 37–41.
Ramanujam, S. (1976). Chickpea, in *Evolution of Crop Plants* (ed. N. W. Simmonds), Longman, London, pp. 157–159.
Ramiah, M. and Narayanasamy, P. (1974). A new graft-transmissible disease of cowpea (*Vigna sinensis* (Torner) Savi), *Curr. Sci.*, **43**, 417–418.

Rands, R. D. and Brotherton, W. (1925). Bean varietal tests for disease resistance, *J. agric. Res.*, **31**, 101–154.

Rangaswami, G. and Prasad, N. N. (1959). A bacterial disease of *Cicer arietinum* L. caused by *Xanthomonas cassiae* Kulkarni *et al.*, *Indian Phytopath.*, **12**, 172–175.

Rao, A. S., Rao, R. D. V. J. P., and Reddy, P. S. (1980). A whitefly-transmitted yellow mosaic disease on groundnut (*Arachis hypogaea* L.), *Curr. Sci.*, **49**, 160.

Rao, K. S. and Tulpule, P. G. (1967). Varietal differences of groundnut in the production of aflatoxin, *Nature, Lond.*, **214**, 738–739.

Rao, M. V. B. and Patel, P. N. (1973a). Studies on resistance in crops to bacterial diseases in India. V. Inheritance of resistance to bacterial pustule in soybean, *Indian Phytopath.*, **26**, 564–567.

Rao, M. V. B. and Patel, P. N. (1973b). Evaluation of chemicals *in vitro* and *in vivo* against the pustule pathogen *Xanthomonas phaseoli* var. *sojense*, *Indian Phytopath.*, **26**, 598–600.

Rao, V. G. (1962). The genus *Phyllosticta* in Bombay-Maharashtra, *Sydowia*, **16**, 275–283.

Rao, V. R. (1980). Groundnut genetic resources of ICRISAT, in *Proc. Intern. Wkshop Groundnuts* (ed. R. W. Gibbons), ICRISAT, Hyderabad, pp. 47–57.

Rathi, Y. P. S. (1979). Temik treatment of pigeon pea seeds for prevention of sterility mosaic, *Acta Bot. Indica*, **7**, 90–91. (*Rev. Pl. Path.*, **59**, 2446.)

Rathi, Y. P. S. and Lal, S. (1977). A wilt disease of pigeon pea caused by *Macrophomina phaseoli*, *Acta Bot. Indica*, **5**, 83–84. (*Rev. Pl. Path.*, **57**, 878.)

Rattray, J. M. (1932). A parasite on cowpeas (*Alectra vogelii* Benth.), *Rhod. agric. J.*, **29**, 791–794.

Rawal, K. M. (1975). Natural hybridization among wild, weedy and cultivated *Vigna unguiculata* (L.) Walp., *Euphytica*, **24**, 699–707.

Rawal, R. D. and Sohi, H. S. (1981). Nature of disease resistance in cowpea to *Septoria vignicola*, *Curr. Sci.*, **50**, 188–189.

Raychaudhuri, S. P. (1942). A disease of pigeon pea (*Cajanus cajan* (L.) Millsp.) caused by *Diplodia cajani* spec. nov., *Indian J. agric. Sci.*, **12**, 837–847.

Razdan, M. I. and Cocking, E. C. (1981). Improvement of legumes by exploring extra-specific genetic variation, *Euphytica*, **30**, 819–833.

Reddick, D. and Stewart, V. B. (1918). Varieties of beans susceptible to mosaic, *Phytopathology*, **8**, 530–534.

Reddick, D. and Stewart, V. B. (1919). Transmission of the virus of bean mosaic in seed and observations on thermal death-point of seed and virus, *Phytopathology*, **9**, 445–450.

Reddy, B. V. S., Green, J. M., and Bisen, S. S. (1978b). Genetic male sterility in pigeon pea, *Crop Sci.*, **18**, 362–364.

Reddy, D. D. R. (1975). Pathogenicity and control of root-knot nematode (*Meloidogyne* spp.) infecting chickpea, *Mysore J. agric. Sci.*, **9**, 434–439.

Reddy, D. V. R., Iizuka, N., Ghanekar, A. M., Murthy, V. K., Kuhn, C. W., Gibbons, R. W., and Chohan, J. S. (1978a). The occurrence of peanut mottle virus in India, *Pl. Dis. Reptr.*, **62**, 978–982.

Reddy, D. V. R. and McDonald, D. (1981). Research on groundnut virus diseases at ICRISAT, *International Conference on Tropical Crop Protection*, Lyon, 8–10 July 1981, abstracts, p. 80.

Reddy, M. V. (1980). Calixin M – an effective fungicide for eradication of *Ascochyta rabiei* in chickpea seed, *Internat. Chickpea Newsl.*, **3**, 12.

Reddy, M. V., Gridley, H. E. and Kaack, H. J. (1980). Major disease problems of chickpea in North Africa, *Internat. Chickpea Newsl.*, **3**, 13–14.

Reddy, M. V. and Nene, Y. L. (1979). Ring spot symptom: a genotypic expression of pigeon pea sterility mosaic, *Trop. Grain Leg. Bull.*, **15**, 27–29.

Reddy, M. V. and Nene, Y. L. (1980). Influence of sterility mosaic resistant pigeon peas on multiplication of mite vector, *Indian Phytopath.*, **33**, 61–63.

Reddy, M. V., Nene, Y. L., and Verma, J. P. (1979). Pea leaf roll virus causes chickpea stunt, *Internat. Chickpea Newsl.*, **1**, 8.

Reed, W., Lateef, S. S., and Sithananthan, S. (1980). Insect pest management on chickpea, *Proc. Internat. Wkshop in Chickpea Improv.*, International Crops Research Institute for the Semi-Arid Tropics, Hyderabad, India, pp. 179–183.

Reeder, B. D., Norton, J. D., and Chambliss, O. L. (1972). Inheritance of bean yellow mosaic virus resistance in southern pea, *Vigna sinensis* (Torner) S., *J. Am. Soc. hort. Sci.*, **97**, 235–237.

Resslar, P. M. (1980). A review of the nomenclature of the genus *Arachis* L., *Euphytica*, **29**, 813–817.

Rey, G. J. V. and Lozano, T. C. J. (1961). Estudios fisiológicas de la roya del fríjol (*Phaseolus vulgaris* L.) causado por el *Uromyces phaseoli* var. *typica* Arth., *Acta agron., Palmira*, **11**, 147–186.

Reyes, G. M. and Romasanta, R. (1940). Varietal susceptibility of peanuts to black spot (*Cercospora personata* (B & C) Ell. and Ev.), *Phillipp. J. Agric.*, **11**, 371–381.

Rhoades, H. L. (1976). Effect of *Indigofera hirsuta* on *Belonolaimus longicaudatus, Meloidogyne incognita*, and *M. javanica* and subsequent crop yields, *Pl. Dis. Reptr.*, **60**, 384–386.

Ribeiro, R. de L. D. and Hagedorn, D. J. (1976). A new bacterial disease of beans and peas in Brazil, *Proc. Amer. Phytopath. Soc.*, **3**, 262–263 (abstr.)

Ribeiro, R. de L. D. and Hagedorn, D. J. (1979a). Screening for resistance to and pathogenic specialization of *Fusarium oxysporum* f.sp. *phaseoli*, the causal agent of bean yellows, *Phytopathology*, **69**, 272–276.

Ribeiro, R. de L. D. and Hagedorn, D. J. (1979b). Inheritance and nature of resistance in beans to *Fusarium oxysporum* f.sp. *phaseoli, Phytopathology*, **69**, 859–861.

Ribeiro, R. de L. D., Hagedorn, D. J., Durbin, R. D., and Uchytil, T. F. (1979). Characterization of the bacterium inciting bean wildfire in Brazil, *Phytopathology*, **69**, 208–212.

Richards, B. L. and Burkholder, W. H. (1943). A new mosaic disease of beans, *Phytopathology*, **33**, 1215–1216 (abstr.).

Richardson, Y. B. and Evans, A. M. (1972). *Phaseolus coccineus* as a source of resistance to anthracnose, *Colletotrichum lindemuthianum, A. Rep. Bean Improv. Co-op.*, **15**, 80–82.

Riggs, R. D. and Hamblen, M. L. (1966). Additional weed hosts of *Heterodera glycines, Pl. Dis. Reptr.*, **50**, 15–16.

Rijks, D. A. and Huxley, P. A. (1964). The empirical relation between solar radiation and hours of bright sunshine near Kampala, Uganda, *J. appl. Ecol.*, **1**, 339–345.

Riley, E. A. (1960). A revised list of plant diseases in Tanganyika Territory. *Mycol. Pap. No. 75*, Commonwealth Mycological Institute, Kew, 42 pp.

Riley, R. (1973). Genetic changes in hosts and the significance of disease, *Ann. appl. Biol.*, **75**, 128–132.

Rios, G. P., Watt, E. E., de Aranjo, J. P. P., and das Neves, B. P. (1980). Identification of sources of resistance to the principal diseases of southern pea (*Vigna unguiculata* (L.) Walp.) in Brazil, *A. Rep. Bean Improv. Co-op.*, **23**, 106.

Rizwi, M. A. (1978). A new leaf spot disease of pigeon pea, *Indian Phytopath.*, **31**, 235.

Robbins, R. T., Dickerson, O. J., and Kyle, J. H. (1972). Pinto bean yield increased by chemical control of *Pratylenchus* spp., *J. Nemat.*, **4**, 28–32. (*Helminth. Abstr.*, B **41**, 892.)

Roberts, E. H. and Ellis, R. H. (1980). See physiology and seed quality in soyabean, in *Advances in Legume Science* (eds R. J. Summerfield and A. H. Bunting), HMSO, London, pp. 297–311.

Roberts, E. H., Summerfield, R. J., Minchin, F. R., and Hadley, P. (1980). Phenology of chickpeas (*Cicer arietinum*) in contrasting aerial environments, *Expl. Agric.*, **16**, 343–360.

Robertson, D. G. (1963). Further studies on the host range of cowpea yellow mosaic virus, *Trop. Agric. (Trin.)*, **40**, 319–324.
Robertson, D. G. (1965). The local lesion reaction for recognizing cowpea varieties immune from and resistant to cowpea yellow mosaic virus, *Phytopathology*, **55**, 923–925.
Robertson, D. G. (1966). Seed-borne viruses of cowpea in Nigeria. BSc thesis, University of Oxford, 111 pp.
Robinson, R. A. (1960). Notes on Kenya agriculture VIII: important plant diseases, *E. Afr. agric. J.*, **25**, 131–146.
Robinson, R. A. (1969). Disease resistance terminology, *Rev. appl. Mycol.*, **48**, 593–606.
Robinson, R. A. (1976). *Plant Pathosystems*, Springer-Verlag, Berlin, 184 pp.
Robinson, R. A. (1980). New concepts in breeding for disease resistance, *A. Rev. Phytopath.*, **18**, 189–210.
Rochow, W. F. (1972). The role of mixed infections in the transmission of plant viruses by aphids, *A. Rev. Phytopath.*, **10**, 101–124.
Roebuck, P., Sexton, R., and Mansfield, J. W. (1978). Ultrastructural observations on the development of the hypersensitive reaction in leaves of *Phaseolus vulgaris* cv. Red Mexican inoculated with *Pseudomonas phaseolicola* (race 1), *Physiol. Pl. Path.*, **12**, 151–157.
Rogers, K. M., Norton, J. D., and Chambliss, O. L. (1973). Inheritance of resistance to cowpea chlorotic mottle virus in southern pea, *Vigna sinensis*, *J. Am. Soc. hort. Sci.*, **98**, 62–63.
Ross, J. P. (1962). Crop rotation effects on the soybean cyst nematode population and soybean yields, *Phytopathology*, **52**, 815–818.
Ross, J. P. (1968a). Effect of single and double infection of soybean mosaic and bean pod mottle viruses on soybean yield and seed characters, *Pl. Dis. Reptr.*, **52**, 344–348.
Ross, J. P. (1968b). Additional physiological races of *Cercospora sojina* on soybeans in North Carolina, *Phytopathology*, **58**, 708–709.
Ross, J. P. (1969). Pathogenic variation among isolates of soybean mosaic virus, *Phytopathology*, **59**, 829–832.
Ross, J. P. (1977). Effect of aphid-transmitted soybean mosaic virus on yields of closely related resistant and susceptible soybean lines, *Crop Sci.*, **17**, 869–872.
Ross, J. P. (1978). A pad inoculator for infecting large numbers of plants with virus, *Pl. Dis. Reptr.*, **62**, 122–125.
Ross, J. P. and Harper, J. A. (1970). Effects of *Endogone* mycorrhiza on soybean yields, *Phytopathology*, **60**, 1552–1556.
Rossel, H. W. (1977). Preliminary investigations on the identity and ecology of legume virus diseases in Northern Nigeria, *Trop. Grain Leg. Bull.*, **8**, 41–46.
Rossel, H. W. and Ferguson, J. M. (1979). A new economical screenhouse for virus research in tropical areas, *FAO Pl. Prot. Bull.* **27**, 74–76.
Rothwell, A. (1962). Diseases of groundnuts in Southern Rhodesia, *Rhod. Agric. J.*, **59**, 199–201.
Rothwell, A. (1975). Peanut rust in Rhodesia, *Pl. Dis. Reptr.*, **59**, 802–803.
Roy, K. W. and Abney, T. S. (1977). Antagonism between *Cercospora kikuchii* and other seedborne fungi of soybeans, *Phytopathology*, **67**, 1062–1066.
Royes, W. V. (1976). Pigeon pea, in *Evolution of Crop Plants* (ed. N. W. Simmonds), Longmans, London, pp. 154–156.
Rubaihayo, P. R. and Leakey, C. L. A. (1970). Introduction, testing and selection of soya beans in Uganda, *E. Afr. agric. For. J.*, **36**, 77–82.
Rudorf, W. (1955). The transference of resistance to bean mosaic viruses 1 (common bean mosaic) and 2 (yellow bean mosaic) to fertile hybrid plants from the cross *P. vulgaris* × *P. coccineus, Naturwissenschaften*, **42**, 19–20.

Ruppel, E. G. and Barnes, D. K. (1965). *Phytophthora* wilt of *Tephrosia vogelii* in Puerto Rico, *Pl. Dis. Reptr.*, **49**, 676–679.
Russell, G. E. (1978). *Plant Breeding for Pest and Disease Resistance*, Butterworths, London, 485 pp.
Russell, P. E. (1975). Variation in the virulence of some streptomycin resistant mutants of *Pseudomonas phaseolicola*, *J. appl. Bact.*, **39**, 175–180.
Russell, P. E. (1976). Potential sources of resistance to halo blight of beans, *A. Rep. Bean Improv. Co-op.*, **19**, 65–67.
Russell, P. E. and Brown, L. (1977). *Alternaria alternata* on *Phaseolus vulgaris, Pl. Path.*, **26**, 47.
Russell, P. E. and Warburg, P. W. (1976). *Fusarium culmorum* on *Phaseolus vulgaris, Pl. Path.*, **25**, 55–56.
Ryker, T. C. (1942). Loss of sporulation in *Cercospora, Phytopathology*, **32**, 16 (abstr.).
Saad, S. and Hagedorn, D. J. (1969). Symptomatology and epidemiology of *Alternaria* leaf spot of bean, *Phaseolus vulgaris, Phytopathology*, **59**, 1530–1533.
Sabet, K. A. and Ishag, F. (1969). Studies on the bacterial diseases of Sudan crops. VIII. Survival and dissemination of *Xanthomonas phaseoli* (E. F. Smith) Dowson, *Ann. appl. Biol.*, **64**, 65–74.
Sabet, K. A., Ishag, F., and Khalil, O. (1969). Studies on the bacterial diseases of Sudan crops. VII. New records, *Ann. appl. Biol.*, **63**, 357–369.
Sachchidananda, J., Singh, S., Prakash, N., and Verma, V. S. (1973). Bean common mosaic virus on cowpea in India, *Z. Pflanzenkrankh. Pflanzenschutz*, **80**, 88–91.
Saettler, A. W. (1971a). Chemical sprays for the control of common-fuscous bacterial blight in navy (pea) beans, *A Rep. Bean Improv. Co-op.*, **14**, 54–55.
Saettler, A. W. (1971b). Seedling injection as an aid in identifying bean blight bacteria, *Pl. Dis. Reptr.*, **55**, 703–706.
Saksena, H. K. and Dwivedi, R. P. (1973). Web blight of black gram caused by *Thanatephorus cucumeris, Indian J. Farm Sci.*, **1**, 58–61.
Saksena, H. K. and Kumar, K. (1972). Some aspects of epidemiology and control of *Phyllosticta* leaf spot of arthar (*Cajanus cajan* (L.) Millsp.), *Proc. Nat. Sci. Acad. India*, B **37**, 399–406. (*Rev. Pl. Path.*, **54**, 5221.)
Saksena, H. K. and Prasad, R. (1955). Studies on gram rust, *Uromyces ciceris-arientini* (Grogn.) Jacz., *Indian Phytopath.*, **8**, 94–98.
Saksena, H. K., Singh, G. P., and Nath, S. (1967). Blight disease of groundnuts in India, *Indian Phytopath.*, **20**, 67–69.
Sanders, F. E., Mosse, B., and Tinker, P. B. (eds.) (1975). *Endomycorrhizas*, Academic Press, London, 626 pp.
Sanders, J. H. (1977). Economic analysis in the bean program of CIAT, *A. Rep. Bean Improv. Co-op.*, **20**, 105–106.
Sanderson, K. E. and Srb, A. M. (1965). Heterokaryosis and parasexuality in the fungus *Ascochyta imperfecta, Am. J. Bot.*, **52**, 72–81.
Santos-Filho, H. P., Ferraz, S., and Vieira, C. (1976). Inheritance of resistance to angular leaf spot in *Phaseolus vulgaris* L., *A. Rep. Bean Improv. Co-op.*, **19**, 69–70.
Sargeant, K., Sheridan, A., O'Kelly, J., and Carnaghan, R. B. A. (1961). Toxicity associated with certain samples of groundnuts, *Nature, Lond.*, **192**, 1096–1097.
Sarojini, T. S. (1951). Soil conditions and root diseases. II. *Fusarium udum* disease of red gram (*Cajanus cajan* (Linn.) Millsp.), *Proc. Indian Acad. Sci.*, B. **33**, 49–68.
Sarojini, T. S. (1954). Soil conditions and root diseases. XI. *Neocosmospora vasinfecta* Smith disease of *Cajanus cajan*, *J. Madras Univ., Sect. B.*, **24**, 137–142. (*Rev. appl. Mycol.*, **34**, 727.)
Sasaki, M. (1936). On the anthracnose of adzuki bean in Hokkaido, *Rep. Hokkaido agric. Exp. Sta.*, **36**, 53–77. (*Rev. appl. Mycol.*, **16**, 12.)

Sasser, J. N. (1980). Root-knot nematodes: a global menace to crop production, *Pl. Dis.*, **64**, 36–41.
Sasser, J. N., Wells, J. C., and Nelson, L. A. (1968). The effect of nine parasitic nematode species on growth, yield and quality of peanuts as determined by soil fumigation and correlation of nematode populations with host response, *Nematologica*, **14**, 15 (abstr.).
Sathe, A. V. (1972). Identification and nomenclature of soybean rust from India, *Curr. Sci.*, **41**, 264–265.
Sattar, A. (1933a). On the occurrence, perpetuation and control of gram (*Cicer arietinum* L.) blight caused by *Ascochyta rabiei* (Pass.) Labrousse, with special reference to Indian conditions, *Ann. appl. Biol.*, **20**, 612–632.
Sattar, A. (1933b). A comparative study of the fungi associated with blight diseases of certain cultivated leguminous plants, *Trans. Br. mycol. Soc.*, **18**, 276–301.
Satya, H. N. (1964). A new species of *Pestalotiopsis* on *Arachis hypogaea* L. from Bhopal, *Curr. Sci.*, **33**, 57.
Savage, E. J., Clayton, C. W., Hunter, J. H., Brenneman, J. A., Laviola, C., and Galleghy, M. E. (1968). Homothallism, heterothallism, and interspecific hybridization in the genus *Phytophthora*, *Phytopathology*, **58**, 1003–1021.
Saxena, K. B., Wallis, E. S., and Byth, D. E. (1980). Development of a photoinsensitive early flowering male sterile line of pigeon pea (*Cajanus cajan* (L.) Millsp.), *Trop. Grain Leg. Bull.*, **20**, 22–23.
Schafer, J. F. (1971). Tolerance to plant disease, *A. Rev. Phytopath.*, **9**, 235–252.
Scherff, R. H. (1973). Bacterial blight of soybeans as influenced by populations of yellow bacteria on leaves and pods, *Phytopathology*, **63**, 752–755.
Schieber, E. and Zentmyer, G. A. (1971). A new bean disease in the Caribbean area, *Pl. Dis. Reptr.*, **55**, 207–208.
Schiller, C. T. and Sinclair, J. B. (1979). Soybean seed lot contamination by *Melanopsichium pennsylvanicum* smut galls, *Phytopathology*, **69**, 605–608.
Schmitthenner, A. F. (1972). Evidence of a new race of *Phytopththora megasperma* var. *sojae* to soybean, *Pl. Dis. Reptr.*, **56**, 536–539.
Schneider, R. W. and Sinclair, J. B. (1975). Inhibition of conidial germination and germ tube growth of *Cercospora canescens* by cowpea leaf diffusates, *Phytopathology*, **65**, 63–65.
Schneider, R. W., Williams, R. J., and Sinclair, J. B. (1976). *Cercospora* leaf spot of cowpea: models for estimating yield loss, *Phytopathology*, **66**, 384–388.
Schnock, M. G., Hoffman, G. M., and Kruger, J. (1975). A new physiological strain of *Colletotrichum lindemuthianum* infecting *Phaseolus vulgaris* L., *Hort. sci.*, **10**, 140.
Schreiber, F. (1932). Resistanzzüchtung bei *Phaseolus vulgaris, Phytopath. Z.*, **4**, 415–454.
Schroder, H. W., Boosalis, M. G., and Moore, M. B. (1953). An improved method of growing inoculum of plant disease fungi, *Phytopathology*, **43**, 401–402.
Schroth, M. N. and Cook, R. J. (1964). Seed exudation and its influence on pre-emergence damping-off of bean, *Phytopathology*, **54**, 670–673.
Schroth, M. N. and Hendrix, F. F. (1962). Influence of nonsusceptible plants on the survival of *Fusarium solani* f. *phaseoli* in soil, *Phytopathology*, **52**, 906–909.
Schroth, M. N. and Snyder, W. C. (1961). Effect of host exudates on chlamydospore germination of the bean root rot fungus, *Fusarium solani* f. *phaseoli, Phytopathology*, **51**, 389–393.
Schroth, M. N., Toussoun, T. A., and Snyder, W. C. (1963). Effect of certain constituents of bean exudate on germination of chlamydospores of *Fusarium solani* f. *phaseoli* in soil, *Phytopathology*, **53**, 809–812.
Schroth, M. N., Vitanza, V. B., and Hildebrand, D. C. (1971). Pathogenic and nutritional variation in the halo blight group of fluorescent pseudomonads of bean, *Phytopathology*, **61**, 852–857.

Schultz, H. K. and Dean, L. L. (1947). Inheritance of curly top disease reaction in the bean, *Phaseolus vulgaris, J. Am. Soc. Agron.*, **39**, 47–51.
Schuster, M. L. (1955). A method for testing resistance of beans to bacterial blights, *Phytopathology*, **45**, 519–520.
Schuster, M. L. (1959). Relation of root-knot nematodes and irrigation water to the incidence and dissemination of bacterial wilt of bean, *Pl. Dis. Reptr.*, **43**, 27–32.
Schuster, M. L. (1977). Survival of bacterial pathogens of soybean, *Indian J. agric. Sci.*, **47**, 270–273.
Schuster, M. L. and Christiansen, D. W. (1957). An orange-colored strain of *Corynebacterium flaccumfaciens* causing bean wilt, *Phytopathology*, **47**, 51–53.
Schuster, M. L. and Coyne, D. P. (1971). New virulent strains of *Xanthomonas phaseoli, Pl. Dis. Reptr.*, **55**, 505–506.
Schuster, M. L. and Coyne, D. P. (1974). Survival mechanisms of phytopathogenic bacteria, *A. Rev. Phytopath.*, **12**, 199–221.
Schuster, M. L. and Coyne, D. P. (1975). Genetic variation in bean bacterial pathogens, *Euphytica*, **24**, 143–147.
Schuster, M. L., Coyne, D. P., and Hoff, B. (1973). Comparative virulence of *Xanthomonas phaseoli* strains from Uganda, Colombia, and Nebraska, *Pl. Dis. Reptr.*, **57**, 74–75.
Schuster, M. L., Vidaver, A. K., and Mandel, M. (1968). A purple pigment-producing bean wilt bacterium, *Corynebacterium flaccumfaciens* var. *violaceum* n.var., *Can. J. Microbiol.*, **14**, 423–427.
Schwartz, H. F. (1980). The international bean rust nursery format, *A Rep. Bean Improv. Co-op.*, **23**, 25–26.
Schwartz, H. F., Correa, V. F., Pineda, D. P. A., Otoya, M. M., and Katherman, M. J. (1981a). Dry bean yield losses caused by ascochyta, angular, and white leaf spots in Colombia, *Pl. Dis.*, **65**, 494–496.
Schwartz, H. F. and Gálvez, G. E. (eds) (1980). *Bean Production Problems: Disease, Insect, Soil and Climatic Constraints of* Phaseolus vulgaris, Centro Internacional de Agricultura Tropical, Cali, Colombia, 424 pp.
Schwartz, H. F., Gálvez, G. E., van Schoonhoven, A., Howeler, R. H., Graham, P. H., and Flor, C. (1978a). *Field Problems of Beans in Latin America*, Centro Internacional de Agricultura Tropical, Cali, Colombia, 136 pp.
Schwartz, H. F., Katherman, M. J., and Thung, M. D. T. (1981b). Yield response and resistance of dry beans to powdery mildew in Colombia, *Pl. Dis.*, **65**, 737–738.
Schwartz, H. F., Pastor Corrales, M. A., and Singh, S. P. (1982). New sources of resistance to anthracnose and angular leaf spot of beans (*Phaseolus vulgaris* L.), *Euphytica*, in press.
Schwartz, H. F. and Sanders, J. H. (1979). Plant diseases of dry beans (*Phaseolus vulgaris* L.) in Latin America and strategies for their control, in *Diseases of Tropical Food Crops* (eds. H. Maraite and J. A. Mayer), U.C.L., Louvain-la-Neuve, Belgium, pp. 171–189.
Schwartz, H. F. and Steadman, J. R. (1980). White mold, in *Bean Production Problems* (eds H. F. Schwartz and G. E. Gálvez), Centro Internacional de Agriculatura Tropical, Cali, Colombia, pp. 111–126.
Schwartz, H. F., Steadman, J. R., and Coyne, D. P. (1977). Resistance of Charlevoix and Valentine to infection by *Sclerotinia sclerotiorum, A. Rep. Bean Improv. Co-op.*, **20**, 71–72.
Schwartz, H. F., Steadman, J. R., and Coyne, D. P. (1978b). Influence of *Phaseolus vulgaris* blossoming characteristics and canopy structure upon reaction to *Sclerotinia sclerotiorum, Phytopathology*, **68**, 465–470.
Scott, H. A. and Fulton, J. P. (1978). Comparison of relationships of southern bean mosaic virus and cowpea strain of tobacco mosaic virus with bean leaf beetle, *Virology*, **84**, 207–209.

Scott, H. A. and Hoy, J. W. (1981). Blackgram mottle virus. *CMI/AAB Descriptions of Plant Viruses No. 237*.

Scott, P. R. and Hollins, T. W. (1980). Pathogenic variation in *Pseudocercosporella herpotrichoides, Ann. appl. Biol.*, **94**, 297–300.

Scott, P. R., Johnson, R., Wolfe, M. S., Lowe, H. J. B., and Bennett, F. G. A. (1980). Host-specificity in cereal parasites in relation to their control, *Appl. Biol.*, **5**, 349–393.

Sechet, M. (1955). Un *Pleospora* parasite des feuilles d'arachide, *Oléagineux*, **10**, 414.

Seinhorst, J. W. (1980). How to interpret Koch's postulates in plant nematology. Contrib. *Plant Disease Etiology: 100 years of Koch's Postulates*, Fed. Br. Pl. Path., London, 17–19 Dec.

Selassie-Gebre, K., Marchoux, G., and Quiot, J. B. (1976). Méthode d'identification du virus du flétrissement de la féve (broad bean wilt virus: BBWV) isolé de *Vigna sinensis* (L.) Savi., *Annls. Phytopath.*, **8**, 331–342.

Semancik, J. S. (1972). Bean pod mottle virus. *CMI/AAB Descriptions of Plant Viruses No. 108*.

Semangoen, H. (1958). Penjakit-penjakit virus pada Katjang-Pandjang (*Vigna sinensis*), *Kongr. Ilmu Penget. Nas. Pertama 1958*, 379–407. (*Rev. appl. Mycol.*, **40**, 327–328.)

Sempio, C., Della Torre, G., Ferranti, F., Barberini, B., and Draoli, R. (1975). Defence mechanism in beans resistant to rust, *Phytopath. Z.*, **83**, 244–266.

Sen Gupta, P. K. (1974). Diseases of major pulse crops in India, *PANS*, **20**, 409–413.

Sen Gupta, P. K. and Das, C. R. (1964). Effects of vitamins, hormones and trace elements on growth and sporulation of *Stemphylium sarcinaeforme*, causal organism of leaf spot of gram, *Pl. Dis. Reptr.*, **48**, 37–40.

Sequeira, L. and Graham, T. L. (1976). Agglutination of avirulent strains of *Pseudomonas solanacearum* by potato lectin, *Proc. Am. Phytopath. Soc.*, **3**, 233 (abstr.).

Seth, M. L. (1962). Transmission of pigeon pea sterility by an eriophyid mite, *Indian Phytopath.*, **15**, 225–227.

Seth, M. L. (1965). Further observations and studies on pigeon pea sterility, *Indian Phytopath.*, **18**, 317–319.

Sethi, C. L. and Sharma, N. K. (1976). Nature of resistance of cowpea to the root-knot nematode, *Meloidogyne incognita*, and the pigeon pea cyst nematode, *Heterodera cajani, Indian J. Nemat.*, **6**, 81–85. (*Helminth. Abstr.*, B. **48**, 1131.)

Sethi, S. C. (1979). Male sterility, *Internat. Chickpea Newsl.*, **1**, 4.

Shakoor, A., Ahsanul-Haq, M., Sadiq, M., and Sarwar, G. (1977). Induction of resistance to yellow mosaic virus in mung bean through induced mutations, in *Induced Mutations Against Plant Diseases*, International Atomic Energy Agency, Vienna, pp. 293–301.

Shands, H. C., Vieira, C., and Zaumeyer, W. J. (1964). Observations on dry bean diseases in Brazil, *Pl. Dis. Reptr.*, **48**, 784–787.

Shanmugasundaram, S. (1979). Varietal development and germplasm utilization in soybeans, *AVRDC Tech. Bull.*, **13**, Asian Vegetable Research and Development Centre, Shanhua, Taiwan.

Shanmugasundaram, S., Kno, G. C., and Nalampang, A. (1980). Adaptation and utilization of soyabeans in different environments and agricultural systems, in *Advances in Legume Science* (eds. R. J. Summerfield and A. H. Bunting), HMSO, London, pp. 265–277.

Sharma, D. and Green, J. M. (1975). Perspective of pigeon pea and ICRISAT's breeding program, *Internat. Wkshop on Grain Legumes*, 13–16 January, 1975, International Crops Research Institute for the Semi-Arid Tropics, Hyderabad, India, pp. 19–29.

Sharma, D., Kannaiyan, J., and Reddy, L. J. (1982). Inheritance of resistance to blight in pigeon peas, *Pl. Dis.*, **66**, 22–25.

Sharma, D., Saxena, K. B., and Green, J. M. (1977). Potential of ratooning in pigeonpea, *Fld. Crops Res.*, **1**, 165–172.
Sharma, D. C. (1966). Studies on bunchy top, chlorosis and ring mottle virus diseases of groundnut (*Arachis hypogaea*), *Phytopath. Z.*, **57**, 127–137.
Sharma, H. C. and Khare, M. N. (1969). Studies on wilt of bengal gram (*Cicer arietinum* L.) at Jabalpur, *Jawaharlal Nehru Krishi Vishwa Vidyalaya Res. J.*, **3**, 122–123.
Sharma, N. D. (1974). A previously unrecorded leafspot of peanut caused by *Phomopsis* sp., *Pl. Dis. Reptr.*, **58**, 640.
Sharma, N. D., Joshi, L. K., and Vyas, S. C. (1977). A new inoculation technique for testing *Fusarium* wilt of pigeon pea, *Indian Phytopath.*, **30**, 406–407.
Sharma, N. K. and Sethi, C. L. (1976a). Reaction of certain cowpea varieties to *Meloidogyne incognita* and *Heterodera cajani, Indian J. Nemat.*, **6**, 99–102. (*Helminth. Abstr.*, B. **48**, 1132.)
Sharma, N. K. and Sethi, C. L. (1976b). Interrelationship between *Meloidogyne incognita, Heterodera cajani* and *Rhizobium* sp. on cowpea (*Vigna sinensis* (L.) Savi), *Indian J. Nemat.*, **6**, 117–123. (*Helminth Abstr.*, B. **48**, 1617.)
Sharma, S. R. and Varma, A. (1975a). Three sap transmissible viruses from cowpea in India, *Indian Phytopath.*, **28**, 192–198.
Sharma, S. R. and Varma, A. (1975b). Natural incidence of cowpea viruses and their effect on yield of cowpea, *Indian Phytopath.*, **28**, 330–334.
Sharma, S. R. and Varma, A. (1976). Cowpea yellow fleck – a whitefly transmitted disease of cowpea, *Indian Phytopath.*, **29**, 421–423.
Sharp, E. L. (1973). Evidence for environmental races of *Puccinia striiformis, Proc. 2nd Int. Congr. Pl. Path.*, Minneapolis, abstr. 0719.
Shaw, F. J. F. (1936a). Studies in Indian pulses. The inheritance of morphological characters and of wilt resistance in rahar (*Cajanus indicus* Spreng.), *Indian J. agric. Sci.*, **6**, 139–189.
Shaw, F. J. F. (1936b). The inheritance of morphological characters and of wilt resistance in rahar (*Cajanus indicus* Spreng.), *Proc. Indian Acad. Sci.*, B **3**, 491–492.
Shekhawat, G. S. and Patel, P. N. (1977). Seed transmission and spread of bacterial blight of cowpea and leaf spot of green gram in summer and monsoon seasons, *Pl. Dis. Reptr.*, **61**, 390–392.
Shekhawat, G. S., Patel, P. N., and Raj, S. (1977). Histology of cowpea plant infected with *Xanthomonas vignicola, Z. Pflanzenkrank. und Pflanzenschutz*, **84**, 547–558.
Sheldrake, A. R., Narayanan, A., and Kannaiyan, J. (1978a). Some effects of the physiological state of pigeon peas on the incidence of the wilt disease, *Trop. Grain Leg. Bull.*, **11/12**, 24–25.
Sheldrake, A. R., Saxena, N. P., and Krishnamurthy, L. (1978b). The expression and influence on yield of the 'double-podded' character in chickpeas (*Cicer arietinum* L.), *Fld. Crops Res.*, **1**, 243–253.
Shepherd, C. J. (1978). Mating behaviour of Australian isolates of *Phytophthora* species. 1. Inter- and intraspecific mating, *Aust. J. Bot.*, **26**, 123–138.
Shepherd, C. J. and Pratt, B. H. (1973). Separation of two ecotypes of *Phytophthora dreschsleri* Tucker occurring in Australian native forests, *Aust. J. biol. Sci.*, **26**, 1095–1107.
Shepherd, R. J. (1964). Properties of a mosaic virus of cowpea and its relationship to the bean pod mottle virus, *Phytopathology*, **54**, 466–473.
Shepherd, R. J. (1970). Pea enation mosaic virus. *CMI/AAB Descriptions of Plant Viruses No. 25*.
Shepherd, R. J. (1971). Southern bean mosaic virus. *CMI/AAB Descriptions of Plant Viruses No. 57*.
Sheriff, N. M., Khan, W. M. A., and Iyemperumal, S. (1977). A note on the study of

red gram mutants for resistance to root-disease under field conditions, *Madras agric. J.*, **64**, 691.
Sherwin, H. S. and Kreitlow, K. W. (1952). Discoloration of soybean seeds by the frogeye fungus, *Cercospora sojina, Phytopathology*, **42**, 568–572.
Sherwin, H. S. and Lefebvre, C. L. (1951). Reaction of cowpea varieties to bacterial canker, *Pl. Dis. Reptr.*, **35**, 303–317.
Sherwood, R. T. (1969). Morphology and physiology in four anastomosis groups of *Thanatephorus cucumeris, Phytopathology*, **59**, 1924–1929.
Shibles, R. (1980). Adaptation of soyabeans to different seasonal durations, in *Advances in Legume Science* (eds R. J. Summerfield and A. H. Bunting), HMSO, London, pp. 279–286.
Shit, S. K. and Sen Gupta, P. K. (1978). Possible existence of physiologic races of *Fusarium oxysporum* f.sp *udum*, the incitant of the wilt of pigeon pea, *Indian J. agric. Sci.*, **48**, 629–632.
Shock, T. L. and Goodman, R. M. (1981). Time-course studies on virus titer and DNA component ratio in beans infected with bean golden mosaic virus, *Phytopathology*, **71**, 80–82.
Shortt, B. J., Grybauskas, A. P., Tenne, F. D., and Sinclair, J. B. (1981). Epidemiology of *Phomopsis* seed decay of soybean in Illinois, *Pl. Dis.*, **65**, 62.
Shoyinka, S. A. (1976). Attempted control of virus incidence in cowpeas by the use of barrier crops, in *Intercropping in Semi-arid Areas* (eds J. H. Monyo, A. D. R. Ker, and M. Campbell), IDRC, Ottawa, p. 46.
Shoyinka, S. A., Bozarth, R. F., Reese, J., and Rossel, H. W. (1978). Cowpea mottle virus: a seed-borne virus with distinctive properties infecting cowpeas in Nigeria, *Phytopathology*, **68**, 693–699.
Shoyinka, S. A. and Okusanya, B. A. O. (1975). Field occurrence and identification of southern bean mosaic virus (cowpea strain) in Nigeria, *Niger. Soc. Pl. Prot. occas. publ.*, **1**, 27–28 (abstr.).
Shukla, D. N. and Bhargava, S. N. (1978). Seed and seedling-rot of arhar, *Proc. Nat. Acad. Sci. India*, B. **48**, 167.
Shukla, D. S. (1975). Incidence of *Fusarium* wilt of pigeonpea in relation to soil composition, *Indian Phytopath.*, **28**, 395–396.
Siddaramaiah, A. L., Desai, S. A., and Jayaramaiah, R. (1981). Occurrence and severity of a mycoparasite on 'tikka' leaf spots of groundnut, *Curr. Res.*, **10**, 14–15.
Sigee, D. C. and Epton, H. A. S. (1975). Ultrastructure of *Pseudomonas phaseolicola* in resistant and susceptible leaves of French bean, *Physiol. Pl. Path.*, **6**, 29–34.
Silbernagel, M. J. (1969). Mexican strain of bean common mosaic virus, *Phytopathology*, **59**, 1809–1812.
Silbernagel, M. J. and Jafri, A. M. (1974). Temperature effects on curly top resistance in *Phaseolus vulgaris, Phytopathology*, **64**, 825–827.
Silbernagel, M. J. and Zaumeyer, W. J. (1973). Beans, in *Breeding Plants for Disease Resistance* (ed. R. R. Nelson), Pennsylvania State Univ. Press, pp. 253–269.
Silvera, G. A. (1967). Evaluacion de la resistencia de variedades de frijol a la mancha angular y otros estudios fisiologicos sobre *Isariopsis griseola, XIIIa Reunion Annual, PCCMCA*, San José, Costa Rica, 60–62.
Simbwa-Bunnya, M. (1972). Resistance of groundnut varieties to bacterial wilt (*Pseudomonas solanacearum*) in Uganda, *E. Af. agric. For. J.*, **37**, 341–343.
Simmonds, J. H. (1955). *Rep. Dep. Agric. Qd. 1953—54*, pp. 55–56; *1954—55*, pp. 51–52. (*Rev. appl. Mycol.*, **35**, 877.)
Simmonds, N. W. (1979). *Principles of Crop Improvement*, Longman, London, 408 pp.
Simons, M. D. (1979). Modification of host-parasite interactions through artificial mutagenesis, *A. Rev. Phytopath.*, **17**, 75–96.
Sinclair, J. B. (1977). Infectious soyabean diseases of world importance, *PANS*, **23**, 49–57.

Sinclair, J. B. (1978). Chemical control of soybean rust, in *Proc. Workshop Soybean Rust in W. Hemisphere* (ed. N. G. Vakili), USDA, pp. 30–33.
Sinclair, J. B. (1982). *Compendium of Soybean Diseases*, 2nd Edn, American Phytopath. Soc., St. Paul, Minnesota.
Sinclair, J. B. and Dhingra, O. D. (1975). *An Annotated Bibliography of Soybean Diseases, 1882—1974*, INTSOY Series No. 7, College of Agric., Univ. Illinois, Urbana-Champaign, 280 pp.
Sinclair, J. B. and Shurtleff, M. C. (eds.) (1975). *Compendium of Soybean Diseases*, American Phytopath, Soc., St. Paul, Minnesota, 69 pp.
Sinclair, J. B. and Walker, J. C. (1955). Inheritance of resistance to cucumber mosaic virus in cowpea, *Phytopathology*, **45**, 563–564.
Sing, V. O. and Schroth, M. N. (1977). Bacteria-plant cell surface interactions: active immobilization of saprophytic bacteria in plant leaves, *Science*, **197**, 759.
Singh, A. B. and Pandey, P. K. (1979). A new strain of tobacco mosaic virus causing mosaic disease of pigeon pea, *Curr. Sci*., **48**, 261.
Singh, A. K., Sastri, D. C., and Moss, J. P. (1980a). Utilization of wild *Arachis* species at ICRISAT, in *Proc. Intern. Wkshop Groundnuts* (ed. R. W. Gibbons), ICRISAT, Hyderabad, pp. 82–90.
Singh, B. B., Gupta, S. C., and Singh, B. D. (1974a). Sources of field resistance to rust and yellow mosaic diseases of soybeans, *Indian J. Genet. Pl. Breed*., **34**, 400–404.
Singh, B. B. and Malick, A. S. (1978). Inheritance of resistance to yellow mosaic in soybean, *Indian J. Genet. Pl. Breed*., **38**, 258–261.
Singh, B. M., Saharan, G. S., Sood, A. K., and Shyam, K. R. (1978). *Asochyta* leaf spot of mash and cowpea, *Indian Phytopath*., **31**, 388–389.
Singh, B. M. and Sharma, Y. R. (1975). Screening of bean lines for resistance to angular leaf spot caused by *Isariopsis griseola, Indian Phytopath*., **28**, 435–436.
Singh, D. and Patel, P. N. (1977). Studies on resistance in crops to bacterial diseases in India. VII. Inheritance of resistance to bacterial blight disease in cowpea, *Indian Phytopath*., **30**, 99–102.
Singh, D. V., Lal, S. and Singh, S. N. (1974b). Breeding gram (*Cicer arietinum* L.) for resistance to wilt, *Indian J. Genet. Pl. Breed*., **34**, 267–270.
Singh, G. and Bedi, P. S. (1974). The relative reaction of different varieties of gram (*Cicer arietinum* L.) to foot-blight caused by *Operculella padwickii* Kheswalla, *J. Res., Punjab agric. Univ.*, **11**, 400–402.
Singh, G. and Bedi, P. S. (1975). The perpetuation of *Operculella padwickii*, the cause of foot blight (rot) of gram in the Punjab, *Indian Phytopath*., **28**, 546–548.
Singh, I. and Chohan, J. S. (1974). Seed-borne fungi on cowpea (*Vigna sinensis*), *Indian Phytopath*., **27**, 239–240.
Singh, K. B., Hawtin, G. C., Nene, Y. L., and Reddy, M. V. (1981a). Resistance in chickpeas to *Ascochyta rabiei*, *Pl. Dis*., **65**, 586–587.
Singh, K. B., Gridley, H. E., and Hawtin, G. C. (1981c). Strategy for breeding *Ascochyta* blight resistant cultivars. *Proc. Wkshop on Ascochyta Blight and Winter Sowing of Chickpea*, 4–7 May 1981, International Center for Agricultural Research in the Dry Areas, Aleppo, Syria.
Singh, K. B., Meyoeci, K., Izgin, N., and Tuwafe, S. (1981d). Screening kabuli chickpea for cold tolerance, *Internat. Chickpea Newsl*., **4**, 11–12.
Singh, K. B., Nene, Y. L., and Reddy, M. V. (1981b). An international screening of chickpea for resistance to *Ascochyta* blight. *Proc Wkshop on Ascochyta Blight and Winter Sowing of Chickpea*, 4–7 May 1981, International Center for Agricultural Research in the Dry Areas, Aleppo, Syria.
Singh, N. and Singh, R. S. (1981). Lysis of mycelium of *Fusarium oxysporum* f.sp. *udum* in soil amended with organic matters, *Pl. Soil*, **59**, 9–15.
Singh, N. D. (1975). Evaluation of nematode population in pigeon pea, in *Tropical Diseases of Legumes* (eds J. Bird and K. Maramorosch), Academic Press, New York, pp. 147–149.

Singh, N. D. and Farrell, K. M. (1972). Occurrence of *Rotylenchulus reniformis* in Trinidad, West Indies, *Pl. Dis. Reptr.*, **56**, 551.

Singh, P., Vasudeva, R. S., and Bajaj, B. S. (1965). Seed bacterization and biological activity of bulbiformin, *Ann. appl. Biol.*, **55**, 89–97.

Singh, P. J. and Mehrotra, R. S. (1980a). Relation between seed exudate and host susceptibility in gram (*Cicer arietinum* L.) to *Rhizoctonia bataticola, Pl. Soil,* **56,** 265–271.

Singh, P. J. and Mehrotra, R. A. (1980b). Biological control of *Rhizoctonia bataticola* on gram by coating seed with *Bacillus* and *Streptomyces* spp. and their influence on plant growth, *Pl. Soil*, **56**, 475–483.

Singh, R. and Mall, T. P. (1976). A new virus disease of arhar (*Cajanus cajan* (L.) Millsp.), *Curr. Sci.*, **45**, 635–636.

Singh, R. A. and Pavgi, M. S. (1972). Some diseases of pigeon pea in Uttar Pradesh, *FAO Pl. Prot. Bull.*, **20**, 116–118.

Singh, R. N. (1979). Natural infection of bean (*Phaseolus vulgaris*) by mung bean yellow mosaic virus, *Indian J. Mycol. Pl. Path.*, **9**, 124–126.

Singh, R. S. and Sinha, R. P. (1955). Studies on the wilt disease of cowpea in Uttar Pradesh. I. Occurrence and symptoms of the disease and identity of the causal organism, *J. Indian bot. Soc.*, **34**, 375–381.

Singh, S. P., White, J. W., and Gutierrez, J. A. (1980b). Male sterility in dry beans. *A. Rep. Bean Improv. Co-op.*, **23**, 55–57.

Singh, S. R. and Allen, D. J. (1979). *Cowpea Pests and Diseases*, Manual series No. 2, International Institute of Tropical Agriculture, Ibadan, Nigeria, 113 pp.

Singh, S. R. and Allen, D. J. (1980). Pests, diseases, resistance and protection in cowpeas, in *Advances in Legume Science* (eds R. J. Summerfield and A. H. Bunting), HMSO, London, pp. 419–443.

Singh, U. B. (1934). Studies on *Cercospora indica* n.sp., parasitic on *Cajanus indicus* Spreng, *Indian J. agric. Sci.*, **4**, 343–360.

Sinha, M. K. and Prasad, T. (1977). Deterioration of 'arhar' seeds by *Aspergillus flavus*, *Indian Phytopath.*, **30**, 70–72.

Sinha, O. K. and Khare, M. N. (1977). Seed-borne fungi of cowpea and their importance, *Indian Phytopath.*, **30**, 469–472.

Sitterly, W. R. and Epps, W. M. (1958). Lima bean scab found in South Carolina, *Pl. Dis. Reptr.*, **42**, 1309.

Sivanesan, A. and Holliday, P. (1971a). *Elsinoe canavaliae. CMI Descriptions of Pathogenic Fungi and Bacteria No. 313*.

Sivanesan, A. and Holliday, P. (1971b). *Elsinoe phaseoli. CMI Descriptions of Pathogenic Fungi and Bacteria No. 314*.

Skidmore, D. I., Shattock, R. C., and Shaw, D. S. (1981). Oospores of *Phytophthora infestans* induced by a British isolate of *Phytophthora drechsleri.* Contrib. *Fed. Br. Pl. Path and Cambridge Pl. Path. Club meeting*, Cambridge, March 1981.

Skiles, R. L. and Cardona-Alvarez, C. (1959). Mancha gris, a new leaf disease of bean in Colombia, *Phytopathology*, **49**, 133–135.

Skipp, R. A. (1975). Progress in understanding the nature of resistance to anthracnose disease in cowpea. *Proc. IITA Collaborators' Meeting on Grain Legume Improvement*, International Institute of Tropical Agriculture, Ibadan, Nigeria, pp. 78–79.

Skipp, R. A. and Deverall, B. J. (1972). Relationships between fungal growth and host changes visible by light microscopy during infection of bean hypocotyls (*Phaseolus vulgaris*) susceptible and resistant to physiological races of *Colletotrichum lindemuthianum, Physiol. Pl. Path.*, **2**, 357–374.

Skipp, R. A. and Deverall, B. J. (1973). Studies on cross-protection in the anthracnose disease of beans, *Physiol. Pl. Path.*, **3**, 299–313.

Skoog, H. A. (1952). Studies on host-parasite relations of bean varieties resistant and susceptible to *Pseudomonas phaseolicola* and toxin production by the parasite, *Phytopathology*, **42**, 475 (abstr.).

Skotland, C. B. (1958). Bean pod mottle virus of soybean, *Pl. Dis. Reptr.*, **42**, 1155–1156.
Slack, S. A. and Scott, H. A. (1971). Hemolymph as a reservoir for the cowpea strain of southern bean mosaic virus in the bean leaf beetle, *Phytopathology*, **61**, 538–540.
Slusarenko, A. J. and Wood, R. K. S. (1981). Differential agglutination of races 1 and 2 of *Pseudomonas phaseolicola* by a fraction from cotyledons of *Phaseolus vulgaris* cv. Red Mexican, *Physiol. Pl. Path.*, **18**, 187–193.
Slusher, R. L. and Sinclair, J. B. (1973). Development of *Phytophthora megasperma* var. *sojae* in soybean roots, *Phytopathology*, **63**, 1168–1171.
Smartt, J. (1960a). Genetic instability and outcrossing in the groundnut variety Mani Pintar, *Nature, Lond.*, **186**, 1070–1071.
Smartt, J. (1960b). A guide to soya bean cultivation in Northern Rhodesia, *Rhod. agric. J.*, **57**, 459–463.
Smartt, J. (1961). The diseases of groundnut in Northern Rhodesia, *Emp. J. exp. Agric.*, **29**, 79–87.
Smartt, J. (1970). Interspecific hybridization between cultivated American species of the genus *Phaseolus, Euphytica*, **19**, 480–489.
Smartt, J. (1976a). *Tropical Pulses*, Longman, London.
Smartt, J. (1976b). Comparative evolution of pulse crops, *Euphytica*, **25**, 139–143.
Smartt, J. (1979). Authentication of interspecific hybrids in *Phaseolus, A. Rep. Bean Improv. Co-op.*, **22**, 12–14.
Smartt, J. (1980). Evolution and evolutionary problems in food legumes, *Econ. Bot.*, **34**, 219–235.
Smith, C. E. (1924). Transmission of cowpea mosaic by bean-leaf beetle, *Science*, **60**, 268.
Smith, D. A. and Ingham, J. L. (1980). Legumes, fungal pathogens, and phytoalexins, in *Advances in Legume Science* (eds R. J. Summerfield and A. H. Bunting), HMSO, London, pp. 207–223.
Smith, D. A., Kuhn, P. J., Bailey, J. A., and Burden, R. S. (1980). Detoxification of phaseollidin by *Fusarium solani* f.sp. *phaseoli, Phytochemistry*, **19**, 1673–1675.
Smith, D. A., van Etten, H. D., and Bateman, D. F. (1975). Accummulation of phytoalexins in *Phaseolus vulgaris* hypocotyles following infection by *Rhizoctonia solani, Physiol. Pl. Path.*, **5**, 51–64.
Smith, D. H. (1971). A simple method for producing *Cercospora arachidicola* conidial inoculum, *Phytopathology*, **61**, 1414.
Smith, D. H. (1980). Groundnut foliar disease in the United States, in *Proc. Intern. Wkshop Groundnuts* (ed. R. W. Gibbons), ICRISAT, Hyderabad, pp. 186–192.
Smith, E. F. and McCulloch, L. (1919). *Bacterium solanacearum* in beans, *Science*, **50**, 238.
Smith, K. M. (1972). *A Textbook of Plant Virus Diseases*, 3rd edn. Longman, London, 684 pp.
Smith, T. E. (1960). Occurrence of *Verticillium* wilt on peanuts, *Pl. Dis. Reptr.*, **44**, 435.
Smith, T. E. (1961). *Verticillium* wilt resistance in peanuts (*Arachis hypogaea*), *Phytopathology*, **51**, 411–412.
Smithson, J. B., Redden, R., and Rawal, K. (1980). Methods of crop improvement and genetic resources in *Vigna unguiculata*, in *Advances in Legume Science* (eds R. J. Summerfield and A. H. Bunting), HMSO, London, pp. 445–457.
Sneep, J. (1945). De Ascochyta-vlekkenziekte van de Boon (*Phaseolus*), *Tijdschr. Plziekt.*, **51**, 1–16.
Sneh, B., Humble, S. J., and Lockwood, J. L. (1977). Parasitism of oospores of *Phytophthora megasperma* var. *sojae, P. cactorum, Pythium* sp., and *Aphanomyces euteiches* by oomycetes, chytridiomycetes, hyphomycetes, actinomyces, and bacteria, *Phytopathology*, **67**, 622–628.
Snowdon, J. D. (1921). Report of the Government Botanist for the period April to

31 December, 1920. *Ann. Rep. Dept. Agric. Uganda* for the nine months ending 31 Dec. 1920, pp. 43–46. (*Rev. appl. Mycol*., **1**, 205 abstr.)

Snyder, W. C. and Hansen, H. N. (1940). Species concept in *Fusarium, Am. J. Bot.*, **27**, 64–67.

Snyder, W. C. and Hansen, H. N. (1941). The species concept in *Fusarium* with reference to section Martiella, *Am. J. Bot.*, **28**, 738–742.

Snyder, W. C., Paulos, A. O., and Gold, A. H. (1956). Virus yellows of garbanzo, *Phytopathology*, **46**, 27 (abstr.).

Sohi, H. S. and Dass, R. (1972). Relative resistance and susceptibility of different varieties of *Phaseolus vulgaris* L. to angular leaf spot disease (*Isariopsis griseola* Sacc.), *Indian J. Hort*., **29**, 109–110.

Sohi, H. S. and Rawal, R. D. (1974). Control cowpea diseases, *Indian Hort*., **19**, 15–19.

Sohi, H. S. and Rawal, R. D. (1975). Note on the relative resistance of cowpea varieties to powdery-mildew, *Indian J. agric. Sci.*, **45**, 89–90.

Sohi, H. S., Sharma, S. L., Verma, B. R., and Sachdeva, K. B. (1965). Floury leaf spot of bean caused by *Ramularia phaseolina* Petrak, *Indian Phytopath*., **18**, 384.

Sokhi, S. S. and Sohi, H. S. (1976). Epidemiological studies on rust of cowpea, *PANS*, **22**, 92–94.

Solel, Z. and Kostrinski, J. (1964). The control of *Ascochyta* anthracnose of chickpea, *Phytopath. Med*., **3**, 119–120. (*Rev. appl. Mycol*., **45**, 1589.)

Solomon, S., Agrikar, G. P., Solanki, M. S., and Morbad, I. R. (1957). A study of heterosis in *Cajanus cajan* (L.) Millsp., *Indian J. Genet. Pl. Breed*., **17**, 90–95.

Somani, R. B., Wangikar, P. D., and Shukla, V. N. (1975). A new stem canker and die back disease of pigeon pea, *Indian Phytopath*., **28**, 436–437.

Soomro, M. P. and Baluch, M. A. (1968). Varietal classification in gram (*Cicer arietinum* L.), *W. Pakistan J. agric. Res*., **6** (4), 1–23.

Sowell, G., Kuhn, C. W., and Brantley, B. B. (1965). Resistance of southern pea, *Vigna sinensis*, to cowpea chlorotic mottle virus, *Proc. Am. Soc. hort. Sci*., **86**, 487–490.

Sowell, G., Smith, D. H., and Hammons, R. O. (1976). Resistance of peanut plant introductions to *Cercospora arachidicola*, *Pl. Dis. Reptr*., **60**, 494–498.

Spegazzini, C. L. (1881). *Anal. Soc. cient. Argent.*, **12**, 67.

Spence J. A. (1975). The importance of diseases in relation to the grain legume research program in the eastern Caribbean, in *Tropical Diseases of Legumes* (eds J. Bird and K. Maramorosch), Academic Press, New York, pp. 151–155.

Spence, J. A. and Williams, S. J. A. (1972). Use of photoperiod response to change plant design, *Crop Sci*., **12**, 121–122.

Spencer, J. A. and Walters, H. J. (1969). Variations in certain isolates of *Corynespora cassiicola, Phytopathology*, **59**, 58–60.

Sprague, R. (1930). Notes on *Phyllosticta rabiei* on chick-pea, *Phytopathology*, **20**, 591–593.

Sprague, R. (1935). *Ascochyta boltshauseri* on beans in Oregon, *Phytopathology*, **25**, 416–420.

Spurling, A. T. (1973). Field trials with Canadian Wonder beans in Malawi. *Expl. Agric*., **9**, 97–105.

Sreenivasulu, P., Iizuka, N., Rajeshwari, R., Reddy, D. V. R., and Nayudu, M. V. (1981). Peanut green mosaic virus – a member of the potato virus Y group infecting groundnut (*Arachis hypogaea*) in India, *Ann. appl. Biol*., **98**, 255–260.

Srinivasan, K. V. (1953). *Isariopsis griseola* Sacc. on *Phaseolus vulgaris* L., *Curr. Sci*., **22**, 20.

Srivastava, A. S., Upadhyay, K. D., and Singh, G. (1974). Effect of root-knot nematode, *Meloidogyne javanica*, on gram crop, *Indian J. Nemat*., **4**, 248–251.

Stace-Smith, R. (1970). Tobacco ringspot virus. *CMI/AAB Descriptions of Plant Viruses No. 17*.
Stadt, S. J. and Saettler, A. W. (1981). Effect of host genotype on multiplication of *Pseudomonas phaseolicola, Phytopathology*, **71**, 1307–1310.
Stanton, W. R. (1966). *Grain Legumes in Africa*, FAO, Rome, 183 pp.
Staples, R. R. (1958). *Rep. Min. Agric. Rhod. Nyasald 1956—7*, pp. 7–86. (*Rev. appl. Mycol.*, **38**, 295–296.)
Starr, G. H. and Kercher, C. J. (1969). Passage of *Pseudomonas phaseolicola* in bean plants through sheep, *Phytopathology*, **59**, 1976.
Steadman, J. R., Kerr, E. D., and Mumm, R. F. (1975). Root rot of bean in Nebraska: primary pathogen and yield loss appraisal, *Pl. Dis. Reptr.*, **59**, 305–308.
Steele, W. M. (1972). Cowpeas in Nigeria. Ph.D thesis, Univ. Reading.
Steele, W. M. (1976). Cowpea, in *Evolution of Crop Plants* (ed. N. W. Simmonds), Longmans, London, pp. 183–185.
Steele, W. M. and Mehra, K. L. (1980). Structure, evolution, and adaptation to farming systems and environments in *Vigna*, in *Advances in Legume Science* (eds R. J. Summerfield and A. H. Bunting), HMSO, London, pp. 393–404.
Steiner, K. G. (1975). *Les maladies des plantes au Togo, liste commentée des hôtes*. Service de la Protection des Vegetaux, Lomé, 57 pp.
Stevens, R. B. (1960). Cultural practices in disease control, in *Plant Pathology, An Advanced Treatise* III (eds J. G. Horsfall and A. E. Dimond), Academic Press, New York, pp. 357–429.
Stevenson, J. A. (1946). Fungi novi denominati – II, *Mycologia*, **38**, 524–533.
Stewart, R. B. and Yirgou, D. (1967). Index of plant diseases in Ethiopia, *Haile Sellassie I Univ. Coll. Agric. Exp. Sta. Bull. No. 3.*, 95 pp.
Stewart, R. D. (1957). An undescribed species of *Pyrenochaeta* on soybean, *Mycologia*, **49**, 115–117.
Stholasuta, P., Bailey, J. A., Severin, V., and Deverall, B. J. (1971). Effect of bacterial inoculation of bean and pea leaves on the accumulation of phaseollin and pisatin, *Physiol. Pl. Path.*, **1**, 177–183.
Storey, H. H. (1935). The rosette disease of peanuts (*Arachis hypogaea* L.), *E. Afr. agric. J.*, **1**, 206–211.
Storey, H. H. and Bottomley, A. M. (1925). Transmission of a rosette disease of the groundnut, *Nature, Lond.*, **116**, 97.
Storey, H. H. and Bottomley, A. M. (1928). The rosette disease of peanuts, *Ann. appl. Biol.*, **15**, 26–45.
Storey, H. H. and Ryland, A. K. (1955). Transmission of groundnut rosette virus, *Ann. appl. Biol.*, **43**, 423–432.
Storey, H. H. and Ryland, A. K. (1957). Viruses causing rosette and other diseases in groundnuts, *Ann. appl. Biol.*, **45**, 318–326.
Stover, R. H. and Malo, S. E. (1972). The occurrence of fusarial wilt in normally resistant 'Dwarf Cavendish' banana, *Pl. Dis. Reptr.*, **56**, 1000–1003.
Sturgess, O. W. and Egan, B. T. (1960). A wilt disease of velvet bean caused by *Phytophthora drechsleri* Tucker, *Tech. Common. Bur. Sug. Exp. Stas Qd 1–4*, pp. 9–13. (*Rev. appl. Mycol.* **39**, 717–718.)
Subrahmanyam, P. (1979). Leaf blight of groundnut caused by *Myrothecium roridum*, *FAO Pl. Prot. Bull.*, **27**, 95–96.
Subrahmanyam, P. and McDonald, D. (1980). Occurrence of *Olpidium brassicae* on groundnut, *Trans. Br. mycol. Soc.*, **75**, 506–509.
Subrahmanyam, P., Mehan, V. K., Nevill, D. J., and McDonald, D. (1980). Research on fungal diseases of groundnut at ICRISAT, in *Proc. Intern. Wkshop Groundnuts* (ed. R. W. Gibbons), ICRISAT, Hyderabad, pp. 193–198.
Subrahmanyam, P., Reddy, D. V. R., Gibbons, R. W., Rao, V. R., and Garren, K. H. (1979). Current distribution of groundnut rust in India, *PANS*, **25**, 25–29.

Subramanian, C. V. (1955). Studies on South India fusaria. IV. The 'wildtype' in *Fusarium udum* Butler, *J. Indian bot. Soc.*, **34**, 29–36.
Subramanian, K. S., Samuel, G. S., Janarthanan, R., and Kandaswamy, T. K. (1973). Studies on the varietal resistance of pigeon pea (*Cajanus cajan*) to sterility mosaic disease, *Madras agric. J.*, **60**, 38–40.
Subramanian, S. (1963). *Fusarium* wilt of pigeon pea. I. Symptomatology and infection studies, *Proc. Indian Acad. Sci.*, B **57**, 134–148.
Sumarno, R. S. and Sudjadi, M. S. (1977). Breeding for soybean rust resistance in Indonesia, in *Rust of Soybean – the Problem and Research Needs* (eds R. E. Ford and J. B. Sinclair), Urbana-Champaign, Illinois, pp. 66–70.
Summerfield, R. J., Minchin, F. R., Roberts, E. H., and Hadley, P. (1980). The effects of photoperiod and air temperature on growth and yield of chickpea (*Cicer arietinum* L.), in *Proc. Internat. Wkshop in Chickpea Improv.*, International Crops Research Institute for the Semi-Arid Tropics, Hyderabad, India, pp. 121–149.
Summerfield, R. J. and Wien, H. C. (1980). Effects of photoperiod and air temperature on growth and yield of economic legumes, in *Advances in legume Science* (eds. R. J. Summerfield and A. H. Bunting), HMSO, London, pp. 17–36.
Sun, M. K. C. and Herbert, T. T. (1972). Purification and properties of a severe strain of peanut mottle virus, *Phytopathology*, **62**, 832.
Suneson, C. A. (1960). Genetic diversity – a protection against plant diseases and insects, *Agron. J.*, **52**, 319–321.
Suryanarayana, D. and Pathak, S. R. (1968). Foot blight – a new disease of gram, *FAO Pl. Prot. Bull.*, **16**, 71–73.
Suteri, B. D. (1974). Occurrence of soybean yellow mosaic in Uttar Pradesh, *Curr. Sci.*, **43**, 689–690.
Sutton, B. C. and Waterston, J. M. (1966). *Ascochyta phaseolorum. CMI Descriptions of Pathogenic Fungi and Bacteria No. 81*.
Swaminathan, M. S. and Jain, H. K. (1973). Food legumes in Indian agriculture, in *Nutritional Improvement of Food Legumes by Breeding* (ed. M. Milner), Wiley, New York, pp. 69–82.
Swinburne, T. R. (1980). Siderophores and iron chelators as factors in the infection and aggression of fungal pathogens. Contrib. *Plant Disease Etiology: 100 years of Koch's postulates*, Fed. Br. Cl. Path., London, 17–19 Dec.
Sydow, H. and Sydow, P. (1913). Novae fungorum species, X, *Annales Mycologici*, **11**, 254–271.
Sydow, H. and Sydow, P. (1914). A contribution to knowledge of the parasitic fungi of the island of Formosa, *Annales Mycologici*, **12**, 105–112.
Szarka, J. and Velich, I. (1980). Classification of isolates of *Pseudomonas phaseolicola* (Burkh.) Dowson by means of the diameter of the water-soaked lesion caused on pods of bean, *A. Rep. Bean Improv. Co-op.*, **23**, 105.
Tachibana, H. and Shih, M. (1965). A leaf-crinkling bacterium of soybean, *Pl. Dis. Reptr.*, **49**, 396–397.
Taiwo, M. A. (1978). An evaluation of advanced cowpea selections for resistance to cowpea aphid-borne mosaic and cowpea mosaic viruses. MS thesis, Cornell Univ., Ithaca.
Taiwo, M. A. and Gonsalves, D. (1982). Serological grouping of isolates of blackeye cowpea mosaic and cowpea aphidborne mosaic viruses, *Phytopathology*, **72**, 583–589.
Taiwo, M. A., Gonsalves, D., Provvidenti, R., and Thurston, H. D. (1982). Partial characterization and grouping of isolates of blackeye cowpea mosaic and cowpea aphidborne mosaic viruses, *Phytopathology*, **72**, 590–596.
Taiwo, M. A., Provvidenti, R., and Gonsalves, D. (1981). Inheritance of resistance to blackeye cowpea mosaic virus in *Vigna unguiculata*, *J. Hered.*, **72**, 433–434.
Takimoto, S. (1934). A new anthracnose of azuki bean, *Ann. phytopath. Soc. Japan*, **4**, 21–24. (*Rev. appl. Mycol.*, **14**, 342.)

Talley, J., Warren, F. H. J. B., Torrance, L., and Jones, R. A. C. (1980). A simple kit for detection of plant viruses by the latex serological test, *Pl. Path.*, **29**, 77–79.

Tamada, T. and Kojima, M. (1977). Soybean dwarf virus. *CMI/AAB Descriptions of Plant Viruses No. 179*.

Tarr, S. A. J. (1955). *The Fungi and Plant Diseases of the Sudan*, Commonwealth Mycological Institute, Kew.

Tarr, S. A. J. (1958). Control of seed-bed losses of groundnuts by seed treatment, *Ann. appl. Biol.*, **46**, 178–185.

Tarr, S. A. J. (1972). *Principles of Plant Pathology*, Macmillan, London, 632 pp.

Taylor, C. E. (1977). Nematological, in *Plant Health and Quarantine in International Transfer of Genetic Resources* (eds W. B. Hewitt and L. Chiarappa), CRC Press, Ohio, pp. 17–23.

Taylor, D. P. (1976). Plant nematology problems in tropical Africa, *Helm. Abstr.*, B **45**, 269–284.

Taylor, D. P., Anderson, R. V., and Haglund, W. A. (1958). Nematodes associated with Minnesota crops. I. Preliminary survey of nematodes associated with alfalfa, flax, peas, and soybeans, *Pl. Dis. Reptr.*, **42**, 195–198.

Taylor, D. P., Netscher, C., and Germani, G. (1978). *Adansonia digitata* (Baobab), a newly discovered host for *Meloidogyne* sp. and *Rotylenchulus reniformis*: agricultural implications, *Pl. Dis. Reptr.*, **62**, 276–277.

Taylor, J. D. (1970). The quantitative estimation of the infection of bean seed with *Pseudomonas phaseolicola* (Burkh.) Dowson, *Ann. appl. Biol.*, **66**, 29–36.

Taylor, J. D. and Dudley, C. L. (1977). Effectiveness of late copper and streptomycin sprays for the control of halo-blight of beans (*Pseudomonas phaseolicola*), *Ann. appl. Biol.*, **85**, 217–221.

Taylor, J. D., Dudley, C. L., and Presly, L. (1979b). Studies of halo-blight seed infection and disease transmission in dwarf beans, *Ann. appl. Biol.*, **93**, 267–277.

Taylor, J. D., Innes, N. L., Dudley, C. L., and Griffiths, W. A. (1978). Sources and inheritance of resistance to halo-blight of *Phaseolus* beans, *Ann. appl. Biol.*, **90**, 101–110.

Taylor, J. D., Phelps, K., and Dudley, C. L. (1979a). Epidemiology and strategy for the control of halo-blight of beans, *Ann. appl. Biol.*, **93**, 167–172.

Tayo, T. O. (1982). Growth, development and yield of pigeon pea (*Cajanus cajan* (L.) Millsp.) in the lowland tropics. 2. Effect of reduced assimilatory capacity, *J. agric. Sci., Camb.*, **98**, 71–77.

Tehon, L. R. (1933). Notes on the parasitic fungi of Illinois, *Mycologia*, **25**, 237–257.

Tehon, L. R. (1937). Notes on the parasitic fungi of Illinois, VI, *Mycologia*, **29**, 434–446.

Tehon, L. R. and Stout, G. L. (1928). An ascomycetous leaf spot of cowpea, *Phytopathology*, **18**, 701–703.

Temple, S. R. and Song, L. (1980). Crop improvement and genetic resources in *Phaseolus vulgaris* for the tropics, in *Advances in Legume Science* (eds R. J. Summerfield and A. H. Bunting), HMSO, London, pp. 365–373.

Tenne, F. D., Prasartsee, C., Machado, C. C., and Sinclair, J. B. (1974). Variation in germination and seedborne pathogens among soybean seed lots from three regions in Illinois, *Pl. Dis. Reptr.*, **58**, 411–413.

Tervet, I. W. (1945). The influence of fungi on storage, on seed viability and seedling vigor of soybeans, *Phytopathology*, **35**, 3–15.

Tervet, I. W. and Cassell, R. C. (1951). The use of cyclone separators in race identification of cereal rusts, *Phytopathology*, **41**, 286–290.

Thapliyal, P. N. and Choudhary, N. N. (1976). Resistance of some soybean varieties to *Phakopsora pachyrhizi*: uredia and uredospore production, *Indian Phytopath.*, **29**, 343–345.

Thaxter, R. (1889). A new American *Phytophthora, Bot. Gaz.*, **14**, 273–274.

Thomas, C. E. (1969). Transmission of tobacco ringspot virus by *Tetranychus* sp., *Phytopathology*, **59**, 633–636.

Thomas, H. R. and Zaumeyer, W. J. (1950a). Red node, a virus disease of beans, *Phtopathology*, **40**, 832–846.

Thomas, H. R. and Zaumeyer, W. J. (1950b). Inheritance of symptom expression of pod mottle virus, *Phytopathology*, **40**, 1007–1010.

Thomas, J. E. and Bowyer, J. W. (1980). Properties of tobacco yellow dwarf and bean summer death viruses, *Phytopathology*, **70**, 214–217.

Thomas, M. D. and Leary, J. V. (1980). A new race of *Pseudomonas glycinea*, *Phytopathology*, **70**, 310–312.

Thomas, P. E. and Mink, G. I. (1979). Beet curly top virus. *CMI/AAB Descriptions of Plant Viruses No. 210*.

Thomason, I. J., Erwin, D. C., and Garber, M. J. (1959). The relationship of the root knot nematode, *Meloidogyne javanica*, to *Fusarium* wilt of cowpea, *Phytopathology*, **49**, 602 606.

Thomason, I. J. and McKinney, H. E. (1960). Reaction of cowpea, *Vigna sinensis*, to root-knot nematode, *Meloidogyne* spp., *Pl. Dis. Reptr*., **44**, 51–53.

Thongmeearkom, P., Honda, Y., Saito, Y., and Syamananda, R. (1981). Nuclear ultrastructural changes and aggregates of viruslike particles in mung-bean cells affected by mungbean yellow mosaic disease, *Phytopathology*, **71**, 41–44.

Thongmeearkom, P., Paschal, E. H., and Goodman, R. M. (1978). Yield reductions in soybeans infected with cowpea mosaic virus, *Phytopathology*, **68**, 1549–1551.

Thottappilly, G. (1969). Untersuchungen über das Blattrollvirus der Erbse und seine Vektoven 1. Übertragungsversuche mit verschiedenen Blattlausarten, die sich an viruskranken *Pisum sativum* – Pflanzen entwickelt hatten, *Phytopath. Z*., **64**, 327–337.

Thouvenel, J-C., Dollet, M., and Fauquet, C. (1976). Some properties of peanut clump, a newly discovered virus, *Ann. appl. Biol*., **84**, 311–320.

Thouvenel, J-C. and Fauquet, C. (1981a). Further properties of peanut clump virus and studies on its natural transmission, *Ann. appl. Biol*., **97**, 99–107.

Thouvenel, J-C. and Fauquet, C. (1981b). Peanut clump virus. *CMI/AAB Descriptions of Plant Viruses No. 235*.

Thouvenel, J-C., Fauquet, C., and Larny, D. (1978). Transmission par la graine du virus du 'clump' du l'arachide, *Oléagineux*, **33**, 503–504.

Thouvenel, J-C., Monsarrat, A., and Fauquet, C. (1982). Isolation of cowpea mild mottle virus from diseased soybeans in the Ivory Coast, *Pl. Dis*., **66**, 336–337.

Thresh, J. M. (1974a). Temporal patterns of virus spread, *A. Rev. Phytopath*., **12**, 111–128.

Thresh, J. M. (1974b). Vector relationships and the development of epidemics: the epidemiology of plant viruses, *Phytopathology*, **64**, 1050–1056.

Thresh, J. M. (1976). Gradients of plant virus diseases, *Ann. appl. Biol*., **82**, 381–406.

Thresh, J. M. (1980). The origin and epidemiology of some important plant virus diseases, *Appl. Biol*., **5**, 1–65.

Thung, T. H. (1937). Phytopathologische waarnemingen, *Meded. Proefst. vorstenl. Tab*., **84**, 25–42. (*Rev. appl. Mycol*., **16**, 779.)

Tiffanay, L. H. and Gilman, J. C. (1954). Species of *Colletotrichum* from legumes, *Mycologia*, **46**, 52–75.

Tinline, R. D. and MacNeill, B. H. (1969). Parasexuality in plant pathogenic fungi, *A. Rev. Phytopath*., **7**, 147–170.

Tinsley, T. W. (1959). Pea leaf roll, a new virus disease of legumes in England, *Pl. Path*., **8**, 17–18.

Tisdale, W. B. (1924). *Rept. Florida Agric. Exp. Sta. for the fiscal year ending June 30, 1924*, pp. 121R–129R. (*Rev. appl. Mycol*., **5**, 8–9.)

Tisdale, W. B. and Williamson, M. M. (1923). Bacterial spot of lima bean, *J. agric. Res*., **25**, 141–154.

Tisselli, O., Sinclair, J. B. and Hymowitz, T. (1980). *Sources of Resistance to Selected Fungal, Bacterial, Viral and Nematode Diseases of Soybeans*, INTSOY series no. 18, Univ. Illinois, 134 pp.

Toler, R. W. and Dukes, P. D. (1965). *Choanephora* pod rot of cowpeas, *Pl. Dis. Reptr.*, **49**, 347–350.

Toler, R. W., Thompson, S. S., and Barker, J. M. (1963). Cowpea (southern pea) diseases in Georgia, 1961–1962, *Pl. Dis. Reptr.*, **47**, 746–747.

Toler, R. W. and Wester, R. E. (1966). A survey of lima bean diseases in the South, 1965, *Pl. Dis. Reptr.*, **50**, 316–317.

Tollenaar, H. and Martin, B. C. (1972). Perchlorate in Chilean nitrate as the cause of leaf rugosity in soybean plants in Chile, *Phytopathology*, **62**, 1164–1166.

Tompkins, C. M. and Gardner, M. W. (1935). Relation of temperature to infection of bean and cowpea seedlings by *Rhizoctonia bataticola, Hilgardia*, **9**, 219–230.

Toussoun, T. A., Nash, S. M., and Snyder, W. C. (1960). The effect of nitrogen sources and glucose on the pathogenesis of *Fusarium solani* f. *phaseoli*, *Phytopathology*, **50**, 137–140.

Toussoun, T. A. and Nelson, P. E. (1975). Variation and speciation in the fusaria, *A. Rev. Phytopath.*, **13**, 71–82.

Toussoun, T. A., Patrick, Z. A., and Snyder, W. C. (1963). Influence of crop residue decomposition products on the germination of *Fusarium solani* f. *phaseoli* chlamydospores in soil, *Nature, Lond.*, **197**, 1314–1316.

Townsend, G. R. (1939). Diseases of beans in Southern Florida, *Bull. Flo Agric. Exp. Sta.*, **336**.

Trenbath, B. R. (1977). Interactions among disease hosts and diverse parasites, in *The Genetic Basis of Epidemics in Agriculture* (ed. P. R. Day), *Ann. NY Acad. Sci.*, **287**, 124–150.

Tribe, H. T. and Ahmed, A. H. M. (1975). Use of autoclavable plastic bags in fungus culture work, *Trans. Br. mycol. Soc.*, **64**, 362–363.

Trutmann, P., Keane, P. J., and Merriman, P. R. (1982). Biological control of *Sclerotinia sclerotiorum* on aerial parts of plants by the hyperparasite *Coniothyrium minitans, Trans. Br. mycol. Soc.*, **78**, 521–529.

Tsai, K-H., Lu, Y-C., and Oka, H. I. (1974). Mutation breeding of soybean for the resistance to rust disease, *SABRAO J.*, **6**, 181–191.

Tsuchizaki, T. (1973). Peanut stunt virus isolated from beans in Japan, *Ann. phytopath. Soc. Japan*, **39**, 67–72.

Tsuchizaki, T., Yora, K., and Asuyama, H. (1970). The viruses causing mosaic of cowpeas and azuki beans, and their transmissibility through seeds. Seed transmission of viruses in cowpea and azuki bean plants. I. Factors influencing seed transmission, *Annls. phytopath. Soc. Japan*, **36**, 112–120.

Tu, J. C. (1978). Protection of soybean from severe *Phytophthora* root rot by *Rhizobium, Physiol Pl. Path.*, **12**, 233–240.

Tu, J. C. (1981). Anthracnose (*Colletotrichum lindemuthianum*) on white bean (*Phaseolus vulgaris* L.) in Southern Ontario: spread of the disease from an infection focus, *Pl. Dis.*, **65**, 477–480.

Tu, J. C., Ford, R. E., and Grau, C. R. (1970b). Some factors affecting the nodulation and nodule efficiency in soybeans infected by soybean mosaic virus, *Phytopathology*, **60**, 1653–1656.

Tu, J. C., Ford, R. E., and Quiniones, S. S. (1970a). Effects of soybean mosaic virus and/or bean pod mottle virus infection on soybean nodulation, *Phytopathology*, **60**, 518–523.

Tu, J. C. and Vaartaja, O. (1981). The effect of the hyperparasite (*Gliocladium virens*) on *Rhizoctonia solani* and on *Rhizoctonia* root rot of white beans, *Can. J. Bot.*, **59**, 22–27.

Tucker, C. M. (1927). Pigeon pea anthracnose, *J. agric. Res.*, **34**, 589–596.

Tudge, C. (1977). *The Famine Business*, Faber and Faber, London, 141 pp.

Turner, G. J. (1971). Fungi and plant disease in Sarawak. *Phytopath. Pap. No. 13*, Commonwealth Mycological Institute, Kew, 55 pp.
Turner, G. J. (1974). Possible transmission of *Puccinia polysora* by bees, *Trans. Br. mycol. Soc.*, **62**, 205–206.
Turner, G. J. and Tribe, H. T. (1975). Preliminary field plot trials on biological control of *Sclerotinia trifoliorum* by *Coniothyrium minitans, Pl. Path.*, **24**, 109–113.
Upadhyay, R. S., Rai, B., and Gupta, R. C. (1981). *Fusarium udum* parasitic on *Aspergillus luchuensis* and *Syncephalastrum racemosum*, *Pl. Soil*, **63**, 407–413.
Uyemoto, J. K. and Grogan, R. G. (1977). Southern bean mosaic virus: evidence for seed transmission in bean embryos, *Phytopathology*, **67**, 1190–1196.
Uyemoto, J. K., Provvidenti, R., and Purcifull, D. E. (1973). Host range and serological properties of a seed-borne cowpea virus, *Phytopathology*, **63**, 208–209 (abstr.).
Vaheeduddin, S. (1958). Evolving wilt resistant strains of red gram, *Andhra agric. J.*, **5**, 163–164.
Vakili, N. G. (1972). Distribution of *Entyloma* smut of beans in Central America, *Phytopathology*, **62**, 794 (abstr.).
Vakili, N. G. (1977). Field screening of cowpeas for *Cercospora* leaf spot resistance, *Trop. Agric. (Trin.)*, **54**, 69–76.
Vakili, N. G. (1978a). Distribution of *Entyloma* smut of beans and cowpeas in Tropical America and its possible centre of origin, *FAO Pl. Prot. Bull.*, **26**, 19–24.
Vakili, N. G. (1978b). Field observation and host range of soybean rust, *Phakospora pachyrhizi*, in Puerto Rico, in *Proc. Workshop Soybean Rust in W. Hemisphere* (ed. N. G. Vakili), USDA, Mayaguez, pp. 4–15.
Vakili, N. G. (1979). Field survey of endemic leguminous hosts of *Phakopsora pachyrhizi* in Puerto Rico, *Pl. Dis. Reptr.*, **63**, 931–935.
Vakili, N. G. (1981). Distribution of *Phakopsora pachyrhizi* on *Lablab purpureus* in Puerto Rico, *Pl. Dis.*, **65**, 817–819.
Vakili, N. G. and Bromfield, K. R. (1976). *Phakopsora* rust on soybean and other legumes in Puerto Rico, *Pl. Dis. Reptr.*, **60**, 995–999.
Vakili, N. G., Kaiser, W. J., Pérez, J. A., and Cortés-Monllor, A. (1975). Bacterial blight of beans caused by two *Xanthomonas* pathogenic types from Puerto Rico, *Phytopathology*, **65**, 401–403.
Vakili, N. G. and Maramorosch, K. (1974). Witch's broom disease caused by mycoplasm-like organisms on pigeon pea (*Cajanus cajan*) in Puerto Rico, *Pl. Dis. Reptr.*, **58**, 96.
Valenta, J. R. and Sisler, H. D. (1962). Evidence for a chemical basis of resistance of lima bean plants to downy mildew, *Phytopathology*, **52**, 1030–1037.
Valladares-Sanchez, N. E., Coyne, D. P., and Schuster, M. L. (1979). Differential reaction of leaves and pods of *Phaseolus* germplasm to strains of *Xanthomonas phaseoli* and transgressive segregation for tolerance from crosses of susceptible germplasm, *J. Am. Soc. hort. Sci.*, **104**, 648–654.
Valverde, R., Moreno, R., and Gamez, R. (1978). Beetle vectors of cowpea mosaic virus in Costa Rica, *Turrialba*, **28**, 90–91.
van Arsdel, E. P. and Harrison, A. L. (1972) Possible origin of peanut rust epidemics in Texas, *Phytopathology*, **62**, 794 (abstr.).
van Breukelen, E. (1979). Latent period as a tool for selecting bean varieties with partial resistance to bean rust. Contrib. *Proc. Symp. Grain Leg. Improv. E. Africa*, Nairobi, Kenya, 20–24 Aug. 1979.
van den Heuval, J. and Vollaard, P. J. (1976). Metabolism of phaseollin by different races of *Colletotrichum lindemuthianum, Neth. J. Pl. Path.*, **82**, 103–108.
van der Maesen, L. J. G. (1972). Cicer *L., a monograph of the genus, with special reference to the chickpea* (Cicer arietinum *L.*), *its ecology and cultivation*. Mededelingen Landouwhogeschool, Wageningen, The Netherlands, 342 pp.
van der Maesen, L. J. G. (1975). Germplasm collection and evaluation in *Cicer* and *Cajanus*, in *Internat. Workshop on Grain Legumes*, International Crops Research

Institute for the Semi-Arid Tropics, Hyderabad, India, pp. 229–237.
van der Maesen, L. J. G. (1979). Wild pigeonpeas in Africa, *Pl. genet. resour. Newsl.*, **40**, 8–10.
van der Maesen, L. J. G. (1980). India is the native home of the pigeonpea, *Landbouwhogeschool Wageningen Misc. Pap.*, **19**, 257–262.
van der Maesen, L. J. G. (1981). Taxonomy of *Cajanus*, in *Proc. Internat. Wkshop in Pigeonpeas*, **2**, 9–13, International Crops Research Institute for the Semi-Arid Tropics, Hyderabad, India.
van der Maesen, L. J. G., Pundir, R. P. S., and Remanandan, P. (1980). The current status of chickpea germplasm work at ICRISAT, in *Proc. Internat. Wkshop on Chickpea Improv.*, International Crops Research Institute for the Semi-Arid Tropics, Hyderabad, India, pp. 28–32.
van der Plank, J. E. (1963). *Plant Diseases: Epidemics and Control*, Academic Press, New York, 349 pp.
van der Plank, J. E. (1968). *Disease Resistance in Plants*, Academic Press, New York, 206 pp.
van Gundy, S. D. (1965). Factors in survival of nematodes, *A. Rev. Phytopath.*, **3**, 43–68.
van Gundy, S. D., Kirkpatrick, J. D., and Golden, J. (1977). The nature and role of metabolic leakage from root-knot nematode galls and infection by *Rhizoctonia solani, J. Nemat.,* **9**, 113–121. (*Helminth. Abstr.*, B **46**, 1355.)
van Hoof, H. A. (1963). Een siekte van Kousebandjes (*Vigna unguiculata* Walp.), veroorzaakt door de schimmel *Elsinoe*, *Surinaam. Landb.*, **11**, 27–30. (*Rev. appl. Mycol.*, **42**, 717.)
van Kammen, A. (1971). Cowpea mosaic virus. *CMI/AAB Descriptions of Plant Viruses No. 47*.
van Kammen, A. and de Jager, C. P. (1978). Cowpea mosaic virus. *CMI/AAB Descriptions of Plant Viruses No. 197*.
van Rheenen, H. A., Hasselbach, O. E., and Muigai, S. G. S. (1981). The effect of growing beans together with maize on the incidence of bean diseases and pests, *Neth. J. Pl. Path.*, **87**, 193–199.
van Rheenen, H. A., Muigai, S. G. S., and Kitivo, D. K. (1979). Male sterility in beans, *Euphytica*, **28**, 761–763.
Vargas, E. (1980). Rust, in *Bean Production Problems* (eds H. F. Schwartz and G. E. Gálvez), Centro Internacional de Agricultura Tropical, Cali, Colombia, pp. 17–36.
Varma, A., Sharma, S. R., and Moharir, A. V. (1978). Witch's broom of cowpea – a mycoplasmal disease, *Curr. Sci.*, **47**, 56–57.
Varma, P. M. (1955). Ability of the white-fly to carry more than one virus simultaneously, *Curr. Sci.*, **24**, 317–318.
Vasudeva, R. S. and Govindaswamy, C. V. (1953). Studies on the effect of associated soil microflora on *Fusarium udum* Butl., the fungus causing wilt of pigeonpea (*Cajanus cajan* (L.) Millsp.), with special reference to its pathogenicity, *Ann. appl. Biol.*, **40**, 573–583.
Vasudeva, R. S., Jain, A. C., and Nema, K. G. (1952). Investigations of the inhibitory action of *Bacillus subtilis* on *Fusarium udum* Butl., the fungus causing wilt of pigeon pea (*Cajanus cajan* (L.) Millsp.), *Ann. appl. Biol.*, **39**, 229–238.
Vasudeva, R. S. and Roy, T. C. (1950). Effect of associated microflora on *Fusarium udum* Butl., the fungus causing wilt of pigeon-pea (*Cajanus cajan* (L.) Millsp.), *Ann. appl. Biol.*, **37**, 169.
Vasudeva, R. S., Singh, G. P., and Iyengar, M. R. S. (1962). Biological activity of bulbiformin in soil, *Ann. appl. Biol.*, **50**, 113–119.
Vasudeva, R. S., Singh, P., Sen Gupta, P. K., Mahmood, M., and Bajaj, B. S. (1963). Further studies on the biological activity of bulbiformin, *Ann. appl. Biol.*, **51**, 415–423.

Vasudeva, R. S. and Srinivasan, K. V. (1952). Studies on wilt disease of lentil (*Lens esculenta* Moench.), *Indian Phytopath*., **5**, 23–32.

Venette, J. R. and Kennedy, B. W. (1975). Naturally produced aerosols of *Pseudomonas glycinea, Phytopathology*, **65**, 737–738.

Venkatakrishnaiya, N. S. (1950). *Phytophthora parasitica* on French bean, *Phaseolus vulgaris* L., *Indian J. agric. Sci*., **20**, 391–394.

Venkataraman, K. (1959). Phyllody in Bengal gram. *Madras agric. J*., **46**, 97–98.

Venkateswarlu, S., Reddy, A. R., Singh, R. M., and Singh, R. B. (1980). Reaction of pigeon pea varieties to wilt and mosaic virus (SMV), *Trop. Grain Leg. Bull*., **17/18**, 25–28.

Venkateswarlu, S. and Singh, R. B. (1980). Natural crossing in pigeon pea (*Cajanus cajan* (L.) Millsp.), *Trop. Grain Leg. Bull.*, **19**, 24–26.

Verdcourt, B. (1970). Studies in the Leguminosae-Papilionoideae for the 'Flora of Tropical East Africa', *Kew Bull*., **24**, 507–569.

Verma, M. M., Singh, G., Sandhu, T. S., Singh, H., Sandhu, S. S., Singh, K., and Bhullar, B. S. (1981). Sources of resistance to gram blight and grey mold, *Internat. Chickpea Newsl*., **4**, 14–16.

Verma, P. R. and Patel, P. N. (1969). Host range, varietal resistance and epidemiological observations on *Cercospora* leaf spot disease of cowpea, *Indian Phytopath*., **22**, 61–66.

Vetten, H. J. and Allen, D. J. (in press). Effects of environment and host on vector biology and incidence of two whitefly-spread diseases of legumes in Nigeria, *Ann. appl. Biol*.

Vieira, C. (1972). Resisténcia horizontal às doencas e diversidade genética no melhoramento do feijoeiro no Brasil, *Revta Ceres*, **19**, 261–279.

Vieira, C. and Wilkinson, R. E. (1972). The importance of field resistance and genetical diversity in bean breeding programs in South-Central Brazil, *A Rep. Bean Improv. Co-op*., **15**, 94–96.

Vignarajah, N. (1978). Mungbean research and production in Sri Lanka, *Proc. 1st Internat. Mungbean Symp*. Asian Vegetable Research and Development Centre, Taiwan, pp. 9–11.

Vir, S. and Grewal, J. S. (1974a). Physiologic specialization in *Asochyta rabiei* the causal organism of gram blight, *Indian Phytopath*., **27**, 355–360.

Vir, S. and Grewal, J. S. (1974b). Evaluation of fungicides for the control of gram blight, *Indian Phytopath*., **27**, 641–643.

Vir, S. and Grewal, J. S. (1974c). Changes in phenolic content of gram plant induced by *Ascochyta rabiei* infection, *Indian Phytopath*., **27**, 524–526.

Vir, S., Grewal, J. S., and Gupta, V. P. (1975). Inheritance of resistance to *Ascochyta* blight in chickpea, *Euphytica*, **24**, 209–211.

Vishwakarma, S. N. and Basu Chaudhary, K. C. (1974). Floral blight of gram incited by *Alternaria alternata, Neth. J. Pl. Path*., **80**, 100–112.

Vovlas, C. and Rana, G. L. (1972). Virus diseases of market garden plants in Apulia. VII. *Lens esculenta* Moench, a natural host of pea enation mosaic virus, *Phytopath. medit*., **11**, 97–102.

Voysest, O., Burga, C., and Burns, R. (1976). 'Canario PF-210': a new rust resistant bean variety for the Peruvian central coast, *A. Rep. Bean Improv. Co-op*., **19**, 87–88.

Vyas, S. C. and Joshi, L. K. (1975). A new record of parasitic dodder on chickpea (*Cicer arietinum* L.), *Curr. Sci*., **44**, 701–702.

Wade, B. L. and Andrus, C. F. (1941). A genetic study of the common bean mosaic under conditions of natural field transmission, *J. agric. Res*., **63**, 389–393.

Walker, C. A. and Chambliss, O. L. (1981). Inheritance of resistance to blackeye cowpea mosaic virus in *Vigna unguiculata* (L.) Walp., *J. Am. Soc. hort. Sci*., **106**, 410–412.

Walker, J. C. and Patel, P. N. (1964a). Spash dispersal and wind as factors in epidemiology of halo blight of bean, *Phytopathology*, **54**, 140–141.

Walker, J. C. and Patel, P. N. (1964b). Inheritance of resistance to halo blight of bean, *Phytopathology*, **54**, 952–954.

Wallace, D. H. (1980). Adaptation of *Phaseolus* to different environments, in *Advances in Legume Science* (eds R. J. Summerfield and A. H. Bunting), HMSO, London, pp. 349–357.

Wallace, D. H. and Wilkinson, R. E. (1965). Breeding for *Fusarium* root rot resistance in beans, *Phytopathology*, **55**, 1227–1231.

Wallace, D. H. and Wilkinson, R. E. (1973). Correlation of seed size and root rot resistance, *A. Rep. Bean Improv. Co-op*., **16**, 57–58.

Wallace, G. B. (1939). French bean diseases and bean fly in East Africa, *E. Afr. agric. J*., **5**, 170–175.

Wallace, G. B. (1941). Yellow bean mosaic and notes on other bean diseases, *E. Afr. agric. J*., **7**, 114–115.

Wallen, V. R. and Jackson, H. R. (1975). Model for yield loss determination of bacterial blight of field beans utilizing aerial infrared photography combined with field plot studies, *Phytopathology*, **65**, 942–948.

Wallen, V. R. and Sutton, M. D. (1965). *Xanthomonas phaseoli* var. *fuscans* (Burkh.) Starr and Burkh. on field bean in Ontario, *Can. J. Bot*., **43**, 437–446.

Waller, J. M. (1976). The influence of climate on the incidence and severity of some diseases of tropical crops, *Rev. Pl. Path*., **55**, 185–194.

Waller, J. M. and Sutton, B. C. (1979). *Mycovellosiella cajani. CMI Descriptions of Pathogenic Fungi and Bacteria No. 628*.

Walters, H. J. (1969). Beetle transmission of plant viruses, *Adv. Virus Res*., **15**, 339–363.

Walters, H. J. and Barnett, O. W. (1964). Bean leaf beetle transmission of Arkansas cowpea mosaic virus, *Phytopathology*, **54**, 911 (abstr.).

Walters, H. J. and Dodd, N. L. (1969). Identification and beetle transmission of an isolate of cowpea chlorotic mottle virus from *Desmodium, Phytopathology*, **59**, 1055 (abstr.).

Walters, H. J. and Henry, D. G. (1970). Bean leaf beetle as a vector of the cowpea strain of southern bean mosaic virus, *Phytopathology*, **60**, 177–178.

Wang, D. (1961). The nature of starch accumulation at the rust infection site in leaves of Pinto bean plants, *Can. J. Bot*., **39**, 1595–1604.

Warren, H. L. (1975). Effect of inoculum concentration on resistance of lima bean to *Rhizoctonia solani, Phytopathology*, **65**, 341–345.

Waterhouse, G. M. and Waterston, J. M. (1966). *Phytophthora megasperma. CMI Descriptions of Pathogenic Fungi and Bacteria No. 115*.

Waterhouse, W. L. (1953). Australian rust studies XIII. Specialization of *Uromyces phaseoli* (Pers.) Wint. in Australia, *Proc. Linn. Soc. N.S.W*., **78**, 226–232.

Waterston, J. M. (1944). *Rep. Dep. Agric. Bermuda for 1943*, pp. 7–8. (*Rev. appl. Mycol*., **24**, 51.)

Waterston, J. M. (1968). Inoculation. *Rev. appl. Mycol*., **47**, 217–222.

Waterworth, H. E. (1981). Bean mild mosaic virus. *CMI/AAB Descriptions of Plant Viruses No. 231*.

Waterworth, H. E., Meiners, J. P., Lawson, R. H., and Smith, F. F. (1977). Purification and properties of a virus from El Salvador that causes mild mosaic in bean cultivars, *Phytopathology*, **67**, 169–173.

Watkins, G. M. (1943). Further note on a cowpea bacterial canker in Texas, *Pl. Dis. Reptr*., **27**, 556.

Watson, I. A. (1970a). Changes in virulence and population shifts in plant pathogens, *A. Rev. Phytopath*., **8**, 209–230.

Watson, I. A. (1970b). The utilization of wild species in the breeding of cultivated crops resistant to plant pathogens, in *Genetic Resources in Plants – their Exploration and Conservation* (eds O. H. Frankel and E. Bennett), pp. 441–457, International Biol. Prog. and Blackwell Scientific Publ., Oxford.

Watson, M. A. and Okusanya, B. A. M. (1967). Studies on the transmission of groundnut rosette virus by *Aphis craccivora, Ann. appl. Biol.*, **60**, 199–208.

Watt, E. E. (1978). *1st Ann. Rep. IITA/EMBRAPA/IICA Cowpea Program in Brazil*, 1978, Goiania.

Weatherwax, P. (1954). *Indian Corn in Old America*, Macmillan, New York, 253 pp.

Webber, H. J. and Orton, W. A. (1902). A cowpea resistant to root-knot (*Heterodera radicicola*), *U.S. Dep. Agric. Bur. Pl. Ind. Bull.*, **17**, 23–36.

Weber, G. F. (1939). Web-blight, a disease of beans caused by *Corticium microsclerotia, Phytopathology*, **29**, 559–575.

Weber, G. F. (1941). Cowpea scab, *Pr. Bull. Fla agric. Exp. Sta.*, **557**. (*Rev. appl. Mycol.*, **21**, 238, abstr.)

Webster, D. M., Temple, S. R., and Schwartz, H. F. (1980). Selection for resistance to *Xanthomonas phaseoli* in dry beans, *Crop Sci.*, **20**, 519–522.

Wei, C. T. (1950). Notes on *Corynespora. Mycol. Pap. No. 34*, Commonwealth Mycological Institute, Kew, 10 pp.

Weimer, J. L. (1949). Red stem canker of cowpea, caused by *Phytophthora cactorum, J. agric. Res.*, **78**, 65–75.

Weimer, J. L. and Luttrell, E. S. (1945). A canker of cowpea and *Macrophomina* leaf spot of cowpea and snap bean, *Pl. Dis. Reptr.*, **29**, 127–129.

Weiss, E. A. (1967). Soya bean trials on the Vasin Gishu (Western Kenya), *E. Afr. agric. For. J.*, **32**, 223–228.

Weller, D. M. and Saettler, A. W. (1976). Chemical control of common and fuscous blights in Michigan navy (pea) beans, *Pl. Dis. Reptr.*, **60**, 793–797.

Wells, D. G. and Deba, R. (1961). Sources of resistance to the cowpea yellow mosaic virus, *Pl. Dis. Reptr.*, **45**, 878–881.

Wells, J. C. (1962). Peanut rust in North Carolina, *Pl. Dis. Reptr.*, **46**, 65.

Weltzien, H. C. (1972). Geophytopathology, *A. Rev. Phytopath.*, **10**, 277–298.

Weltzien, H. C. (1978). Geographical distribution of powdery mildews, in *The Powdery Mildews* (ed. D. M. Spencer), Academic Press, London, pp. 39–49.

Weltzien, H. C. and Kaack, H. J. (1981). Epidemiological aspects of chickpea *Ascochyta* blight. *Proc. Wkshop on Ascochyta Blight and Winter Sowing of Chickpea*, 4–7 May 1981, International Center for Agricultural Research in the Dry Areas, Aleppo, Syria.

Wernham, C. C. (1948). The species value of pathogenicity in the genus *Xanthomonas, Phytopathology*, **38**, 283–291.

Weste, G. (1974). *Phytophthora cinnamomi*: the cause of severe disease in certain native communities in Victoria, *Aust. J. Bot.*, **22**, 1–8.

Wester, R. E. (1968). New lima beans resistant to 'B' strain of *Phytophthora phaseoli, Pl. Dis. Reptr.*, **52**, 563–564.

Wester, R. E., Goth, R. W., and Dreschsler, C. (1966). Overwintering of *Phytophthora phaseoli, Phytopathology*, **56**, 95–97.

Wester, R. E. and Jorgensen, H. (1959). A new race of *Phytophthora phaseoli* from lima beans, *Pl. Dis. Reptr.*, **43**, 184–186.

Westerlund, F. V., Campbell, R. N., and Kimble, K. A. (1974). Fungal root rots and wilt of chickpea in California, *Phytopathology*, **64**, 432–436.

Westphal, E. (1974). *Pulses in Ethiopia, their Taxonomy and Agricultural Significance*, College of Agric., Haile Sellassie I University, Ethiopia and Agric. Univ., Wageningen, The Netherlands, 261 pp.

Wharton, A. L. (1967). Detection of infection by *Pseudomonas phaseolicola* (Burkh.) Dowson in white-seeded dwarf bean seed stocks, *Ann. appl. Biol.*, **60**, 305–312.

Wheeler, B. E. J. (1969). *An Introduction to Plant Diseases*, Wiley, London, 374 pp.

Whitehead, A. G. (1969). The distribution of root-knot nematodes (*Meloidogyne* spp.) in tropical Africa, *Nematologica*, **15**, 315–333.

Whitehead, M. D. and Thirumalachar, M. J. (1960). An undescribed smut disease on soybeans, *Mycologia*, **52**, 189–192.
Whitney, P. J. (1978). Broomrape (*Orobanche*) seed germination inhibitors from plant roots, *Ann. appl. Biol.*, **89**, 475–478.
Whitney, W. K. and Gilmer, R. M. (1974). Insect vectors of cowpea mosaic virus in Nigeria, *Ann. appl. Biol.*, **77**, 17–21.
Wien, H. C. and Roesingh, C. (1980). Ethylene evolution by thrips-infested cowpea provides a basis for thrips resistance screening with ethephon sprays, *Nature, Lond.*, **283**, 192–193.
Wien, H. C. and Summerfield, R. J. (1980). Adaptation of cowpeas in West Africa: effects of photoperiod and temperature responses in cultivars of diverse origins, in *Advances in Legume Science* (eds R. J. Summerfield and A. H. Bunting), HMSO, London, pp. 405–417.
Wien, H. C. and Tayo, T. O. (1978). The effect of defoliation and removal of reproductive structures on growth and yield of tropical grain legumes, in *Pests of Grain Legumes: Ecology and Control* (eds S. R. Singh, H. F. van Emden, and T. Ajibola Taylor), Academic Press, London, pp. 241–252.
Wienk, J. F. (1968). *Phytophthora nicotianae*: a cause of zebra disease in *Agave* hybrid no. 11648 and other agaves, *E. Afr. agric. For. J.*, **33**, 261–268.
Wilcox, J. R., Laviolette, F. A., and Martin, R. J. (1975). Heritability of purple seed stain resistance in soybeans, *Crop Sci.*, **15**, 525–526.
Wild, H. (1954). Rhodesian witchweeds, *Rhod. Agric. J.*, **51**, 330–342.
Wilding, N. and Brobyn, P. J. (1980). Effects of fungicides on development of *Entomophthora aphidis, Trans. Br. mycol. Soc.*, **75**, 297–302.
Wilkinson, R. E. and Vieira, C. (1972). A standardized system for surveying and reporting of races of bean rust is urgently needed, *A. Rep. Bean Improv. Co-op.*, **15**, 98–99.
Willey, R. W., Rao, M. R., and Natarajan, M. (1981). Traditional cropping systems with pigeonpea and their improvement, in *Proc. Internat. Wkshop on Pigeonpeas I*, International Crops Research Institute for the Semi-Arid Tropics, Hyderabad, India, pp. 11–25.
Williams, C., Birchfield, W., and Hartwig, E. E. (1973). Resistance in soybeans to a new race of root-knot nematode, *Crop Sci.*, **13**, 299–301.
Williams, C. N. (1961a). Growth and morphogenesis of *Striga* seedlings, *Nature, Lond.*, **189**, 378–381.
Williams, C. N. (1961b). Tropism and morphogenesis of *Striga* seedlings in the host rhizosphere, *Ann. Bot.*, n.s. **25**, 407–415.
Williams, D. J. and Nyvall, R. F. (1980). Leaf infection and yield losses caused by brown spot and bacterial blight diseases of soybean, *Phytopathology*, **70**, 900–902.
Williams, F. J., Amin, K. S., and Baldev, B. (1975). *Phytophthora* stem blight of *Cajanus cajan, Phytopathology*, **65**, 1029–1030.
Williams, F. J., Grewal, J. S., and Amin, K. S. (1968). Serious and new diseases of pulse crops in India in 1966, *Pl. Dis. Reptr.*, **52**, 300–304.
Williams, R. J. (1974a). Plant pathology in the Grain Legume Improvement Program at IITA. *Proc. 1st IITA Grain Legume Improv. Wkshop*, International Institute of Tropical Agriculture, Ibadan, Nigeria, pp. 261–268.
Williams, R. J. (1974b). Comparison of varietal reactions to cowpea anthracnose in field plot and detached petiole inoculations, *Proc. Am. Phytopath. Soc.*, **1**, 87 (abstr.).
Williams, R. J. (1975a). Control of cowpea seedling mortality in Southern Nigeria, *Pl. Dis. Reptr.*, **59**, 245–248.
Williams, R. J. (1975b). Diseases of cowpea (*Vigna unguiculata* (L.) Walp.) in Nigeria, *PANS*, **21**, 253–267.
Williams, R. J. (1975c). The control of cowpea diseases in the IITA Grain Legume

Improvement Program, in *Tropical Diseases of Legumes* (eds J. Bird and K. Maramorosch), Academic Press, New York, pp. 139–146.

Williams, R. J. (1975d). The international testing program: aims, progress and problems seen from IITA. *Proc. IITA Collab. Meet. Grain Legume Improv.*, International Institute of Tropical Agriculture, Ibadan, Nigeria, pp. 66–72.

Williams, R. J. (1976). A whitefly transmitted golden mosaic of lima beans in Nigeria, *Pl. Dis. Reptr.*, **60**, 853–857.

Williams, R. J. (1977a). The identification of multiple disease resistance in cowpea, *Trop. Agric. (Trin.)*, **54**, 53–60.

Williams, R. J. (1977b). The identification of resistance to cowpea (yellow) mosaic virus, *Trop. Agric. (Trin.)*, **54**, 61–68.

Williams, R. J. (1982). Important factors in finding and using disease resistance, *Seminar, Pl. Breed. Inst.*, Cambridge, 21 May 1982.

Williams, R. J. and Ayanaba, A. (1975). Increased incidence of *Pythium* stem rot in cowpea treated with benomyl and related fungicides, *Phytopathology*, **65**, 217–218.

Williams, T. D. (1968). Plant-parasitic nematodes, in *Plant Pathologists' Pocketbook*, Commonwealth Mycological Institute, Kew, pp. 119–136.

Williams, T. H. and Liu, P. S. W. (1976). A host list of plant diseases in Sabah, Malaysia. *Phytopath, Pap. No. 19*, Commonwealth Mycological Institute, Kew, 67 pp.

Williamson, C. E. (1950). Ethylene, a metabolic product of diseased or injured plants, *Phytopathology*, **40**, 205–208.

Wilsie, C. P. and Takahashi, M. (1934). Natural crossing in the pigeonpea, *J. agric. Res.*, **49**, 923–927.

Wilson, C. (1947). Concealed damage of peanuts in Alabama, *Phytopathology*, **37**, 657–668.

Wilson, D. M., Mixon, A. C., and Troeger, J. M. (1977). Aflatoxin contamination of peanuts resistant to seed invasion by *Aspergillus flavus, Phytopathology*, **67**, 922–924.

Wilson, E. M. (1958). Rust-TMV cross-protection and necrotic-ring reaction in bean, *Phytopathology*, **48**, 228–231.

Wilson, K. I. and Jose, P. C. (1968). Pod rot of *Vigna sinensis* (L.) Savi caused by *Choanephora cucurbitarum* (Berk. and Rav.) Thaxter, *Indian Phytopath.*, **21**, 130–131.

Wilson, M. and Henderson, D. M. (1966). *British Rust Fungi*, Cambridge Univ. Press. 384 pp.

Wilson, R. D. (1936). A bacterial disease of snake beans, *J. Proc. Roy. Soc. N.S.W.*, **69**, 215–223. (*Rev. appl. Mycol.*, **15**, 772.)

Wilson, R. D. (1947). Rainfall in relation to the production of bean seed free of the bacterial blight diseases, *Agric. Gaz. N.S.W.*, **58**, 15–20.

Wilson-Jones, K. (1953a). Relations of witchweed (*Striga*) to fertility in tropical soils, *Nature, Lond.*, **172**, 128.

Wilson-Jones, K. (1953b). The witchweeds of Africa, *Wld. Crops*, **5**, 263–266.

Wingard, S. A. (1933). Nature of rust resistance in beans, *Phytopathology*, **23**, 38 (abstr.).

Wingard, S. A. (1935). Host-parasite relationship in bean rust, *Phytopathology*, **25**, 39 (abstr.).

Winstead, N. N., Skotland, C. B., and Sasser, J. N. (1955). Soybean cyst nematode in North Carolina, *Pl. Dis. Reptr.*, **39**, 9–11.

Wolf, F. A. and Lehman, S. G. (1926). Diseases of soybean which occur both in North Carolina and the Orient, *J. agric. Res.*, **33**, 391–396.

Wolfe, M. S., Barrett, J. A., and Jenkins, J. E. E. (1981). The use of cultivar mixtures for disease control, in *Strategies for the Control of Cereal Disease* (eds J. F. Jenkin and R. T. Plumb), Blackwell Scientific Publ.

Woodbury, W. and Stahmann, M. A. (1970). Role of surface films in the germination or rust uredospores, *Can. J. Bot.*, **48**, 499–511.

Woodhead, T. (1966). Empirical relations between cloud amount, insolation and sunshine duration in East Africa: 1, *E. Afr. agric. for. J.*, **32**, 211–213.

Woodroof, J. G. (1966). *Peanuts: Production, Processing, Products*, Avi, Westport, Connecticut, 291 pp.

Woodroof, N. C. (1933). Two leaf spots of the peanut (*Arachis hypogaea* L.), *Phytopathology*, **23**, 627–640.

Woolley, J. N. (1980). A survey of the distribution and production of cowpea and prospects for the crop in Central America and the Caribbean. Unpubl. rep., International Institute of Tropical Agriculture, Ibadan, Nigeria, 139 pp.

Wong, E. Mortimer, P. I., and Geissman, T. A. (1965). Flavonoid constituents of *Cicer arietinum, Phytochemistry*, **4**, 89–95.

Wyatt, J. E. and Fassuliotis, G. (1979). Methods for screening snap beans for resistance to root knot nematodes, *Hort. sci.*, **14**, 27–28.

Wyatt, S. D. and Kuhn, C. W. (1979). Replication and properties of cowpea chlorotic mottle virus in resistant cowpeas, *Phytopathology*, **69**, 125–129.

Wynn, W. K. (1976). Appressorium formation over stomates by the bean rust fungus: response to a surface contact stimulus, *Phytopathology*, **66**, 136–146.

Wynne, J. C., Rowe, R. C., and Beute, M. K. (1975). Resistance of peanut genotypes to *Cylindrocladium crotalariae, Peanut Sci.*, **2**, 54–56.

Wyse, D. L., Meggitt, W. F., and Penner, D. (1976). Effect of herbicides on the development of root rot on navy bean, *Weed Sci.*, **24**, 11–15.

Yamamoto, W. (1934). *Cercospora* from Formosa I. *Trans. Sapporo Nat. Hist. Soc.*, **13**, 139–144. (*Rev. appl. Mycol.*, **13**, 805.)

Yang, C. Y. (1978). Soybean rust in Asia, in *Proc. Workshop Soybean Rust in W. Hemisphere* (ed. N. G. Vakili), USDA, pp. 34–43.

Yang, C. Y. (1980). Developments in crop protection in *Glycine*, in *Advances in Legume Science* (eds R. J. Summerfield and A. H. Bunting), HMSO, London, pp. 323–335.

Yang, S. L. and Zettler, F. W. (1975). Effects of alarm pheromones on aphid probing behaviour and virus transmission efficiency, *Pl. Dis. Reptr.*, **59**, 902–905.

Yarwood, C. E. (1951). Associations of rust and virus infections, *Science*, **114**, 127–128.

Yarwood, C. E. (1954). Mechanism of acquired immunity to a plant rust, *Proc. Nat. Acad. Sci.*, **40**, 374–377.

Yarwood, C. E. (1956a). Simultaneous self-stimulation and self-inhibition of uredospore germination, *Mycologia*, **48**, 20–24.

Yarwood, C. E. (1956b). Cross-protection with two rust fungi, *Phytopathology*, **46**, 540–544.

Yarwood, C. E. (1961). Uredospore production by *Uromyces phaseoli, Phytopathology*, **51**, 22–27.

Yarwood, C. E. (1968). Selective retention of uredospores by leaves, *Phytopathology*, **58**, 359–361.

Yarwood, C. E. and Hecht-Poinar, E. (1973). Viruses from rusts and mildews, *Phytopathology*, **63**, 1111–1115.

Yayock, J. Y. (1977). An epidemic of rosette disease and its effect on growth characteristics and yield of groundnuts in Nigeria, *Oléagineux*, **32**, 113–115.

Yeh, C. C., Tschanz, A. T., and Sinclair, J. B. (1981). Induced teliospore formation by *Phakopsora pachyrhizi* on soybeans and other hosts, *Phytopathology*, **71**, 1111–1112.

Yen, D. E. and Brien, R. M. (1960). French bean rust (*Uromyces appendiculatus*). Studies on resistance and determination of rust races present in New Zealand, *N.Z. J. agric. Res.*, **3**, 358–363.

Yen, W. Y., Chen, M. J., and Huang, K. T. (1956). Leaf scorch of peanut (a new disease), *J. Agric. For., Taiwan*, **10**, 145–168. (*Rev. appl. Mycol.*, **37**, 438.)
Yerkes, W. D. (1956). *Chaetoseptoria wellmanii* in Mexico, *Mycologia*, **48**, 738–740.
Yerkes, W. D. (1958). Additional new races of *Colletotrichum lindemuthianum* in Mexico, *Pl. Dis. Reptr.*, **42**, 329.
Yerkes, W. D. and Ortiz, M. F. (1956). New races of *Colletotrichum lindemuthianum* in Mexico, *Phytopathology*, **46**, 564–567.
Yerkes, W. D. and Patiño, G. (1960). The severe bean mosaic virus from Mexico, *Phytopathology*, **50**, 334–338.
Yorinori, J. T. and Homechin, M. (1978). Races of *Cercospora sojina* in Parana, Brazil, *3rd Int. Congr. Pl. Path.*, Munich, Germany, p. 304 (abstr.).
York, D. W., Dickson, M. H., and Abawi, G. S. (1977). Inheritance of resistance to seed decay and pre-emergence damping-off in snap beans caused by *Pythium ultimum, Pl. Dis. Reptr.*, **61**, 285–289.
Yoshii, K. (1980). Common and fuscous blights, in *Bean Production Problems* (eds H. F. Schwartz and G. E. Gálvez), Centro Internacional de Agricultura Tropical, Cali, Colombia, pp. 157–172.
Yoshii, K. and Aamodt, D. (1978). Evaluation of bean varieties for resistance to *Pseudocercosporella albida* in highland Guatemala, *Phytopath. News*, **12**, 269 (abstr.).
Yoshii, K., Gálvez, G. E., and Alvarez, G. (1976). Estimation of yield losses in beans caused by common blight, *Proc. Am. Phytopath. Soc.*, **3**, 298–299.
Yoshii, K., Gálvez, G. E., and Alvarez, G. (1978). Screening bean germplasm for tolerance to common blight caused by *Xanthomonas phaseoli* and the importance of pathogenic variation to varietal improvement, *Pl. Dis. Reptr.*, **62**, 343–347.
Yoshii, K., Gálvez, G. E., Temple, S. R., Orozco, S. H., Masaya, P., and Aldana, L. F. (1980). Release of three new Guatemalan bean varieties tolerant to golden mosaic virus, *A. Rep. Bean Improv. Co-op.*, **23**, 122–125.
Yuwa, J. A. (1964). Introduction of soyabeans into Abuja (Nigeria), *Samaru Agric. Newsl.*, **63/64**, 100–102.
Zachos, D. G., Panagopoulos, C. G., and Makris, S. A. (1963). Recherches sur la biologie, l'epidémiologie, et la lutte contre l'anthracnose du pois-chiche, *Annls. Inst. phytopath. Benaki N.S.*, **5**, 167–192.
Zambettakis, C. and Bokelee-Morvan, A. (1976). Recherches sur la structure du tégument séminal de la graine d'arachide et son influence sur la pénétration de l'*Aspergillus flavus, Oléagineux*, **31**, 219–228.
Zambettakis, C., Bockelee-Morvan, A., Waliyar, F., and Rossion, J. (1977). Différences variétales de sensibilité de l'arachide à la contamination par *A. flavus* aux champs et en conditions artificielles, *Oléagineux*, **32**, 377–385.
Zaumeyer, W. J. (1946). Field control of bean rust with sulfur, *Phytopathology*, **36**, 689 (abstr.).
Zaumeyer, W. J. (1953). Alfalfa yellow mosaic virus systemically infectious to beans, *Phytopathology*, **43**, 38–42.
Zaumeyer, W. J. (1960). A new race of bean rust in Maryland, *Pl. Dis. Reptr.*, **44**, 459–462.
Zaumeyer, W. J. (1963). Two new strains of alfalfa mosaic virus systemically infectious to bean, *Phytopathology*, **53**, 444–449.
Zaumeyer, W. J. (1972a). Genetic vulnerability in snap beans, *A. Rep. Bean Improv. Co-op.*, **15**, 37–40.
Zaumeyer, W. J. (1972b). Genetic vulnerability in dry beans, *A. Rep. Bean Improv. Co-op.*, **15**, 40–43.
Zaumeyer, W. J. and Goth, R. W. (1964). A new severe symptom-inducing strain of common bean mosaic virus, *Phytopathology*, **54**, 1378–1385.
Zaumeyer, W. J. and Harter, L. L. (1941). Inheritance of resistance to six physiologic races of bean rust, *J. agric. Res.*, **63**, 599–622.

Zaumeyer, W. J. and Meiners, J. P. (1975). Disease resistance in beans, *A. Rev. Phytopath.*, **13**, 313–334.
Zaumeyer, W. J. and Thomas, H. R. (1948). Pod mottle, a virus disease of beans, *J. agric. Res.*, **77**, 81–96.
Zaumeyer, W. J. and Thomas, H. R. (1950). Yellow stipple, a virus disease of bean, *Phytopathology*, **40**, 847–859.
Zaumeyer, W. J. and Thomas, H. R. (1957). A monographic study of bean diseases and methods for their control, *Tech. Bull. U.S. Dep. Agric.*, **868**, 2nd edn.
Zentmyer, G. A. (1966). Role of amino acids in chemotaxis of zoospores of three species of *Phytophthora, Phytopathology*, **56**, 907 (abstr.).
Zentmyer, G. A. (1977). Origin of *Phytophthora cinnamomi*: evidence that is not an indigenous fungus in the Americas, *Phytopathology*, **67**, 1373–1377.
Zettler, F. W. and Evans, I. R. (1972). Blackeye cowpea mosaic virus in Florida: host range and incidence in certified cowpea seed, *Proc. Florida State hort. Soc. 1972*, **85**, 99–101. (*Rev. Pl. Path.*, **53**, 4671.)
Zimmerman, A. (1907). Ueber eine Krankheit der Erdnüsse (*Arachis hypogaea*), *Der Pflanzer*, **3**, 129–133.
Zink, F. W., Grogan, R. G., and Welch, J. E. (1956). The effect of the percentage of seed transmission upon subsequent spread of lettuce mosaic virus, *Phytopathology*, **46**, 662–664.
Zummo, N. (1977). Diseases of giant witchweed, *Striga hermonthica*, in West Africa, *Pl. Dis. Reptr.*, **61**, 428–430.
Zúñiga de Rodríguez, J. E. and Victoria, K., J. I. (1975). Determinación de las razas fisiológicas de la roya del fríjol (*Uromyces phaseoli* var. *typica* Arth.) en el Valle de Cauca, *Acta agron. Palmira*, **25**, 75–85.

Index